T0173329

White Rose Maths
Key Stage 3
Student Book 3

Ian Davies, Caroline Hamilton and Sahar Shillabeer

William Collins' dream of knowledge for all began with the publication of his first book in 1819. A self-educated mill worker, he not only enriched millions of lives, but also founded a flourishing publishing house. Today, staying true to this spirit, Collins books are packed with inspiration, innovation and practical expertise.

They place you at the centre of a world of possibility and give you exactly what you need to explore it.

Collins. Freedom to teach.

Published by Collins
An imprint of HarperCollins*Publishers*
The News Building
1 London Bridge Street
London
SE1 9GF

1st Floor
Watermarque Building
Ringsend Road
Dublin 4
Ireland

Browse the complete Collins catalogue at www.collins.co.uk

British Library Cataloguing-in-Publication Data
A catalogue record for this publication is available from the British Library.

Authors: Ian Davies, Caroline Hamilton and
    Sahar Shillabeer
Series editors: Ian Davies and Caroline Hamilton
Publisher: Katie Sergeant
Product manager: Jennifer Hall
Product developer: Natasha Paul
Content editor: Tina Pietron
Editors: Karl Warsi, Julie Bond, Phil Gallagher,
    Tim Jackson
Proofreader: Catherine Dakin
Answer checkers: Laurice Suess and Tim Jackson
Project manager: Karen Williams
Cover designer: Kneath Associates Ltd
Internal designer and illustrator:
    Ken Vail Graphic Design Ltd
Typesetter: Ken Vail Graphic Design Ltd
Production controller: Katharine Willard
Printed and bound by Grafica Veneta in Italy

p 137 Nenov Brothers Images/Shutterstock, p 202 Aleksandar Grozdanovski/Shutterstock, p 209 PHOTO JUNCTION/Shutterstock, pp 223 and 239 Inked Pixels/Shutterstock, p 278t Evgenia.B/Shutterstock, p 278b Vector things/Shutterstock, p 323tl mikiekwoods/Shutterstock, p 323tr cobalt88/ Shutterstock, p 323bl urfin/Shutterstock, p 323br nikiteev_konstantin/Shutterstock, p 331 creatOR76/ Shutterstock, p 348 VectorHot/Shutterstock, p 370c logistock/Shutterstock, pp 417tl and 418b d1sk/ Shutterstock, p 424t Vitalliy/Shutterstock, p 424b fokke baarssen/Shutterstock, pp 482 and 483r Catharina van Delden/Shutterstock, p 483l topseller/ Shutterstock, p 532tl Eva Speshneva/Shutterstock, p 532cl MaryValery/Shutterstock, p 601 tl cTermit/ Shutterstock, p 601bl kavalenkava/Shutterstock, p 671 Rattanapon Ninlapoom/Shutterstock, p 676 Edwin Remsberg/Alamy.

Photo acknowledgements
The publishers wish to thank the following for permission to reproduce photographs. Every effort has been made to trace copyright holders and to obtain their permission for the use of copyright materials. The publishers will gladly receive any information enabling them to rectify any error or omission at the first opportunity.

MIX
Paper from
responsible sources
FSC™ C007454

This book is produced from independently certified FSC™ paper to ensure responsible forest management.

For more information visit: **www.harpercollins.co.uk/green**

Contents

Contents

Introduction

How to use this book

Welcome to the **Collins White Rose Maths Key Stage 3** course. We hope you enjoy your learning journey. Here is a short guide to how to get the most out of this book.

Ian Davies and Caroline Hamilton, authors and series editors

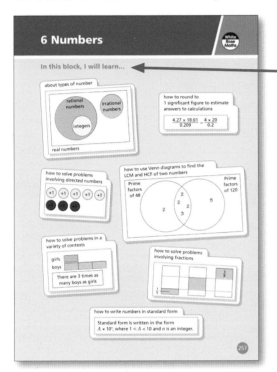

Block overviews Each block of related chapters starts with a visual introduction to the key concepts and learning you will encounter.

Small steps The learning for each chapter is broken down into small steps to ensure progression and understanding. The 🄗 symbol against a chapter or small step indicates more challenging content. The 🄡 symbol against a small step indicates revision content.

Key words Important terms are defined at the start of the chapter, and are highlighted the first time that they appear in the text. Definitions of all key terms are provided in the glossary at the back of the book.

Are you ready? Remind yourself of the maths you already know with these questions, before you move on to the new content of the chapter.

Models and representations Familiarise yourself with the key visual representations that you will use through the chapter.

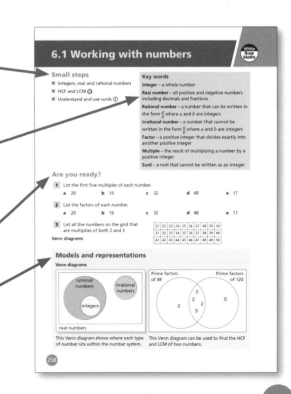

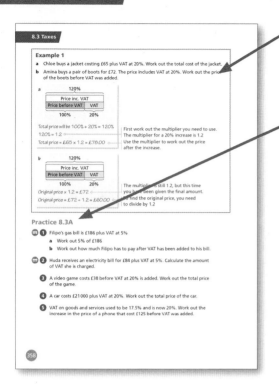

Worked examples Learn how to approach different types of questions with worked examples that clearly walk you through the process of answering, using lots of visual representations.

Practice Put what you've just learned into practice. Icons suggest tools or skills to help you approach a question:

- use manipulatives such as multi-link cubes or Cuisenaire rods
- draw a bar model
- draw a diagram
- discuss with a partner
- think deeply

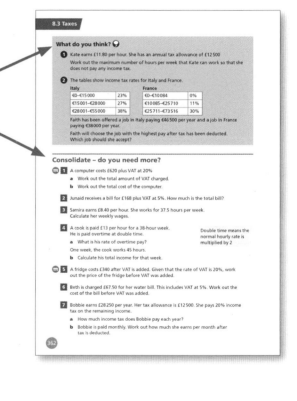

There is a **What do you think?** section at the end of every practice exercise.

Consolidate Reinforce what you've learned in the chapter with additional practice questions.

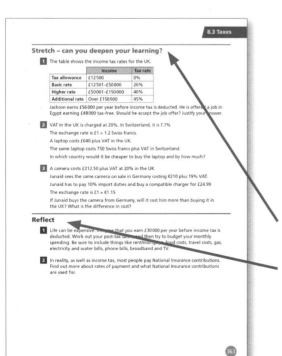

Stretch Take the learning further and challenge yourself to apply it in new ways.

Reflect Look back over what you've learned to make sure you understand and remember the key points.

Block summary Bring together the learning at the end of the learning block with fluency, reasoning and problem-solving statements and review questions.

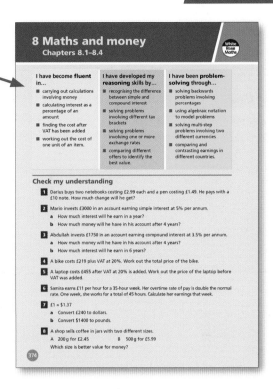

Answers Check your work using the answers provided at the back of the book.

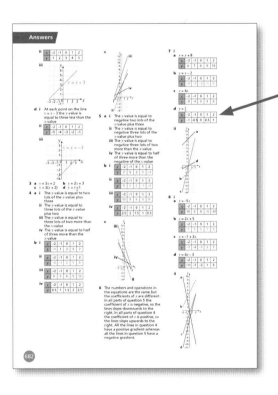

1 Straight line graphs

In this block, I will learn...

how to recognise the graphs of $y = x$ and $y = -x$

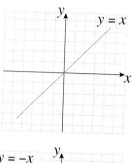

how to plot graphs from a table of values

x	−2	−1	0	1	2
y	−10	−7	−4	−1	2

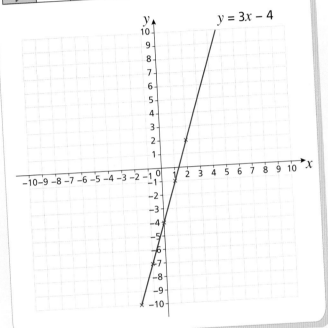

about perpendicular lines ⒣

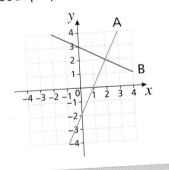

how to interpret the gradient and y-intercept of a straight line

$$y = 2x + 1$$

Gradient y-intercept
$(m) = 2$ $(c) = +1$

about inverse proportion in real-life situations ⒣

Number of bakers	1	2	5	10	15	20	30
Time taken (mins)	60	30	12	6	4	3	2

1.1 Plotting and reading graphs

Small steps

- Recognise lines parallel to the axes, and also the lines $y = x$ and $y = -x$ (R)
- Use tables of values (R)

Key words

Parallel – always the same distance apart and never meeting

Axis (plural: **axes**) – a reference line on a graph

Coordinate – an ordered pair used to describe the position of a point

Table of values – this lists pairs of values that represent a relationship between two variables and that can be used to plot a graph

Are you ready?

1 a Here are five pairs of coordinates.

   (3, 6)   (3, 0)   (3, 1)   (3, 3)   (3, –10)

   i What do all the coordinate pairs have in common?

   ii Copy and complete this sentence to describe the coordinates.

   In each pair of coordinates, the _____ value is equal to _____.

b Here are five pairs of coordinates.

   (2, –7)   (3, –7)   (0, –7)   (–9, –7)   (1000, –7)

   i What do all the coordinates have in common?

   ii Copy and complete the sentence to describe the coordinates.

   In each pair of coordinates, the _____ value is equal to _____.

2 a Write down three pairs of coordinates whose x-value is equal to –1

   b Write down three pairs of coordinates whose y-value is equal to 4

3 Describe each expression in words. The first one has been done for you.

   a $2x + 1$ *"two lots of x plus one"*

   b $3x + 5$ **c** $5x - 4$ **d** $-x + 1$

4 Evaluate each expression for $x = 5$

   a $2x + 1$ **b** $3x + 5$ **c** $5x - 4$ **d** $-x + 1$

Models and representations

Table of values

$y = 3x$ ○— The y-value is 3 times the x-value.

x	0	1	2	3
y	0	3	6	9

Function machine

This function machine can be used to work out the **coordinates** of points that lie on the line $y = 2x + 5$

Example 1

Draw the graph of $y = 5$

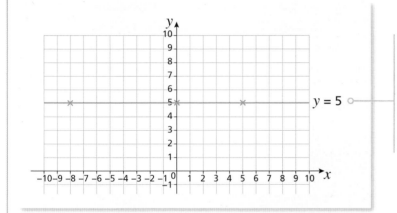

The equation of the line is "y is equal to 5" so the y-value is equal to 5 in the pair of coordinates for all points on the line.

Examples are (–8, 5), (0, 5) and (5, 5)

Practice 1.1A

1   **a**   Draw a coordinate grid with x- and y-axes that go from –10 to 10

  b   **i**   Write down three pairs of coordinates in which the y-value is equal to 3

    ii   Plot your coordinates and join them to draw the line with equation $y = 3$

    iii Choose another point on the line and write down its coordinates.

      How does this confirm that your line has the equation $y = 3$?

  c   On your coordinate grid, draw and label each of these straight lines.

    i $y = -5$ **ii** $y = 1$ **iii** $y = -9$ **iv** $4 = y$

  d   What do you notice? Copy and complete this sentence to describe each line.

    Each line of the form $y = a$ is a horizontal line through point _____ on the _____-axis.

2 Draw a coordinate grid with x- and y-axes that go from −10 to 10

    a   **i**   Write down three pairs of coordinates in which the x-value is equal to 3

        ii   Plot your coordinates and join them to draw the line with equation $x = 3$

        iii Choose another point on the line and write down its coordinates.
           How does this confirm that your line has the equation $x = 3$?

    b On your coordinate grid, draw and label each of these straight lines.

        i $x = -7$ **ii** $x = -1$ **iii** $x = 9$ **iv** $2 = x$

    c What do you notice? Copy and complete this sentence to describe each line.

       Each line of the form $x = b$ is a vertical line through point _____ on the _____-axis.

3 Draw a coordinate grid with x- and y-axes that go from −10 to 10

   Draw and label each of these lines on your grid.

    a $y = 2$ **b** $x = -3$ **c** $x = 9$

    d $y = 10$ **e** $y = -10$ **f** $x = 0$

4 Write down the equation of each line.

    a   **b**   **c**   **d**

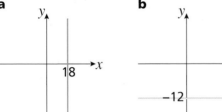
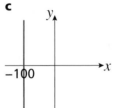
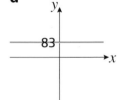

   Discuss with your partner how you know.

5 Draw a coordinate grid with x- and y-axes that go from −10 to 10

    a   **i**   Write down three pairs of coordinates in which the y-value is equal to
           the x-value.

        ii   Plot your coordinates and join them to draw the line with equation $y = x$

        iii Choose another point on the line and write down its coordinates.
           How does this confirm that your line has equation $y = x$?

    b On the same coordinate grid, draw the graph of $y = -x$

What do you think?

1 Any line of the form "$x = \ldots$" is parallel to the x-axis.

Do you agree with Beca? Explain your answer.

2 Find the equation of each line being described.

a A vertical line through –17 on the x-axis.

b A horizontal line through 12.4 on the y-axis.

c A line parallel to the x-axis through (15, 4)

d A straight line through (2, –7), (2.4, –7) and (2.8, –7)

e A straight line on which the x- and y-values are equal at all points.

3 Abdullah is trying to find the equation of this line.

This can't be $y = x$ because the line isn't at a 45 degree angle to the axes.

Do you agree with Abdullah? Explain your answer.

Example 2

a Complete the table of values for $y = 3x - 4$

x	–2	–1	0	1	2
y					

b Draw the graph of $y = 3x - 4$

a

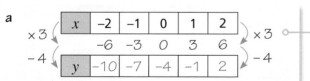

You can use this function machine to help you to work out the values in the table.

Input   Output

x → × 3 → – 4 → y

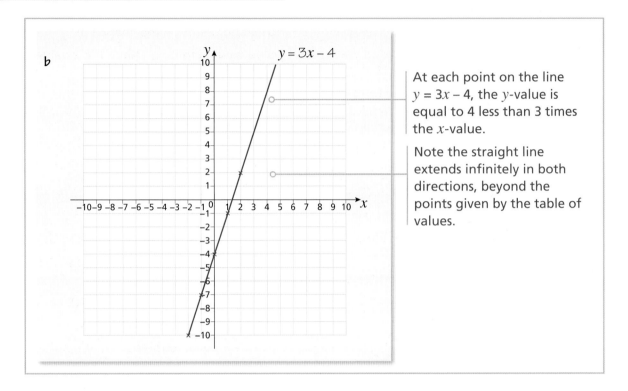

b

$y = 3x - 4$

At each point on the line $y = 3x - 4$, the y-value is equal to 4 less than 3 times the x-value.

Note the straight line extends infinitely in both directions, beyond the points given by the table of values.

Practice 1.1B

1 Match each function machine with the correct equation.

a Input → x → × 3 → y ← Output

$y = -3x$

b Input → x → + 3 → y ← Output

$y = x - 3$

c Input → x → × –3 → y ← Output

$y = 3 + x$

d Input → x → – 3 → y ← Output

$y = 3x$

2 **a** **i** Copy and complete this sentence to describe the points on the line $y = 3x$

At each point on the line $y = 3x$, the y-value is equal to _____ times the x-value.

ii Complete a table of values for $y = 3x$ using values of x from –2 to 2

iii Draw the graph of $y = 3x$ for values of x from –2 to 2

b **i** Copy and complete the sentence to describe the points on the line $y = -3x$

At each point on the line $y = -3x$, the y-value is equal to _____ times the x-value.

ii Complete a table of values for $y = -3x$ using values of x from –2 to 2

iii Draw the graph of $y = -3x$ for values of x from –2 to 2

c **i** Copy and complete the sentence to describe the points on the line $y = x + 3$

At each point on the line $y = x + 3$, the y-value is equal to three _____ than the x-value.

ii Complete a table of values for $y = x + 3$ for values of x from –2 to 2

iii Draw the graph of $y = x + 3$ for values of x from –2 to 2

d **i** Copy and complete the sentence to describe the points on the line $y = x - 3$

At each point on the line $y = x - 3$, the y-value is equal to three _____ than the x-value.

ii Complete a table of values for $y = x - 3$ using values of x from –2 to 2

iii Draw the graph of $y = x - 3$ for values of x from –2 to 2

3 Match each function machine with the correct equation.

a Input   Output

x → × 3 → + 2 → y

$y = \dfrac{x + 3}{2}$

b Input   Output

x → × 2 → + 3 → y

$y = 3x + 2$

c Input   Output

x → + 2 → × 3 → y

$y = 3(x + 2)$

d Input   Output

x → + 3 → ÷ 2 → y

$y = 2x + 3$

4 Here are the equations of four straight lines.

i $y = 2x + 3$ **ii** $y = 3x + 2$ **iii** $y = 3(x + 2)$ **iv** $y = \dfrac{x + 3}{2}$

For each line

a describe the relationship between the x- and y-values at each point

b complete a table of values using values of x from –2 to 2

c draw the graph for values of x from –2 to 2

Draw all four graphs on the same coordinate grid. Remember to label each line with its equation.

5 Here are the equations of four lines.

i $y = -2x + 3$　　　　**ii** $y = 2 - 3x$　　　　**iii** $y = 3(-x + 2)$　　　　**iv** $y = \dfrac{3 - x}{2}$

For each line

a describe the relationship between the x- and y-values at each point

b complete a table of values using values of x from –2 to 2

c draw the graph for values of x from –2 to 2

Draw all four graphs on the same coordinate grid. Remember to label each line with its equation.

6 What's the same and what's different about the lines in questions **4** and **5**? Why does this happen?

7 For each line

　i complete a table of values using values of x from –2 to 2

　ii draw the graph for values of x from –2 to 2

a $y = x + 8$　　　　**b** $y = x - 2$　　　　**c** $y = 4x$　　　　**d** $y = \dfrac{x}{2}$

8 For each line

　i complete a table of values using values of x from –2 to 2

　ii draw the graph for values of x from –2 to 2

a $y = -5x$　　　　**b** $y = 2x + 5$　　　　**c** $y = -1 + 3x$　　　　**d** $y = 4x - 3$

9 For each line

　i complete a table of values using values of x from –2 to 2

　ii draw the graph for values of x from –2 to 2

a $y = \dfrac{5 - x}{4}$　　　　**b** $y = 2 - 6x$　　　　**c** $y = 3(x + 1)$　　　　**d** $y = 3(2x + 1)$

What do you think?

1 Zach and Flo have each sketched the graph of $y = 3x + 7$. Explain why each of them must have made a mistake.

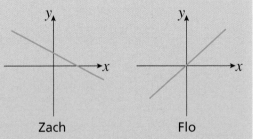

Zach　　　　　　　　　Flo

2 Chloe has completed a table of values for $y = 15x - 8$ using values of x from –2 to 2. She says, "I can see a sequence in my y-values." What is the term-to-term rule for the sequence that Chloe has spotted?

3 Seb has completed a table of values for $y = 5x - 1$. He says, "The equation starts with $5x - 1$ so I only need to find the first value then add 5 each time."

a Explain why Seb's method hasn't worked.

b Will Seb's method ever work?

x	0	1	2	4	20
y	–1	4	9	14	19

Consolidate – do you need more?

1 Write down the equation of each line.

a

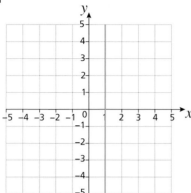

b

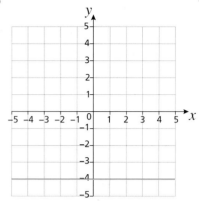

c

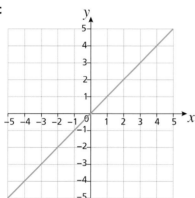

d

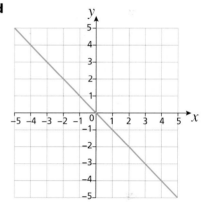

2 a Copy and complete each table of values for the given equations.

i $y = 2x$

x	−2	−1	0	1	2
y					

ii $y = 4x$

x	−2	−1	0	1	2
y					

iii $y = x + 2$

x	−2	−1	0	1	2
y					

iv $y = x + 4$

x	−2	−1	0	1	2
y					

b Draw a coordinate grid with x- and y-axes that go from −8 to 8.
Plot the graph of each equation on your grid.

    i $y = 2x$ **ii** $y = 4x$ **iii** $y = x + 2$ **iv** $y = x + 4$

c What's the same and what's different about the graphs?

3 **a** Copy and complete each table of values for the given equations.

i $y = 2x - 1$

x	-2	-1	0	1	2
y					

ii $y = 2x + 2$

x	-2	-1	0	1	2
y					

b Draw a coordinate grid with x- and y-axes that go from -8 to 8. Plot the graph of each equation on your grid.

i $y = 2x - 1$ **ii** $y = 2x + 2$

c What's the same and what's different about the graphs?

Stretch – can you deepen your learning?

1 Ed has drawn the line with equation $y = -x$

Explain how you know that Ed must have made a mistake.

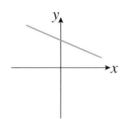

2 Could this be the graph of $y = 7x + 3$? Explain your answer.

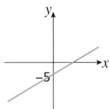

3 Which of these pairs of values lie on the line $y = -x$? How do you know?

$(17, -\frac{34}{2})$ $(12, \sqrt{144})$ $(2ab, 2ba)$ $(5 + p, p + 5)$ $(7 - q, q - 7)$

4 Without drawing, write down the coordinates of the point where the line $y = 17$ meets the line $x = -19.8$

5 Do these three lines meet at a single point? Show working to support your answer.

$7 = x$ $-2 = y$ $y = 3 - \frac{5x}{7}$

6 A quadrilateral is bordered by these four lines.

$y = 2$ $y = -7$ $x = 6$ $y = 2x + 4$

a Write down the coordinates of a point

i on the perimeter of the quadrilateral

ii inside the quadrilateral.

b Write the coordinates of each vertex of the quadrilateral.

c Find the area of the quadrilateral.

Reflect

1 How can you tell from its equation whether a line will be horizontal, vertical or diagonal?

2 How does the equation of a line describe the relationship between the coordinates at any point on the line?

1.2 Gradients and intercepts

Small steps

- Compare gradients
- Compare intercepts
- Understand and use $y = mx + c$

Key words

Gradient – the steepness of a line

Intercept – the point at which a graph crosses, or intersects, a coordinate axis

Are you ready?

1 Copy and complete the table of values for $y = 4x - 1$

x	−2	−1	0	1	2
y					

2 Draw x- and y-axes numbered from −8 to 8. Draw the graph of each of these lines on your grid.

    a $x = -4$ **b** $y = 6$ **c** $y = x$ **d** $y = 4x - 1$

3 **i** Generate the first five terms of the sequences given by the following nth terms.

    ii Write down the term-to-term rule of each sequence.

    a $3n + 1$ **b** $3n - 4$

    c What do you notice? Why does this happen?

Models and representations

Table of values

$y = 3x$ ○─┤ The y-value is 3 times the x-value.

x	0	1	2	3
y	0	3	6	9

Function machine

Input   Output

x ⟶ × 2 ⟶ + 5 ⟶ y

This function machine can be used to work out the values of points that lie on the line $y = 2x + 5$

Example 1

a Draw each of these lines on the same coordinate grid.

   i $y = 2x + 1$ **ii** $y = 2x - 4$ **iii** $y = -x + 1$

b Which two lines are parallel? Why does this happen?

c Which two lines have the same y-intercept? Why does this happen?

a

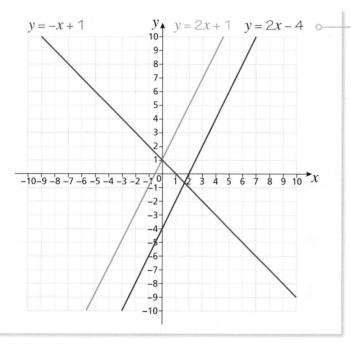

Remember to label each line when you have drawn it.

You can use a table of values to help you to work out the values of some points on each line.

b $y = 2x + 1$ and $y = 2x - 4$ are parallel because they have the same gradient.

This happens because the equations of these lines have the same coefficient of x. The number before x is 2 in both equations.

c $y = 2x + 1$ and $y = -x + 1$ have the same y-intercept.

This is because they cut the y-axis at the same point, (0, 1)

This happens because the equations of these lines have the same constant term (in both equations it is +1), so when $x = 0$ they have the same value.

Practice 1.2A

1   **a**   Draw each of these lines on the same coordinate grid.

     $y = x$ $y = 2x$ $y = 3x$ $y = 6x$

     b   What do you notice?

     c   In your own words, explain what happens to a line when you increase its gradient.

2 **a** Draw each of these lines on the same coordinate grid.

$y = -x$ $\qquad$ $y = -2x$ $\qquad$ $y = -3x$ $\qquad$ $y = -6x$

b What do you notice?

c In your own words, explain what it means for the gradient of a line to be negative.

3 State whether each line has a positive or a negative gradient.

a 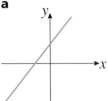 **b** 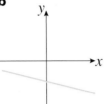 **c** 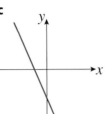 **d**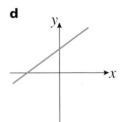

Discuss with your partner how you know.

4 **a** Draw each of these lines on the same coordinate grid.

$y = 2x$ $\qquad$ $y = 2x + 1$ $\qquad$ $y = 2x + 5$ $\qquad$ $y = 2x - 4$

b All of the lines are parallel.

Explain how you can tell this from

i the graphs $\qquad\qquad$ **ii** the equations.

c Write down the equation of another line that is parallel to these lines.

5 **a** Draw each of these lines on the same coordinate grid.

$y = x + 2$ $\qquad$ $y = x + 3$ $\qquad$ $y = x + 5$ $\qquad$ $y = x + 8$

b What do you notice?

c All of these lines have a positive y-intercept. Explain what this means.

6 **a** Draw each of these lines on the same coordinate grid.

$y = x - 2$ $\qquad$ $y = x - 3$ $\qquad$ $y = x - 5$ $\qquad$ $y = x - 8$

b What do you notice?

c All of these lines have a negative y-intercept. Explain what this means.

7 State whether each of these lines has a positive or negative y-intercept.

a 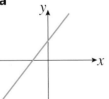 **b** 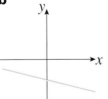 **c** 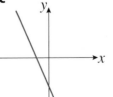 **d**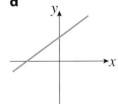

Discuss with your partner how you know.

8 **a** Draw each of these lines on the same coordinate grid.

$y = 2x + 3$ $y = 5x + 3$ $y = 3 + 4x$ $y = 3 - x$

b All of the lines have the same y-intercept.

Explain how you can tell this from

i the graphs   **ii** the equations.

c Write down the equation of another line that has the same y-intercept.

What do you think?

1

$y = 2 + 3x$ and $y = 2x + 1$ are parallel because they each have 2 after the "equal to" symbol.

Is Marta correct? Give reasons to support your answer.

2 Line A has equation $y = 3x + 5$. Give the equation of a line that is

a steeper than A   **b** parallel to A

c less steep than A   **d** sloping in the opposite direction to A

3 **a** Three of the four lines shown in the diagram are parallel.

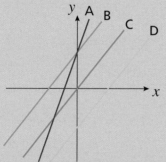

Which line is not parallel to the others? Explain how you know.

b The equation of line B is $y = 5x + 9$. Suggest equations for the other lines.
Compare your answers with a partner.

4 The line with equation $y = 5x - 7$ intersects the y-axis at the same point as the line $y = b - x$. Work out the value of b.

The general equation of a straight line is $y = mx + c$, where m is the **gradient** and c is the y-**intercept**.

Example 2

A straight line has equation $y = 4x - 9$

Write down

a the gradient of the line

b the coordinates of the y-intercept.

a 4 ○—| The gradient of the straight line is given by the coefficient of x in the equation.

b $(0, -9)$ ○—| The y-intercept of a straight line is given by the constant in the equation.

You are asked for the coordinates of the y-intercept, so you need to give the value of x as well as the value of y

Practice 1.2B

1 Here are the equations for eight lines.

    i $y = 5x + 3$ **ii** $y = 2x - 7$ **iii** $y = -4x + 1$

    iv $y = -3x - 9$ **v** $y = \frac{3}{5}x$ **vi** $y = -\frac{x}{3} + 8$

    vii $7x - 11 = y$ **viii** $-100x + 134 = y$

    For each line, state

a the gradient

b the coordinates of the y-intercept.

2 Here are the equations for four lines.

    i $y = 10x + 19$ **ii** $y = -10x + 19$

    iii $y = 10x - 19$ **iv** $y = -10x - 19$

    For each of these lines, state

a the gradient

b the coordinates of the y-intercept.

c What's the same and what's different about the lines?

3 Here are the equations for eight lines.

    i $y = 5 + 3x$ **ii** $y = 7 - 2x$ **iii** $y = 0.5 - 4x$

    iv $y = 19 + 3x$ **v** $y = 11 + \frac{3}{5}x$ **vi** $y = 54 - \frac{2x}{3}$

    vii $7 - 11x = y$ **viii** $-100 - 134x = y$

    For each line, state

a the gradient

b the coordinates of the y-intercept.

4 Here are the equations for eight lines.

i $y = 20x - 17$ **ii** $y = 32x + 11$ **iii** $y = 19 - 32x$

iv $y = 19 - 20x$ **v** $y = 11 - 20x$ **vi** $y = -17 + 32x$

vii $y = 0.75 + 20x$ **viii** $\frac{3}{4} - 32x = y$

Identify

a four pairs of parallel lines

b four pairs of lines that have the same y-intercept.

c Explain how you identified the lines in each case.

5 Write down the equations of three lines that

a are parallel to $y = 23x + 9$

b are parallel to $y = 1 - 7x$

c intersect the y-axis at the same point as the line $y = 23x + 9$

d intersect the y-axis at the same point as the line $y = 1 - 7x$

Compare your answers with a partner. What do you notice?

6 Without drawing, decide whether each pair of lines will be parallel. Explain your reasoning.

a $y = 4x + 5$ and $y = 3x + 5$ **b** $y = 7 - x$ and $y = -x + 4$

c $y = \frac{2}{3}x - 9$ and $y = 4 + \frac{2x}{3}$ **d** $y = 18x - 11$ and $y = 11 - 18x$

e $y = 4(3x + 1)$ and $y = 3x - 8$ **f** $y = 2x + 7$ and $y = 2(x - 9)$

7 Without drawing, decide whether each pair of lines will intersect the y-axis at the same point. Explain your reasoning.

a $y = 4x + 5$ and $y = 3x + 5$ **b** $y = 0.5 - x$ and $y = -x + \frac{1}{2}$

c $y = \frac{2}{3}x - 9$ and $y = 9 + \frac{2x}{3}$ **d** $y = 18x + 11$ and $y = 11 - 8x$

e $y = 4(3x + 1)$ and $y = 3x + 1$ **f** $y = 2x - 7$ and $y = 2(x - 3.5)$

8 Write the equations of the lines with these gradients and y-intercepts.

a gradient 7 and y-intercept (0, 2)   **b** gradient 4 and y-intercept (0, –9)

c gradient –5 and y-intercept (0, 0)   **d** gradient –1 and y-intercept (0, 11)

9 A straight line has gradient 72 and intersects the y-axis at (0, –5). Write the equation of the line.

10 A straight line is parallel to $y = 10x + 14$ and intersects the y-axis at (0, –21). Write the equation of the line.

What do you think?

1 The line $y = 5 + 2x$ has gradient 5 and y-intercept 2.

    a   Explain the mistake Jackson has made.

    b   Correct Jackson's answer.

2 $y = 3x + 1$ and $y = 7 - 3x$ both have a gradient of 3, so they're parallel.

    Do you agree with Faith? Explain your answer.

3 The gradient of $y = \dfrac{119x}{21} + 5$ is $\dfrac{119x}{21}$

    Explain why Ed is wrong.

4 Identify the gradient and y-intercept of the straight line with equation $y = \dfrac{4x + 25}{5}$

Consolidate – do you need more?

1 Write down the gradient of each of these lines.

    a $y = 8x - 3$
    b $y = 10 + x$
    c $y = 15 - 6x$

2 Write down the coordinates of the y-intercept of each of these lines.

    a $y = 17 + 2x$
    b $y = \dfrac{1}{5}x + \dfrac{2}{3}$
    c $y = x - 12.5$

3 Emily says that $y = 3x + 5$ and $y = 5 + 3x$ will produce the same graph.

Do you agree with Emily? Explain your answer.

4 Write down the equations of three lines that

    a   are steeper than the line $y = 2x$

    b   intersect the y-axis at (0, 4)

    c   are parallel to the line $y = 7x - 3$

Stretch – can you deepen your learning?

1 Beca says that $y = 8x + \sqrt{25}$ and $y = 5 + 8x$ are parallel. Marta says they are not.

Who do you agree with? Explain your answer.

2 The equation of line A is $y = 19x + 13$

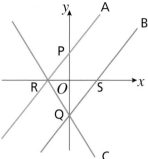

    a The distance from O to P is the same as the distance from O to Q. The lines A and B are parallel. What is the equation of line B?

    b The distance from O to R is the same as the distance from O to S. What is the equation of line C?

3 Lines l_1 and l_2 are parallel. The equation of l_1 is $y = \dfrac{5x + 9}{4}$. l_2 passes through the point $(0, 34)$. Find an equation for l_2

4   **a** A straight line has equation $y = 5x + 2a + 3$. The line passes through the point $(0, 19)$. Work out the value of a

    b A straight line has equation $y = jx + 3j + k$. The gradient of the line is $-\dfrac{1}{2}$

    The line passes through the point $(0, 10.5)$. Work out the values of j and k

5 Here is the graph of $y = a - \dfrac{4}{3}x$

    a Beca says, "The equation should be $y = -a - \dfrac{4}{3}x$"

      i Suggest why Beca may think this.

      ii Explain why Beca is not correct.

    b On a copy of the graph, sketch the graphs of

      i $y = -a - \dfrac{4}{3}x$

      ii $y = a - x$

      iii $y = a + \dfrac{4}{3}x$

6 Faith and Zach have each written the equation of a line parallel to $y = 12x + 7$

    a They have different answers, so Zach thinks that one of them must be wrong. Explain why Zach's thinking is incorrect.

    b Faith says, "10000 people could each get a different answer and still be correct."

    Do you agree? Explain your reasoning.

Reflect

Explain how you can tell from their equations whether two lines

a are parallel

b have the same y-intercept.

1.3 Equations of lines

Small steps

- Write an equation in the form $y = mx + c$
- Find the equation of a line from a graph
- Interpret gradient and intercepts of real-life graphs

Key words

Equation – a statement with an equal sign, which states that two expressions are equal in value

Gradient – the steepness of a line

Intercept – the point at which a graph crosses, or intersects, a coordinate axis

Are you ready?

1 Use the fact that $5x + 5y = 70$ to work out the value of each expression.

   a $10x + 10y$ **b** $20x + 20y$ **c** $x + y$ **d** $3x + 3y$ **e** $5x$

2 A straight line has equation $y = -4x + 9$. Write down

   a the gradient of the line   **b** the coordinates of the y-intercept.

3 A straight line has gradient 7 and intersects the y-axis at $(0, -3)$. What is the equation of the line?

4 10 identical books cost £62.50. How much do 15 of the same book cost?

Models and representations

Coordinate grid

Always check the scale. A square may represent a different value on each axis.

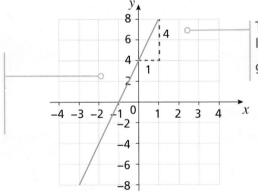

To work out the **gradient** of a line you can use the formula

$$\text{gradient} = \frac{\text{change in } y}{\text{change in } x}$$

20

Example 1

A straight line has equation $7y = 21x + 63$

a Write the equation in the form $y = mx + c$

b Find

    i the gradient of the line

    ii the coordinates of the y-intercept.

a $7y = 21x + 63$ Your **equation** currently starts with $7y = ...$, so you need to adjust it to make it in the form $y = ...$

     $\div 7 \quad \div 7 \quad \div 7$ To get from $7y$ to y, you need to divide by 7. Remember to divide everything by 7 to keep the equation balanced.

     $y = 3x + 9$

b **i**   3   When the equation of the line is in the form $y = mx + c$ you can see that the gradient is 3

Remember: the gradient is given by the value of m, which is the coefficient of x

     ii $(0, 9)$ The value of c is 9, which means that the coordinates of the y-**intercept** are (0, 9)

Remember: the y-intercept is given by the value of c, which is the constant term in the equation.

Practice 1.3A

1 Write each of these equations in the form $y = mx + c$

     a $2y = 18x + 20$ **b** $3y = 6x + 21$ **c** $10y = 40x + 50$

     d $6y = 24x + 6$ **e** $5y = 90 + 10x$ **f** $12x + 24 = 3y$

     g $35 + 77x = 7y$ **h** $4y = 10 + 8x$ **i** $2y = 7x + 8$

     j $10y = 15 + 60x$ **k** $15x + 45 = 30y$ **l** $21y = 14x + 7$

2 For each line in question **1**, find

     i the gradient

     ii the coordinates of the y-intercept.

3

> The line with equation $y - 7 = 5x$ intersects the y-axis at (0, −7)

Explain why Beca is incorrect.

4 Write each of these equations in the form $y = mx + c$

a $y + 9 = 2x$ b $y - 8x = 11$ c $y + x = 1$ d $y - 11 = 3x$

e $12 + y = 4x$ f $y - \dfrac{1}{2} = 10x$ g $y - 5x - 1 = 0$ h $0 = 11x - 7 - y$

5 For each line in question **4**, find

     i the gradient

     ii the coordinates of the y-intercept.

6 Write the equation of a line parallel to each of these.

a $7y = 21x - 35$ b $y - 10 = 6x$ c $11x + y = 15$ d $20 + 4x = 2y$

e $y + 20 = x$ f $20y = 5 - 10x$ g $x + y = 16$ h $0.5y = 2x + 1$

7 Write the equation of a line with the same y-intercept as each of these.

a $x + y = 0$ b $2y = 8 - 4x$ c $100y = 20x + 300$ d $7 = 15x + y$

e $60y = x + 60$ f $-15x + y = 17$ g $3(2x + 5) = 6y$ h $2y = 5(11 - 6x)$

What do you think? 💭

1 Two straight lines have equations $4y = 12(x + 1)$ and $y - 3 = 3x$

     a Show that these points lie on both of the lines.

         i (2, 9)   **ii** (−1, 0)

     b Chloe says, "Any point that lies on one line will lie on the other."

        Explain why Chloe is correct.

2 Match the equations which represent the same straight lines.

a

$y - 3x = 11$ $20y = 220 - 20x$

b

$x + y = 11$ $y - \dfrac{x}{3} = \dfrac{11}{3}$

c

$3y - x = 11$ $\dfrac{y}{8} = 1.375 - \dfrac{3}{8}x$

d

$11 - 3x = y$ $121 + 33x = 11y$

3 For each line, find

     i the gradient   **ii** the coordinates of the y-intercept.

     a $\dfrac{1}{4}y = 5x - 7$ **b** $\dfrac{y}{3} = \dfrac{3}{2}x + 1$ **c** $x + 7 = \dfrac{3}{5}y$ **d** $\dfrac{5y}{8} = \dfrac{4}{7}x - 3$

The gradient of a straight line can be calculated using gradient = $\dfrac{\text{change in } y}{\text{change in } x}$

Example 2

The diagram shows a straight line.

a Write down the coordinates of the y-intercept.

b Calculate the gradient.

c Find the equation of the line.

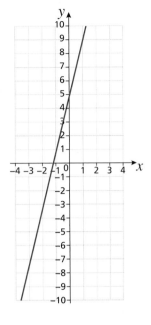

a $(0, 5)$ ⊙──────────── The line intersects the y-axis at $(0, 5)$

b

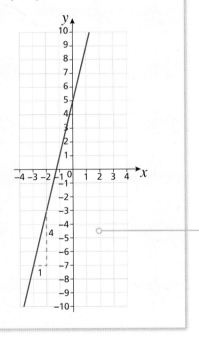

⊙──────── The gradient is how far up or down the line travels for every one unit across. For each 1 square across, the line goes up 4 squares, so the gradient of the line is 4

gradient $= \dfrac{\text{change in } y}{\text{change in } x}$ ○————| Use this formula to calculate the gradient.

$= \dfrac{4}{1}$

$= 4$

the gradient is 4

c $y = 4x + 5$ ○————| You can write the equation in the form $y = mx + c$, where m is the gradient and c is where the line cuts the y-axis.

Practice 1.3B

1 The gradient of each line is given. Write the equation of each line.

a gradient = 1

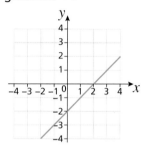

b gradient = –2

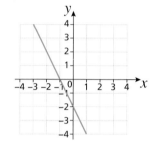

c gradient = 3

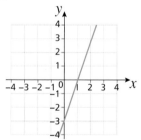

d gradient $= -\dfrac{1}{2}$

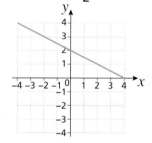

2 Find the gradient of each line.

a

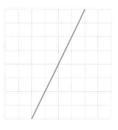

b

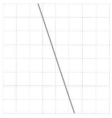

c

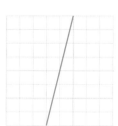

d

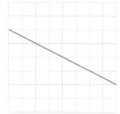

3 Find an equation for each line.

a

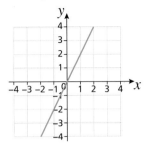

b

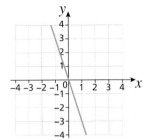

c

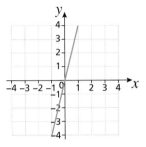

d

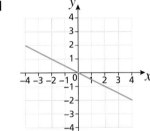

4 Find an equation for each line.

a

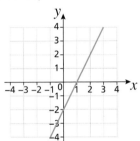

b

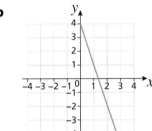

c

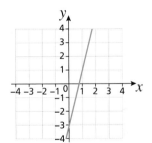

d

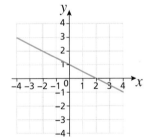

5 Find an equation for each line.

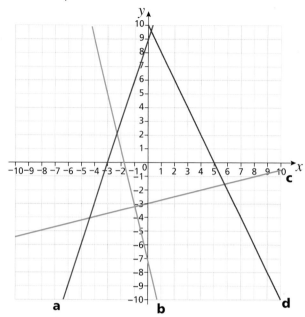

6 The graph shows the cost of buying a given number of books.

    a Use the graph to find the cost of

        i 7 books

        ii 1 book.

    b **i** Find an equation of the line.

        ii Interpret the gradient and
y-intercept of the line in
the context of the question.

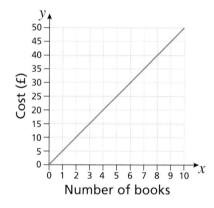

7 The graph shows the cost of a taxi journey for a given number of miles.

    a Use the graph to find the cost of

        i a 1-mile journey

        ii a 6-mile journey.

    b Work out the cost per extra mile
travelled.

    c Interpret the gradient and
y-intercept of the line in the
context of the question.

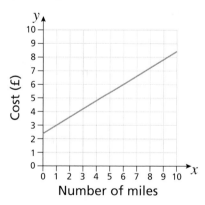

8 The graph shows the mass of a box when some identical tins are put into it.

  a Use the graph to estimate the mass of

    i one tin

    ii the empty box.

  b **i** Find an equation of the line.

    ii Interpret the gradient and y-intercept of the line in the context of the question.

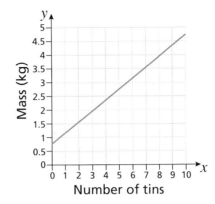

What do you think? 💭

1 **a** Ed has found the gradient of this line to be 5
    Without calculation, explain why Ed must be incorrect.

  b Emily has found the gradient of the line to be $-\dfrac{1}{5}$. This is incorrect. What mistake do you think Emily has made?

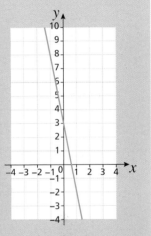

2 Beca says that the equation of this line is $y = \dfrac{1}{2}x$

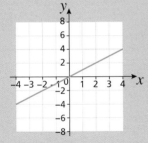

Explain why Beca is incorrect.

3 Line P is parallel to the line shown. Line P passes through the point $(0, -7)$. Find an equation of line P.

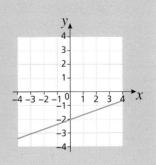

27

Consolidate – do you need more?

1 Write each of these equations in the form $y = mx + c$

   a $2y = 4x + 6$ **b** $3y = 12x - 9$ **c** $10y = 60x + 70$

2 Ed says, "I can write $\frac{1}{2}y = 3x + 2$ in the form $y = mx + c$ by multiplying it by 2"

   Show that Ed is correct.

3 Write $\frac{1}{4}y = 5x + 1$ in the form $y = mx + c$

4 Work out the gradient and the coordinates of the y-intercept of these lines.

   a $4y = 8x + 20$ **b** $9y = 27x - 9$ **c** $11y = 33 + 55x$

5 Find an equation of each line.

a

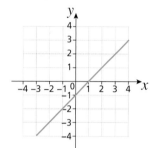

b

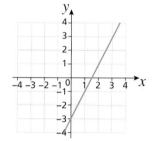

c

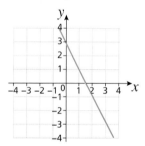

d

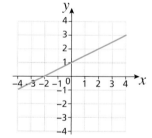

Stretch – can you deepen your learning?

1 True or false? The gradient of the line with equation $by = 7bx + 18b$ is 7

2 Find an equation of each line.

a

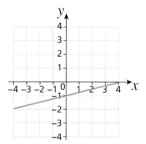

b

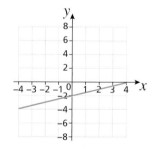

c 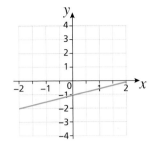 **d**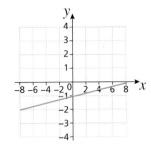

3 The coordinates of point A are (18, –20).
Find an equation of the line shown.

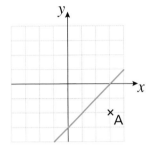

4 On a coordinate grid, plot the points (–3, 5) and (2, –5)

    a Draw the straight line that passes through these points.

    b Find an equation of your line in part **a**

5 The cost of hiring a bouncy castle is £65 plus £20 per hour. Chloe is going to draw a graph to represent this information.

    a Explain why Chloe's graph will be a straight line.

    b **i** Find an equation of Chloe's straight line.

       ii Interpret the gradient and y-intercept of the line in the context of the question.

6 **a** The straight line with equation $2y = ax + 11$ has gradient 18. Work out the value of a

    b The straight line with equation $y + b = 5x$ intersects the y-axis at (0, 3).
Work out the value of b

    c The straight line with equation $cy = 24x + 5$ has gradient 3. Find the coordinates of the y-intercept.

Reflect

What do you think are the most common mistakes when finding an equation of a straight line?

Small steps

- Explore perpendicular lines Ⓗ
- Model real-life graphs involving inverse proportion Ⓗ

Key words

Inverse proportion – if two quantities are in inverse proportion, when one quantity increases, the other decreases at the same rate

Perpendicular – at right angles to

Reciprocal – the result of dividing 1 by a given number. The product of a number and its reciprocal is always 1

Are you ready?

1 The area of each rectangle is 48 cm². The lengths are not drawn to scale. Work out the length of the unknown side in each shape.

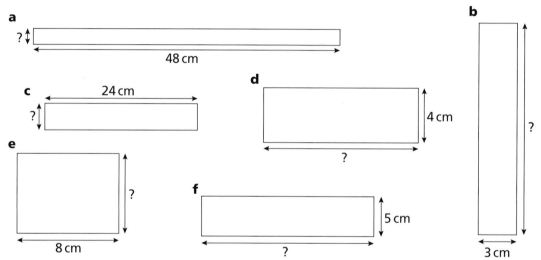

2 Write the missing number in each of these calculations.

a $\frac{1}{2} \times \boxed{} = 1$

b $\frac{1}{3} \times \boxed{} = 1$

c $\boxed{} \times \frac{1}{9} = 1$

d $11 \times \boxed{} = 1$

e $100 \times \boxed{} = 1$

f $\boxed{} \times \frac{1}{5} = 1$

g $\frac{2}{5} \times \boxed{} = 1$

h $\boxed{} \times \frac{3}{4} = 1$

3 Write the reciprocal of each number.

a 3   b 7   c $\frac{1}{4}$ d $\frac{1}{10}$ e $\frac{3}{2}$ f $\frac{7}{8}$

Models and representations

Table of values

x	1	2	5	10	100
y	100	50	20	10	1

If x and y are **inversely proportional** to one another, then the product of the two numbers in each column is the same. This fact can help you to find other values.

Two lines are **perpendicular** if the product of their gradients is –1
The gradient of one line will be the negative **reciprocal** of the other.

Example 1

a A and B are perpendicular lines. Explain how this can be seen from the graph.

b Find the equation of

   i line A   **ii** line B

c Show numerically that lines A and B are perpendicular.

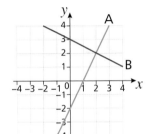

a A and B meet at a right angle.

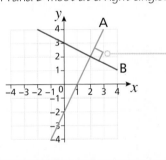

Two lines are perpendicular when they meet at a right angle.

b

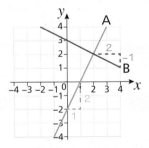

Find the gradient and y-intercept of each line.

  i   gradient $= \dfrac{\text{change in } y}{\text{change in } x} = \dfrac{2}{1} = 2$

     The y-intercept is at $(0, -2)$

     The equation of A is $y = 2x - 2$ Write the equation in the form $y = mx + c$

ii   gradient $= \dfrac{\text{change in } y}{\text{change in } x} = -\dfrac{1}{2}$

The y-intercept is at $(0, 3)$

The equation of B is $y = -\dfrac{1}{2}x + 3$

c   Let the gradient of lines A and B be m_A and m_B respectively.

Then $m_A \times m_B = 2 \times -\dfrac{1}{2}$ ○——┤ Always define any notation that you use.

$\qquad\qquad\quad = -\dfrac{2}{2}$

$\qquad\qquad\quad = -1$ ○——┤ Two lines are perpendicular if the product of their gradients is −1

Lines A and B are perpendicular.

Practice 1.4A

① By finding the product of the gradients, show that each pair of lines are perpendicular.

a

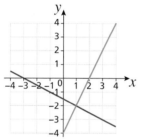

b

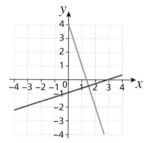

c

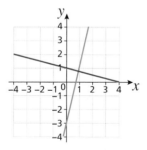

d

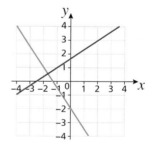

② Write the missing number in each of these calculations.

a $\dfrac{1}{2} \times \boxed{} = -1$ b $\dfrac{1}{3} \times \boxed{} = -1$ c $\boxed{} \times \dfrac{1}{9} = -1$

d $-11 \times \boxed{} = -1$ e $-100 \times \boxed{} = -1$ f $\boxed{} \times \dfrac{1}{5} = -1$

g $\dfrac{2}{5} \times \boxed{} = -1$ h $\boxed{} \times -\dfrac{3}{4} = -1$

3 Write the negative reciprocal of each number.

a 3 **b** 7 **c** $-\frac{1}{4}$ **d** $-\frac{1}{10}$ **e** $\frac{3}{2}$ **f** $-\frac{7}{8}$

4 Find the equation of the line perpendicular to each of the following that passes through the origin.

a $y = 3x$ **b** $y = 7x + 3$ **c** $y = -\frac{1}{4}x - 1$ **d** $-\frac{1}{10}x = y$

e $y = \frac{3x}{2} + 5$ **f** $y = 10 - \frac{7}{8}x$ **g** $y = 9x + 7$ **h** $8 - x = y$

5 Show that $y = 5x - 7$ is perpendicular to $5y = 10 - x$

6 Write the equation of a line that is perpendicular to each of the following.

a $y = \frac{3}{4}x + 9$ **b** $y = 10 - 7x$ **c** $y = \frac{x - 5}{3}$ **d** $y = 3(4x + 1)$

e $4y = 20 - 8x$ **f** $y - 11x = 17$ **g** $7y = 32 - 5x$ **h** $10y = x + 5$

7 Find the equation of the line that is perpendicular to $y = 8x + 9$ and passes through the point $(0, -1)$

8 Find the equation of the line that is perpendicular to $y = 5 - \frac{1}{4}x$ and passes through the point $(0, 28)$

What do you think?

1 Line P has equation $y = 4x + 3$. Line Q is perpendicular to P.

The gradient of line Q is 0.25

Explain the mistake that Zach has made.

2 Lines A and B are parallel. Lines B and C are perpendicular. The equation of line B is $y = 0.6x$. Show that lines A and C are perpendicular.

3 **a** Write an equation of a line

   i perpendicular to $y = 0.75x + 5$

   ii perpendicular to $y = 0.75x + 5$ that passes through the point $(0, 18)$

   b Explain why there are infinitely many answers to part **a i**, but only one answer to part **a ii**.

Example 2

It takes one waitress 30 minutes to clean the tables in a restaurant.

a How long will it take two waitresses working at the same rate?

b Complete the table.

Number of waitresses	1	2	5	10	15
Time taken (minutes)	30				

c Draw a graph to represent this.

a 15 minutes

If there is another waitress working, the job will take less time.

One waitress takes 30 minutes. With twice as many waitresses it will take half the time:

30 minutes ÷ 2 = 15 minutes

This is an example of inverse proportion; when one quantity doubles, the other halves.

The fact that the waitresses are working at the same rate is important.

b

Number of waitresses	1	2	5	10	15
Time taken (minutes)	30	15	6	3	2

The more waitresses that are working, the quicker it will be to clean the tables.

Notice that the product of each column is 30. This is typical of an inverse proportion relationship, and can help when calculating missing values.

c

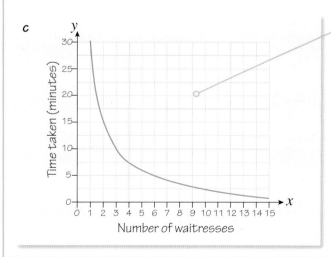

The graph forms a curve with an asymptote at each axis.

The x-axis is an asymptote because the task could not be completed in 'zero' time.

The y-axis is an asymptote because if there were no waitresses, the job would never get done!

Practice 1.4B

1 It takes one baker 60 minutes to ice some cakes.

    a How long would it take two bakers to ice the same number of cakes?

    b Copy and complete the table.

Number of bakers	1	2	5	10	15	20	30
Time taken (minutes)	60						

    c Draw a graph to represent this information.

    d What assumptions have you made?

2 It takes two painters 9 hours to paint a house.

    a How long would the same job take

      i one painter

      ii three painters?

    b Copy and complete the table.

Number of painters	1	2	3	5	6	9	18
Time taken (hours)		9					

    c Draw a graph to represent this information.

    d What assumptions have you made?

3 It takes one machine 15 hours to print some leaflets.

    a How long would it take five machines working at the same rate?

    b Copy and complete the table.

Number of machines	1	2	3	4	5	10	20
Time taken (hours)	15						

    c Draw a graph to represent this information.

4 The time taken to build a shopping centre is inversely proportional to the number of builders working on the task. If there are 20 builders, the shopping centre can be built in 100 days.

    a Draw a graph to represent this information.

    b The project manager wants the shopping centre to be completed in 45 days. How many builders should she hire?

What do you think? 💭

① Explain why there is no column containing zero in any of the tables in the questions in Practice 1.4B.

② The time taken to landscape a garden is inversely proportional to the number of gardeners. If there are four gardeners, the job takes 6 hours. Seb has drawn a graph to represent this information.

Identify

a one thing that Seb has done well

b one mistake that Seb has made.

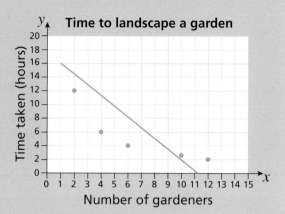

Time to landscape a garden

③ It takes 20 minutes to bake a batch of cupcakes in one oven. Jakub says, "It will take two ovens 10 minutes to bake the same number of cupcakes."

Do you agree? Explain your answer.

Consolidate – do you need more?

1 Write the negative reciprocal of each number.

a 5   **b** $\frac{1}{3}$ **c** −7   **d** $-\frac{4}{5}$

2 Find the equation of the line perpendicular to each of the following and that passes through the origin.

a $y = 5x$ **b** $y = -3x$ **c** $y = 8x$

d $-6x = y$ **e** $y = 2x + 5$ **f** $y = 10 - 4x$

3 Write down the equation of a straight line that is

a perpendicular to $y = 2x + 7$ and passes through the point (0, 1)

b perpendicular to $y = -5x - 6$ and passes through the point (0, 3)

c perpendicular to $y = -\frac{1}{2}x + 1$ and passes through the point (0, −4)

4 It takes two painters 7 hours to paint a house.

a If there were more painters, would you expect it to take more time or less time to paint the house?

b To paint the same house, how long would it take

    i one painter   **ii** four painters?

c What assumptions have you made in your answers to part **b**?

Stretch – can you deepen your learning?

1 Lines J and K are perpendicular. The gradient of J is $\frac{3a}{4}$. Write an expression for the gradient of K.

2 It takes x cleaners y hours to clean a building. Write an expression for how long it would take 7 cleaners working at the same rate to clean the same building.

3 The equation of l_1 can be written as $y = 5x + px - 17.2$. The equation of l_2 can be written as $15y = x + 8$. l_1 and l_2 are perpendicular. Work out the value of p

4 Lines A and B are perpendicular. Line A has equation $y = \frac{2}{3}x + 5$. Line B passes through (4, 10). Find the equation of line B

5 L_1 and L_2 are perpendicular. The equation of L_1 is $y = \frac{1}{2}x - 5$. Find the equation of L_2

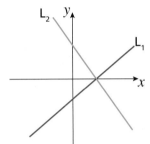

Reflect

1 How can you tell by looking at a graph whether two lines are perpendicular? How can you tell by looking at the equations of the lines?

2 Explain what it means when two quantities are inversely proportional.

I have become **fluent in...**	I have developed my **reasoning** skills by...	I have been **problem-solving** through...
■ recognising and plotting lines parallel to the axes ■ recognising and plotting the lines $y = x$ and $y = -x$ ■ using tables of values ■ drawing lines of the form $y = mx + c$ ■ identifying gradients and intercepts of straight lines.	■ comparing gradients of straight lines using graphs and equations ■ comparing intercepts of straight lines using graphs and equations ■ interpreting gradients and intercepts of real-life graphs ■ explaining why a point does or does not lie on a straight line.	■ modelling real-life graphs involving inverse proportion ⓗ ■ exploring perpendicular lines ⓗ ■ using prior knowledge to solve multi-step problems.

Check my understanding

1   **a**   Complete a table of values for $y = 3x - 5$ using values of x from -2 to 2

     b   Draw the graph of $y = 3x - 5$ for values of x from -2 to 2

2   A straight line has equation $y = 2x + 7$

     a   What is the gradient of the line?

     b   What are the coordinates of the y-intercept?

3   A straight line has equation $8y = 15x + 12$ ⓗ

     a   What is the gradient of the line?

     b   What are the coordinates of the y-intercept?

4   By drawing a sketch, find the equation of the straight line that passes through $(0, 6)$ and $(1, 3)$

5   The cost of a taxi journey per mile can be modelled using the straight line $y = 1.5x + 4$

     An 8-mile journey costs £16. Find the

     a   minimum taxi fare   **b**   cost per mile.

6   It takes 5 machines 12 hours to print some leaflets.

     How long would it take 15 machines working at the same rate to print the same leaflets? ⓗ

7   Write an equation of the straight line that is perpendicular to $y = 10 - 3x$ and passes through the point $(0, 4)$ ⓗ

2 Forming and solving equations

In this block, I will learn...

how to solve one-step and two-step equations and inequalities

$a + 5 > -3$ $20 = 3b + 2$ $\dfrac{c - 7}{4} \leqslant 10$

how to solve equations with brackets

$3(x + 5) = 20$

how to solve inequalities with negative coefficients

$-y > 4$
$y < -4$

$3 \leqslant 10 - p$
$-3 \geqslant p - 10$
$10 - 3 \geqslant p$
$p \leqslant 7$

how to form equations and inequalities

$(4b + 40)°$
$(6b - 12)°$

$4b + 40 = 6b - 12$

The perimeter of the square is greater than the perimeter of the isosceles triangle. Find the possible values of x

$5x - 2$

$2x + 4$

$3x - 1$

how to solve equations and inequalities with unknowns on both sides

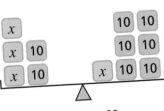

$3x + 20 = x + 60$

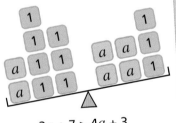

$2a + 7 > 4a + 3$

how to rearrange simple formulae

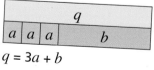

$q = 3a + b$
$a = \dfrac{q - b}{3}$

how to rearrange complex formulae **H**

Make x the subject of $\dfrac{5(2x^2 - 9)}{4}$

Small steps

- Solve one- and two-step equations and inequalities ®
- Solve equations and inequalities with brackets ®

Key words

Equation – a statement with an equal sign, which states that two expressions are equal in value

Inequality – a comparison between two quantities that are not equal to each other

Solve – find a value that makes an equation true or find a set of values that make an inequality true

Solution – a value you can substitute in place of the unknown in an equation or inequality to make it true

Expand – multiply to remove brackets from an expression

Are you ready?

1 Write, in words, the meaning of each expression.

    a $4n$ **b** $\frac{n}{4}$ **c** n^2 **d** $3n + 4$ **e** $3(n + 4)$

2 Find the output of each of these function machines for the given input.

    a $10 \longrightarrow \boxed{\times 8} \longrightarrow \Box$ **b** $10 \longrightarrow \boxed{\div 2} \longrightarrow \Box$

    c $9 \longrightarrow \boxed{-7} \longrightarrow \Box$ **d** $12 \longrightarrow \boxed{+ 18} \longrightarrow \Box$

3 Find the input of each of these function machines for the given output.

    a $? \longrightarrow \boxed{\times 2} \longrightarrow 60$ **b** $? \longrightarrow \boxed{\div 2} \longrightarrow 60$

    c $? \longrightarrow \boxed{+ 12} \longrightarrow 30$ **d** $? \longrightarrow \boxed{- 12} \longrightarrow 50$

4 Find the inverse of each operation.

    a $+ 9$ **b** $- 7$ **c** $\div 3$ **d** $\times 9$

5 Expand the brackets.

    a $3(x + 4)$ **b** $5(y - 7)$ **c** $7(2 + z)$ **d** $2(10 - 3x)$

Models and representations

In Books 1 and 2 you met several different ways of representing **equations** and **inequalities**.

Bar models

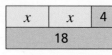

    $2x + 4 = 18$ $2x + 4 > 18$ $2x + 4 < 18$

Balances

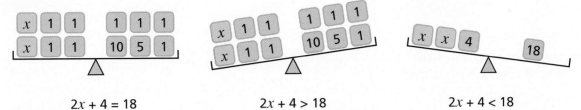

$$2x + 4 = 18 \qquad\qquad 2x + 4 > 18 \qquad\qquad 2x + 4 < 18$$

Function machines

$$x \longrightarrow \boxed{\times 2} \longrightarrow \boxed{+ 4} \longrightarrow 18$$

$$2x + 4 = 18$$

Other representations include algebra tiles or cups and counters.

Example 1

Solve these equations.

a $4x + 5 = 20$ **b** $\dfrac{m}{3} - 5 = 20$

a

$$4x + 5 = 20$$
$-5 \qquad\qquad -5$
$$4x = 15$$
$\div 4 \qquad\qquad \div 4$
$$x = 3.75$$

This is a two-step equation.

Firstly, subtract 5 from each side of the equation to isolate a term in x

Then divide both sides by 4 to find x

Remember: the **solution** to an equation does not have to be an integer.

You can also use bar models or other representations to help you find the solution.

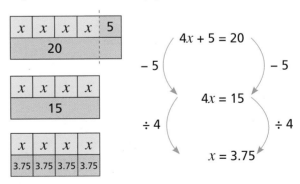

x	x	x	x	5
\multicolumn{5}{c	}{20}

x	x	x	x
\multicolumn{4}{c	}{15}

x	x	x	x
3.75	3.75	3.75	3.75

$$4x + 5 = 20$$
$-5 \qquad\qquad -5$
$$4x = 15$$
$\div 4 \qquad\qquad \div 4$
$$x = 3.75$$

b

$$\frac{m}{3} - 5 = 20$$

$+5$ $+5$ ○── This time, you need to start by adding 5 to each side of the equation.

$$\frac{m}{3} = 25$$

$\times 3$ $\times 3$ ○── Then multiply both sides by 3 to find m

$$m = 75$$

Here is the bar model illustration for this equation.

$\frac{m}{3}$ start with $\frac{m}{3}$

$\frac{m}{3}$ 5   this portion is $\frac{m}{3} - 5$

$\frac{m}{3}$ 20   5 $\frac{m}{3} - 5 = 20$

$\frac{m}{3}$ 25 $\frac{m}{3} = 25$

$\frac{m}{3}$ $\frac{m}{3}$ $\frac{m}{3}$ 25   25   25   You need three lots of $\frac{m}{3}$ to make m

m 75 $m = 75$

Remember: you can check the solution to an equation by substituting the answer back into the original equation.

For example, in part **a**, the equation was $4x + 5 = 20$. Substitute $x = 3.75$ into the left-hand side:

$4 \times 3.75 + 5 = 15 + 5 = 20$, which is the value on the right-hand side.

So $x = 3.75$ is the correct solution to the equation.

Example 2

a Solve these inequalities.

   i $7 + a > 3$ **ii** $6b + 2 \leqslant 44$

b State

   i   the smallest possible integer value of a **ii**   the greatest possible integer value of b

a   **i**

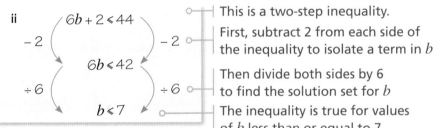

$$7 + a > 3$$
$$-7 \quad\quad -7$$
$$a > -4$$

By subtracting 7 from both sides of the **inequality**, you can see that it is true for any value of a which is greater than -4

You can check the solution by testing some values, for example a value greater than -4 (such as -1) and a value less than -4 (such as -10).

Remember: when you **solve** an inequality, you get a set of solutions for the unknown, not just a single value.

When $a = -1$, $7 + a = 7 + -1 = 6$, which is greater than 3, so this a valid solution.

When $a = -10$, $7 + a = 7 + -10 = -3$, which is not greater than 3, so this is not a valid solution.

You cannot check all possible values of a, as there are an infinite number.

ii

$$6b + 2 \leqslant 44$$
$$-2 \quad\quad -2$$
$$6b \leqslant 42$$
$$\div 6 \quad\quad \div 6$$
$$b \leqslant 7$$

This is a two-step inequality.

First, subtract 2 from each side of the inequality to isolate a term in b

Then divide both sides by 6 to find the solution set for b

The inequality is true for values of b less than or equal to 7

b   **i** -3 The first integer greater than -4 is -3

ii   7 $b \leqslant 7$, so the greatest possible value of b is 7   Remember that $\leqslant$ means "less than or equal to".

Practice 2.1A

1   Solve these equations.

   a $\dfrac{x}{3} = 10$ **b** $2y = 9$ **c** $z + 6 = 28$ **d** $q - 3 = 19$

   e $5 + d = 2$ **f** $-12 = 3g$ **g** $\dfrac{b}{5} = -10$ **h** $11 = h - 17$

2   Solve these inequalities.

   a $n + 5 > 12$ **b** $3p < 60$ **c** $2f \geqslant 10$ **d** $6 < 2x$

   e $6 < d - 7$ **f** $-6 < d - 7$ **g** $\dfrac{m}{2} > 5$ **h** $\dfrac{m}{2} > -5$

③ Here are Beca, Jackson and Chloe's methods for solving the equation $2x + 9 = 35$

Beca

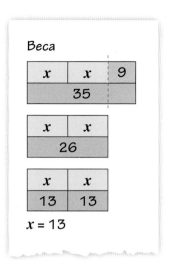

Jackson

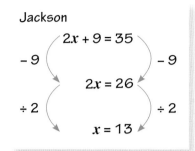

Chloe

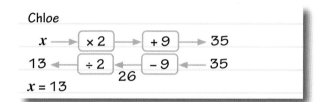

a Whose method do you prefer?

b Solve these equations.

   i $3c + 8 = 29$ 　　　　　**ii** $4n - 3 = 19$ 　　　　　**iii** $10 + 5k = 80$

   iv $12 = 6k - 18$ 　　　　**v** $8 + 10q = 2$

c Which of these equations will have a different solution to the others?
How can you tell?

A $\boxed{4b + 7 = 20}$ 　　　　B $\boxed{20 = 7 + 4b}$ 　　　　C $\boxed{20 + 4b = 7}$

D $\boxed{7 + 4b = 20}$ 　　　　E $\boxed{20 = 4b + 7}$

④ Here are Zach and Marta's methods for solving the equation $8 + \frac{m}{4} = 3$

Zach

$8 + \frac{m}{4} = 3$

$-8 \qquad\qquad -8$

$\frac{m}{4} = -5$

$\times 4 \qquad\qquad \times 4$

$m = -20$

Marta

$8 + \frac{m}{4} = 3$

$\frac{m}{4} + 8 = 3$

$m \rightarrow \boxed{\div 4} \rightarrow \boxed{+ 8} \rightarrow 3$

$-20 \leftarrow \boxed{\times 4} \leftarrow -5 \leftarrow \boxed{- 8} \leftarrow 3$

$m = -20$

a Whose method do you prefer?

b Explain why it would be difficult to represent the equation with a bar model.

c   Solve these equations.

i $\dfrac{k}{5} - 3 = 10$ **ii** $6 + \dfrac{p}{2} = 8$ **iii** $4 + \dfrac{t}{3} = 1$

iv $5 = \dfrac{a}{9} - 2$ **v** $3 = \dfrac{n}{6} + 4$

d   What's the same and what's different about the equations $\dfrac{a}{3} + 5 = 7$ and $\dfrac{a+5}{3} = 7$?

5 **a**   Solve the equation $3m + 8 = 44$

b   Solve the inequality $3m + 8 > 44$

c   What's the same and what's different about your answers to parts **a** and **b**?

6   Solve these inequalities.

a $2t + 5 < 11$ **b** $2t - 7 \leqslant 11$ **c** $\dfrac{t}{2} - 8 > 11$

d $11 < \dfrac{t}{3} + 8$ **e** $\dfrac{5t - 6}{2} < 7$

7 $x = 7$ is a member of the solution set of some of these inequalities. Which inequalities?

A $3x + 5 > 26$ B $3x + 5 \geqslant 26$ C $26 \leqslant 5 + 3x$

D $26 < 3x + 5$ E $26 < 5 + 3x$ F $5 + 3x \leqslant 26$

8   Jakub is solving the equation $\dfrac{a - 3}{4} = 6$

a   Show by substitution that Jakub is wrong.

b   Explain Jakub's mistake.

c   Find the correct solution to the equation.

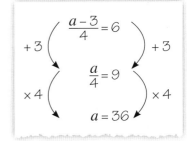

9 **a**   Solve these equations.

i $\dfrac{n}{4} + 3 = 5$ **ii** $\dfrac{n + 3}{4} = 5$ **iii** $n + \dfrac{3}{4} = 5$ **iv** $\dfrac{2n + 3}{4} = 5$ **v** $\dfrac{2n + 3}{4} = -5$

b   Solve these inequalities.

i $6 + \dfrac{p}{5} > 2$ **ii** $\dfrac{6}{5} + p < 2$ **iii** $\dfrac{6 + p}{5} \geqslant 2$ **iv** $\dfrac{p + 6}{5} \leqslant -2$ **v** $-2 > \dfrac{2p - 5}{6}$

What do you think?

1

All equations have exactly one solution.

Faith

All inequalities have an infinite number of solutions.

Seb

Do you agree with either Faith or Seb? Discuss why or why not with a partner.

2 Jackson finds the solutions to two inequalities. He writes $x > 2$ and $y \geqslant 2$

He represents his solutions on number lines like this:

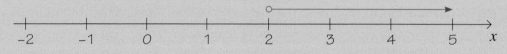

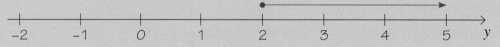

a Why do you think the top number line has an open circle at the left-hand end of the arrow while the bottom number line has a filled-in circle? What is different about the two inequalities?

b Match each inequality below with its corresponding number line.

A 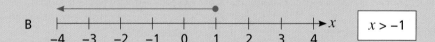 $x < 1$

B 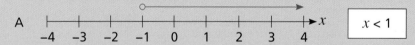 $x > -1$

C 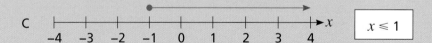 $x \leqslant 1$

D 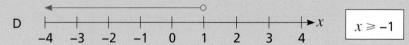 $x \geqslant -1$

c Write some inequalities of your own and represent them on number lines.

3 **a** **i** What do you think $-2 \leqslant x \leqslant 5$ means?

   ii How is $-2 \leqslant x \leqslant 5$ different from $-2 < x < 5$?

b How many possible integer values of x are represented by each of the inequalities in part **a**?

c Write some inequalities of the form $a \leqslant x \leqslant b$, $a < x \leqslant b$ and so on, that have

   i exactly three integer solutions

   ii only one integer solution

   iii no integer solutions.

4 Solve each of these equations.

   a $\dfrac{a}{5} = 6$ **b** $\dfrac{a+3}{5} = 6$ **c** $\dfrac{2a+3}{5} = 6$ **d** $\dfrac{2a-3}{5} = 6$ **e** $2a - \dfrac{3}{5} = 6$

   Compare your methods. What's the same and what's different?

Here is a reminder of how to solve equations and inequalities that involve brackets.

Example 3

a Find the value of x for which $5(x + 3) = 36$

b Find the set of values of y for which $4(y - 7) > 100$

a

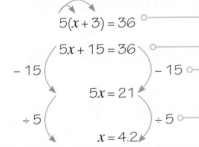

$5(x + 3) = 36$ ○——— You could start by **expanding** the brackets.

$5x + 15 = 36$ ○——— Now you have an equation that is like those you solved earlier.

-15 -15 ○—— Subtract 15 from each side to isolate the term in x

$5x = 21$

$\div 5$ $\div 5$ ○—— Then divide each side by 5 to find the value of x

$x = 4.2$

You can also use bar models to represent equations like this; see question **1** in Practice 2.1B.

b

$4(y - 7) > 100$ ○——— You could start by expanding the brackets.

$\div 4$ $\div 4$ However, as 4 is a factor of 100 you can divide both sides of the inequality by 4 first.

$y - 7 > 25$ ○——— Now you have a simple inequality in y

$+7$ $+7$ ○—— Add 7 to both sides of the inequality to find the solution set.

$y > 32$

Practice 2.1B

1 **a** Compare the following two methods of solving $4(x + 7) = 40$, which are illustrated by bar models.

Method 1

$4(x + 7) = 40$

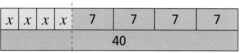

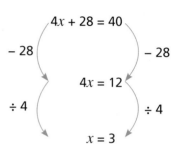

$4x + 28 = 40$

$- 28$ $- 28$

$4x = 12$

$\div 4$ $\div 4$

$x = 3$

Method 2

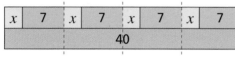

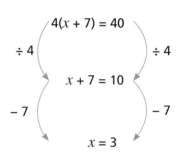

$4(x + 7) = 40$

$\div 4$ $\div 4$

$x + 7 = 10$

$- 7$ $- 7$

$x = 3$

 Which method do you prefer?

    b Use your preferred method to solve

      i $5(a + 12) = 100$ **ii** $3(b + 9) = 36$

    c

> I'd choose one of the methods for $7(a + 8) = 56$ but the other method for $7(b + 5) = 30$

Chloe

    Which method do you think is better for each of Chloe's equations? Explain why.

2 Solve these equations.

    a $2p + 5 = 60$ **b** $2(p + 5) = 60$ **c** $2(2p + 5) = 60$

    d $2(5 + 4p) = 60$ **e** $2p - 5 = 11$ **f** $2(p - 5) = 11$

    g $2(2p - 5) = 11$

3 Solve these equations.

a $4(x - 5) = 40$

b $4(x - 5) = 30$

c $4(x - 5) + 20 = 40$

d $4(x - 5) + 4x = 20$

4 a Solve the equation $3(p - 2) = 6$

b Solve the inequality $3(p - 2) < 6$

c What's the same and what's different about your solutions to parts **a** and **b**?

5 Solve these inequalities.

a $5(x + 3) < 10$

b $5(x - 3) < 10$

c $5(x - 3) < -10$

d $10 > 5(x - 3)$

e $-10 < 5(3 + x)$

f $-10 \leqslant 5(3x + 2)$

g $-10 \geqslant 5(3x - 2)$

What do you think? 💡

1 Benji is finding other equations that will have the same solution as $4a + 13 = 5$

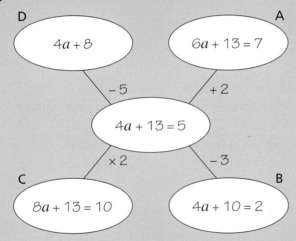

a Benji has made errors in three of his new equations. Find the errors and correct them.

b Write three more equations with the same solution as $4a + 13 = 5$

2 A rectangle has width x cm. The length of the rectangle is 4 cm longer than the width and the perimeter of the rectangle is 38 cm.

a Show that this information leads to the equation $2(2x + 4) = 38$

b Work out the area of the rectangle.

Consolidate – do you need more?

1 Solve these equations.

   a $4x = 12$ **b** $4x + 2 = 12$ **c** $4x - 2 = 12$ **d** $4(x + 2) = 12$ **e** $4(x - 2) = 12$

2 Solve these inequalities.

   a $4x < 12$ **b** $4x + 2 < 12$ **c** $4x - 2 \geqslant 12$ **d** $4(x + 2) < 12$ **e** $4(x - 2) \leqslant 12$

3 Solve these equations and inequalities.

   a $5c + 3 = 38$ **b** $4 + x < 19$ **c** $21 > 2y - 9$

   d $\dfrac{t}{10} = 3$ **e** $6(1 + p) \geqslant 480$ **f** $12 > 5(m - 1)$

   g $7(2g + 11) = 84$ **h** $\dfrac{5h + 1}{8} < 4.5$

Stretch – can you deepen your learning?

1

> $3x + 2y = 12$ has more than one solution.

Ed

a Show that Ed is correct by finding some solutions to the equation $3x + 2y = 12$. Compare your solutions with a partner's. What's the same and what's different? How many solutions are there?

b Draw the graph of $3x + 2y = 12$. What is the connection between the coordinates of the points on the graph of $3x + 2y = 12$ and the solutions to the equation $3x + 2y = 12$?

> I also know that $x = y$. There is only one solution to $3x + 2y = 12$ where $x = y$

Marta

c Explain how you can find Marta's solution

   i using a graph

   ii using an algebraic method.

   Which method is more likely to be accurate?

2 What's the same and what's different about solving these equations?

A $\quad 5 - 2y = 1$

B $\quad 5 - 2y = -1$

C $\quad -5 - 2y = -1$

D $\quad 5 - 2y = y$

E $\quad 5 - 2y = -y$

You will explore solving equations such as $5 - 2y = y$, where the letter appears on both sides of the equation, in Chapter 2.3

3 Each of these equations has more than one solution. Find all the possible solutions to each equation.

a $\quad x^2 = 16$

b $\quad 2y^2 = 50$

c $\quad (3z)^2 = 144$

d $\quad 4m^2 + 3 = 12$

e $\quad 4(m + 3)^2 = 16$

f $\quad (4m + 3)^2 = 16$

Reflect

What's the same and what's different about solving an equation and solving an inequality?

2.2 Inequalities with negative numbers

Small steps

■ Solve inequalities with negative numbers

Key words

Inequality – a comparison between two quantities that are not equal to each other

Solve – find a value that makes an equation true or find a set of values that make an inequality true

Solution – a value you can substitute in place of the unknown in an equation or inequality to make it true

Are you ready?

1 Copy and complete using > or <

  a 5 ◯ 3   **b** 5 ◯ –3   **c** –5 ◯ 3   **d** –5 ◯ –3

2 Put these numbers in order of size, starting with the lowest.

  10   –6   –10   –4   12   –3   7   –1.5

3 a Solve the equation $10 + 3x = 31$

  b Solve the inequality $8 + 3y < 44$

4 a Solve the equation $5a + 20 = 5$

  b Solve the inequality $3b - 1 > -4$

Models and representations

Algebra tiles

Algebra tiles are a great way to model equations and **inequalities** with negative numbers. Remember the red side of an algebra tile represents negatives.

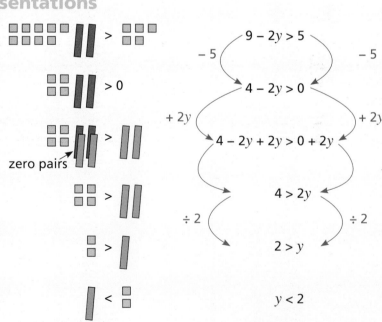

Look at what happens when you perform operations to both sides of an equation.

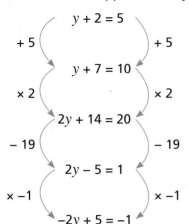

$y = 3$ is the **solution**

$3 + 7 = 10$, so $y = 3$ is still the solution

$2 \times 3 + 14 = 6 + 14 = 20$, so $y = 3$ is still the solution

$2 \times 3 - 5 = 6 - 5 = 1$, so $y = 3$ is still the solution

$-2 \times 3 + 5 = -6 + 5 = -1$, so $y = 3$ is still the solution

If you add or subtract the same number to both sides of an equation, or multiply/divide both sides of an equation by any number, the new equation still has the same solution.

Now look at what happens when you perform operations to both sides of an inequality.

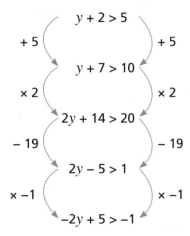

The solution set is $y > 3$, so $y = 4$ is in the solution set

$4 + 7 = 11$ and $11 > 10$, so $y = 4$ is still in the solution set

$2 \times 4 + 14 = 22$ and $22 > 20$, so $y = 4$ is still in the solution set

$2 \times 4 - 5 = 3$ and $3 > 1$, so $y = 4$ is still in the solution set

$-2 \times 4 + 5 = -8 + 5 = -3$, but $-3 < -1$ so $y = 4$ is not in the solution set

If you add or subtract the same number to both sides of an inequality, or if you multiply/divide both sides of an inequality by a positive number, the new inequality still has the same solution set.

If you multiply/divide both sides of an inequality by a negative number, the new inequality has a different solution set. See Worked example 1 to see what happens to the solution set.

Example 1

Solve the inequality $-2y + 5 > -1$

Method A

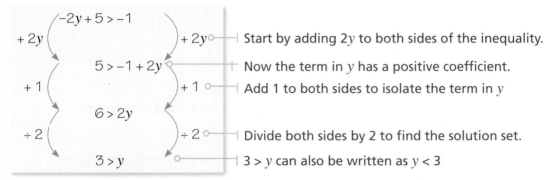

Start by adding $2y$ to both sides of the inequality.

Now the term in y has a positive coefficient.
Add 1 to both sides to isolate the term in y

Divide both sides by 2 to find the solution set.

$3 > y$ can also be written as $y < 3$

Notice that the inequalities you looked at earlier all had solution set $y > 3$
After multiplying by -1, the solution set is $y < 3$

If you multiply/divide both sides of an inequality by a negative number, the direction of the inequality sign needs to be reversed to keep the statement true.

Method B

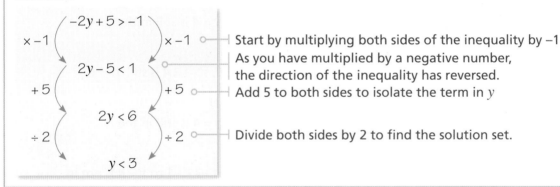

Start by multiplying both sides of the inequality by -1
As you have multiplied by a negative number, the direction of the inequality has reversed.
Add 5 to both sides to isolate the term in y

Divide both sides by 2 to find the solution set.

Practice 2.2A

1 Compare these ways of solving the equation $10 - 4x = 2$

Method 1

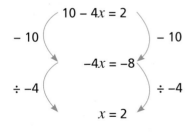

Method 2

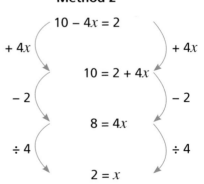

Method 3

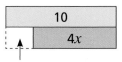

start with 10

subtract $4x$

This must be 2

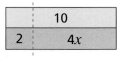

$10 - 4x = 2$ is the same as $10 - 2 = 4x$

$4x = 10 - 2$

$4x = 8$
$\div 4 \qquad \div 4$
$x = 2$

2 Solve these equations.

a $30 - x = 12$ **b** $30 - 2x = 12$ **c** $30 - 2(x + 5) = 12$

d $6(5 - x) = 18$ **e** $6(5 - \frac{x}{2}) = 18$ **f** $5(6 - 2x) = 18$

3 Given that $a > 5$, which of the following statements are true?

A $a + 3 > 5 + 3$ B $a - 3 > 5 - 3$

C $a \times 2 > 5 \times 2$ D $a \times (-1) > 5 \times (-1)$

4 Solve these inequalities.

a $3y > 30$ **b** $-3y > 30$ **c** $\frac{y}{3} > 30$

d $\frac{-y}{3} > 30$ **e** $y + 3 > 30$ **f** $y - 3 > 30$

g $3 - y > 30$

Compare your answers with a partner's, testing at least one value in each of your solution sets.

5 Solve these inequalities.

a $2y + 3 > 11$ **b** $2y - 3 > 11$ **c** $-2y + 3 > 11$

What's the same and what's different?

6 Solve these inequalities.

a $-3x + 5 < 12$ **b** $10 - \frac{t}{2} \geqslant 7$ **c** $8 < 2 - \frac{p}{5}$

d $6 < 12 - \frac{q}{3}$ **e** $10 + 4g \geqslant 7$ **f** $8 - \frac{2p}{3} \leqslant 6$

g $8 - \frac{2q}{3} \leqslant -6$

7 **a** **i** Solve the equations $2(3x - 8) = 20$ and $2(8 - 3x) = 20$

ii Compare the methods you used.

b Solve these inequalities.

i $5(2p - 3) \leqslant 40$ **ii** $5(3 - 2p) \leqslant 40$

What do you think?

1 **a** What's the same and what's different about solving these two equations?

$8 - 3x + 4 = 9$ and $8 - (3x + 4) = 9$

b Now compare the methods for solving these equations.

$8 - 3x - 4 = 9$ and $8 - (3x - 4) = 9$

2 Form and solve inequalities to answer these number puzzles.

a I think of a number, double it and subtract 10. The answer is greater than 15. Find the set of possible values that my number could take.

b I think of a number, double it and subtract the answer from 10. The answer is greater than 15. Find the set of possible values that my number could take.

c I think of a number, halve it and subtract the answer from 10. The answer is not more than 8. Find the set of possible values that my number could take.

d I think of a number, subtract 4, double the result and then subtract this result from 10. The answer is greater than 7. Find the set of possible values that my number could take.

Consolidate – do you need more?

1 Spot the mistakes in these answers.

a

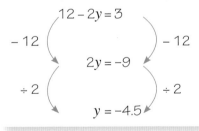

$12 - 2y = 3$
-12 -12
$2y = -9$
$\div 2$ $\div 2$
$y = -4.5$

b

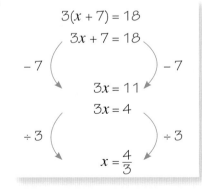

$3(x + 7) = 18$
$3x + 7 = 18$
-7 -7
$3x = 11$
$3x = 4$
$\div 3$ $\div 3$
$x = \dfrac{4}{3}$

c

$10 - 4x > 2$
-10 -10
$-4x > -8$
$\div -4$ $\div -4$
$x > 2$

2 Solve these inequalities.

a $-t > 3$ **b** $-t < -3$ **c** $2t > -6$ **d** $-2t > -6$

e $\dfrac{-t}{4} < 5$ **f** $\dfrac{-t}{6} < -10$ **g** $t - 7 \geqslant 10$ **h** $7 - t \geqslant 10$

3 a Solve these equations.

i $3 + 2x = 17$ **ii** $3 - 2x = 17$

b Solve these inequalities.

i $3 + 2x > 17$ **ii** $3 - 2x > 17$

Stretch – can you deepen your learning?

1 a Explain why the equation $x^2 = 16$ has two solutions.

b Which of these numbers satisfy the inequality $x^2 \geqslant 25$?

10, −10, 4, −4, 4.9, −4.9, 5, −5, 5.1, −5.1

c

If $x^2 \geqslant 25$ then either $x \geqslant 5$ or $x \leqslant -5$

Ed

Explain why Ed is right.

d Find the two regions that represent the solutions to these inequalities.

i $x^2 > 9$ **ii** $x^2 > 100$ **iii** $x^2 + 5 > 149$ **iv** $2x^2 - 3 > 95$

2 a

If $x^2 < 36$ then $-6 < x < 6$

Seb

Explain what is meant by $-6 < x < 6$

b Express as a single region the solutions to these inequalities.

i $x^2 < 16$ **ii** $x^2 < 49$ **iii** $2x^2 - 3 \leqslant 29$ **iv** $3x^2 + 8 < 200$

c Which of these inequalities have solutions that can be written as a single region?

A $y^2 > 100$ B $t^2 < 64$ C $b^2 \geqslant 4$ D $g^2 \leqslant 1$ E $5m^2 > 45$

3 a Discuss with a partner how you can represent the solutions to inequalities with squared terms on a number line.

b Challenge each other to find inequalities with squared terms, given the solutions in the form of number lines.

Look back to Chapter 2.1, Exercise 2.1A, What do you think? question **2** to remind yourself how to show solutions to inequalities on number lines.

Reflect

Explain your approaches to solving equations and inequalities where the unknown has a negative coefficient.

2.3 Unknowns on both sides

Small steps

- Solve equations with unknowns on both sides
- Solve inequalities with unknowns on both sides

Key words

Equation – a statement with an equal sign, which states that two expressions are equal in value

Solve – find a value that makes an equation true or find a set of values that make an inequality true

Solution – a value you can substitute in place of the unknown in an equation or inequality to make it true

Are you ready?

1 Solve these equations.

   a $10x = 30$ **b** $x + 10 = 30$ **c** $x - 10 = 30$ **d** $10 - x = 30$

2 Solve these equations.

   a $3y + 7 = 40$ **b** $3y - 7 = 41$ **c** $40 - 3y = 16$ **d** $40 + 3y = 16$

3 What's the same and what's different about solving the equations $4x - 3 = 17$ and $17 = 4x - 3$?

4 Solve these inequalities.

   a $t - 7 > 10$ **b** $2t - 7 > 10$ **c** $t + 10 > 7$ **d** $10 - t > 7$ **e** $10 - 2t > 7$

5 Simplify these expressions.

   a $7x - 3x$ **b** $3x - 7x$ **c** $-3x + 7x$ **d** $-3x - 7x$

Models and representations

You can use the same models to represent **equations** and inequalities with unknowns on both sides as for those with unknowns on one side only.

Bar model

x	x	x	5
x	11		

Balance

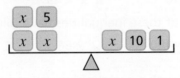

Cups and counters

These all show the equation

$$3x + 5 = x + 11$$

The first step to **solving** equations with unknowns on both sides is to adjust the equation so that there are unknowns on one side only. Bar models are very useful for illustrating this.

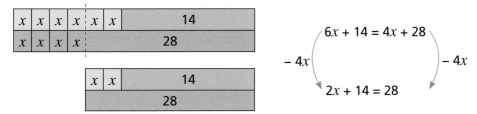

$$6x + 14 = 4x + 28$$
$$- 4x \qquad - 4x$$
$$2x + 14 = 28$$

You can also use a balance method using just symbols, as shown alongside the bar model.

A balance method is particularly useful if the coefficient of one of the unknowns is negative.

$$4y + 8 = 18 - 6y$$
$$+ 6y \qquad + 6y$$
$$10y + 8 = 18$$

The equation now has an unknown on one side only, and you already know how to solve equations like this.

Example 1

Solve the equation $4x + 2 = x + 17$ Think of the equation as a set of balanced scales.

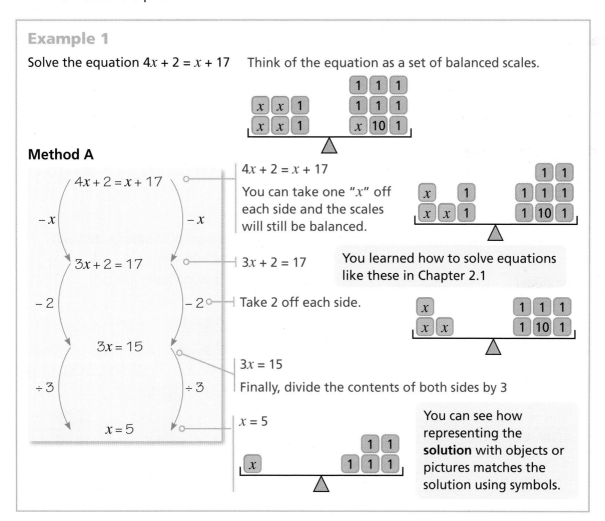

Method A

$$4x + 2 = x + 17$$
$$-x \qquad -x$$
$$3x + 2 = 17$$
$$-2 \qquad -2$$
$$3x = 15$$
$$\div 3 \qquad \div 3$$
$$x = 5$$

$4x + 2 = x + 17$

You can take one "x" off each side and the scales will still be balanced.

$3x + 2 = 17$

You learned how to solve equations like these in Chapter 2.1

Take 2 off each side.

$3x = 15$

Finally, divide the contents of both sides by 3

$x = 5$

You can see how representing the **solution** with objects or pictures matches the solution using symbols.

Check:

$$4x + 2 = 4 \times 5 + 2 \qquad x + 17 = 5 + 17$$
$$ = 20 + 2 \qquad\qquad = 22$$
$$ = 22$$

You can check your answer by substituting it into both sides of the equation.

Both sides of the equation have the same value so the solution is correct.

Method B

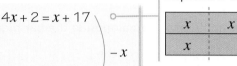

Represent the equation as a bar model. $4x + 2 = x + 17$

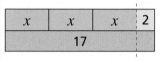

take x from both bars

$3x + 2 = 17$

take 2 from both bars

$3x = 15$

divide into 3 equal parts

$x = 5$

Example 2

Solve

a $4p + 6 = 8p$

b $2y + 7 = 5y - 2$

a

$$4p + 6 = 8p$$
$$-4p \qquad\qquad -4p$$
$$6 = 4p$$
$$\div 4 \qquad\qquad \div 4$$
$$1.5 = p$$

Start by subtracting $4p$ from both sides, as this will leave unknowns on one side only.

Using a balance, the equation looks like this:

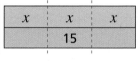

You could also show these steps on a bar model – try it!

Taking $4p$ from each side of the balance leaves:

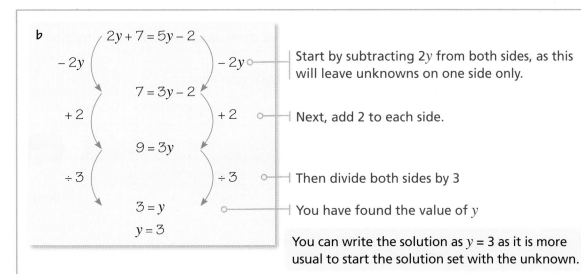

It is not as easy to represent equations with negative values using balances and bar models, but you can use the same methods as you have used with positive numbers.

Practice 2.3A

1. Copy and complete the solution to the equation $6a + 3 = 2a + 19$
   Use the bar model to help you.

   | a | a | a | a | a | a | 3 |
   | a | a | 19 | | | | |

   $$-2a \left(\begin{array}{c} 6a + 3 = 2a + 19 \\ 4a + 3 = 19 \end{array} \right) -2a$$

2. **a** What equation does this bar model show?

   | b | b | b | b | b | 2 |
   | b | b | b | 14 | | |

   b A section of length $3b$ can be "cut off" each bar. What equation is shown now?

   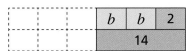

   c Solve the equation you wrote for part **b** to find the value of b

   d Check that your value of b gives the same value for both the left-hand side and the right-hand side of the original equation.

3. **a** **i** Draw a bar model to show the equation $3x = x + 20$

   ii How can you adjust the equation so that there are only terms involving x on one side?

   iii Solve the equation $3x = x + 20$

   iv Check your answer.

b i Draw a bar model to show the equation $16 + 2q = 6q$

ii How can you adjust the equation so that there are only terms involving q on one side?

iii Solve the equation $16 + 2q = 6q$ and check your answer.

4 Abdullah and Benji are solving the equation $8x + 3 = 2x + 21$

I'm going to subtract $2x$ from both sides of the equation.

I'm going to subtract $6x$ from both sides of the equation.

Abdullah   Benji

a Whose strategy is most sensible? Explain why.

b Solve the equation $8x + 3 = 2x + 21$ and check your answer.

5 a Explain what would be your first step for solving each of these equations.

    i $3g + 8 = 7g + 4$ **ii** $5m + 3 = 3m + 19$ **iii** $6 + 4t = t + 27$

b Now solve the equations in part **a**.

6 Solve these equations.

a $6x + 4 = 4x + 10$

b $6x - 4 = 4x + 10$

c What's the same and what's different about your methods?

7 Solve these equations.

    a $5n - 3 = 4n + 2$ **b** $c + 11 = 7c - 7$ **c** $2h - 3 = 15 + 4h$

8 Bobbie is solving the equation $5x - 6 = 2 + 7x$

$$5x - 6 = 2 + 7x$$
$-5x$ $-5x$
$$6 = 2 + 2x$$
-2 -2
$$4 = 2x$$
$\div 2$ $\div 2$
$$2 = x$$

She checks her solution and discovers that it is incorrect.

$$5 \times 2 - 6 = 10 - 6 = 4 \qquad\qquad 2 + 7 \times 2 = 2 + 14 = 16$$
$$4 \neq 16$$

Identify Bobbie's error and find the correct solution.

9 Solve these equations. The answers may be positive or negative integers or fractions.

a $3m + 7 = 8m + 2$ **b** $7t - 6 = 4t + 5$

c $6h - 8 = 2h - 7$ **d** $12 + 3w = 6w - 4$

e $5p = 2p + 8$ **f** $5p = 2p - 8$

g $k - 19 = 6k - 3$ **h** $6t = 4t + 8 - 3t$

What do you think?

1 Here is how Marta created an equation with unknowns on both sides that has the solution $x = 4$

$$x = 4$$
$$\times 2 \qquad\qquad \times 2$$
$$2x = 8$$
$$+ 7 \qquad\qquad + 7$$
$$2x + 7 = 15$$
$$+ 3x \qquad\qquad + 3x$$
$$5x + 7 = 3x + 15$$

a Create another equation with unknowns on both sides with the solution $x = 4$

b Create an equation with the unknown on both sides for each of these solutions.

   i $a = 3$ **ii** $b = -3$ **iii** $c = \dfrac{1}{2}$ **iv** $d = -\dfrac{1}{2}$

2 a Chloe and Jakub are solving the equation $10 - 3t = 2 + 5t$

 I'm going to subtract $5t$ from both sides of the equation.

I'm going to add $3t$ to both sides of the equation.

Chloe   Jakub

   i Whose strategy is most sensible? Explain why.

   ii Solve the equation $10 - 3t = 2 + 5t$

b Solve these equations.

   i $2t + 4 = 9 - t$ **ii** $6 - 2m = 12 + m$ **iii** $8 - 3x = 5 - 2x$

   iv $3g - 6 = 10 - g$ **v** $6 - 2v = 11 - 7v$ **vi** $7 - 4f = 9 + 6f$

3 Form and solve equations to solve these number puzzles.

a I think of a number, double it and add 12. The answer is three times the number I started with. What was my starting number?

b I think of a number, double it and add 12. The answer is the number I started with. What was my starting number?

c I think of a number, add 12 and then double the result. The answer is four times the number I started with. What was my starting number?

d Make up some number puzzles of your own and challenge a partner to answer them.

You can use the same methods for solving inequalities with unknowns on both sides as you use for equations.

Example 3

Find the range of values of y for which $2y + 7 \leqslant 5y - 2$

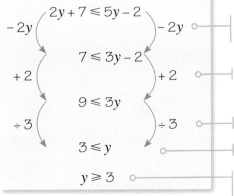

Start by subtracting $2y$ from both sides as this will leave unknowns on one side only.

Next, add 2 to each side.

Then divide both sides by 3

This gives the solution set for y

You can write "3 is less than or equal to y" or "y is greater than or equal to 3" as these mean the same thing. It is more usual to start your solution set with the unknown.

It is not as easy to represent inequalities using balances and bar models, but you use the same methods as you have learned for equations.

Compare this solution with that of Example 1, Method B.

Practice 2.3B

 1 a Explain what your first step would be for solving each of these inequalities.

    i $2x + 2 > 10$ **ii** $10 < 2x + 2$

b Copy and complete the working to solve these inequalities.

    i $5x > 3x + 12$
    $-3x \Big(\quad \Big) -3x$
    $2x > 12$

    ii $6x + 2 \leqslant 8x$
    $-6x \Big(\quad \Big) -6x$
    $2 \leqslant \square$

    iii $7x + 3 \geqslant 3x + 19$
    $-3x \Big(\quad \Big) -3x$
    $\square \geqslant 19$

2 a Explain what your first step would be for solving each of these inequalities.

    i $5x + 2 > x + 14$ **ii** $5x + 2 > 2x + 14$

    iii $5x + 2 > 3x + 14$ **iv** $5x + 2 > 4x + 14$

b Solve the inequalities in part **a**

3 Solve these inequalities.

a $6x + 5 > 2x + 21$ b $6x + 5 \leqslant 2x + 21$

c $6x + 5 \geqslant 2x + 21$ d $2x + 21 < 6x + 5$

What's the same and what's different?

4 Solve these inequalities.

a $6y + 2 > 2y + 22$ b $6y - 2 > 2y + 22$

What's the same and what's different?

5 Solve these inequalities.

a $5w + 3 > 2w + 12$ b $5w - 3 > 2w + 12$

c $5w + 3 > 2w - 12$ d $5w - 3 > 2w - 12$

e $10 + 2q \leqslant q - 7$ f $4m - 8 < 12 + 2m$

g $2 + 5y \leqslant 2y - 6$ h $10p + 4 < 8p - 7$

6 Find the smallest integer value of x that satisfies each of these inequalities.

a $6x + 4 > 2x + 7$ b $5x - 6 \geqslant 2x - 1$

c $6x - 3 < 10x - 19$ d $3x + 7 \geqslant x - 12$

What do you think?

1 a   Here is how Chloe starts to solve the inequality $10 + 2y > 35 - 3y$

     i   Why do you think she starts by adding $3y$ to each side of the inequality instead of subtracting $2y$ from each side of the inequality?

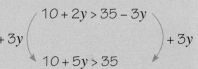

     ii   Copy and complete Chloe's working to solve the inequality.

   b   Solve these inequalities.

     i $3a - 5 < 11 - a$ ii $15 - 2b \leqslant 9 + b$ iii $5 + 4c > 10 - 6c$

2 a   Ed and Faith are solving the inequality $6 - 3y > 10 - 5y$

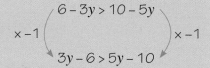

I'm going to add $3y$ to both sides of the inequality.   I'm going to add $5y$ to both sides of the inequality.

     Ed   Faith

     i   Whose strategy is most sensible? Explain why.

     ii   Solve the inequality $6 - 3y > 10 - 5y$

   b   Seb starts to solve this inequality by multiplying both sides by -1

     What mistake has Seb made?

   c   Solve these inequalities.

     i $11 - 2d \geqslant 6 - 4d$ ii $15 - 3d < 7 - d$ iii $8 - 6d > 6 - 8d$

Consolidate – do you need more?

1 a What equation does this bar model show?

a	a	a	a	3

a	a	11

b A section of length $2a$ can be "cut off" each bar. What equation is shown now?

	a	a	3

	11

c Solve the equation to find the value of a

d Check that your value of a gives the same value for both the left-hand side and the right-hand side of the original equation.

2 a Explain what would be your first step for solving each of these equations.

   i $7a + 5 = 3a + 13$ **ii** $6m + 7 = 3m + 19$

   iii $6 + 6t = t + 21$ **iv** $3f + 4 = 2f + 6$

b Solve the equations in part **a**

3 Solve these equations.

a $3a + 5 = 11$ **b** $3a + 11 = 5$

c $3a + 11 = 5a$ **d** $3a + 5 = 11a$

What's the same and what's different?

4 Solve these equations.

a $3v - 4 = 2v + 8$ **b** $3v + 4 = 2v - 1$ **c** $8b + 6 = 10b + 4$

d $8b - 6 = 10b + 4$ **e** $8b - 6 = 10b - 4$

5 Solve these inequalities.

a $4x + 3 > 7$ **b** $4x + 3 > 7x$ **c** $4x - 3 > 7$

d $4x - 3 > 7x$ **e** $4x + 3 > 7x - 6$ **f** $4x - 6 > 7x - 3$

Stretch – can you deepen your learning?

1 a Jackson and Jakub are solving the equation $\frac{1}{2}x + 3 = \frac{1}{4}x + 5$

Jackson

I'm going to subtract $\frac{1}{2}x$ from both sides of the equation first.

I'm going to multiply both sides of the equation by 4 first.

Jakub

Use both methods to solve the equation. Which method do you prefer?

b Solve these equations.

   i $\frac{1}{3}x + 5 = \frac{1}{4}x + 7$ **ii** $\frac{3}{5}x + 8 = \frac{3}{4}x + 2$ **iii** $10 - \frac{1}{2}x = \frac{3}{4}x - 10$

c Solve these inequalities.

   i $\frac{1}{5}y - 3 > \frac{7}{10}y$ **ii** $\frac{5}{6}y - 10 \leqslant \frac{2}{3}y - 5$ **iii** $\frac{7}{9}y - 2 < \frac{2}{3}y + 6$

2 Solve these equations.

   a $3(x + 4) = 2x - 1$ **b** $3(x + 4) = 2(x - 1)$

   c $3 - (x + 4) = 2(x - 1)$ **d** $3 - (x + 4) = 2 - (x - 1)$

   e $3x - (x + 4) = 2x - (x - 1)$

3 Solve these inequalities.

   a $5(t - 1) < 3(t - 2)$ **b** $5(t - 1) \geqslant 3(2 - t)$

   c $5(1 - t) \leqslant 3(t - 2)$ **d** $5(1 - t) < 3(2 - t)$

4 **a** Find all the integer values of x that satisfy the inequality $7 < 4x + 3 < 27$

       Hint: think of this as two separate inequalities $7 < 4x + 3$ and $4x + 3 < 27$

   b Find the range of values of x that satisfy the inequality $2x + 5 < 4x + 3 < 2x + 11$

   c Explain why no values of x satisfy the inequality $3x + 4 < 2x + 5 < 5x + 1$

5 I can see the solution to the equation $8 - 7x = 8 + 7x$ without having to do any calculations.

Explain how Zach can do this.

Reflect

1 What's the same and what's different about solving an equation or an inequality with an unknown on one side compared to solving an equation or an inequality with an unknown on both sides?

2 Write a series of instructions to support another student who is solving an equation or an inequality with an unknown on both sides and with brackets.

Small steps

- Solve equations and inequalities in context
- Substitute into formulae and equations

Key words

Variable – a numerical quantity that might change, often denoted by a letter, for example x or t

Formula (plural: **formulae**) – a rule connecting variables written with mathematical symbols

Equation – a statement with an equal sign, which states that two expressions are equal in value

Are you ready?

1 Describe any pairs or sets of angles that

   a add up to 180°

   b add up to 360°

   c are always equal to each other.

2 I think of a number and call it n. Form an expression for each of these sets of instructions.

   a I double my number and then add 40

   b I double my number and then subtract 40

   c I halve my number and then add 40

   d I halve my number and then subtract 40

   e I double my number and then subtract it from 40

   f I add 40 to my number and then I double the result.

   g I subtract 40 from my number and then I halve the result.

3 Solve these equations.

   a $3g + 12 = 33$

   b $40 = 2p - 7$

   c $10(q - 3) = 80$

   d $\frac{m}{4} - 3 = 12$

   e $\frac{n + 3}{4} = 12$

   f $6a + 8 = 2a + 24$

4 The probability that it will rain next week is 0.84. What is the probability that it won't rain next week?

Models and representations

Bar models

a	a	a	a	a	7
a	a	a		20	

$5a + 7 = 3a + 20$

b	b	b	b
21		9	

$4b - 9 = 21$

Function machines

$y \longrightarrow \boxed{\div 2} \longrightarrow \boxed{+ 7} \longrightarrow 40$

$\frac{y}{2} + 7 = 40$

$t \longrightarrow \boxed{+ 3} \longrightarrow \boxed{\times 5} \longrightarrow 60$

$(t + 3) \times 5 = 60$

or $5(t + 3) = 60$

You could also use balances, algebra tiles, cups and counters or any of the other representations you have used in earlier chapters.

In this chapter, you will use your knowledge of other areas of mathematics to form **equations** and then use the techniques you have learned in this block to solve them.

You can use whichever method you think is most appropriate to solve each equation; for example, symbols, function machines or bar models.

Example 1

Find the size of the largest angle in this triangle.

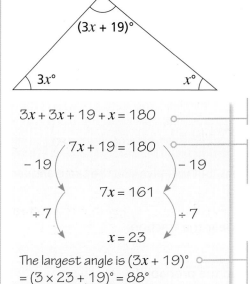

$3x + 3x + 19 + x = 180$

Form an equation using the fact that the angles in a triangle add up to 180°

$7x + 19 = 180$

$-19 \qquad\qquad -19$

Simplify the expression on the left-hand side and then solve the equation using your preferred method.

$7x = 161$

$\div 7 \qquad\qquad \div 7$

$x = 23$

The largest angle is $(3x + 19)°$
$= (3 \times 23 + 19)° = 88°$

Don't forget to answer the question. You are asked to find the size of the largest angle, not just the value of x

Example 2

The probability that a biased coin lands on heads is four times the probability that it lands on tails. Work out the probability that the coin lands on heads.

Method A

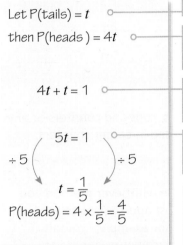

Let P(tails) = t

then P(heads) = $4t$

$4t + t = 1$

$5t = 1$

$\div 5 \left(\qquad \right) \div 5$

$t = \dfrac{1}{5}$

P(heads) = $4 \times \dfrac{1}{5} = \dfrac{4}{5}$

Use a letter to represent the probability of getting tails.

Form an expression for P(heads) using the information in the question.

The probabilities of all possible events add up to 1. Use this fact to form an equation.

Simplify the expression on the left-hand side and then solve the equation.

You could also have solved this using a bar model:

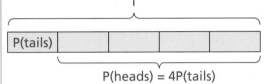

P(heads) = 4P(tails)

It is easier to write the probabilities in terms P(tails) because it avoids writing the probability of heads as a fraction. See Method B below for the alternative way of labelling the probabilities.

Remember: you can give probabilities as fractions, decimals or percentages.

Method B

let P(heads) = h

then P(tails) = $\dfrac{h}{4}$

$h + \dfrac{h}{4} = 1$

$\dfrac{4h}{4} + \dfrac{h}{4} = 1$

$\times 4 \left(\dfrac{5h}{4} = 1 \right) \times 4$

$5h = 4$

$\div 5 \left(\qquad \right) \div 5$

$h = \dfrac{4}{5}$

The probability that the coin lands on heads is 4 times the probability that it lands on tails, so the probability that the coin lands on tails is one-quarter of the probability that it lands on heads.

As in Method A, P(heads) + P(tails) = 1

This method involves fractions, but will give you the same answer.

Multiply both sides by 4 to clear the fraction.

Divide both sides by 5 to find the probability of the coin landing on heads.

Practice 2.4A

1 **a** Write the sum of the adjacent angles on a straight line.

b Form and solve equations to find the sizes of the unknown angles in these diagrams.

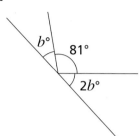

i   ii   iii

2 Zach, Faith and Flo are each thinking of a number.

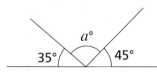 6 times my number is 504

Zach

18 more than my number is 161

Faith

5 less than a quarter of my number is 30

Flo

a Form and solve equations or inequalities to find the numbers that Zach, Faith and Flo are thinking of.

b Make up your own number puzzles like these and challenge a partner.

3 The total price of three pens and one pencil is 99p. The pencil costs 18p.

Call the price of one pen xp

a Which of these equations represents the information given?

A $\frac{x}{3} + 18 = 99$ B $3(x + 18) = 99$

C $3x + 18 = 99$ D $x + 3 \times 18 = 99$

b Solve the correct equation to find the price of one pen.

4 A theme park charges £18 for a child ticket. The total cost of two adult tickets and three child tickets is £110

a Form an equation to find the price, £a, of an adult ticket.

b Find the value of a

c Find the cost for a party of 6 adults and 13 children to visit the theme park.

5 The cost of a helicopter ride is £59 plus £12 per mile.

a Form an inequality to find the greatest number of whole miles you can ride in the helicopter with a budget of £150

b Solve your inequality.

6 Use your knowledge of angle rules to find the values of the letters.

a

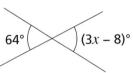

64° $(3x - 8)°$

b

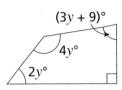

$(3y + 9)°$ $4y°$ $2y°$

c

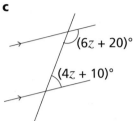

$(6z + 20)°$ $(4z + 10)°$

d

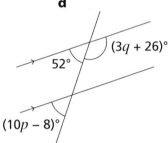

52° $(3q + 26)°$ $(10p - 8)°$

e

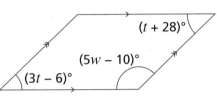

$(t + 28)°$ $(5w - 10)°$ $(3t - 6)°$

7 Benji goes shopping. He has £120. He wants to buy a pair of jeans for £45 and as many T-shirts, priced at £12 each, as possible.

    a Form an inequality to find the maximum number of T-shirts Benji can buy.

    b Solve your inequality.

8 The spinner is spun once. The table shows the probabilities that it lands on 1, 2, 3 or 4

Score	1	2	3	4
Probability	0.3	0.3	x	$3x$

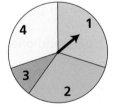

    a Work out the value of x

    b Find the probability that the spinner lands on an even number.

9 The perimeter of the square is greater than the perimeter of the equilateral triangle. Find the range of possible values of x

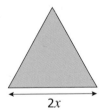

$x + 2$ $2x$

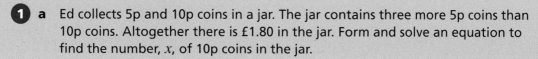

What do you think? 💭

1 **a** Ed collects 5p and 10p coins in a jar. The jar contains three more 5p coins than 10p coins. Altogether there is £1.80 in the jar. Form and solve an equation to find the number, x, of 10p coins in the jar.

b The price of a frying pan is £5 more than the price of a saucepan. The frying pan and saucepan together cost less than £60. Form and solve an inequality that represents the price, £p, of the saucepan.

2 Abdullah has four packets of mints and seven more mints. Beca has two packets of mints and 31 more mints. Each packet contains the same number of mints.

a If Abdullah and Beca each have the same number of mints, how many mints do they have altogether?

b What can you say about the number of mints in a packet if

i Abdullah has more mints than Beca   **ii** Beca has more mints than Abdullah?

c Explain why Abdullah cannot have 97 mints, but Beca can.

3 A dice is biased so that the probabilities of it landing on 1, 2, 3, 4, 5 and 6 are in the ratio 1:2:3:4:5:6

a Find the probability that the dice lands on 4   **b** Find an event with probability $\frac{1}{3}$

Remember: a **formula** is a rule connecting **variables** written with mathematical symbols. Sometimes, depending on which variables you know or don't know, when you substitute numbers into a formula you get an equation to solve.

Example 3

The formula for the area of a circle is $A = \pi r^2$

a Find the area of circle with diameter 12 cm. Give your answer in terms of π

b The area of a circle is 100 cm². Find the radius of the circle.
Give your answer to 3 significant figures.

a If $d = 12$ cm, then $r = 12 \div 2$ ○——┤ Remember: the radius of a circle is half of its diameter.

$\qquad = 6$ cm

$A = \pi r^2$ ○————————┤ Write down the formula for the area.

$A = \pi \times 6^2$ ○————————┤ Substitute the value of r.

$A = 36\pi$ cm² ○————————┤ Leave your answer in terms of π, as requested.

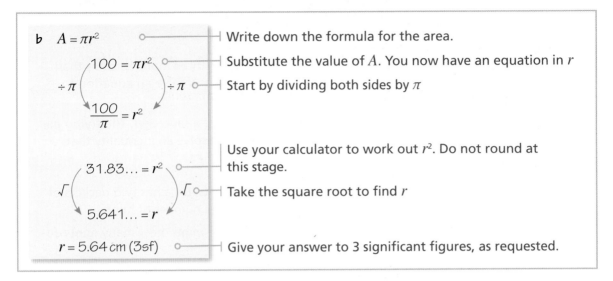

b $A = \pi r^2$ ———— Write down the formula for the area.

$100 = \pi r^2$ ———— Substitute the value of A. You now have an equation in r

$\div \pi$ $\div \pi$ ———— Start by dividing both sides by π

$\dfrac{100}{\pi} = r^2$

$31.83\ldots = r^2$ ———— Use your calculator to work out r^2. Do not round at this stage.

$\sqrt{}$ $\sqrt{}$ ———— Take the square root to find r

$5.641\ldots = r$

$r = 5.64\,\text{cm}$ (3sf) ———— Give your answer to 3 significant figures, as requested.

Practice 2.4B

1   A rectangle has length l and width w. The formula for the area is $A = lw$

   a   Use the formula to find A when $l = 40$ and $w = 17$

   b   The area of a different rectangle is $720\,\text{cm}^2$. The length of this rectangle is $45\,\text{cm}$.

     i   Use $A = lw$ to form an equation in w

     ii   Solve your equation to find the width of this rectangle.

2   The rule for the nth term of a sequence is $4n + 17$

   a   Find the 12th term of the sequence.

   b   Find the position in the sequence of the term 109

   c   Explain why 304 is not a term in the sequence.

   d   Find the first term in the sequence that is more than 1000

3   Use the formula $a^2 = c^2 - b^2$ to find the value of

   a a when $c = 10$ and $b = 8$

   b c when $a = 15$ and $b = 8$

   c b when $a = 3$ and $c = 5$

   d   Why can't you find a when $c = 8$ and $b = 9$?

4   A taxi driver charges £2.80 for the first mile and 90p for each mile after that.

   a   Which of these is the correct formula for the total cost, C, in pounds, of a journey of m miles?

     A $C = 2.80 + 90m$ B $C = (2.8 + 0.9)m$ C $C = 0.9m + 2.8$

   b   Find the cost of a 12-mile journey.

   c   Seb has £10. How far can he travel in the taxi?

5 **a** Write down the formula for the area of a triangle.

    b The area of a triangle is $8\,\text{cm}^2$

        i Find the height of the triangle if the base is $10\,\text{cm}$.

        ii Find the height of the triangle if it is equal to the base.

6 Use the formula $s = \frac{1}{2}(u + v)t$ to work out

    a s when $u = 10$, $v = 15$ and $t = 40$

    b t when $s = 20$, $u = 0$ and $v = 8$

    c v when $s = 30$, $t = 12$ and $u = 8$

What do you think? 💭

1 The formula for the volume, V, of a cuboid with length l, width w and height h is $V = lwh$

> You will explore this formula in Block 4

How many cuboids with integer lengths can you find with volume $24\,\text{cm}^3$? State their side lengths.

2 The gradient, m, of the line segment joining the points with coordinates (a, b) and (c, d) can be found using the formula $m = \dfrac{d - b}{c - a}$

    a Use the formula to find the gradient of the line segment joining

        i (2, 3) and (6, 15)   **ii** (5, –2) and (–3, –5)

    b A line segment with gradient $\frac{1}{2}$ joins the points with coordinates (4, 7) and (8, p). Compare finding the value of p by using the formula with drawing a sketch or any other method you can think of. Investigate with other gradients, including negative ones, and other coordinates.

Consolidate – do you need more?

1 The angles in a quadrilateral are $x°$, $2x°$, $3x°$ and $4x°$. Find the size of each of the angles in the quadrilateral.

2 Five times a number added to 11 gives the result 56. What is the number?

3 Ed's dad is three times as old as Ed. Their total age is 48 years. How old is Ed?

4 There are 12 more girls in the choir than boys. There are 48 students altogether in the choir.

Use the bar model to form and solve an equation to find the number of girls in the choir.

Boys

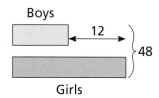

Girls

5 Emily has m marbles. Jakub had three times as many marbles as Emily, but lost 17 of them. Jakub still has more marbles than Emily. Form and solve an inequality to find the smallest number of marbles Emily could have.

6 The perimeter of a nonagon is given by the formula $P = 4c + 5d$

   a Find P when $c = 12$ and $d = 20$

   b Find c when $P = 80$ and $d = 12$

   c Find d when $P = 60$ and $c = 9$

7 The total number of points a team has scored in a football league is found using the formula $T = 3w + d$, where w is the number of games won and d is the number of games drawn. A team draws 7 games and has 43 points. How many games have they won?

Stretch – can you deepen your learning?

1 **a** The sum of three consecutive integers is 261. Find the greatest of the three integers.

   b The sum of three consecutive integers is −120. Find the smallest of the three integers.

   c Find the sums of several sets of three consecutive integers. Compare the sums to the middle integer. What do you notice?

     You will explore looking for and testing relationships like this in Chapters 3.3 and 3.4

2 **a** Flo has two cats, Misty and Smokey. Misty is twice as old as Smokey. In four years' time, the sum of their ages will be 17. How old is Misty now?

   b Zach has two dogs, Autumn and Titch. Four years ago, Autumn was three times as old as Titch. The sum of their ages now is 12. What will be the sum of their ages next year?

3 The formula $D = b^2 - 4ac$ is used in A-level Mathematics.

   a Find D when $b = -5$, $a = 3$ and $c = 5$

   b Find a when $D = -44$, $b = -2$ and $c = 4$

   c **i** Find some values of a, b and c for which $D = 0$

     ii Can you find values of a, b and c that are all positive?

     ii Can you find values of a, b, and c that are all negative?

Reflect

1 How do you go about forming an equation? What information do you need to know?

2 What is the difference between a formula and an equation?

Small steps

- Rearrange formulae (one-step)
- Rearrange formulae (two-step)
- Rearrange complex formulae including brackets and squares (H)

Key words

Variable – a numerical quantity that might change, often denoted by a letter, for example x or t

Formula (plural: **formulae**) – a rule connecting variables written with mathematical symbols

Subject – the variable in a formula that is expressed in terms of the other variables

Equation – a statement with an equal sign, which states that two expressions are equal in value

Are you ready?

1 Write down the inverse of

   a -10 b $\times 12$ c $\div 10$ d $+15$

2 The formula for the area of a parallelogram is $A = bh$
   a Find A when $b = 6.7$ and $h = 3.8$
   b Find b when $A = 32.4$ and $h = 7.2$
   c Find h when $A = 600$ and $b = 12.5$

3 Solve these equations.

   a $\frac{x}{4} = 10$ b $\frac{x}{4} - 3 = 10$ c $6 + \frac{x}{4} = 10$

4 a The square of a number is 64. What is the number?
   b The square root of a number is 64. What is the number?

Models and representations

Bar model

t				
p	p	p	p	p

This bar model shows $t = 5p$ and $p = \frac{t}{5}$

Function machines

$p \longrightarrow \boxed{\times 5} \longrightarrow t$

inverse

$p \longleftarrow \boxed{\div 5} \longleftarrow t$

$t = 5p$

$p = \frac{t}{5}$

You learned in Books 1 and 2, and Chapter 2.4 of this book, that a **formula** is a rule connecting **variables**, written using mathematical symbols.

For example, $A = \pi r^2$ is the formula for finding the area of a circle of radius r. Because the formula is written to help you to find A, we say that A is the **subject** of the formula.

Another example is $P = 2(l + w)$, which is a formula for finding the perimeter of a rectangle with length l and width w. Here P is the subject of the formula.

Sometimes it is useful to rearrange a formula to make a different letter the subject. When you need to do this depends on which variables you know and which you want to calculate.

Example 1

An equilateral triangle has side length l

a Write a formula for the perimeter, P, of the equilateral triangle in terms of l

b Find P when $l = 60$

c Find l when $P = 60$

d Find a formula for l in terms of P

a $P = 3l$ — Each side of the equilateral triangle has length l, so the perimeter P is given by $P = l + l + l$, which simplifies to $P = 3l$

b When $l = 60, P = 3 \times 60 = 180$ — Substitute the value of l into the formula.

c When $P = 60$

$$60 = 3l$$
$\div 3 \left(\right) \div 3$ — Substitute the value of P into the formula.
— Divide both sides by 3 to find l
$$20 = l$$

So $l = 20$

d
$$P = 3l$$
$\div 3 \left(\right) \div 3$
$$\frac{P}{3} = l$$
$$l = \frac{P}{3}$$

— Start with the formula you worked out in part **a**

— The term involving l is $3l$. To find l in terms of P, you need to divide both sides of the **equation** by 3

— Now you have a formula with l instead of $3l$

— It is usual to write the formula with the subject first.

> Remember: if two things are equal, then the order in which you state them doesn't matter, $l = \frac{P}{3}$ is the same as $\frac{P}{3} = l$

The bar model shows both

$P = 3l$ and $l = \frac{P}{3}$

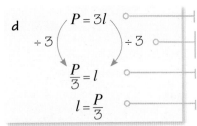

— $\frac{P}{3}$ and $\frac{1}{3}P$ mean the same, so you could also write $l = \frac{1}{3}P$

Example 2

An electrician charges £40 per hour plus a callout charge of £60

a Write a formula for the total cost (£C) of hiring the electrician for t hours.

b The electrician charges one client £220. For how many hours did they work?

c Rearrange your formula from part **a** to make t the subject.

a 1 hour costs £40

so t hours costs $t \times 40 = £40t$ — Multiply t by the cost per hour.

total cost = charge for the — Add the callout charge to the cost of t hours' work.
hours worked + £60

so $C = 40t + 60$ — Write the formula in algebraic form.

You could also have written $C = 60 + 40t$

b
$220 = 40t + 60$ — Substitute the values you know into the formula. Here you know that $C = 220$

-60 -60 — Solve the equation using the methods you learned in Chapter 2.1

$160 = 40t$

$\div 40$ $\div 40$

$4 = t$

The electrician worked for 4 hours. — Write a statement to answer the question.

c
$C = 40t + 60$ — Rearrange the formula in the same way as you would solve the equation to find the value of t

-60 -60

$C - 60 = 40t$

$\div 40$ $\div 40$

$\dfrac{C - 60}{40} = t$ — Although $\dfrac{C - 60}{40} = t$ can be easily used to find t, it is more common to write a formula with the subject first.

$t = \dfrac{C - 60}{40}$

Remember: $\dfrac{C - 60}{40} = t$ and $t = \dfrac{C - 60}{40}$ are the same.

Practice 2.5A

1 Write down the subject of each formula.

a $P = 5l$ **b** $l = \dfrac{P}{5}$ **c** $F = ma$ **d** $a = \dfrac{F}{m}$

e $m = \dfrac{v - u}{F}$ **f** $s = \dfrac{1}{2}(u + v)t$ **g** $I = \dfrac{PTR}{100}$ **h** $t = ar^n$

2 The formula for finding the perimeter, P, of a square of side length x is $P = 4x$

Compare these methods of rearranging the formula to make x the subject.
Which do you prefer?

Draw a bar model

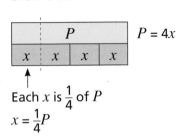

Function machine

Symbols

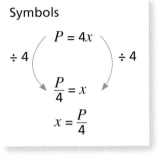

3 Make y the subject of each formula. You may use the bar models provided, or any other method, to help you.

a $x = y + 3$

| y | 3 |

b $x = y - 3$

| x | 3 |

c $x = 3y$

| y | y | y |

d $x = \dfrac{y}{3}$

| $\frac{y}{3}$ | $\frac{y}{3}$ | $\frac{y}{3}$ |
| x | x | x |

4 Beca and Marta are rearranging $m = 3n + 4$ to make n the subject.

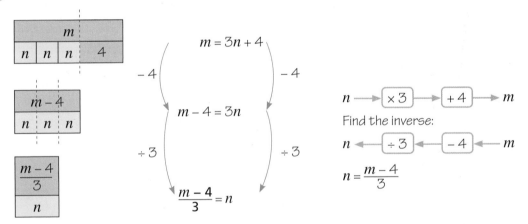

Use your preferred method to rearrange these formulae to make t the subject.

a $p = 3t + 7$ **b** $p = 3t - 7$ **c** $p = 3t + a$ **d** $p = 3t - a$

e $p = \dfrac{t}{2} + 7$ **f** $p = \dfrac{t}{2} - 7$ **g** $p = \dfrac{t}{2} + a$ **h** $p = \dfrac{t}{2} - a$

5 The circumference of a circle is 40 cm. Find, to 3 significant figures, the diameter of the circle by

a substituting $C = 40$ into the formula $C = \pi d$ and solving the resulting equation

b rearranging $C = \pi d$ to make d the subject and then substituting $C = 40$

What's the same and what's different?

6 Given that $t = 2m + 3n$

   a express m in terms of t and n

   b express n in terms of t and m

> This is another way of saying "Make m the subject of the formula $t = 2m + 3n$". This will make m equal to an expression involving t, n and some numbers.

7 Which of these is the correct rearrangement of the formula $a = 2bc$ to make b the subject?

A $b = \dfrac{2a}{c}$

B $b = \dfrac{a}{c} - 2$

C $b = \dfrac{a}{c/2}$

D $b = \dfrac{a}{2c}$

E $b = \dfrac{a}{2/c}$

F $b = \dfrac{a}{2} - c$

What do you think? 💭

1 Zach and Emily are rearranging $a = b - 2c$ to make c the subject.

Here is Zach's working.

  a Explain Zach's mistake.

$$a = b - 2c$$
$$-b \curvearrowright \qquad \curvearrowleft -b$$
$$a - b = 2c$$
$$\div 2 \curvearrowright \qquad \curvearrowleft \div 2$$
$$\frac{a - b}{2} = c$$

b Here is the start of Emily's working. Copy and complete it.

$$a = b - 2c$$
$$+2c \curvearrowright \qquad \curvearrowleft +2c$$
$$a + 2c = b$$
$$-a \curvearrowright \qquad \curvearrowleft -a$$
$$= $$
$$= $$

c Chloe uses a bar model.

start with b

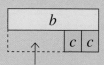

subtract $2c$

This must be a

subtract a from both bars

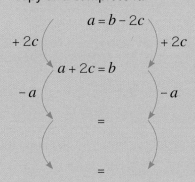

divide by 2

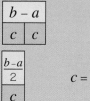

$$c = \frac{b - a}{2}$$

Whose method do you prefer? Discuss this with a partner.

d Make c the subject of each of these formulae.

    i $x = y + c$ **ii** $x = y - c$ **iii** $x = y - 4c$ **iv** $x = y - \dfrac{c}{2}$

2 Jakub is making x the subject of the formula $t = 10 - 3x$ using function machines.

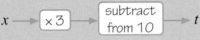

$x \longrightarrow \boxed{\times 3} \longrightarrow \boxed{\begin{array}{c}\text{subtract} \\ \text{from } 10\end{array}} \longrightarrow t$

Find the inverse:

$\boxed{?} \longleftarrow t$

> I don't know what the inverse of "subtract from 10" is.

Jakub

Help Jakub to complete the solution.

3 The equation of a straight line is $x + y = 10$

    a Rearrange $x + y = 10$ so that it is in the form $y = \ldots$

    b Does the straight line $x + y = 10$ have a positive or a negative gradient? How do you know?

    c Find the gradients of the straight lines with these equations.

        i $4x + y = 10$ **ii** $4x + 2y = 10$ **iii** $4x - 2y = 10$

    d

> $y = 3x - 5$ is an equation.

> $y = 3x - 5$ is a formula.

Seb   Flo

Who do you agree with?

4

> I can check that I have rearranged a formula correctly by substituting values into the original and rearranged forms.

Benji

Benji is correct. Work with a partner to check your answers to some of the earlier questions using substitution.

In the next section, you will learn how to rearrange some more complex formulae, including some with brackets, squares and square roots.

Example 3

Ed rearranges $a = bc^2$ to make c the subject.

Ed is incorrect.

Identify Ed's mistake and work out the correct answer.

$$a = bc^2$$
$$\div b \Big(\qquad \Big) \div b$$
$$\frac{a}{b} = c^2$$
$$\div 2 \Big(\qquad \Big) \div 2$$
$$\frac{a}{2b} = c$$

The first step is correct.

In the second step, Ed has divided by 2 instead of taking the square root.

$$a = bc^2$$
$$\div b \Big(\qquad \Big) \div b$$
$$\frac{a}{b} = c^2$$
$$\sqrt{} \Big(\qquad \Big) \sqrt{}$$
$$\sqrt{\frac{a}{b}} = c$$

$$c = \sqrt{\frac{a}{b}}$$

Explain the error clearly.

The inverse of squaring is taking the square root, not dividing by 2

The correct second step is to take the square root of both sides.

Remember to write the final answer with the new subject on the left-hand side.

Example 4

The area of a trapezium is given by the formula $A = \frac{1}{2}(a + b)h$

a Express h in terms of A, a and b

b Express a in terms of A, b and h

a

$$A = \frac{1}{2}(a + b)h$$
$$\times 2 \Big(\qquad \Big) \times 2$$
$$2A = (a + b)h$$
$$\div (a + b) \Big(\qquad \Big) \div (a + b)$$
$$\frac{2A}{a + b} = h$$

$$h = \frac{2A}{a + b}$$

Start by writing the formula.

Multiply by 2 to clear the fraction.

It is always good to clear any fractions first if you can. Here, multiplying both sides by 2 gives easier expressions to work with.

Divide both sides by $(a + b)$

The sum of a and b is multiplied by h on the right-hand side, so dividing both sides by $a + b$ leaves you with h, which is what you want.

You don't need the brackets around a and b on the denominator. Why not?

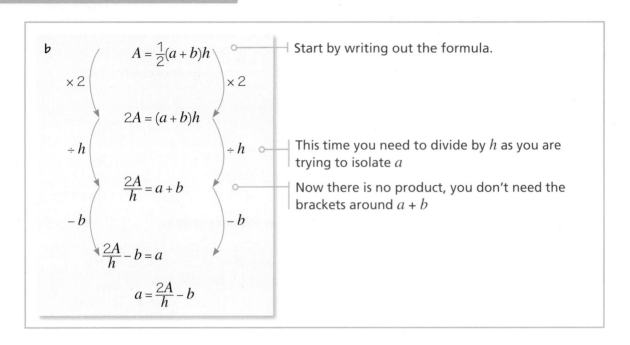

b

$A = \frac{1}{2}(a + b)h$ ○ ──── Start by writing out the formula.

$\times 2$ $\times 2$

$2A = (a + b)h$

$\div h$ $\div h$ ○ ── This time you need to divide by h as you are trying to isolate a

$\frac{2A}{h} = a + b$ ○ ── Now there is no product, you don't need the brackets around $a + b$

$- b$ $- b$

$\frac{2A}{h} - b = a$

$a = \frac{2A}{h} - b$

Practice 2.5B

1 Abdullah uses function machines to rearrange $x = 2y^2$ and $x = (2y)^2$ to make y the subject of each.

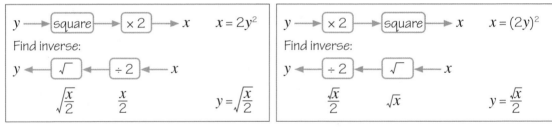

$y \rightarrow \boxed{\text{square}} \rightarrow \boxed{\times 2} \rightarrow x$ $\quad x = 2y^2$

Find inverse:

$y \leftarrow \boxed{\sqrt{}} \leftarrow \boxed{\div 2} \leftarrow x$

$\sqrt{\frac{x}{2}} \qquad \frac{x}{2} \qquad\qquad y = \sqrt{\frac{x}{2}}$

$y \rightarrow \boxed{\times 2} \rightarrow \boxed{\text{square}} \rightarrow x$ $\quad x = (2y)^2$

Find inverse:

$y \leftarrow \boxed{\div 2} \leftarrow \boxed{\sqrt{}} \leftarrow x$

$\frac{\sqrt{x}}{2} \qquad \sqrt{x} \qquad\qquad y = \frac{\sqrt{x}}{2}$

Copy and complete the working to rearrange these formulae using a different method.
Check that you get the same answers as Abdullah.

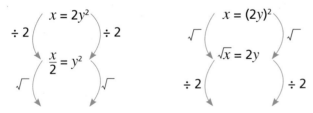

$\div 2$ $\left(\begin{array}{c} x = 2y^2 \\ \\ \frac{x}{2} = y^2 \end{array}\right) \div 2$

$\sqrt{}$ $\sqrt{}$

$\sqrt{} \left(\begin{array}{c} x = (2y)^2 \\ \\ \sqrt{x} = 2y \end{array}\right) \sqrt{}$

$\div 2$ $\div 2$

2 Rearrange these to make q the subject each time.

    a $\quad p = 2q + 4$ **b** $\quad p = q^2 + 4$ **c** $\quad p = 2q^2 + 4$

    What's the same and what's different?

3 **a** A rectangle has length $l = 16.4\,$cm and width $w = 9.3\,$cm. The perimeter, P, is $51.4\,$cm. Use the formula $P = 2(l + w)$ to verify the value of the perimeter.

b Copy and complete these methods for making w the subject of the formula $P = 2(l + w)$

expand $\left\langle \begin{array}{c} P = 2(l + w) \\ P = 2l + 2w \\ P - 2l = 2w \end{array} \right\rangle$ expand
$-2l$ $-2l$

$\div 2$ $\left\langle \begin{array}{c} P = 2(l + w) \\ \dfrac{P}{2} = l + w \end{array} \right\rangle$ $\div 2$

c Use both of your answers to part **b** to verify that $w = 9.3$ when $P = 51.4$ and $l = 16.4$

Why are there two different formulae for w? Why do they give the same answer?

4 Make t the subject of each of these formulae.

a $x = 3(t + 4)$ **b** $x = 3(2t + 4)$ **c** $x = 3(2t - 4)$

d $x = 3(4 + 5t)$ **e** $x = 3(4 - 5t)$ **f** $x = 3(t^2 + 4)$

Compare your answers with a partner's. Are they the same or are they different?

5 Rearrange each of these formulae to make x the subject.

a $y = \dfrac{x}{5}$ **b** $y = \dfrac{x}{5} + 1$ **c** $y = \dfrac{x + 1}{5}$ **d** $y = x + \dfrac{1}{5}$ **e** $y = \dfrac{3x - 1}{5}$

f $y = \dfrac{3x^2 - 1}{5}$ **g** $y = \dfrac{1 - 3x}{5}$ **h** $y = \dfrac{1 - 3x^2}{5}$ **i** $y = \dfrac{(3x - 1)^2}{5}$ **j** $y = 3 + \dfrac{2(x + 1)}{5}$

6 **a** Identify the correct working for making a the subject of $y = \sqrt{3a}$

A

$\div 3$ $\left\langle \begin{array}{c} y = \sqrt{3a} \\ \dfrac{y}{3} = \sqrt{a} \end{array} \right\rangle$ $\div 3$

square $\left\langle \begin{array}{c} \dfrac{y^2}{9} = a \end{array} \right\rangle$ square

$a = \dfrac{y^2}{9}$

B

square $\left\langle \begin{array}{c} y = \sqrt{3a} \\ y^2 = 3a \end{array} \right\rangle$ square

$\div 3$ $\left\langle \begin{array}{c} \dfrac{y^2}{3} = a \end{array} \right\rangle$ $\div 3$

$a = \dfrac{y^2}{3}$

Explain how you can check your answer.

b Make a the subject of each of these formulae.

i $y = 4\sqrt{a}$ **ii** $y = 4 + \sqrt{a}$ **iii** $y = \sqrt{4 + a}$ **iv** $y = \sqrt{a - 4}$ **v** $y = \sqrt{4 - a}$

vi $y = \sqrt{\dfrac{a}{4}}$ **vii** $y = \sqrt{\dfrac{a}{4}} + 1$ **viii** $y = \sqrt{\dfrac{a}{4} + 1}$ **ix** $y = \sqrt{\dfrac{a^2}{4} + 1}$ **x** $y = \sqrt{a^2 + \dfrac{1}{4}}$

What do you think? 💭

1 $V = IR$ is a formula used in physics.

Rearrange the formula to make

    a I the subject   **b** R the subject.

2 $P = \dfrac{F}{A}$ is a formula that connects pressure (P), force (F) and area (A)

    a   Express F in terms of P and A **b**   Express A in terms of P and F

3 **a**   Rearrange the formula $K = \dfrac{1}{2}mv^2$ to make

       i m the subject   **ii** v the subject.

   b   Use the formula $K = \dfrac{1}{2}mv^2$ to find an expression with value

       i $\dfrac{1}{2}$ **ii**   2

4 Another formula used in physics is $\dfrac{1}{f} = \dfrac{1}{v} + \dfrac{1}{u}$

    a   Express f in terms of v and u

    b   Express v in terms of f and u Hint: write $\dfrac{1}{v} + \dfrac{1}{u}$ as a single fraction first.

Consolidate – do you need more?

1 Write down two formulae shown by each bar model.

    a

a			
b	b	b	b

    b

x	
y	10

    c

t	
$\frac{1}{2}t$	$\frac{1}{2}t$
m	m

2 Rearrange these formulae to make y the subject.

    a $p = 6y$ **b** $q = \dfrac{y}{6}$ **c** $m = y + 6$ **d** $n = y - 6$

3

c				
d	d	d	d	12

   Use the bar model to express

    a c in terms of d **b** d in terms of c

4

a Use the bar model to express d in terms of c

b Copy and continue the working to make d the subject of the formula.

$$\times 3 \left(\begin{array}{c} \frac{c}{3} = d - 4 \end{array} \right) \times 3$$

5 Make t the subject of these formulae.

a $y = 2t$

b $y = t - 2$

c $y = \dfrac{t}{2}$

d $y = t + 2$

e $y = 2t + 5$

f $y = \dfrac{t}{2} + 5$

g $y = \dfrac{t}{2} - 5$

h $y = 2t - 5$

6 Here are the equations of some straight lines. Rearrange them into the form $y = mx + c$

a $y + 3x = 7$

b $2y + 8 = 4x$

c $x = 2y + 10$

d $3x = 5y - 10$

Why is it useful to write the equation of a straight line in the form $y = mx + c$?

7 Einstein worked out that energy (E), mass (m) and the speed of light (c) are connected by the formula $E = mc^2$. Use this formula to

a express m in terms of E and c

b express c in terms of E and m

Stretch – can you deepen your learning?

1 a An approximate formula for converting a temperature in degrees Celsius (C) to degrees Fahrenheit (F) is given in words as "Double the temperature in degrees Celsius and add 30".

   i Write this formula in symbols.

   ii Rearrange the formula to make C the subject.

b A more accurate formula is $F = \dfrac{9}{5}C + 32$. Rearrange this formula to make C the subject.

c i For what value of C do both the approximate and more accurate formulae give the same value of F?

   ii Find the value of F in this case.

2 In maths and physics, the "equations of motion" relate the displacement of an object (s) with its initial velocity (u), final velocity (v), acceleration (a) and time (t).

   a Acceleration is defined as "change in velocity divided by time". Form an equation for a in terms of v, u and t

   b Show that your answer to part **a** can be rearranged to give $v = u + at$

   c Another equation of motion is $v^2 = u^2 + 2as$. Rearrange this formula to make

      i a the subject   **ii** s the subject   **iii** u the subject.

   d Yet another equation of motion is $s = \frac{1}{2}(u + v)t$. Use the formula in part **b** to show that this leads to the equation $s = ut + \frac{1}{2}at^2$

   e Discuss with a partner whether s, u, v, a and t can or cannot be negative.

3 **a** Why is the formula $p + 3c = q - 4c$ different from most of the formulae you have met in this chapter?

   b Make c the subject of the formula $p + 3c = q - 4c$

   c Make t the subject of the formula $3(x + t) = 5(2x - y)$

Reflect

1 What's the same and what's different about solving an equation and rearranging a formula?

2 When rearranging a formula, how do you know which step to take first?

I have become **fluent** in…	I have developed my **reasoning** skills by…	I have been **problem-solving** through…
<ul><li>solving one-step equations and inequalities</li><li>solving two-step equations and inequalities</li><li>solving equations and inequalities with brackets</li><li>solving two-step equations and inequalities with unknowns on both sides</li><li>rearranging formulae.</li></ul>	<ul><li>identifying variables and expressing relations between variables algebraically</li><li>recognising the difference between formulae and equations</li><li>using formulae to form equations</li><li>interpreting the structure of a problem</li><li>recognising the subject of a formula.</li></ul>	<ul><li>modelling problems from other areas of maths by writing them as equations and inequalities</li><li>solving multi-step problems by representing them as equations and inequalities</li><li>selecting appropriate methods and techniques to apply to problems</li><li>representing equations and inequalities in a wide variety of forms.</li></ul>

Check my understanding

1 Find the equations with the solution $y = 4.5$

$4y = 18$ $\qquad$ $2y - 7 = 2$ $\qquad$ $3\left(y - \frac{1}{2}\right) = 12$ $\qquad$ $2y - 5 = 4y - 15$

2 Solve these inequalities.

a $3p - 7 > 20$ $\qquad$ **b** $10 + 2q \leqslant 5$ $\qquad$ **c** $8t - 3 < 5t - 7$

3 I think of a number. Three less than double my number is equal in value to my number subtracted from 18. Work out the value of my number.

4 A sequence is given by the rule $6n - 3$. Is 879 a term in the sequence? How do you know?

5 **a** Write down a formula, in terms of a and b, for the perimeter of this octagon.

$\quad$ **b** Find the value of b if $a = 1.5$ and P is 17

$\quad$ **c** Express a in terms of b and P

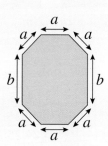

6 Rearrange these formulae to make x the subject.

a $m = 5x$ $\qquad$ **b** $n = x^2$ $\qquad$ **c** $q = 5x^2$

d $g = 10 - 7\sqrt{x}$ Ⓗ $\qquad$ **e** $V = \dfrac{2p - 5x^2}{7}$ Ⓗ

3 Testing conjectures

In this block, I will learn...

how to work with factors, multiples and primes

$21 = 3 \times 7$

The factors of 21 are 1, 3, 7 and 21

The prime factors of 21 are 3 and 7

21 is a multiple of both 3 and 7

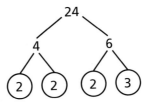

$24 = 2 \times 2 \times 2 \times 3$

$24 = 2^3 \times 3$

how to test conjectures

"When you add two primes the answer is even."

$3 + 5 = 8$ ✓

$11 + 2 = 13$ ✗

The statement is sometimes true.

how to prove conjectures about numbers using diagrams and symbols

"The sum of two even numbers is even."

$2m + 2n = 2(m + n)$

how to expand a pair of binomials

$(x + 1)(x + 3)$

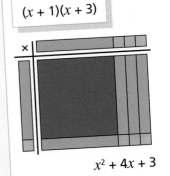

$x^2 + 4x + 3$

how to use algebra to look for patterns

1	2	3	4	5	6	7	8	9	10
11	12	13	14	15	16	17	18	19	20
21	22	23	24	25	26	27	28	29	30
31	32	33	34	35	36	37	38	39	40
41	42	43	44	45	46	47	48	49	50
51	52	53	54	55	56	57	58	59	60
61	62	63	64	65	66	67	68	69	70

x	
?	?
?	

Small steps

■ Explore factors, multiples and primes ®

Key words

Factor – a positive integer that divides exactly into another positive integer

Multiple – the result of multiplying a number by a positive integer

Prime number – a positive integer with exactly two factors, 1 and itself

Prime factor decomposition – writing numbers as a product of their prime factors

Are you ready?

1 Work out

  a 4×1 **b** 4×2 **c** 4×3 **d** 4×4 **e** 4×5

2 **a** What mathematical term is used for multiples of 2?

  b How can you tell by looking at a number whether it is a multiple of 2?

3 Copy and complete these tables of multiplication facts.

×	2	4	6
3			
5			
8			

×		5	7
		15	
4	8		
9			

Models and representations

Array

This array shows

$12 = 2 \times 6$ $12 \div 2 = 6$

$12 = 6 \times 2$ $12 \div 6 = 2$

Bar model

12			
3	3	3	3

This bar model shows

$12 \div 4 = 3$

$12 \div 3 = 4$

Factor tree

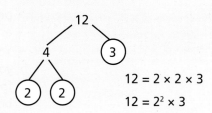

$12 = 2 \times 2 \times 3$

$12 = 2^2 \times 3$

You worked with **factors, multiples** and primes in Books 1 and 2. In this chapter, you will review your learning as you will be using these concepts in the rest of this block.

Example 1

Here is a list of numbers: 21   22   23   24   25   26   27

a   Write down all the prime numbers in the list.

b   Write down all the multiples of 3 in the list.

c   Write down all the factors of 48 in the list.

a   23 ⊸ — 23 is the only **prime number** as all the other numbers have factors other than 1 and themselves, for example:

$21 = 3 \times 7$

$22 = 2 \times 11$

$24 = 4 \times 6$

$25 = 5 \times 5$

$26 = 2 \times 13$

$27 = 3 \times 9$

b   21, 24 and 27 ⊸ — 3 divides exactly into these numbers with no remainder
$21 \div 3 = 7, 24 \div 3 = 8, 27 \div 3 = 9$

Another way of saying this is all the numbers are multiples of 3
$21 = 3 \times 7, 24 = 3 \times 8, 27 = 3 \times 9$

c   24 ⊸ — 24 is a factor of 48 as $24 \times 2 = 48$ (or $48 \div 2 = 24$)

Example 2

Express 54 as a product of its prime factors.

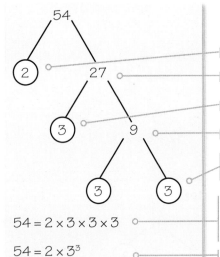

$54 = 2 \times 3 \times 3 \times 3$ ⊸

$54 = 2 \times 3^3$ ⊸

One way is to draw a factor tree.

Start by identifying a pair of factors: $54 = 2 \times 27$

2 is prime so you can circle it; this branch ends here

27 is not prime, so you need to continue: $27 = 3 \times 9$

3 is prime so you can circle it; this branch ends here.

9 is not prime, so you need to continue: $9 = 3 \times 3$

Again 3 is prime, so you can circle both 3s; the factor tree is complete.

Now you can write the product of prime factors in ascending order.

You can also write your answer in index form as $3 \times 3 \times 3$ can be written as 3^3

Remember: this result is called the **prime factor decomposition** of 54

Check that you get the same final answer if you start your tree with $54 = 6 \times 9$

Practice 3.1A

1 **a** What multiplication facts do these arrays show?

   b Use your answers to part **a** to list the factors of 12

2 **a** In how many ways can you arrange 15 counters in an array? List the multiplications that the arrays show.

   b List the factors of 15

   c List the factors of

   i 20   **ii** 24   **iii** 30   **iv** 27

3 Here is a list of numbers.

   15   16   17   18   19   20   21   22   23   24

   a From the list, write down the multiples of

   i 2   **ii** 3   **iii** 4   **iv** 6   **v** 8

   b How can you tell by looking at the list that there are exactly two multiples of 5?

4 **a** Find all the ways in which can you arrange 16 counters in an array.

   b Find all the ways in which can you arrange 17 counters in an array.

   c Explain how your answers to parts **a** and **b** illustrate that 17 is prime but 16 is not.

   d Which of the numbers in this list are prime?

   6   7   9   13   18   19   21   25   29   31   39

5 **a** Explain how you can use a calculator to find out whether a number is a multiple of another number.

   Here is a list of numbers

   90   91   92   93   94   95   96   97   98   99

   b From the list, write down all the multiples of

   i 3   **ii** 7

   c Write down all the prime numbers in the list.

6 **a** Copy and complete this factor tree.

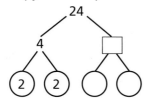

   b Use your answer to part **a** to write 24 as a product of its prime factors.

   c Express each of these numbers as a product of its prime factors.

   i 30   **ii** 40   **iii** 66   **iv** 28   **v** 81   **vi** 120

7 Express each of these numbers as the product of two prime factors.

    a 26   **b** 38   **c** 46   **d** 55   **e** 65   **f** 93

8   **a** Write 108 as a product of its prime factors.

    b Use your answer to part **a** to write 216 as a product of its prime factors.

    c Use you answer to part **b** to write 2160 as a product of its prime factors.

9 Another way of finding the prime factor decomposition of a number is called the "ladder method".

    a Discuss with a partner how ladder method works.

    b Use the ladder method to find the prime factor decomposition of

        i 500   **ii** 252   **iii** 264

Do you prefer the factor tree method or the ladder method?

2	180
2	90
3	45
3	15
5	5
	1

$180 = 2 \times 2 \times 3 \times 3 \times 5$

10   **a** $5a^3 = 135$. Find the value of a

    b $x = 2^3 \times 3^2 \times 5$. Calculate the value of $10x$

What do you think?

1 Do you agree with Marta?

> If a is a factor of b, then b is a multiple of a

2 This book was written in the year 2021

$2021 = 43 \times 47$

Write down the prime factorisations of

    a 4042   **b** 20210   **c** 606300

3 Ali factorises the expression $12x + 18y$

> $12x + 18y \equiv 3(4x + 6y)$

How many other ways can you find to factorise the expression?

4   **a** What percentage of the integers from 1 to 10 are prime?

    b What percentage of the integers from 1 to 100 are prime?

    c Estimate the percentage of the integers from 1 to 1000 that are prime. Check how close you are by finding the correct answer from an internet search.

    d Why do you think the percentage of prime numbers decreases as you include more numbers?

Consolidate – do you need more?

1 **a** What multiplication facts does this rectangle show?

   b What other rectangles can you draw with the same area?

2 List the factors of

   a 10   **b** 28   **c** 40

3 Here is a list of numbers.

   25   26   27   28   29   30   31   32   33   34

   a From the list, write down the multiples of

      i 3   **ii** 4

   b Which numbers in the list have

      i the most factors   **ii** the fewest factors?

4 Write down all the prime numbers between

   a 10 and 20   **b** 35 and 45

5 Seb is thinking of a number.

My number is odd. It is a multiple of 3 and a factor of 30

   What are the possible numbers that Seb could be thinking of?

6 Express each of these numbers as the product of two prime factors.

   a 33   **b** 34   **c** 35   **d** 46   **e** 51   **f** 85

7 Express each of these numbers as a product of its prime factors.

   a 75   **b** 48   **c** 60   **d** 600   **e** 96   **f** 225

Stretch – can you deepen your learning?

1 The number $2x$ will have twice as many factors as the number x

   a Find a pair of numbers for which Zach's statement is

      i correct   **ii** incorrect.

   b Find a general rule to decide when Zach's statement is correct and when it isn't correct.

2 **a** Substitute $n = 1, 2, 3$ and 4 into the expression $n^2 + n + 41$

   b What do you notice about your answers?

   c Explain why your discovery will not be true for $n = 41$

   d Explore substituting other numbers into the expressions $n^2 + n + 41$ and $n^2 - n + 41$

3 **a** Find the five numbers under 100 that have exactly 12 factors.

   b Find the prime factorisations of the numbers from part **a**. Explain why they must have 12 factors.

> You will explore this relationship again in Chapter 3.3

4 I can tell by looking at the prime factorisation of 1225 that it is a square number.

   a Find the prime factorisation of 1225 and explain how Faith knows it is a square number.

   b Use prime factorisation to determine which of these are square numbers.

      1764   729   3969   1980   2744   1 000 000

   c Are any of the numbers in part **b** cube numbers? How do you know?

   d Use your learning to find numbers that are both square and cube numbers.

Reflect

Describe how you could find all the factors of a number such as 14 280

Small steps

- Explore whether a statement is true or false
- Explore whether a statement is always, sometimes or never true

Key words

Conjecture – a statement that might be true that has not yet been proved

Counterexample – an example that disproves a statement

Are you ready?

1 **a** List the first six multiples of 5

    b What is true about the final digits of all multiples of 5?

2 List the factors of

    a 35   **b** 55   **c** 70

3 A fair straight line has the equation $y = 3x - 5$

    a What is the gradient of the line?

    b Where does the line intersect the y-axis?

4 A fair spinner has five equal sections labelled 1, 2, 3, 4 and 5

Write down the probability that the spinner lands on

    a 1   **b** 3   **c** a prime number.

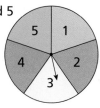

Models and representations

Depending on the area of maths you are working with in this chapter, you can use any of the models and representations used throughout Books 1, 2 and 3. Here are some particularly useful ones.

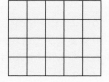

4 is a factor of 20

5 is a factor of 20

20 is a multiple of 4

20 is a multiple of 5

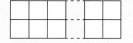

Even number

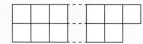

Odd number

In this first part of the chapter, you will explore whether statements are true or false. In each case, the emphasis will be on explaining how you know.

Here are some strategies you might use.

- ▦ Identifying a factual error
- ▦ Performing a calculation
- ▦ Looking at the structure of the mathematics that the statement is about
- ▦ Giving a **counterexample**

Example 1

> All multiples of 6 are even.

Is Chloe's claim true or false? Explain how you know.

True

6 times table	6	12	18	24
	× 3	× 3	× 3	× 3
2 times table	2	4	6	8

All the numbers in the 6 times table are 3 times a number in the 2 times table. All the numbers in the 2 times table have 2 as a factor, so all the numbers in the 6 times table will as well.

Explain how you know that the statement is true.

Notice that it's not enough to say something like "6 = 3 × 2 and 12 = 6 × 2 so it's true" as the claim is about **all** multiples of 6

Example 2

> You find factors of numbers by looking for factor pairs such as 10 = 1 × 10, 10 = 2 × 5. Because factors come in pairs, all numbers except 1 will have an even number of factors.

Is Zach's claim true or false?

Zach's claim is false.

For example, the factors of 25 are 1, 5 and 25 So 25 has 3 factors, which is an odd number.

This is called a counterexample – an example that shows that the statement is incorrect.

You only need one counterexample to show that a statement is false.

Practice 3.2A

1 Are these number statements true or false? How do you know?

a 86 + 24 = 100 **b** 100 − 47 = 63 **c** 848 ÷ 2 = 424

d 3.7 × 10 = 3.70 **e** −2 + −3 = 5

Compare your reasoning with a partner's.

2 Are these fraction, decimal and percentage statements true or false? How do you know?

a $\frac{1}{3} = 30\%$ **b** $\frac{1}{2} > \frac{1}{3}$ **c** $0.17 = \frac{17}{10}$ **d** $\frac{6}{8} = \frac{15}{20}$

e 31% = 0.31 **f** $\frac{3}{5} = 35\%$ **g** $\frac{4}{5} = 80\%$

Compare your reasoning with a partner's.

3 $\frac{1}{2}$ of 846 = 393

Emily: I'm going to divide 846 by 2 to see if the statement is true.

Jackson: I know just by dividing 800 by 2

a Explain how Jackson's calculation will tell him whether the statement is true or false.

b Find out whether these statements are true or false. For each, try to find a way that doesn't involve doing complex calculations.

i 49% of 88 = 45 **ii** 27 + 38 > 27 + 39

iii $\frac{7}{15}$ of 800 < $\frac{11}{20}$ of 800 **iv** 2652 ÷ 12 > 2652 ÷ 13

4 Are these statements true or false? Explain how you know.

a There are more primes between 1 and 10 inclusive than between 11 and 20 inclusive.

b Multiples of 4 are even.

c All the factors of odd numbers are odd.

d Positive integers have at least one factor.

5 Are these statements true or false? Explain how you know.

    a The probability of rolling a 2 on a fair dice is twice the probability of rolling a 1

    b If four people enter a tennis tournament, the probability of any one of them winning is $\frac{1}{4}$

    c If I flip a fair coin twice, the probability of getting two tails is $\frac{1}{4}$

6

> 91 isn't in the times tables, so it's a prime number.

Is Faith's claim true or false? Show calculations to justify your answer.

7 Here are some statements about equations. Which are true and which are false? Justify your answers.

    a $x = 5$ is the solution to the equation $12 - 2x = 5$

    b $x = 3$ is a solution to the equation $x^2 + 4x = 21$

    c $x = -4$ is the solution to the equation $12 - 2x = 20$

    d $x = -4$ is the solution to the equation $2x - 12 = -20$

8

> The area of a triangle $= \frac{1}{2} \times$ base $\times$ height. $7 \times 4 = 28$, so the height of the triangle must be $\frac{1}{2}$ of $4 = 2\,\text{cm}$

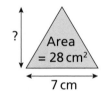

Area $= 28\,\text{cm}^2$

7 cm

Is Abdullah correct? Explain how you know.

9 Here are some statements about the equations of straight-line graphs. Which are true and which are false? Justify your answers.

    a $y = 2 - 5x$ has gradient 5   **b** $y = 2 - 5x$ cuts the y-axis at $(0, 2)$

    c $2y = 6x + 1$ has gradient 6   **d** $2y = 6x + 1$ cuts the y-axis at $(0, 1)$

    e The gradient of $x = 3$ is 0   **f** The gradient of $y = 3$ is 0

What do you think?

1 In Example 2, you discovered that this claim by Zach was false.

> You find factors of numbers by looking for factor pairs such as $10 = 1 \times 10$, $10 = 2 \times 5$. Because factors come in pairs, all numbers except 1 will have an even number of factors.

Most numbers do have an even number of factors, but some have an odd number of factors.

    a Which numbers under 50 have an odd number of factors?

    b What do the numbers in part **a** have in common?

2 Investigate whether these statements are true or false. Explain your answers.

a If $x > y$ then $\dfrac{1}{x} < \dfrac{1}{y}$

b If $a > b$ then $-a > -b$

c Straight lines with equations of the form $x + y = a$ have positive gradients.

In the next exercise, you will explore statements that could be always true, sometimes true or never true.

Example 3

Statements which you don't yet know are true are called conjectures.

"Square numbers have three factors." Is this statement always true, sometimes true or never true?

9 is a square number, and the factors of 9 are 1, 3 and 9. So it has 3 factors.

> Look for examples that support the **conjecture**.

16 is a square number and the factors of 16 are 1, 2, 4, 8 and 16, so it has 5 factors not 3

> Also look for examples that do not support the conjecture.

The statement is sometimes true.

> If you can find examples and counterexamples, then the statement is sometimes true.

Example 4

Are these statements about multiples of 10 always true, sometimes true or never true?

a Multiples of 10 are odd.

b Multiples of 10 are greater than or equal to 10.

a Never true. 10 itself is even, so any multiple of 10 will also be even.

> Explain briefly how you know that the statement is never true.

b Always true. The first multiple of 10 is 10, which is equal to 10, and all the other multiples are greater than 10, for example 20, 30, 40… and get greater as the list continues.

> Explain briefly how you know that the statement is always true.

Practice 3.2B

1 For each statement, give an example of when it is true and an example of when it is false.

    a   Prime numbers are odd.

    b   All factors of even numbers are even.

    c   Multiples of 2 are multiples of 4

    d   Adding three consecutive numbers results in an even number.

2 Are these statements about 5 always true, sometimes true or never true?

    a   Multiples of 5 are odd.

    b   Multiples of 5 are multiples of 10

    c   If you add a multiple of 5 to another multiple of 5, then the answer is a multiple of 5

3 **a**   Find the first three terms in the sequence given by the rule $4n + 1$

    b   Are these statements about the sequence in part **a** always true, sometimes true or never true?

        i   The numbers in the sequence are multiples of 4

        ii The numbers in the sequence are positive.

        iii The numbers in the sequence are odd.

        iv The numbers in the sequence are three less than multiples of 4

        v The numbers in the sequence are prime.

4 Are these statements always true, sometimes true or never true? If a statement is sometimes true, give an example to show when it is true and when it is false.

    a   3-D shapes have at least four faces.

    b   The angles in a triangle add up to 180°

    c   The angles in a polygon add up to a multiple of 360°

    d   Triangles have lines of symmetry.

    e   Straight-line graphs go through the origin.

    f   Quadrilaterals can be split into two triangles of equal size.

    g   A triangle contains two right angles.

    h   Isosceles trapezia have two pairs of equal angles.

5 Decide whether these algebraic statements are always true, sometimes true or never true. Don't forget to consider negative numbers and 0

    a $y + 3 > y + 2$ **b** $2x > x$ **c** $x^2 = x$

    d $p + 3 = p - 3$ **e** $(a + b)^2 = a^2 + b^2$

    Justify your answers.

6   Investigate whether these statements are always true, sometimes true or never true. Explain your answers.

   a   The product of two even numbers is a multiple of an odd number.

   b   The product of two odd numbers is a multiple of an even number.

   c   An even number has an even number of factors.

   d   An odd number has an odd number of factors.

   e   The sum of two prime numbers is even.

   f   Multiples of 7 are 1 more or 1 less than prime numbers.

   g   An even number that is divisible by 3 is also divisible by 6

What do you think?

1   Investigate whether these statements are always true, sometimes true or never true. Explain your answers.

   a   If n is even, $n + 2$ is even.

   b   If m is odd, $m + 1$ is even.

   c   If p is prime $p + 1$ is even.

   d   If $x > a$ then $x > -a$

2   The mean of a set of four numbers is greater than their mode.

   a   Show that Chloe's statement is only sometimes true.

   b   Investigate whether these statements are always true, sometimes true or never true.

   i   The mean of a set of four different numbers is greater than their median.

   ii   The range of a set of four different numbers is greater than 3

   iii The median of a set of four different numbers is greater than the mode.

3   Lydia is investigating the conjecture: "The area of a rectangle is numerically greater than its perimeter."

   a   Why does the conjecture say "numerically greater than" rather than "greater than"?

   b   Is the conjecture always true, sometimes true or never true?

4   Is it always true, sometimes true or never true that when you cut a piece out of a shape, you reduce its area and perimeter? Give examples to justify your answer.

Consolidate – do you need more?

1 Are these number statements true or false? How do you know?

a $\frac{1}{2}$ of 20 = $\frac{1}{4}$ of 40

b 30% of 50 = 50% of 30

c 12 × 6 = 10 × 6 + 2 × 6

d 7.8 ÷ 100 = 0.0078

e −2 × −3 = −6

Compare your reasoning with a partner's.

2 Are these statements true or false? Give a reason for each answer.

a 20% = $\frac{1}{20}$

b 58 is a multiple of 8

c 20 has exactly 6 factors

d 604 − 60 = 544

e It's harder to roll a 6 on a fair dice than it is to roll a 1

f 1 km = 100 m

3 Mr A gets a pay rise of 20%. Ms B gets a pay rise of 25%.

Is it always true, sometimes true or never true that Ms B's pay rise is greater than Mr A's?

4 Are these statements always true, sometimes true or never true? Justify your answers.

a When you fold a square in half you get a rectangle.

b If you subtract a multiple of 10 from any number, the ones digit of that number stays the same.

c If the sides in a shape are all equal in length, the angles in the shape are equal.

d The two diagonals of a parallelogram meet at right angles.

e If you double the perimeter of a square, you double its area.

Stretch – can you deepen your learning?

1 Investigate whether these statements are true or false. Explain your answers.

a For any integer n, 2^n has $n + 1$ factors.

b The product of two prime numbers will have four factors.

c If $p = q - r$ then $q = p - r$

d When you add two even numbers, the answer is even.

e If a number is the product of three different prime numbers, then it will have eight factors.

2 Beca makes two conjectures about the straight line with equation $y = ax + b$

"The straight line meets the y-axis at the point $(0, b)$"

"The straight line meets the x-axis at the point $(-\frac{b}{a}, 0)$"

Show that both of these conjectures are always true.

3 Investigate the following claims.

a A square number has an odd number of factors.

b A cube number has an odd number of factors.

c If the prime factorisation of n is $n = a^x b^y$, then n has $(x + 1)(y + 1)$ factors.

Investigate other prime factorisations.

Reflect

How do you decide whether a statement is true, false or sometimes true?
How can you illustrate that your decision is correct?

Small steps

- Show that a statement is true
- Conjectures about number

Key words

Conjecture – a statement that might be true that has not yet been proved

Prove – to show that something is always true

Counterexample – an example that disproves a statement

Are you ready?

1 Which of these numbers are odd and which are even? How do you know?

18   56   371   90   4005   40 008   123 456   1 234 567

2 Expand the brackets.

  a $4(a + 7)$ **b** $8(3 - b)$ **c** $x(x - 5)$ **d** $3(2a + 5b)$

3 Factorise these expressions.

  a $15b + 10c$ **b** $12 - 6t$ **c** $p^2 + 12p$

4 Work out

  a $2 + -4$ **b** 2×-4 **c** $-4 - 2$ **d** $-4 \div -2$ **e** $-2 - -4$

Models and representations

Counters   **Bars**   **Cubes**

These images could all represent even numbers as they show that they can be split into two equal rows.

Algebra tiles

 Here the tiles are representing three consecutive numbers.

In this chapter, you will explore **conjectures** about number and test whether they are always true, sometimes true or never true. You will learn how to use numbers, words, diagrams and symbols to show whether or not statements or conjectures are true.

You will learn more about using algebra to show whether conjectures are true in the next chapter.

Example 1

Show that $\frac{2}{3}$ of 90 = 50% of 120

$\frac{1}{3}$ of 90 = 90 ÷ 3 = 30

So $\frac{2}{3}$ of 90 = 2 × 30 = 60

50% of 120 = $\frac{1}{2}$ of 120 = 120 ÷ 2 = 60

Both calculations give the answer 60

So $\frac{2}{3}$ of 90 = 50% of 120

You can show this by performing the calculations.

First, work out $\frac{2}{3}$ of 90

Next work out 50% of 120

Use your answers to justify the fact that the two calculations are equal.

Example 2

Show that $\frac{2}{5}$ of 60 = $\frac{4}{5}$ of 30

Method A

$\frac{1}{5}$ of 60 = 60 ÷ 5 = 12

So $\frac{2}{5}$ of 60 = 2 × 12 = 24

$\frac{1}{5}$ of 30 = 30 ÷ 5 = 6

So $\frac{4}{5}$ of 30 = 4 × 6 = 24

Both calculations give the answer 24

So $\frac{2}{5}$ of 60 = $\frac{4}{5}$ of 30

Again you can show this by performing the calculations.

Method B

$\frac{2}{5}$ of 60 = $\frac{2}{5}$ × 60

$= \frac{2 \times 60}{5}$

$= \frac{2 \times 2 \times 30}{5}$

$= \frac{4 \times 30}{5}$

$= \frac{4}{5}$ × 30

Alternatively, you could show that the calculations are the same without working out the answers.

In this step, you can use the connection between 30 and 60 to show that the calculations are equivalent.

The numerical expressions have been kept equal to each other at every step, so this shows that $\frac{2}{5}$ of 60 = $\frac{4}{5}$ of 30

Explain how this diagram shows that $\frac{2}{5}$ of 60 = $\frac{4}{5}$ of 30

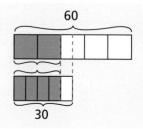

Example 3

Show that $5(a + 2b) + 3(a - 2b) \equiv 4(2a + b)$

Method A

LHS $= 5(a + 2b) + 3(a - 2b)$

$\equiv 5a + 10b + 3a - 6b$

$\equiv 5a + 3a + 10b - 6b$

$\equiv 8a + 4b$

$\equiv 4(2a + b)$, which is the RHS

So $5(a + 2b) + 3(a - 2b) \equiv 4(2a + b)$

To show that an algebraic identity is true, you need to show that both sides of the identity are the same. One way is to start by expanding on the left-hand side (LHS).

Collect like terms – you might be able to do this mentally.

Then simplify.

Factorise $8a + 4b$ to give the expression on the right-hand side (RHS).

State your conclusion.

Method B

LHS $= 5(a + 2b) + 3(a - 2b)$

$\equiv 5a + 10b + 3a - 6b$

$\equiv 5a + 3a + 10b - 6b$

$\equiv 8a + 4b$

RHS $4(2a + b) \equiv 8a + 4b$

Both expressions are equivalent to $8a + 4b$
So $5(a + 2b) + 3(a - 2b) \equiv 4(2a + b)$

You could also work out the LHS and RHS separately and show that the two expressions are identical each other. This is similar to the solution of Example 1.

Practice 3.3A

1 **a** Show by calculation that $80 \times 4 = 40 \times 8$

 b Explain how these diagrams show that $80 \times 4 = 40 \times 8$

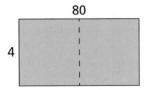
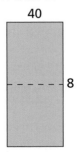

 c What other ways can you find to show that $80 \times 4 = 40 \times 8$?

2 Show that the calculations in each pair are equal.

a 16×75 and 3×400

b $\frac{1}{3}$ of 900 and $\frac{3}{4}$ of 400

c $20 \div 2$ and $40 \div 4$

Compare your methods with a partner's.

3 Use equivalent fractions to show that

a $\frac{4}{5} = 0.8$ **b** $\frac{9}{20} = 0.45$

4 Show that $\frac{5}{6} > \frac{3}{4}$ in as many ways as you can.

5 **a** Show that 40% of 80 is greater than 30% of 90

b Show that

i 60% of 90 = 90% of 60   **ii** 20% of 75 = 75% of 20   **iii** 150% of 30 = 30% of 150

c What do you notice about the questions in part **b**?
Can you make a general conjecture?
Can you explain why it works?

You will see a proof of this result in the next chapter.

6 Look at the red rectangle.

Show that

a the perimeter of the rectangle is $6a + 10$

b the area of the rectangle is $2a^2 + 10a$

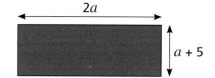

7 Show that

a $3(p + 2q) + 2(p + 3q) \equiv 5p + 12q$

b $3(p + 2q) - 2(p + 3q) \equiv p$

c $4(3p + 6q) + 6(4p - q) \equiv 18(2p + q)$

8 Show that $y = 3$ is a solution to each of these equations.

a $\boxed{6y + 4 = 22}$ **b** $\boxed{40 - 7y = 19}$ **c** $\boxed{5(2y - 1) = 25}$ **d** $\boxed{2y^2 = 18}$

e $\boxed{(y + 5)^2 = 64}$ **f** $\boxed{\dfrac{60}{y} = 20}$ **g** $\boxed{\dfrac{12}{5 - y} = 6}$ **h** $\boxed{y^3 - y^2 - 6y = 0}$

i $\boxed{20 - 4y = 35 - 9y}$

Which equations can you solve using the methods you have learned? How can you use substitution to show that $y = 3$ is a solution to the equations which you don't know how to solve?

9 Show that

a $24 \div \frac{1}{4} > 24 \div \frac{1}{3}$ **b** $16 \div \frac{1}{4} = 32 \div \frac{1}{2}$ **c** When $x > 0$, $x \div \frac{1}{3} > 5x \div 2$

What do you think?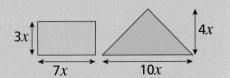

1 Show that the area of the rectangle is greater than the area of the triangle.

2 Show that none of these numbers are prime.

117　187　299　841　1271

3 Use prime factorisation to show that

　a 2304 is a square number but not a cube number

　b 8000 is a cube number but not a square number

　c 46656 is both a square number and a cube number.

4 **a** Show that increasing a number by 50% and then decreasing the result by 50% does not get you back to the original number.

　b Show that increasing a number by $\frac{1}{3}$ and then decreasing the result by $\frac{1}{4}$ gets you back to the original number.

　c Find other fractional increases and decreases which return a number to its original value as in part **b**.

Example 4

> When you add two even numbers, the answer is always even.

Show that Marta's conjecture is true.

Here are two even numbers

m		n
m		n

You can show this using bar models.

Both numbers are even as each bar is made up of two equal parts.

$$\begin{array}{|c|} m \\ \hline m \end{array} + \begin{array}{|c|} n \\ \hline n \end{array} = \begin{array}{|c|c|} m & n \\ \hline m & n \end{array}$$

Adding these gives a bar with two equal rows.

$$\begin{array}{|c|c|} m & n \\ \hline m & n \end{array} = \begin{array}{|c|} m+n \\ \hline m+n \end{array}$$

This is the same as one large bar with two equal rows.

$m + n$
$m + n$

The bar is split into two equal parts, so it represents an even number.

This bar represents an even number

so the sum of two even numbers is even.

In the next chapter, you will explore how to use algebra to show that conjectures like this are true.

Practice 3.3B

1

When you add two odd numbers, the answer is always odd.

Use a counterexample to show that Zach's conjecture is incorrect.

2 **a** Explain why this diagram shows an even number.

b Explain why this diagram shows an odd number.

c Explain how these diagrams show that when you add two odd numbers, the answer is always even.

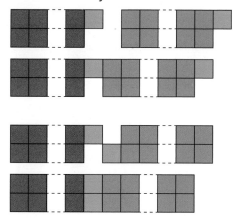

3 Use objects or diagrams to test each of these conjectures.

a One more than an odd number is an even number.

b Five more than an even number is an odd number.

c Two more than an even number is an even number.

d One less than an odd number is an even number.

e The sum of three even numbers is even.

f The sum of three odd numbers is odd.

4 Ali is trying to show that when you add two multiples of 3 the answer is a multiple of 3
He writes:

> $12 = 4 \times 3$ is a multiple of 3
>
> $30 = 10 \times 3$ is a multiple of 3
>
> $12 + 30 = 42$ and $42 = 14 \times 3$ is a multiple of 3
>
> So if you add two multiples of 3 the answer is a multiple of 3

a Why is Ali's solution incomplete?

b Using bar models or otherwise, show that the conjecture "When you add two multiples of 3 the answer is a multiple of 3" is always true.

c Form and test a conjecture about two multiples of 4

d Form and test a conjecture about two multiples of n

e Form and test a conjecture about m multiples of n

5 **a** Multiply any pair of even numbers. Is the answer even or odd?

b Explore with other pairs of even numbers. Make a conjecture about the product of a pair of even numbers.

c Use the diagram to explain why your conjecture is true.

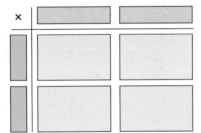

6 Test these conjectures.

a The sum of two negative numbers is negative.

b The product of two negative numbers is negative.

What do you think?

1 **a** Investigate the sum of

    i 2 odd numbers   **ii** 3 odd numbers

    iii 4 odd numbers   **iv** 5 odd numbers, and so on.

  b Can you see a general pattern? Can you make a conjecture? Can you prove that your conjecture is true?

2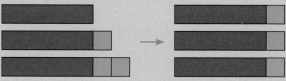

> The mean of three consecutive numbers is equal to the middle number.

  a Test Abdullah's conjecture with several sets of three consecutive numbers.

  b Explain how the pictures show that Abdullah's conjecture is true.

  c Explore the conjecture: "The sum of three consecutive numbers is a multiple of 3"

  d Investigate conjectures about the sum and mean of four or more consecutive numbers.

Consolidate – do you need more?

1 **a** Show by calculation that $200 \times 5 = 100 \times 10$

  b Show that $200 \times 5 = 100 \times 10$ in as many ways as possible.

2 Show that the calculations in each pair are equal.

  a 36×50 and 60×30 **b** $\frac{3}{5}$ of 300 and $\frac{2}{3}$ of 270   **c** $80 \div 10$ and $40 \div 5$

Compare your methods with a partner's.

3 Show that $\frac{3}{5} = 0.6$ using

  a equivalent fractions   **b** a diagram   **c** division.

4 Show that 40% of 70 is equal to 35% of 80

5 **a** Expand

    i $4(2x + 3y)$ **ii** $2(4x + 6y)$

  b What do you notice about your answers to part **a**? Explain why this happens.

6 **a** Show that the areas of these shapes are equal.

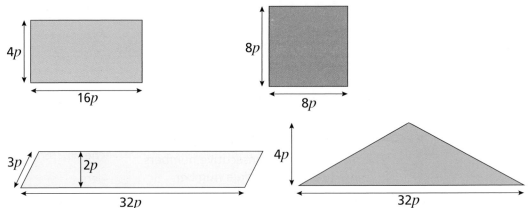

b For which of these shapes can you find the perimeter? Show that none of these perimeters are equal.

7 Use objects or diagrams to test each of these conjectures.

a One more than an even number is an odd number.

b Three more than an odd number is an even number.

c Two more than an odd number is an even number.

d Half of a multiple of 4 is a multiple of 2

Stretch – can you deepen your learning?

1 Here are illustrations of the first four triangular numbers.

a Use diagrams to show that the sum of two consecutive triangular numbers is a square number.

b The nth triangular number is given by the formula $T(n) = \frac{1}{2}n(n + 1)$
Verify that the formula is correct.

c

There is only one prime triangular number.

Is Flo's conjecture correct? How do you know?

2 The digit sum of a number is found by adding its digits, for example, the digit sum of 423 is $4 + 2 + 3 = 9$

 If the digit sum of a number is divisible by 9, then the number itself is divisible by 9

   a Verify that Benji's conjecture is true for the number 423 and test the conjecture for some other numbers of your choice.

   b Show that the conjecture is true for two-digit numbers. (Hint: a two-digit number with a in the tens column and b in the ones column can be written as $10a + b$)

   c Show that Benji's conjecture is true for three-digit numbers.

   d Will the conjecture be true for numbers with more than three digits? Explain why or why not.

3 The converse of an "If… then… " statement is made by swapping the "if" and "then" parts of the statement.

   a Write down the converse of Benji's conjecture in question **2**

   b Is the converse of Benji's conjecture also true?

4 Make and test a conjecture about the connection between the digit sum of a number and whether it is a multiple of 3. Can you prove your conjecture? Is the converse also true?

Reflect

1 How do you "show" that a conjecture is true?

2 Why is it easier to show that a conjecture is false than to show that it is true?

3 What's the difference between a demonstration and a proof?

Small steps

- Expand a pair of binomials
- Explore conjectures with algebra

Key words

Expand – multiply to remove brackets from an expression

Factorise – find the factors you need to multiply to make an expression

Binomial – an expression with two terms

Conjecture – a statement that might be true that has not yet been proved

Prove – to show that something is always true

Are you ready?

1 Expand the brackets.

    a $4(a + 4)$ **b** $a(a + 4)$ **c** $4(a - 4)$ **d** $a(a - 4)$

    e $b(a + 3)$ **f** $b(3 + a)$ **g** $b(3 + b)$ **h** $b(3 - b)$

2 Simplify

    a $3y + 4y$ **b** $3y - 4y$ **c** $-3y + 4y$ **d** $-3y - 4y$

3 Sort these into groups of like terms.

x^2 2 $2x$ $7x$ $2x^2$ 7 $-2x$ -2

4 Expand and simplify

    a $3(p + 2q) + 4(3p + 5q)$

    b $3(p + 2q) + 4(3p - 5q)$

    c $3(p + 2q) - 4(3p - 5q)$

5 Factorise

    a $2a + 2b$ **b** $3p + 3q + 6$ **c** $4m + 8n + 2$

Models and representations

You can use algebra tiles and area models to represent multiplying more than one set of brackets. There are interactive versions of algebra tiles available online.

Algebra tiles

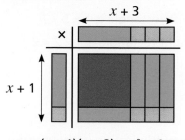

$$(x + 1)(x + 3) \equiv x^2 + 4x + 3$$

Area models

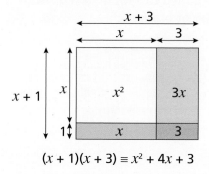

$$(x + 1)(x + 3) \equiv x^2 + 4x + 3$$

Cubes or counters

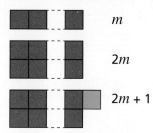

m

$2m$

$2m + 1$

You can use **counters** and **cubes** alongside **symbols** to represent different types of number.

An expression with two terms is called a **binomial**. $x + 3$ and $y - 4$ are examples of binomials.

Later in the chapter, you will combine this with earlier algebra skills to look at some **conjectures** and how to **prove** them.

You may have learnt about **expanding** a pair of binomials in the Higher chapter 7.3 of Book 2. If so, the first part of this chapter will be revision.

Example 1

Expand $(a + 1)(b + 2)$

Method A

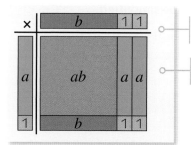

You can partition $a + 1$ into one a and one 1, and you can partition $b + 2$ into one b and two 1s.

You can then find the product of each section and add up to find the total product.

$(a + 1)(b + 2) \equiv ab + 2a + b + 2$

The two rectangles of area a can be simplified to $2a$ but you cannot simplify the expression any further as all the terms are unlike terms.

Method B

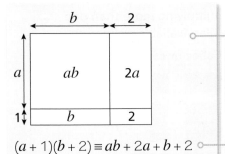

The area of the top left rectangle is $a \times b = ab$

The area of the top right rectangle is $a \times 2 = 2a$

The area of the bottom left rectangle is $1 \times b = b$

The area of the bottom right rectangle is $1 \times 2 = 2$

$(a + 1)(b + 2) \equiv ab + 2a + b + 2$ ⊸ Total area is $ab + 2a + b + 2$

Method C

$$(a + 1)(b + 2) \equiv a(b + 2) + 1(b + 2)$$
$$\equiv ab + 2a + b + 2$$

You can partition the multiplication without using a diagram by thinking of "$a + 1$" lots of "$b + 2$" as "a" lots of "$b + 2$" added to 1 lot of "$b + 2$" and expanding each pair of brackets.

This can also be shown using arrows:

$(a + 1)(b + 2) \equiv ab + 2a + b + 2$

Example 2

Expand $(x + 2)(x + 3)$

Method A

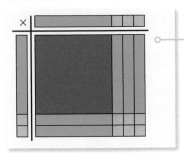

You can use algebra tiles to partition the multiplication.

$$(x + 2)(x + 3) \equiv x^2 + 3x + 2x + 6$$
$$\equiv x^2 + 5x + 6$$

This time the answer can be simplified as $3x$ and $2x$ are like terms.

Method B

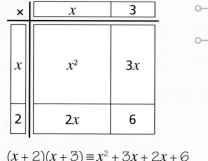

You can also represent the product as an area model.

The area of the top left rectangle is $x \times x = x^2$

The area of the top right rectangle is $x \times 3 = 3x$

The area of the bottom left rectangle is $2 \times x = 2x$

The area of the bottom right rectangle is $2 \times 3 = 6$

$$(x + 2)(x + 3) \equiv x^2 + 3x + 2x + 6$$
$$\equiv x^2 + 5x + 6$$

The total area is $x^2 + 5x + 6$

Method C

$$(x + 2)(x + 3) \equiv x(x + 3) + 2(x + 3)$$
$$\equiv x^2 + 3x + 2x + 6$$
$$\equiv x^2 + 5x + 6$$

You can partition the multiplication without using a diagram by thinking of "$x + 2$" lots of "$x + 3$" as "x" lots of "$x + 3$" added to 2 lots of "$x + 3$" and expanding each pair of brackets.

You can simplify the $2x + 3x$ to a single term $5x$ but you cannot simplify further.

This can also be shown using arrows:

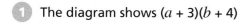

$$(x + 2)(x + 3) \equiv x^2 + 3x + 2x + 6$$
$$\equiv x^2 + 5x + 6$$

Practice 3.4A

1 The diagram shows $(a + 3)(b + 4)$

   a Copy and complete the diagram to show the area of each of section. One is filled in for you.

   b Use the diagram to expand $(a + 3)(b + 4)$

   c Use your expression in part **b** to work out 23×64

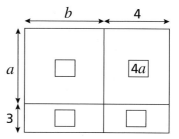

2 Using diagrams, or otherwise, expand these pairs of binomials.

   a $(a + 2)(b + 5)$ **b** $(p + 4)(q + 4)$

   c $(c + 7)(d + 1)$ **d** $(f + 6)(g + 7)$

3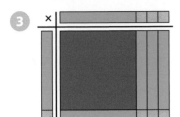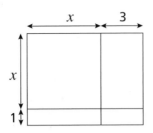

The diagrams use algebra tiles and an area model to represent $(x + 1)(x + 3)$

a Write down the four terms in the expansion of $(x + 1)(x + 3)$

b Write and simplify the expansion of $(x + 1)(x + 3)$

 c Which diagram do you find most useful?

4 Use diagrams, algebra tiles or a written method to expand these pairs of binomials. Remember to simplify your answers.

a $(x + 4)(x + 2)$ **b** $(x + 3)(x + 5)$

c $(x + 7)(x + 2)$ **d** $(x + 5)(x + 3)$

e $(x + 2)(x + 4)$ **f** $(x + 2)(x + 7)$

g $(x + 6)(x + 1)$ **h** $(x + 1)(x + 6)$

5 Which parts of question **4** have the same answers? Why?

6

$(x + 4)^2 \equiv x^2 + 4^2 \equiv x^2 + 16$

a By writing $(x + 4)^2$ as $(x + 4)(x + 4)$, show that Jackson is wrong.

b Expand and simplify

   i $(x + 5)^2$ **ii** $(x + 3)^2$ **iii** $(7 + x)^2$

c Use your answers to part **b** to work out

   i 25^2 **ii** 33^2 **iii** 57^2

7 The diagram uses algebra tiles to represent $(x - 1)(x + 3)$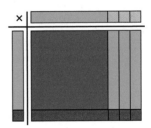

a Identify the correct expansion of $(x - 1)(x + 3)$

   A $x^2 + 3x + x - 3$ B $x^2 - 3x - x - 3$

   C $x^2 + 3x - x + 3$ D $x^2 + 3x - x - 3$

b Write the expansion of $(x - 1)(x + 3)$ in its simplest form.

8 Mario is expanding $(x - 1)(x + 5)$. He writes

$$(x - 1)(x + 5) \equiv x^2 + 5x - x - 5 \equiv x^2 + 5 - 5 \equiv x^2$$

a Explain why Mario is wrong.

b Find the correct expansion of $(x - 1)(x + 5)$

9 Expand and simplify.

a $(x + 3)(x - 2)$ b $(x - 3)(x + 2)$ c $(y - 4)(y + 5)$

d $(p - 3)(p + 5)$ e $(x - 2)(x - 5)$ f $(g - 2)(g - 6)$

g $(t + 5)^2$ h $(t - 5)^2$ i $(x + 4)(5 - x)$

What do you think?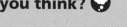

1 a Ali thinks that the expansion of $(a + b + 3)(c + 2)$ will have six terms.
Use a diagram to show that Ali is correct.

b Can the expression be simplified? Why or why not?

c Expand

     i $(a + b - 3)(c + 2)$ **ii** $(a - b + 3)(c - 2)$ **iii** $(a - b - 3)(c - 2)$

d Expand

     i $(a + b + 3)(a + 2)$ **ii** $(a + b + 3)(a + b)$ **iii** $(a - b - 3)(3 - a)$

2 a Expand and simplify

     i $(x + 5)(x - 5)$ **ii** $(x + 4)(x - 4)$ **iii** $(x + 2)(x - 2)$ **iv** $(x + y)(x - y)$

     What do you notice about your results?

b Use your answers to part **a** to write down the answers to

     i $(x + 9)(x - 9)$ **ii** $(y - 3)(y + 3)$ **iii** $(p - 6)(p + 6)$ **iv** $(6 + p)(6 - p)$

c **i** Write $x^2 - 144$ as the product of two binomials.

     ii Write $53^2 - 47^2$ as the product of two binomials.

     iii Use your answer to part **ii** to show that $53^2 - 47^2 = 600$

     iv Work out $7.3^2 - 2.7^2$

3 a Which is the correct expansion of $(2x + 3)(x + 4)$?

     A $2x^2 + 7x + 12$ B $2x^2 + 27x + 12$ C $2x^2 + 11x + 12$

b Investigate expansions of the type

     i $(ax + b)(x + d)$ **ii** $(ax + b)(cx + d)$

Start with a, b, c and d all positive and then move on to include negative numbers in your expressions.

In Practice 3.3A question **5**, you showed facts such as 60% of 90 = 90% of 60. You will now prove a general result for this type of result using algebra. You will also prove other results algebraically.

Example 3

$a\%$ of $b = b\%$ of a

Show that Ed's conjecture is true.

$a\%$ of $b = \dfrac{a}{100} \times b = \dfrac{ab}{100}$ ○—| First, find an expression for $a\%$ of b

$b\%$ of $a = \dfrac{b}{100} \times a = \dfrac{ba}{100}$ ○—| Then, find an expression for $b\%$ of a

One way of showing that two things are equal is to show they are both equal to the same thing.

$ab = ba$, so $\dfrac{ab}{100} = \dfrac{ba}{100}$ ○—| We know that the expressions are equal because multiplication is commutative.

So $a\%$ of $b = b\%$ of a

You could use this result to find the answer to calculations such as 84% of 25 by working out 25% of 84 instead.

Example 4

Show that the sum of two even numbers is always even.

If a is even then $a = 2m$ where m is an integer. ○—| Any even number is a multiple of 2

If b is even then $b = 2n$ where n is an integer. ○—| The two even numbers might be different, so you cannot use the same letter for them.

$a + b = 2m + 2n = 2(m + n)$ ○—| Add the expressions for a and b. You can **factorise** because 2 is a common factor.

So $a + b$ is a multiple of 2, which means that $a + b$ is even. ○—| Remember to write a statement to explain your answer.

Any number that has 2 as a factor is even.

Compare this with the method that used bar models which you explored in Chapter 3.3

Here are two even numbers.

m / m n / n These bars represent $2m$ and $2n$

m/m + n/n = m n / m n The total is 2 bars of length $m + n$

m n / m n = $m + n$ / $m + n$ Each green bar is $m + n$ long.

$m + n$ / $m + n$ This is also even.
      So the sum of two even numbers is even.

Practice 3.4B

 1 Copy and complete these sentences using either the word "odd" or "even".

    a If x is even then $x + 1$ is _____
    b If x is odd then $x + 1$ is _____

    c If x is even then $x + 2$ is _____
    d If x is odd then $x + 2$ is _____

    e If x is an integer then $2x$ is _____
    f If x is an integer then $2x + 1$ is _____

2 p is an even number.

    a What can you say about the numbers represented by these expressions?

       i $p + 10$ **ii** $p - 1$ **iii** $p + 5$

    b

You can't tell whether $3p$ is even or odd.

Jakub

2p is definitely a multiple of 4

Faith

    Investigate Jakub's and Faith's claims.

3 Write an algebraic expression to represent each of these.

    a 1 greater than n **b** 2 less than n **c** 2 multiplied by n

    d 2 more than n **e** n multiplied by 4

4   If k is an integer, then $6k$ is a multiple of 6

a Explain why Beca is correct.

b If k is an integer, write an expression in terms of k for

    i a multiple of 5   **ii** a multiple of 8   **iii** 1 less than a multiple of 7

c Beca writes

$6k = 2 \times 3k$, so $6k$ must be even.

Use a similar strategy to show that $6k$ is a multiple of 3

d If k is an integer, which of these expressions are always even?

$4k$	$6k + 1$	$3k$	$5k - 2$	$3k + 2$	$10k - 2$

Can you determine anything about the other expressions?

5 Filipo and Darius are trying to prove the conjecture: "When you add two odd numbers, the answer is even."

Filipo starts

Darius starts

$2n + 1 + 2n + 1 \equiv 4n + 2$

$2n + 1 + 2m + 1 \equiv 2m + 2n + 2$

a Why is Darius' approach better than Filipo's?

b Factorise Darius' expression to show that the sum of two odd numbers is always even.

6 Use algebra to prove that all these conjectures are true.

a The sum of three even numbers is even. (Hint: call the numbers $2a$, $2b$ and $2c$)

b The sum of two multiples of 5 is a multiple of 5

c The sum of two consecutive integers is odd. (Hint: call the smallest integer n)

d The sum of three consecutive integers is a multiple of 3. (Hint: call the smallest integer n)

7 Use a counterexample to show that all of these conjectures are false.

a If p is prime and x is even, then $p + x$ is odd.

b If m is even, then $\frac{m}{2}$ is even.

c If x is an integer, then $5x$ is always a multiple of 10

8 **a** Is this conjecture true or false?

"All the terms in the sequence given by the rule $4n + 2$ are even."

Explain how you know.

b Make up and test conjectures about the sequences given by these rules.

i $4n + 1$ **ii** $3n - 1$ **iii** $6n - 1$

Explain your findings.

9 **a** Multiply a pair of even numbers. Is the answer even or odd?

b Use algebra to show that the product of two even numbers is even.

c Use algebra to show that the product of two even numbers is a multiple of 4

d Make and test a conjecture about the product of three even numbers.

e Make and test a conjecture about the product of two odd numbers.

10 Make up and test your own conjectures about adding, subtracting, multiplying and dividing odd numbers, even numbers, multiples of 3, and so on. Can you prove your results using algebra?

What do you think?

1 Marta is testing the conjecture: "The square of an even number is always even."

She writes:

> An even number can be written as $2n$.
>
> Squaring $2n$ gives $2n^2 = 2 \times n^2$, which is a multiple of 2, so the square of an even number is even.

a What mistake has Marta made?

b Correct Marta's proof.

c Is the square of an even number always, sometimes or never a multiple of 4?

d By expanding a pair of binomials, prove that the square of an odd number is always odd. What else can you say about the square of an odd number?

2 Use algebra to show that the sum of three odd numbers is always odd but the sum of four odd numbers is always even.

3 In Chapter 3.3, you used objects to prove this conjecture:

"The mean of three consecutive numbers is equal to the middle number."

a Use algebra to prove the conjecture.

b Use algebra to investigate conjectures about the sum and mean of four or more consecutive numbers.

Consolidate – do you need more?

1 The diagram shows $(a + 5)(b + 2)$

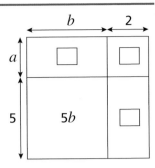

a Copy and complete the diagram to show the area of each section. One is filled in for you.

b Use the diagram to expand $(a + 5)(b + 2)$

c Use your expansion from part **b** to work out 25×32

2 Using diagrams, or otherwise, find the expansion of each of these pairs of binomials.

a $(a + 3)(b + 4)$ **b** $(c + 6)(d + 3)$ **c** $(g + 7)(f + 2)$

3 **a** Expand

i $(x + 4)(y + 5)$ **ii** $(x + 4)(x + 5)$

What's the same and what's different?

b Expand and simplify

i $(y + 3)(y + 10)$ **ii** $(p + 6)(p + 4)$ **iii** $(m + 4)(m + 9)$ **iv** $(n + 1)^2$

4 **a** Expand and simplify

i $(t + 3)(t + 4)$ **ii** $(t + 3)(t - 4)$ **iii** $(t - 3)(t + 4)$ **iv** $(t - 3)(t - 4)$

What's the same and what's different?

b Expand and simplify

i $(p + 5)(p - 6)$ **ii** $(q - 2)(q - 6)$ **iii** $(g - 4)(g + 7)$ **iv** $(f + 3)^2$

5 Write algebraic expressions that represent the following.

a 1 less than q **b** 2 greater than q **c** q multiplied by 3

d 2 less than q **e** 3 more than q

6 Is the sum of an even number and an odd number even or odd? Prove your answer by using $2n$ to represent an even number and $2m + 1$ to represent an odd number.

7 **a** Write an expression for a multiple of 3

b Write an expression for a multiple of 5

c Ali thinks that the sum of a multiple of 3 and a multiple of 5 must be a multiple of 8. Do you agree with Ali? Justify your answer.

d Use algebra to show that the sum of a multiple of 6 and a multiple of 9 is always a multiple of 3.

8 Show that all multiples of 10 are

a even   **b** multiples of 5

9 Investigate the product of two multiples of 3. Make a conjecture and test it. Can you prove your result?

Stretch – can you deepen your learning?

1 **a** Show that $(x^2 + 5x + 7)(x + 2) \equiv x^3 + 7x^2 + 17x + 14$

 b Expand $(x + 1)(x + 2)(x + 3)$ by first expanding $(x + 1)(x + 2)$ and then multiplying the result by $x + 3$. Give your answer in its simplest form.

 c Check your answer to part **b** by expanding the three brackets in a different order.

 d Expand and simplify

 i $(x + 1)^3$ **ii** $(x - 1)^3$ **iii** $(x + 2)^3$ **iv** $(x - 2)^3$

 e Work out $(x + 3)^3$ and use your answer to expand $(x - 3)^3$

2 **a** Write down three consecutive integers.

 Find the product of the first and third numbers.

 Find the square of the second number.

 Compare your answers. What do you notice? Repeat for other sets of three consecutive integers.

 b Use algebra to show that the result that you obtained in part **a** is always true by

 i calling the smallest of the consecutive numbers x

 ii calling the greatest of the numbers x

 Discuss with a partner the different approaches taken in parts **bi** and **ii**.

3 In 1742, the German mathematician Christian Goldbach conjectured: "Every even integer can be written as the sum of two prime numbers." The conjecture remains unproved. Investigate the conjecture for numbers up to 100. What patterns can you see?

Reflect

1 How do you expand a pair of binomials? When can you and when can't you simplify your answer?

2 How can you use algebra to describe odd and even numbers? What results about combining odd and even numbers can you prove?

3.5 Searching for pattern

Small steps

- Explore the 100 square
- Expand a pair of binomials
- Explore conjectures with algebra

Key words

Expand – multiply to remove brackets from an expression

Conjecture – a statement that might be true that has not yet been proved

Are you ready?

1 n is an integer. Write an expression for each of the following integers.

    a 1 more than n **b** 10 more than n **c** 11 less than n

2 Expand and simplify

    a $x(x + 2)$ **b** $y(y + 10)$ **c** $t(t - 4)$ **d** $(p + 3)(p + 10)$

3 Simplify these expressions.

    a $a + a + 1 + a + 10 + a + 11$

    b $b + b + 6 + b + 10 + b + 20$

    c $c(c + 10) - c(c + 5)$

Models and representations

Hundred square

1	2	3	4	5	6	7	8	9	10
11	12	13	14	15	16	17	18	19	20
21	22	23	24	25	26	27	28	29	30
31	32	33	34	35	36	37	38	39	40
41	42	43	44	45	46	47	48	49	50
51	52	53	54	55	56	57	58	59	60
61	62	63	64	65	66	67	68	69	70
71	72	73	74	75	76	77	78	79	80
81	82	83	84	85	86	87	88	89	90
91	92	93	94	95	96	97	98	99	100

In this chapter, you will explore patterns in the hundred square and other grids. You will make, test and prove **conjectures**.

Example 1

Here are two shaded squares in a hundred square

1	2	3	4	5	6	7	8	9	10
11	12	13	14	15	16	17	18	19	20
21	22	23	24	25	26	27	28	29	30
31	32	33	34	35	36	37	38	39	40
41	42	43	44	45	46	47	48	49	50
51	52	53	54	55	56	57	58	59	60
61	62	63	64	65	66	67	68	69	70
71	72	73	74	75	76	77	78	79	80
81	82	83	84	85	86	87	88	89	90
91	92	93	94	95	96	97	98	99	100

This square is called S_{23}

This square is called S_{46}

a Find the total of the numbers in S_{23}

b Find the total of the numbers in S_{46}

c Chloe thinks that the total of the numbers in S_{46} will be double the total of the numbers in S_{23}. Is Chloe correct?

d Find an expression for the total of the numbers in S_x. (S_x is the square with x in its top left-hand corner.)

e Prove that the total of the numbers in any 2 by 2 square is always even.

a $23 + 24 + 33 + 34 = 114$ —— Add the four numbers in the square.

b $46 + 47 + 56 + 57 = 206$

c Total of $S_{46} = 46 + 47 + 56 + 57 = 206$ —— Work out the totals and compare them.

$2 \times$ total of $S_{23} = 2 \times 114 = 228$

$206 \neq 228$, so Chloe is wrong —— State your conclusion clearly, justifying your answer.

d

x	$x + 1$
$x + 10$	$x + 11$

Total $= x + x + 1 + x + 10 + x + 11 = 4x + 22$

Find an expression for the total by adding these and simplifying the answer.

Find an expression, in terms of x, for each of the cells in the square.

The number to the left of x is one more than x, so is $x + 1$

The number directly below x is ten more than x, so is $x + 10$

The number to the left of $x + 10$ is one more than $x + 10$, so is $x + 11$

e $4x + 22 = 2(x + 11) = 2 \times$ an integer so is always even

Factorise to show that the number represented by the expression is always a multiple of 2

Practice 3.5A

1 Here is part of a hundred square.

The highlighted blocks are R_{14} and R_{32}. The value of each block is found by adding the numbers in each cell.

a Show that $R_{14} = 45$

b Find the value of R_{32}

c Explain why you cannot find R_{29}

d Copy and complete the table to show the value of R_x

x	$x + 1$	

e Find an expression, in terms of x, for the value of R_x
Check that your expression gives the correct values of R_{14} and R_{32}

f Use your expression for R_x to work out R_{79}

1	2	3	4	5	6	7	8	9	10
11	12	13	14	15	16	17	18	19	20
21	22	23	24	25	26	27	28	29	30
31	32	33	34	35	36	37	38	39	40
41	42	43	44	45	46	47	48	49	50
51	52	53	54	55	56	57	58	59	60

2 Here is part of another hundred square.

The highlighted blocks are J_{15} and J_{28}

The value of each block is found by adding the numbers in each cell.

Find the value of

a J_{15}

b J_{28}

c Copy and complete the table to show the values of J_x

1	2	3	4	5	6	7	8	9	10
11	12	13	14	15	16	17	18	19	20
21	22	23	24	25	26	27	28	29	30
31	32	33	34	35	36	37	38	39	40
41	42	43	44	45	46	47	48	49	50
51	52	53	54	55	56	57	58	59	60

	x	
$x + 19$		

d Find an expression, in terms of x, for the value of J_x
Check that your expression gives the correct values of J_{15} and J_{28}

e Use your expression for J_x to work out J_{60}

f Explain why J_x is always odd.

g Find the value of x if $J_x = 341$

3 On this hundred square, the highlighted block is called T_4 and the value of the block is found by adding the numbers in each cell.

a Find an expression for the value of T_x

b Is there a block with a value of 428? Justify your answer.

c Find the smallest value of x for which the value of T_x is greater than 200

d Show that the value of these blocks is always a multiple of 6

1	2	3	4	5	6	7	8	9	10
11	12	13	14	15	16	17	18	19	20
21	22	23	24	25	26	27	28	29	30
31	32	33	34	35	36	37	38	39	40
41	42	43	44	45	46	47	48	49	50

4 A 2 by 4 rectangle is drawn on a hundred square.

The products of the opposite corners of the rectangle are $15 \times 28 = 420$ and $18 \times 25 = 450$

a Find the difference between the products of the opposite corners of the rectangle shown.

b Use algebra to show that the difference between the products of the opposite corners of any 2 by 4 rectangle on the hundred square is always the same.

1	2	3	4	5	6	7	8	9	10
11	12	13	14	15	16	17	18	19	20
21	22	23	24	25	26	27	28	29	30
31	32	33	34	35	36	37	38	39	40
41	42	43	44	45	46	47	48	49	50

c Find the difference between the opposite corners of the 4 by 2 rectangle on the hundred square below. What's the same and what's different about this answer and your answer to part **a**?

1	2	3	4	5	6	7	8	9	10
11	12	13	14	15	16	17	18	19	20
21	22	23	24	25	26	27	28	29	30
31	32	33	34	35	36	37	38	39	40
41	42	43	44	45	46	47	48	49	50
51	52	53	54	55	56	57	58	59	60

What do you think? 💭

1. Change the size of the rectangles used in question **4** above.
   Investigate the following:

   ▥ How does the size of the rectangle affect the difference between the products of the opposite corners of the rectangle?

   ▥ Does the orientation of the rectangle affect the outcome?

   ▥ What's the same and what's different if you consider squares instead of rectangles?

2. How do the expressions you found in question **2** above change if you rotate or reflect the shapes? Is there a connection between the expressions?

1	2	3	4	5	6	7	8	9	10
11	12	13	14	15	16	17	18	19	20
21	22	23	24	25	26	27	28	29	30
31	32	33	34	35	36	37	38	39	40
41	42	43	44	45	46	47	48	49	50
51	52	53	54	55	56	57	58	59	60

1	2	3	4	5	6	7	8	9	10
11	12	13	14	15	16	17	18	19	20
21	22	23	24	25	26	27	28	29	30
31	32	33	34	35	36	37	38	39	40
41	42	43	44	45	46	47	48	49	50
51	52	53	54	55	56	57	58	59	60

1	2	3	4	5	6	7	8	9	10
11	12	13	14	15	16	17	18	19	20
21	22	23	24	25	26	27	28	29	30
31	32	33	34	35	36	37	38	39	40
41	42	43	44	45	46	47	48	49	50
51	52	53	54	55	56	57	58	59	60

Consolidate – do you need more?

1. Here is part of a hundred square.

   The highlighted blocks are I_6 and I_{13}
   The value of each block is found by
   adding the numbers in each cell.

   a Show that $I_6 = 48$

   b Find the value of I_{13}

   c Explain why you cannot find I_{86}

   d Copy and complete the table to
   show the values of I_x

1	2	3	4	5	6	7	8	9	10
11	12	13	14	15	16	17	18	19	20
21	22	23	24	25	26	27	28	29	30
31	32	33	34	35	36	37	38	39	40
41	42	43	44	45	46	47	48	49	50
51	52	53	54	55	56	57	58	59	60

x
$x + 20$

   e Find an expression, in terms of x, for the value of I_x
   Check that your expression gives the correct the values of I_6 and I_{13}

   f Use your expression for I_x to work out I_{50}

2 Here is part of another hundred square.

The highlighted blocks are t_{12} and t_{28}
The value of each block is found by
adding the numbers in each cell.

1	2	3	4	5	6	7	8	9	10
11	12	13	14	15	16	17	18	19	20
21	22	23	24	25	26	27	28	29	30
31	32	33	34	35	36	37	38	39	40
41	42	43	44	45	46	47	48	49	50
51	52	53	54	55	56	57	58	59	60

a Find the value of t_{12}

b Find the value of t_{28}

c Copy and complete the table to
show the values of t_x

x

$x + 20$

d Find an expression, in terms of x, for the value of t_x Check that your expression
gives the correct the values of t_{12} and t_{28}

e Use your expression for t_x to work out t_{81}

f Is t_x always even, always odd, or sometimes odd and sometimes even? Explain how
you know.

g Find shapes in the grid whose total is

    i always even   **ii** always odd   **iii** sometimes odd and sometimes even.

Stretch – can you deepen your learning?

1 Here is a C-shape drawn on part of a 10 by 10 grid and on a 9 by 9 grid.

1	2	3	4	5	6	7	8	9	10
11	12	13	14	15	16	17	18	19	20
21	22	23	24	25	26	27	28	29	30
31	32	33	34	35	36	37	38	39	40
41	42	43	44	45	46	47	48	49	50

1	2	3	4	5	6	7	8	9
10	11	12	13	14	15	16	17	18
19	20	21	22	23	24	25	26	27
28	29	30	31	32	33	34	35	36
37	38	39	40	41	42	43	44	45

a Compare the expressions for the totals of the numbers in the shapes on
the two grids.

b How would the expressions change for different sizes of grids?

c How would the expressions change if the shapes were rotated or reflected?

2 a Investigate the difference between the products of opposite corners of rectangles
drawn on different-sized grids.

b What other shapes could you investigate? Write a summary of your results.

Reflect

1 What does it mean to generalise a result?

2 Explain how working systematically helps you to find a solution to a problem.

I have become **fluent** in...	I have developed my **reasoning** skills by...	I have been **problem-solving** through...
▦ recognising factors and multiples ▦ expressing a number as a product of prime factors ▦ simplifying algebraic expressions ▦ expressing one number in terms of another ▦ expanding a pair of binomials.	▦ exploring whether statements are true or false ▦ determining if statements are always true, sometimes true or never true ▦ finding examples and counterexamples to prove or disprove a conjecture ▦ conjecturing relationships and generalisations ▦ developing an argument, justification or proof.	▦ applying different areas of mathematics ▦ breaking down problems into smaller parts ▦ looking at non-routine problems ▦ using diagrams and algebra to represent situations.

Check my understanding

1 **a** Explain why 80 is a multiple of 8 but a factor of 800

    b Express 80 as product of its prime factors.

2 Are these statements true or false? Justify your answers.

    a None of the numbers in the 7 times table are in the 11 times table.

    b $3\,km > 300\,000\,mm$

3 Show that

    a the statement "numbers of the form $6k + 1$ are prime" is only sometimes true

    b $160 \times 25 = 100 \times 40$

    c the product of a multiple of 9 and a multiple of 4 is always a multiple of 12

4 **a** Expand and simplify if possible.

        i $(x + 5)(x - 8)$ **ii** $(x + 5)(y - 8)$ **iii** Expand and simplify $(p - 1)^2$

    b Use your answer to part **iii** to work out 99^2

5 Here is part of a hundred square.

    The highlighted block is B_4 The value of the block is found by adding the numbers in the cells.

    a Find the value of x if $B_x = 429$

    b Test the conjecture "If x is even, then B_x is even, but if x is odd, then B_x is odd". Explain your answer.

1	2	3	4	5	6	7	8	9	10
11	12	13	14	15	16	17	18	19	20
21	22	23	24	25	26	27	28	29	30
31	32	33	34	35	36	37	38	39	40
41	42	43	44	45	46	47	48	49	50

4 Three-dimensional shapes

In this block, I will learn...

the names of 3-D shapes

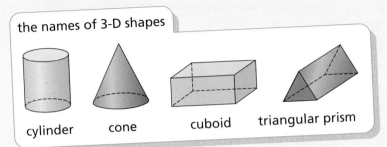

cylinder cone cuboid triangular prism

how to draw nets of 3-D shapes

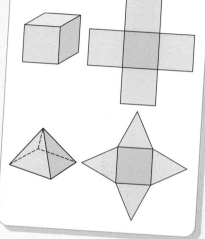

how to draw plans and elevations of 3-D shapes

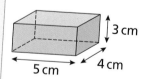

Plan view

Side elevation

Front elevation

how to find the areas of 2-D shapes

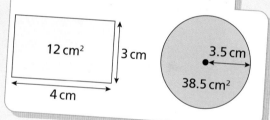

12 cm² 3 cm

4 cm

3.5 cm

38.5 cm²

how to find the surface areas of cubes, cuboids, prisms and cylinders

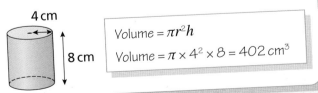

3 cm

5 cm 4 cm

$3 \times 4 = 12$ $12 \times 2 = 24\,cm^2$

$4 \times 5 = 20$ $20 \times 2 = 40\,cm^2$

$5 \times 3 = 15$ $15 \times 2 = 30\,cm^2$

Total surface area $= 24 + 40 + 30 = 94\,cm^2$

how to find the volumes of cubes, cuboids, prisms and cylinders

4 cm

8 cm

Volume $= \pi r^2 h$

Volume $= \pi \times 4^2 \times 8 = 402\,cm^3$

about the volumes of spheres, pyramids and cones **H**

5 cm

Volume $= \frac{4}{3}\pi r^3$

Volume $= \frac{4}{3} \times \pi \times 5^3$

Volume $= 523.6\,cm^3$

Small steps

- Know names of 2-D and 3-D shapes
- Recognise prisms (including language of edges/vertices)

Key words

2-D shape – a flat shape with two dimensions such as length and width

3-D shape – a shape with three dimensions: length, width and height

Edge – a line segment joining two vertices of a 3-D shape; it is where two faces of a 3-D shape meet

Vertex (plural: **vertices**) – a point where two line segments meet; a corner of a shape

Face – a flat surface of a 3-D shape

Prism – a solid shape with polygons at its ends and flat surfaces

Are you ready?

1 Write down the mathematical name for each of these shapes.

a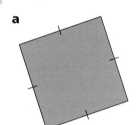
b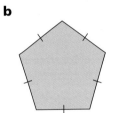
c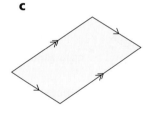
d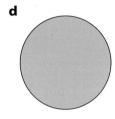

2 Match each triangle to its correct name.

a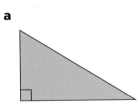
b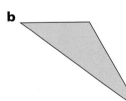
c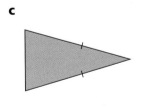
d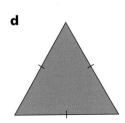

| scalene triangle | isosceles triangle | equilateral triangle | right-angled triangle |

3 Which of these shapes are hexagons?

A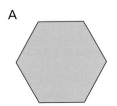
B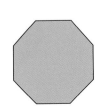
C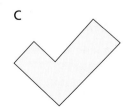
D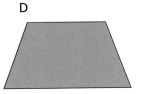

4 Match each 3-D shape to its correct name.

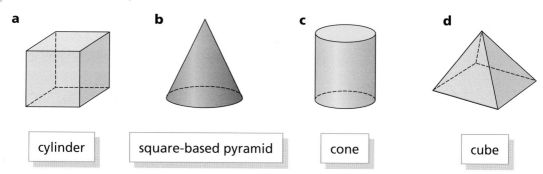

| a | b | c | d |

cylinder | square-based pyramid | cone | cube

Models and representations

Geoboards

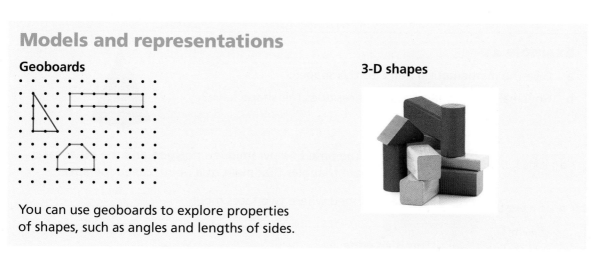

You can use geoboards to explore properties of shapes, such as angles and lengths of sides.

3-D shapes

In this section, you will review your knowledge of the properties of **2-D shapes** and explore the properties of some **3-D shapes**.

Example 1

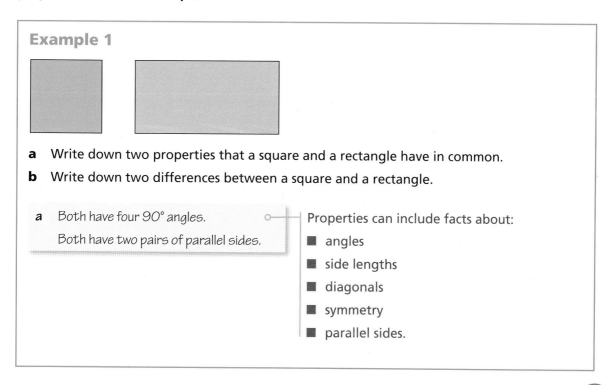

a Write down two properties that a square and a rectangle have in common.

b Write down two differences between a square and a rectangle.

a Both have four 90° angles.

Both have two pairs of parallel sides.

Properties can include facts about:

■ angles

■ side lengths

■ diagonals

■ symmetry

■ parallel sides.

b A square has four sides of equal length but other rectangles do not.

A square has four lines of reflection symmetry but a rectangle has two lines of reflection symmetry.

Can you find any other differences?

Remember: Other than rectangles that are squares, the diagonals of a rectangle are not lines of symmetry. If you fold a rectangle along a diagonal, the two halves do not match up.

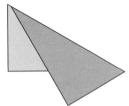

Example 2

a Give the mathematical name of this shape.

b How many edges, faces and vertices does this shape have?

a Square-based pyramid

The base of a pyramid is a polygon and the other faces are triangles that meet at a point, called the apex.

b Edges: 8

Faces: 5

Vertices: 5

Edges are formed where two faces meet.

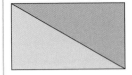

This is an **edge** ⟶

Faces are the flat surfaces or shapes that make up the surface of the 3-D shape.

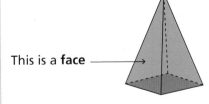

This is a **face** ⟶

Vertices are the corners where two or more edges meet.

This is a **vertex** ⟶

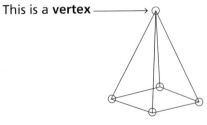

Practice 4.1A

1 What could be the name of a quadrilateral with four angles of equal size?

2 Give the name of a 5-sided shape with sides of equal length.

3

> I am thinking of a quadrilateral in which all sides are equal in length.

Ed

> You must be thinking of a square!

Seb

Is Seb right? Is there another shape that Ed could be thinking of?

4 Match each shape to its description.

   a A quadrilateral with one pair of parallel sides.

   b A shape with one pair of parallel sides and two equal sides.

   c A shape with three lines of symmetry.

   d A quadrilateral whose diagonals meet at 90°

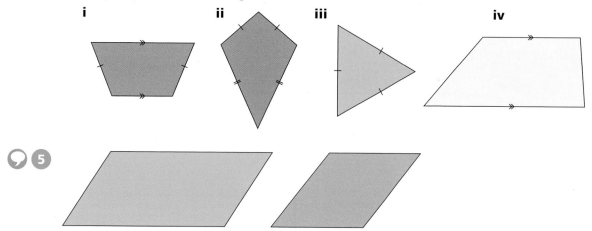

5

   a List two properties that a parallelogram and a rhombus have in common.

   b List two differences between a parallelogram and a rhombus.

6 Darius draws a parallelogram. He says that two of the angles are 60° and 110° Explain why Darius must be wrong.

7 Write down the mathematical name for each of these 3-D shapes.

a **b** **c** **d**

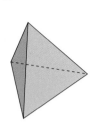
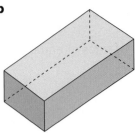

 Where might you see these objects in everyday life?

8 For each shape, write down the number of faces, edges and vertices.

a **b** **c** **d**

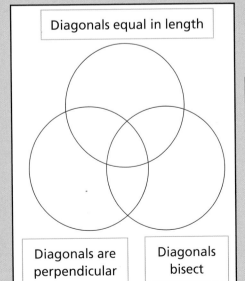

9 Flo draws a 3-D shape with two triangular faces and three rectangular faces.
What shape has she drawn?

What do you think?

1

A square is a rectangle. A rectangle is a parallelogram.

Zach   Beca

Do you agree with Zach and Beca? Explain why or why not.

2 Copy and complete the Venn diagram by sorting the shapes into the correct regions.

Diagonals equal in length

Shapes to sort:

rectangle

square

isosceles trapezium

kite

rhombus

parallelogram

Diagonals are perpendicular

Diagonals bisect

3 The diagram shows a tetrahedron.

a How many faces, edges and vertices does a tetrahedron have?

b Benji sticks two tetrahedra together as shown.

My shape now has double the number of faces, edges and vertices.

The plural of tetrahedron is tetrahedra.

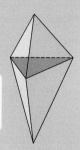

Is Benji correct? How do you know?

In this section, you are going to take a closer look at the properties of **prisms**.

A prism is a shape with a uniform cross-section. This means that if you slice the prism at different points along the length, each cross-section would be the same size and shape.

The shape of the cross-section gives the prism its name.

This is a triangular prism.

This is a pentagonal prism.

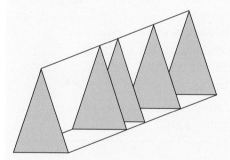

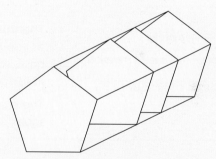

Each slice creates an identical triangular face.   Each slice creates an identical pentagonal face.

The shape of the cross-section is not always a regular polygon, as shown below.

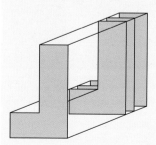

Example 3

Which of these shapes are

   i   prisms

   ii not prisms?

Name the shapes you choose.

a    **b** 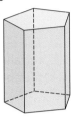   **c** 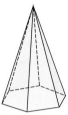   **d** 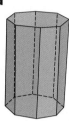   **e**

  i   Prisms: **a**, **b** and **d**   ○───── A prism must have a uniform cross-section.

     a   triangular prism   The cross-sections are identical octagons. ─○

     b   pentagonal prism

     d   octagonal prism

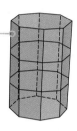

  ii Not prisms: **c** and **e**   ○───── The thickness of shapes **c** and **e** change

     c   hexagonal-based pyramid   as you move from the top vertex to the

     e   square-based pyramid   base, so they would not produce

                             identical cross-sections.

                             The cross-sections are squares. But the

                             squares are different in size.

Practice 4.1B

  1 Give the name of each of these shapes.

     a 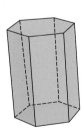   **b** 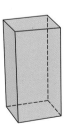   **c**    **d**

2 Which of these shapes are prisms?

A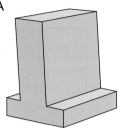

B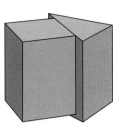

C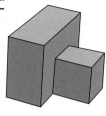

3 **a** Copy and complete the table.

Name	Number of faces	Number of edges	Number of vertices
cuboid			
triangular prism			
hexagonal prism			
pentagonal prism			

b

The number of faces plus the number of vertices is always two greater than the number of edges.

Is Faith right?

c Does Faith's conjecture apply to

   i a square-based pyramid   **ii** a tetrahedron?

d Which of these formulae correctly describes the relationship between the edges (E), the vertices (V) and the faces (F) of a prism?

$$F + V + 2 = E$$ $$F + V = E + 2$$ $$F + V + E = 2$$

4 A shape has 9 faces and 9 vertices. How many edges does it have?

5

I've drawn a shape with 4 faces, 4 vertices and 7 edges.

Explain why Emily is wrong.

6 Marta draws a prism with eight rectangular faces. What is the shape of the cross-section of the prism?

What do you think?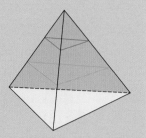

1 Marta slices this shape along the orange and green lines.

She notices that both cross-sections are triangles, so she says that this shape must be a prism.
Is Marta correct? How do you know?

2 Is a cuboid a prism?

3 Why is a cylinder not a prism? In what ways is a cylinder similar to a prism? In what ways is a cylinder different?

4 A prism has a cross-section that is a polygon with n sides. Write expressions, in terms of n, for the number of faces, edges and vertices that it will have.

Consolidate – do you need more?

1 Give the mathematical name of each of these shapes.

a    b    c 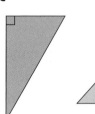   d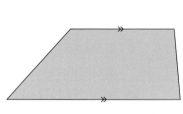

2 Mario draws a quadrilateral with sides of equal length. Which shapes could Mario have drawn?

3 Amina draws a triangle with exactly one line of symmetry. What is the mathematical name for the triangle that Amina has drawn?

4 Which of the following statements is true?

A │ A triangle can be both right-angled and isosceles.

B │ A rectangle is also a type of square.

C │ A triangle can be both isosceles and scalene.

5 Look at these shapes.

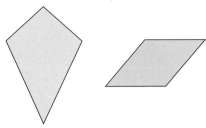

   a List two properties that a kite and a rhombus have in common.

   b List two differences between a kite and a rhombus.

6 Write down the mathematical name for each of these 3-D shapes.

   a   **b**   **c**   **d**

7 For each shape, write down the numbers of faces, edges and vertices.

   a   **b**   **c**   **d**

8 Sketch two different 3-D shapes that have exactly six faces.

Stretch – can you deepen your learning?

1 Give the names of two 3-D shapes that have no vertices.

2

       A cone only has 1 edge.

   Do you agree with Abdullah? Explain your answer.

3 a Sketch and name two different 3-D shapes with seven faces.

   b For each shape, work out the number of edges and vertices.

4 **a** For each of these pyramids, state the number of faces, edges and vertices.

Hint: you could set out your answer in a table.

i 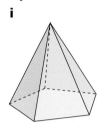 **ii**  **iii** 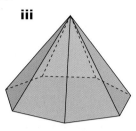 **iv**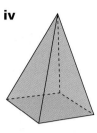

b What do you notice?

c A pyramid has a base with n sides. Write expressions, in terms of n, for the number of faces and edges it has.

5 These five shapes are known as the Platonic solids. What is special about them?

tetrahedron   cube   octahedron   dodecahedron   icosahedron

Reflect

1 Which properties define a prism? How can you tell that a shape is not a prism?

2 What is the relationship between the number of edges, faces and vertices of a 3-D shape? Is this relationship true for *all* 3-D shapes?

Small steps

- Construct accurate nets of cuboids and other 3-D shapes
- Sketch and recognise nets of cuboids and other 3-D shapes
- Explore plans and elevations

Key words

Net – a 2-D shape that can be folded to make a 3-D shape

Plan view – when an object is viewed from above

Front/side elevation – when an object is viewed from the front or side

Isometric drawing – a method of drawing 3-D shapes in two dimensions using special dotty paper

Are you ready?

1 How many faces does a cuboid have?

2 a How many faces does a triangular prism have?

   b What is the shape of each face of a triangular prism?

3 a Accurately draw a rectangle measuring 4 cm by 3 cm.

   b Calculate the area of the rectangle.

4 a Make an accurate drawing of this triangle.

   b Measure the length of the side AC.

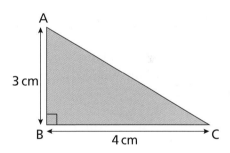

Models and representations

Household objects

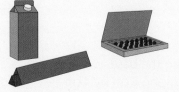

3-D shapes

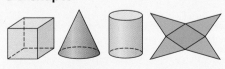

Interlocking cubes

These are useful for helping with **isometric drawing** and when drawing **plans** and **elevations**.

In this chapter, you will explore different 2-D representations of 3-D shapes.

In the first section, you will look at **nets** of 3-D shapes.

This is a net for making a cube. It is made up of six squares.

> A net is a 2-D shape that can be folded to make a 3-D shape.

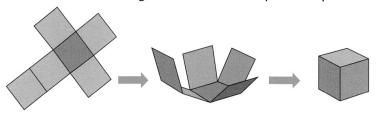

This is a net for a square-based pyramid. It is made up of one square (the base) and four triangles.

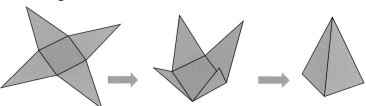

Example 1

Draw an accurate net for this cuboid.

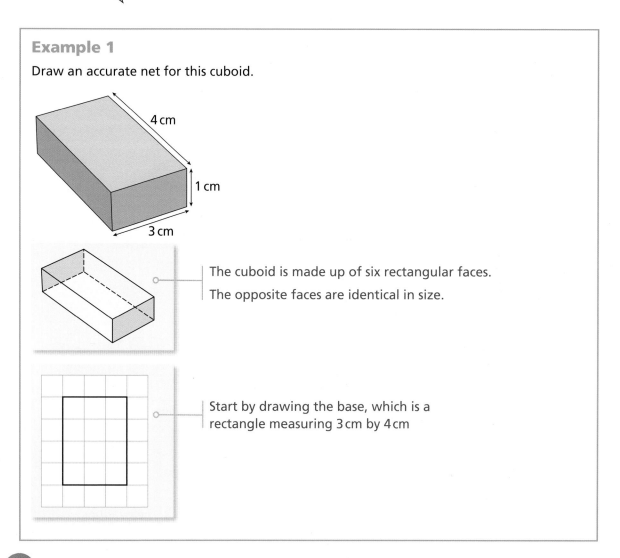

The cuboid is made up of six rectangular faces.

The opposite faces are identical in size.

Start by drawing the base, which is a rectangle measuring 3 cm by 4 cm

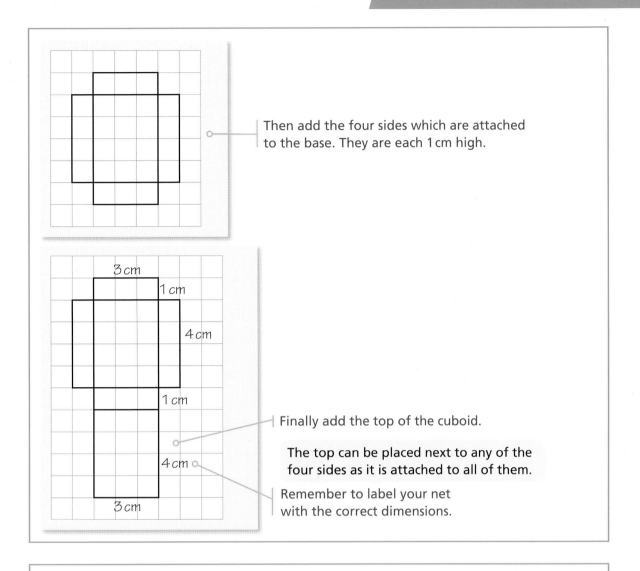

Then add the four sides which are attached to the base. They are each 1 cm high.

Finally add the top of the cuboid.

The top can be placed next to any of the four sides as it is attached to all of them.

Remember to label your net with the correct dimensions.

Example 2

Draw an accurate net for this triangular prism.

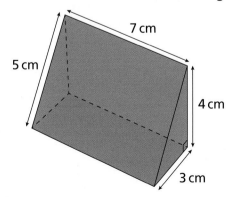

This prism is made up of five faces.

There are two identical right-angled triangles.

There are three rectangles.

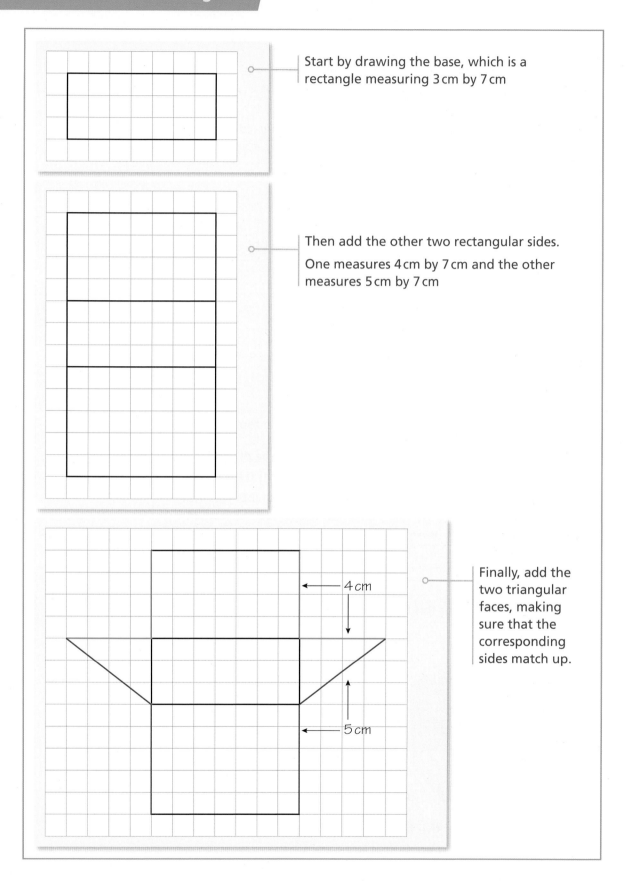

Start by drawing the base, which is a rectangle measuring 3 cm by 7 cm

Then add the other two rectangular sides.

One measures 4 cm by 7 cm and the other measures 5 cm by 7 cm

4 cm

5 cm

Finally, add the two triangular faces, making sure that the corresponding sides match up.

Practice 4.2A

1 Draw an accurate net for a cube measuring 2 cm by 2 cm by 2 cm

2 Draw an accurate net for this cuboid.

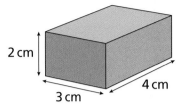

3 This net makes a cuboid. Each square measures 1 cm by 1 cm

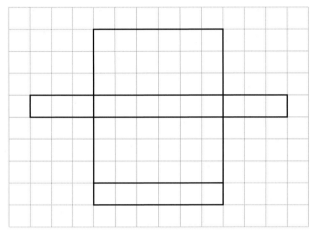

Write down the dimensions of the cuboid.

4 Flo makes a cube using this net.

    a Which edge will meet

       i edge FE   **ii** edge GH?

    b Which vertices will meet vertex D?

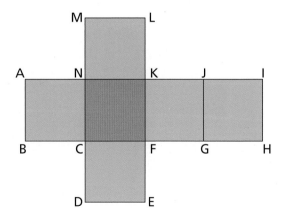

5 Draw an accurate net of this square-based pyramid.

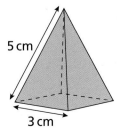

6 Which of these nets could be used to make a closed cube?

A 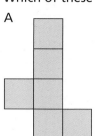 B 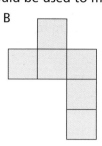 C 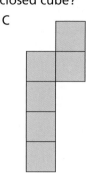 D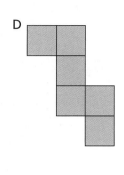

7 Jakub draws this net. He says that it will make a cuboid.

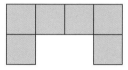

Explain and correct his mistake.

8 The dots on opposite faces of a dice add up to seven.

This is a net of a dice. How many dots will there be on faces **a**, **b** and **c**?

 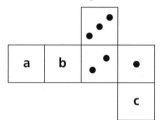

9 Give the name of each of the 3-D shapes that can be made from these nets.

a 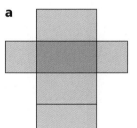   **b** 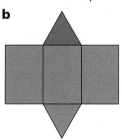   **c** 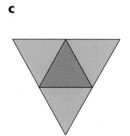   **d**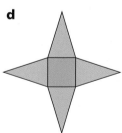

What do you think?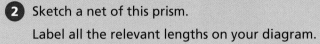

1 Sketch as many different nets of a cube as you can. How many possible nets are there?

2 Sketch a net of this prism.

Label all the relevant lengths on your diagram.

> "Sketch" means that you do not have to draw the net to scale but you still need to label your diagram with the correct lengths.

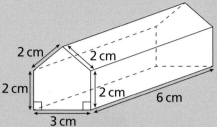

2 cm   2 cm

2 cm   2 cm   6 cm

3 cm

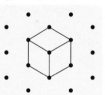

2 cm

5 cm

3 **a** Sketch a net for this cylinder.

b What other measurement can you work out for this cylinder?

4 Why would it be difficult to draw the net of a sphere?

In this section, you will explore isometric drawings, and plans and elevations. These are different methods of representing a 3-D shape.

Isometric drawings use a grid of dots that are arranged in a triangular pattern.

> Isometric paper must be oriented the correct way. Always start by drawing the edge closest to you. Vertical edges are drawn vertically but horizontal edges are drawn at an angle.

Plans and elevations show an object from three different perspectives.

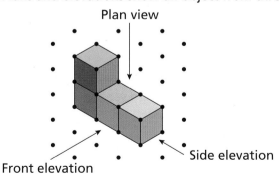

Plan view

Side elevation

Front elevation

The plan view is the view when looking down on the object.

The side elevation is the view from one side.

The front elevation is the view from the front.

Example 3

Mario makes a shape using some cubes.

a Draw Mario's shape on isometric paper.

b Draw the plan view, side elevation and front elevation of the shape.

a

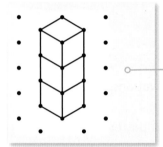

You can make the shape from cubes to help.

Start from the closest edge.

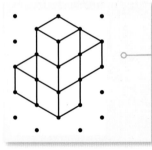

Then build up the three cubes on top of each other.

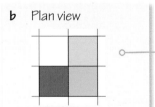

Finally add in the last two cubes, keeping the lines parallel.

b Plan view

You can make the shape from cubes to help.

Plan view – looking from the top. You would see the blue, green and purple cubes. The other cubes would be hidden from view.

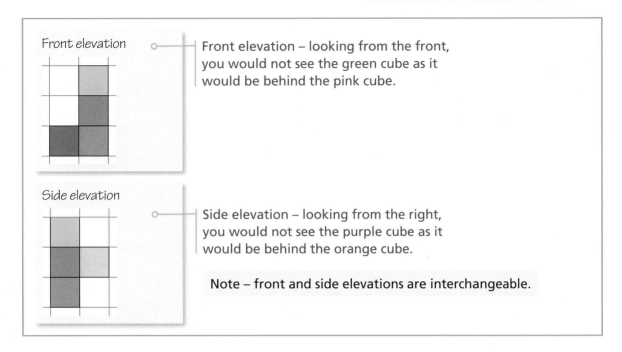

Front elevation – looking from the front, you would not see the green cube as it would be behind the pink cube.

Side elevation – looking from the right, you would not see the purple cube as it would be behind the orange cube.

Note – front and side elevations are interchangeable.

Practice 4.2B

1. Seb makes a cube measuring 2 cm by 2 cm by 2 cm

    Draw Seb's cube on isometric paper.

2. Make some shapes using exactly four cubes. Draw each of your shapes on isometric paper.

3. For each of these shapes draw

    a the plan view   **b** the front elevation   **c** the side elevation.

    i   **ii**   **iii**

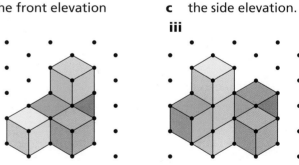

4. Here are the plan view, front elevation and side elevation of some 3-D shapes.

    Draw each 3-D shape on isometric paper. You may find it helpful to make the shapes from cubes.

    a plan   front   side
           view   elevation   elevation

    b plan   front   side
           view   elevation   elevation

5 Each of these shapes is extended by adding a cube to each shaded face.

Draw the new shapes on isometric paper.

a   **b**   **c**

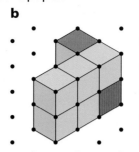
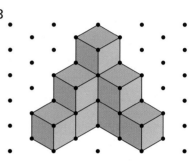

6 These 3-D shapes form a pattern. Shape 1 consists of one cube and shape 2 consists of four cubes.

1   2   3

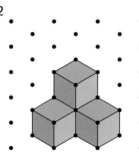

a How many cubes are there in shape 3?

b How many cubes would there be in shape 4?

c Write an expression for the number of cubes used in the nth shape.

d Draw a plan view, side elevation and front elevation for shape 3

What do you think? 💡

1 Make as many shapes as possible using only six multilink cubes.

Use isometric paper to record your results.

2 The cuboid shown is cut in half to make two triangular prisms. Draw one of these prisms on isometric paper.

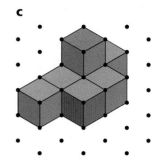

3

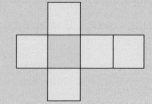

I count 19 straight lines in this picture. This means that a cube has 19 edges.

Is Abdullah correct? Explain your answer.

Consolidate – do you need more?

1 Draw an accurate net for a cuboid that measures 4 cm by 6 cm by 2 cm

2 Chloe uses this net to make a 3-D shape.

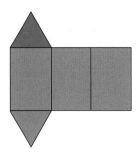

    a What is the name of the shape she makes?

    b Sketch Chloe's 3-D shape.

3 Draw an accurate net for this open box.

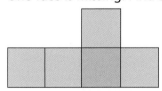

4 Here is an incomplete net for a cube.

One face is missing. Find all the different ways of completing the net.

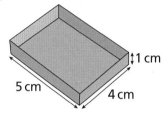

5 Draw an accurate net for this tetrahedron.

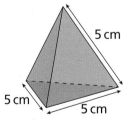

6 Sketch a net for each of these shapes.

    a

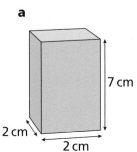

    b

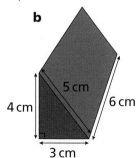

7 Draw each of these shapes on isometric paper.

a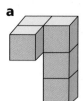

b

c

8 For each of these shapes draw

a the plan view   **b** the front elevation   **c** the side elevation.

i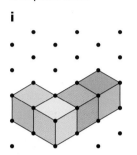

ii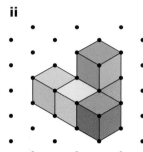

iii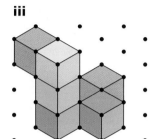

Stretch – can you deepen your learning?

1 Flo draws this side elevation.

 I think that's from a cone.

I think that's from a square-based pyramid.

Flo

Ed

a Explain why both Flo and Ed could be right.

b Draw the plan view of

   i a cone   **ii** a square-based pyramid.

c Are there any other shapes that would have the same side elevation?

2 Here is the net of a dice. Which of the five dice would it make when folded?

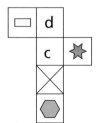

A

B

C

D

E

3 Sketch a net for

   a a cone   **b** an octahedron.

4 Which of these nets will make a triangular prism?

   A   B   C   D

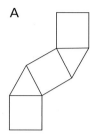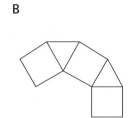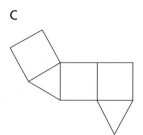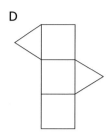

Reflect

Nets, isometric drawings and plans and elevations are all methods of representing 3-D shapes on two-dimensional paper. What are the advantages and disadvantages of each approach? When might each one be useful?

Small steps

- Find areas of 2-D shapes
- Find the surface area of cubes and cuboids

Are you ready?

1   Find the area of these shapes.

Each square measures 1 cm by 1 cm

a   b   c   d

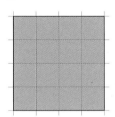 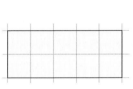  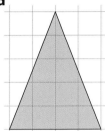

2   Copy the circle and label the parts with these words.

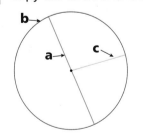

radius

diameter

circumference

3   Round each of these numbers to 2 decimal places.

a   24.354   b   5.6572   c   7.998   d   0.059   e   104.1001

4   Solve these equations.

a $4x = 48$ b $3x = 6.9$ c $\frac{1}{2}x = 23$ d $x^2 = 25$ e $x^2 = 0.36$

5   What are the missing numbers in these statements?

a   1 m = ___ cm   b   3 cm = ___ mm   c   342 cm = ___ m   d   76 mm = ___ cm

Models and representations

Geoboards

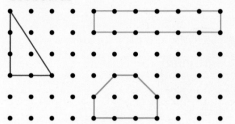

Geoboards show how the dimensions of a shape are linked to its **area**.

Nets

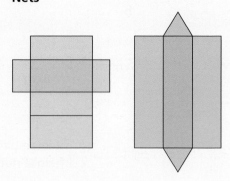

In this chapter, you will review your knowledge of finding areas of 2-D shapes before extending this to find the **surface area** of 3-D shapes.

You will need to use the formulae that you learned in Books 1 and 2

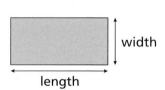

width

length

Area of rectangle = length × width

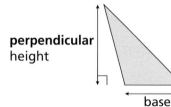

perpendicular height

base

Area of triangle = $\frac{1}{2}$ × base × perpendicular height

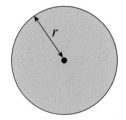

r

Area of circle = $\pi \times r^2$

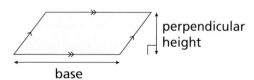

perpendicular height

base

Area of parallelogram = base × perpendicular height

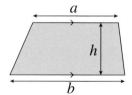

a

h

b

Area of trapezium = $\frac{1}{2}(a + b) \times h$

Example 1

Work out the area of each shape.

a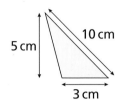

b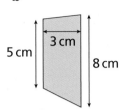

c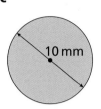

a Area = $\frac{1}{2}$ × base × height

Remember: you need the perpendicular height not the slant height.

Area = $\frac{1}{2}$ × 3 × 5 = 7.5 cm²

Always remember to give the correct units.

This is the height that meets the base at 90°

b Area = $\frac{1}{2}$ × $(a + b)$ × h

You need to use the formula for the area of a trapezium.

Area = $\frac{1}{2}$ × (5 + 8) × 3 = $\frac{1}{2}$ × 13 × 3

First identify a, b and h in the diagram.

= 19.5 cm²

a and b are the two parallel sides. h is the perpendicular distance between them.

c Area = π × r^2

Use the formula for the area of a circle.

Diameter = 10 mm, so the radius = 5 mm

You have been given the diameter of the circle. The radius is half of the diameter.

Area = π × 5^2 = 78.5 mm²

Remember to round your answer sensibly.

Example 2

Work out the area of each of these compound shapes.

a

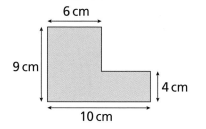

b

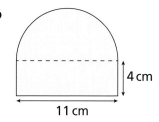

a

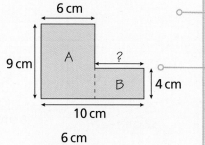

You need to break the shape down into rectangles. You can do this in different ways

To find the area of rectangle B, you need to find the missing length.

The length of the overall shape is 10 cm and the length of A is 6 cm
So the length of B must be 10 cm – 6 cm = 4 cm

Area A = 6 × 9 = 54 cm²

Area B = 4 × 4 = 16 cm²

Now you can work out the area of A and the area of B.

Total area = 54 + 16 = 70 cm²

Finally, add your answers together.

b

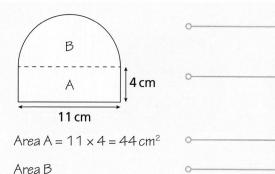

This shape is made up of a rectangle and a semicircle.

You need to work out the area of A and the area of B and add them together.

Area A = 11 × 4 = 44 cm²

Use the formula for the area of a rectangle.

Area B

B is a semicircle.

Diameter = 11 cm, so radius = 5.5 cm

The diameter is 11 cm, so the radius is 11 cm ÷ 2

Area of full circle = $\pi \times 5.5^2 = 95.03\ldots$

Area of B = 95.03… ÷ 2 = 47.5 cm² (1 d.p.)

A semicircle is half of a circle, so you need to divide the area by 2

The total area is 44 + 47.5 = 91.5 cm²

Finally, add the two areas together.

Practice 4.3A

1 Work out the area of each shape.

a

6 cm

b

3 cm

8 cm

c

24.5 mm

16 mm

d

6 cm

13 cm

e

3.1 m   2.4 m

f

9 cm

6.5 cm

g

21 cm

32 cm

h

10 cm

6 cm

8 cm

i

88 mm

55 mm

40 mm

j

12 cm

6 cm

14 cm

k

1.5 m

l

1.4 m

0.65 m

0.8 m

What assumptions have you made?

2 Work out the area of each shape. Round your answers to 1 decimal place.

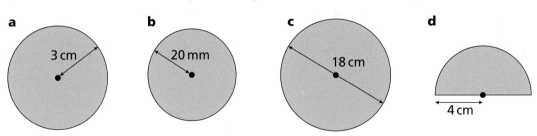

a

3 cm

b

20 mm

c

18 cm

d

4 cm

3 Work out the area of each shape.

a

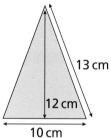

3 cm
6 cm
2 cm
8 cm

b

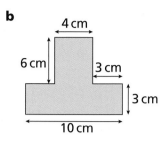

4 cm
6 cm
3 cm
3 cm
10 cm

c

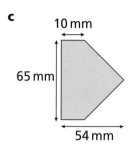

10 mm
65 mm
54 mm

d

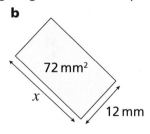

7 m
3 m
3 m

e

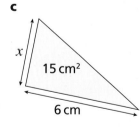

2 cm
16 cm

f

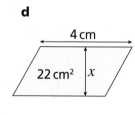

4 cm
9 cm

4 Faith works out the area of this triangle.

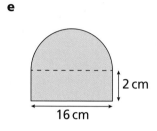

13 cm
12 cm
10 cm

$$A = \frac{1}{2} \times b \times h$$

$$A = \frac{1}{2} \times 10 \times 13 = 65 \, cm^2$$

Explain the mistake Faith has made and work out the correct answer.

5 Find the missing lengths in these shapes.

a

x
49 cm²

b

72 mm²
x
12 mm

c

x
15 cm²
6 cm

d

4 cm
22 cm²
x

6 Find the area of this shape in two different ways.

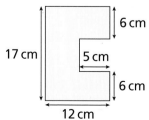

6 cm
17 cm
5 cm
6 cm
12 cm

7 The area of the parallelogram is equal to the area of the trapezium.

Work out the value of x

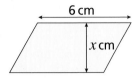
6 cm

x cm

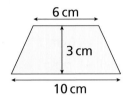
6 cm

3 cm

10 cm

8 A circular dinner plate has area 452 cm². Work out the radius of the circle.

9 Write an expression for the area of each of these shapes.

a

d cm

b

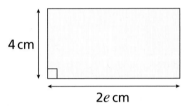
4 cm

2e cm

c

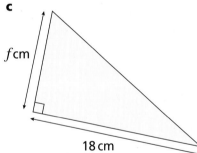
f cm

18 cm

d

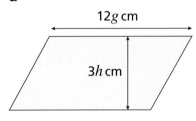
12g cm

3h cm

10 The shaded region of the diagram represents Jakub's lawn.

Grass seed costs £13 per box. Each box covers an area of 20 m².

How much will it cost Jakub to cover his lawn with grass?

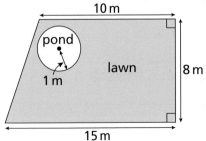
10 m

pond

lawn

1 m

8 m

15 m

What do you think?

1 Which of these shapes has the greater area?

8 cm

9 cm

2 Work out the area of the shaded region in each diagram.

a

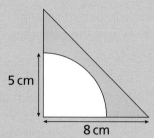

5 cm

8 cm

b

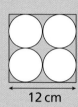

12 cm

3 A circle has circumference 140 cm. Calculate the area of the circle.

4 Work out the area of the shaded region.

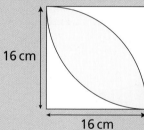

16 cm

16 cm

5 Draw three different trapezia that each have an area of 30 cm²

In this section, you will learn how to find the surface area of cubes and cuboids.

The surface area is the total area of all the **faces** of a 3-D shape.

Example 3

Find the total surface area of this cuboid.

3 cm

4 cm

5 cm

5 cm

3 cm   5 cm   3 cm

3 cm

4 cm

The **net** of a cuboid is made from six rectangles.

You can use the net to help you to find the area of each face. Then you need to add them together.

The faces come in pairs: top and base, front and back, and the two sides.

Pink faces: $3 \times 4 = 12\,cm^2$

$12 \times 2 = 24\,cm^2$

The pink faces form the two sides.

Green faces: $5 \times 4 = 20\,cm^2$

$20 \times 2 = 40\,cm^2$

The green faces are the top and base.

You can find the area of one face from each pair and then double your answer.

Blue faces: $5 \times 3 = 15\,cm^2$

$15 \times 2 = 30\,cm^2$

The blue faces are the front and back.

Total surface area = $24 + 40 + 30 = 94\,cm^2$

Finally, add your answers and include the correct unit for area.

Practice 4.3B

1. Here is a cuboid.

   a   Work out the area of

      i   the front face, F

      ii the side face, S

      iii the top face, T

   b   Work out the total surface area of the cuboid.

   T

   S   F   2 cm

   8 cm   5 cm

2. Work out the surface area of each cuboid.

   a

   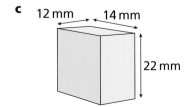

   2 cm

   8 cm   6 cm

   b

   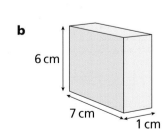

   6 cm

   7 cm   1 cm

   c

   12 mm   14 mm

   22 mm

   d

   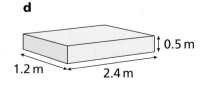

   0.5 m

   1.2 m   2.4 m

3 Which cuboid has the larger surface area? Show working to explain how you know.

A

5 cm

3 cm   8 cm

B

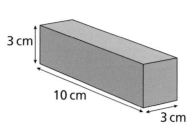

3 cm

10 cm

3 cm

4 Ed stacks three identical cuboids on top of each other as shown.

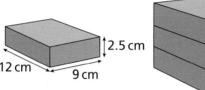

2.5 cm

12 cm   9 cm

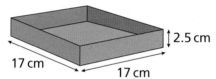

> The surface area of my shape will be three times as large as the surface area of one of the cuboids.

Explain why Ed is wrong.

5 Beca makes an open box. She is going to paint the outside of the box.

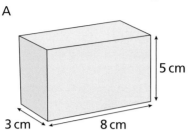

2.5 cm

17 cm   17 cm

Work out the total surface area to be painted.

6 A cube has sides of length x cm. Write an expression, in terms of x, for the surface area of the cube.

7 A cube has surface area 150 cm²

   a Work out the area of each face.

   b Find the length of each side of the cube.

8 A cuboid-shaped storage crate measures 0.8 m by 1 m by 1.5 m

Abdullah paints all the faces of the crate.

One tin of paint costs £8. Each tin covers 4 m²

Work out the total cost of the paint that Abdullah needs.

What do you think? 💭

1. Write an expression for the total surface area of each cuboid.

   a

   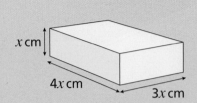

   b

   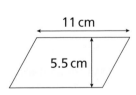

2. The surface area of this cuboid is 376 cm².

   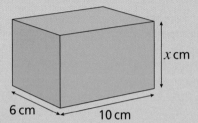

   Work out the value of x

3. Rhys has eight identical cubes.

   He can connect them together along any face. Rhys uses all eight cubes to make a 3-D shape.

   Investigate the different configurations and determine which would minimise the total surface area of Rhys's shape.

Consolidate – do you need more?

1. Find the area of each shape.

   a

   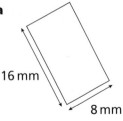

   b

   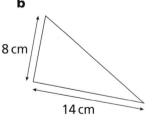

   c

   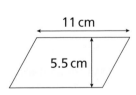

   d

   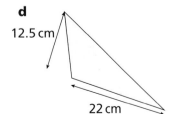

   e

   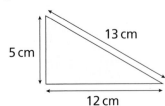

   f

   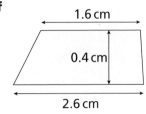

2 Find the area of each shape, leaving your answers in terms of π

a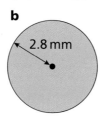
13 cm

b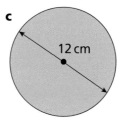
2.8 mm

c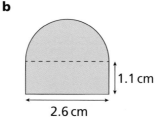
12 cm

d
6 cm

3 Find the area of each shape.

a
6 cm
5 cm
2 cm
11 cm

b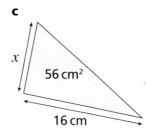
1.1 cm
2.6 cm

c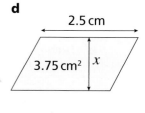
5 cm
17 cm

4 Find the missing length in each of these shapes.

a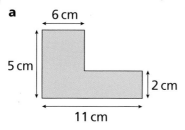
x
144 cm²

b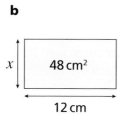
x
48 cm²
12 cm

c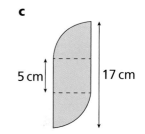
x
56 cm²
16 cm

d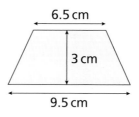
2.5 cm
3.75 cm² x

5 This is Mario's method for working out the area of the trapezium.

$A = \frac{1}{2} \times (6.5 \times 9.5) \times 3$

$A = \frac{1}{2} \times 61.75 \times 3 = 92.625\,cm^2$

6.5 cm
3 cm
9.5 cm

What mistake has Mario made? Work out the correct answer.

6 A cube has sides of length 3 cm. Work out the surface area of the cube.

7 Find the surface area of each of these cuboids.

a

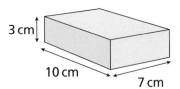

3 cm
10 cm
7 cm

b

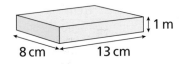

4.5 cm
5 cm
2 cm

c

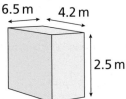

6.5 m   4.2 m
2.5 m

d

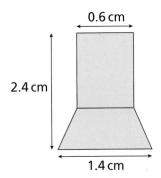

1 m
8 cm   13 cm

8 Huda is tiling her bathroom wall.

The wall is a rectangle measuring 2 m by 6 m

Each tile is a square with side length 50 cm

a What is the minimum number of tiles that Huda needs?

Tiles are sold in boxes of 10. Each box costs £12.99

b How much will it cost Huda to tile her bathroom wall if she uses the minimum number of tiles?

Stretch – can you deepen your learning?

1 The diagram shows a rectangle.

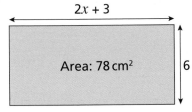

2x + 3

Area: 78 cm²   6

Write and solve an equation to find the value of x

2 Amina makes a pendant by joining a rectangle and trapezium, as shown.

The ratio of the height of the rectangle to the height of the trapezium is 5 : 3

Work out the area of the pendant.

0.6 cm
2.4 cm
1.4 cm

3 Work out the value of x in each of these cuboids.

a

surface area = 248 m²

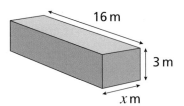

16 m

3 m

x m

b

surface area = 90 cm²

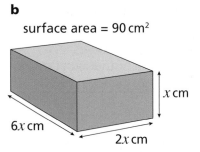

x cm

$6x$ cm

$2x$ cm

4 An A4 piece of paper measures approximately 21 cm by 30 cm. Investigate the largest cube that can be made from a single sheet of A4 paper and calculate its surface area.

Reflect

1 Sketch a parallelogram, a trapezium, a rectangle and a triangle that each have an area of 72 cm²

2 For a cuboid with dimensions x cm, y cm and z cm, the surface area can be found using the formula area = $2(xy + yz + xz)$. Explain why this formula works and how it links to your earlier learning in this chapter.

Small steps

■ Find the surface area of triangular prisms

■ Find the surface area of a cylinder

Key words

Surface area – the sum of the areas of all the faces of a 3-D shape

Prism – a solid shape with polygons at its ends and flat surfaces

Are you ready?

1 Work out the area of each shape.

a

9 cm

b
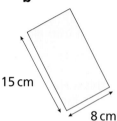
15 cm
8 cm

c
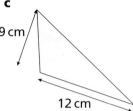
9 cm
12 cm

d
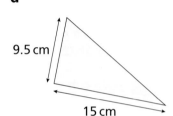
9.5 cm
15 cm

2 Work out the circumference of each circle. Round your answers to 1 decimal place.

a
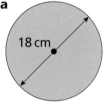
18 cm

b

8.4 mm

c
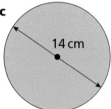
14 cm

d

1.7 m

3 Work out the area of each circle, giving your answers in terms of π

a
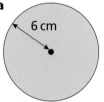
6 cm

b

2.5 m

c

14 m

d
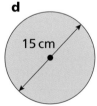
15 cm

4 Round each number to 1 decimal place.

   a   3.256   **b**   0.365   **c**   0.0957   **d**   28.956

5 For each of these nets, state the mathematical name of the 3-D shape that it makes.

a

b

c

Models and representations

Household objects

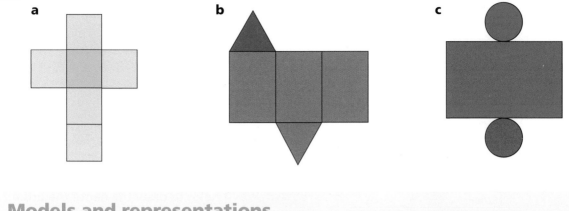

In this chapter, you will extend your knowledge of **surface area** to shapes such as **prisms** and cylinders.

Remember: to find the surface area of a 3-D shape you need to find the sum of the areas of all of its faces. You can draw a net to help you to identify the different faces and their areas.

Example 1

Work out the surface area of this triangular prism.

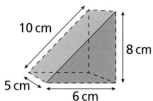

10 cm

8 cm

5 cm

6 cm

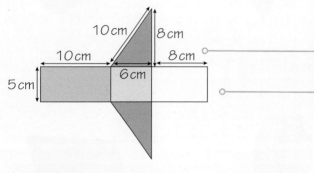

10 cm 8 cm

10 cm 8 cm

6 cm

5 cm

It helps to draw the net of the prism.

There are two identical triangular faces and three rectangular faces.

Purple triangles

$A = \frac{1}{2}b \times h = \frac{1}{2} \times 6 \times 8 = 24 \, cm^2$

Total area of both triangles: $24 \times 2 = 48 \, cm^2$

Make sure you use the correct dimensions: the base and the perpendicular height.

Rectangles

Yellow area = $8 \times 5 = 40 \, cm^2$

Blue area = $6 \times 5 = 30 \, cm^2$

Pink area = $10 \times 5 = 50 \, cm^2$

Notice that all of the rectangles have a width of 5 cm

Total surface area = $48 + 40 + 30 + 50 = 168 \, cm^2$

Add up the areas of all the faces. Remember to include the correct units with your answer.

Practice 4.4A

1 Zach sketches the net of a triangular prism.

  a Copy the net and label each side with the correct length.

  b Work out the surface area of the prism.

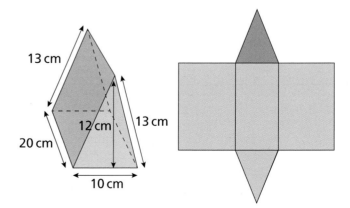

13 cm

12 cm   13 cm

20 cm

10 cm

2 Work out the surface area of these prisms.

a

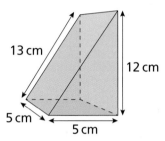
15 cm
9 cm
10 cm
12 cm

b

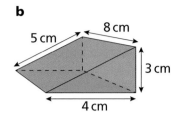
8 cm
5 cm
3 cm
4 cm

c

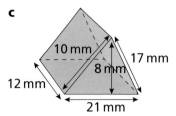
10 mm
17 mm
8 mm
12 mm
21 mm

3 **a** Work out the surface area of this prism.

13 cm
12 cm
5 cm
5 cm

b

If you double the length of a prism, the surface area doubles.

Is Marta correct? Explain your answer.

4 Here is a triangular prism.

Abdullah works out the surface area of the prism.

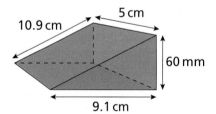
10.9 cm
5 cm
60 mm
9.1 cm

Area of triangle $= \frac{1}{2}b \times h = \frac{1}{2} \times 9.1 \times 60 = 273\,\text{cm}^2$

Two triangles $= 273 \times 2 = 546\,\text{cm}^2$

Area of rectangles $= (9.1 \times 5) + (10.9 \times 5) + (60 \times 5) = 400\,\text{cm}^2$

Total area $= 546 + 400 = 946\,\text{cm}^2$

Identify Abdullah's mistake.

5 The diagram shows a triangular prism.

Express the area of the triangular faces as a percentage of the total surface area.

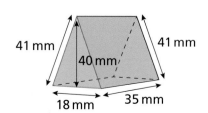
41 mm
41 mm
40 mm
18 mm
35 mm

6 This prism has surface area 222 cm²

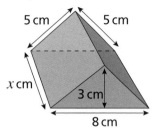

5 cm   5 cm

x cm   3 cm

8 cm

Work out the value of x

What do you think?

1 Here is a triangular prism.

Find the ratio of the total area of the triangular faces to the total area of the rectangular faces.

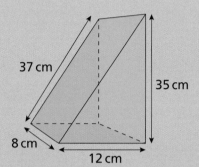

37 cm   35 cm

8 cm   12 cm

2 Flo cuts a cube in half, as shown.

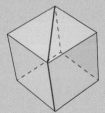

> The surface area of each prism is half of the surface area of the cube.

Explain why Flo is wrong.

3 Amina cuts a regular hexagonal prism into six, identical triangular prisms.

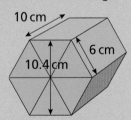

10 cm   6 cm

10.4 cm

Work out the surface area of each triangular prism.

In this section, you will learn how to work out the surface area of a cylinder.

Here is the net of a cylinder.

It is made up of two circles and a rectangle.

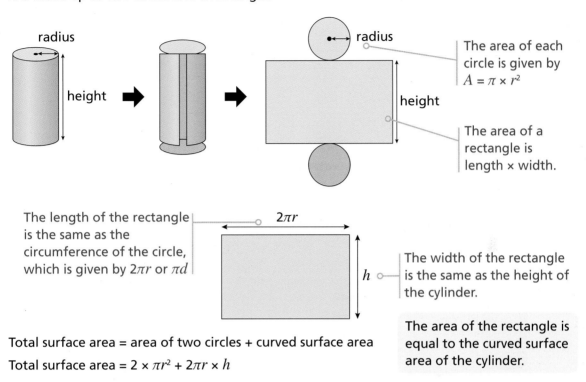

radius

height

radius

The area of each circle is given by $A = \pi \times r^2$

height

The area of a rectangle is length × width.

The length of the rectangle is the same as the circumference of the circle, which is given by $2\pi r$ or πd

$2\pi r$

h

The width of the rectangle is the same as the height of the cylinder.

The area of the rectangle is equal to the curved surface area of the cylinder.

Total surface area = area of two circles + curved surface area

Total surface area = $2 \times \pi r^2 + 2\pi r \times h$

Example 2

Work out the total surface area of this cylinder.

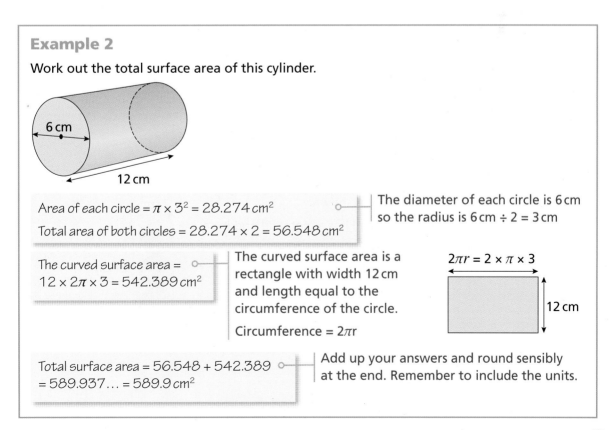

6 cm

12 cm

Area of each circle = $\pi \times 3^2 = 28.274$ cm^2

Total area of both circles = $28.274 \times 2 = 56.548$ cm^2

The diameter of each circle is 6 cm so the radius is 6 cm ÷ 2 = 3 cm

The curved surface area = $12 \times 2\pi \times 3 = 542.389$ cm^2

The curved surface area is a rectangle with width 12 cm and length equal to the circumference of the circle.

Circumference = $2\pi r$

$2\pi r = 2 \times \pi \times 3$

12 cm

Total surface area = $56.548 + 542.389$ = $589.937... = 589.9$ cm^2

Add up your answers and round sensibly at the end. Remember to include the units.

Practice 4.4B

1 Beca sketches a net for this cylinder.

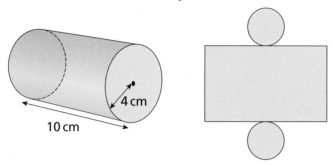

a Copy the sketch and label it with the correct lengths.

b Use your diagram from part **a** to work out the total surface area of the cylinder.

2 Work out the total surface area of each cylinder. Round your answers to 1 decimal place.

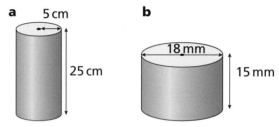

a 5 cm, 25 cm

b 18 mm, 15 mm

3 Work out the total surface area of each cylinder. Give your answers in terms of π

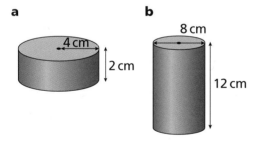

a 4 cm, 2 cm

b 8 cm, 12 cm

4 Tennis balls are sold in tubes, as shown.

Each tube is made of plastic and has a metal lid and base.

Find the total surface area of the plastic tube.

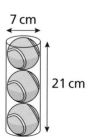

7 cm

21 cm

5 Which shape has the larger surface area? Justify your answer.

A

B

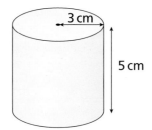

6 A cylinder has a radius of 12 cm and a height of 20 cm

The height of the cylinder is increased by 10%

a Work out the new total surface area of the cylinder.

b Work out the percentage increase in the total surface area.

7 Cat food is sold in cylindrical tins, as shown.

The label covers the entire height of the tin.

The label has a 1 cm overlap vertically so that it can be glued.

Work out the area of the label.

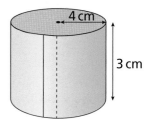

8 A cylinder has radius 4 m and height h m

The total surface area of the cylinder is 128π m²

Work out the value of h

What do you think? 💭

1 This cylinder has a surface area of 972π cm²

Work out the value of x

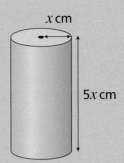

x cm

$5x$ cm

2 Work out the surface area of this semicylinder.

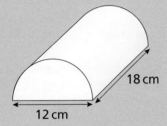

18 cm

12 cm

3 Work out the total surface area of this shape.

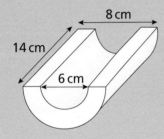

8 cm

14 cm

6 cm

Consolidate – do you need more?

1 Work out the total surface area of each shape.

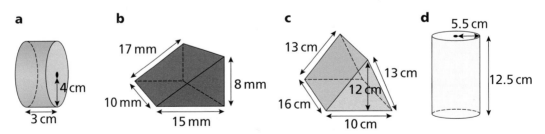

a

4 cm

3 cm

b

17 mm

10 mm

8 mm

15 mm

c

13 cm

13 cm

16 cm

12 cm

10 cm

d

5.5 cm

12.5 cm

2 Work out the total surface area of each of these cylinders. Give your answers in terms of π

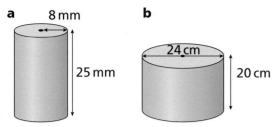

a 8 mm

25 mm

b

24 cm

20 cm

3 Express the curved surface area of this cylinder as a percentage of the total surface area.

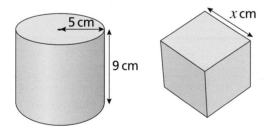

2.2 cm

8.5 cm

4 A cylindrical vase has diameter 10 cm and height 32 cm

Filipo paints the outside of ten of these vases.

One tin of paint covers 2000 cm². How many tins of paint will Filipo need to buy?

5 The cylinder and cube have the same surface area.

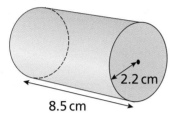

5 cm

9 cm

x cm

Work out the value of x

6 A cylinder has radius 6 m and height of h

The curved surface area of the cylinder is 754 cm²

Work out the value of h

Stretch – can you deepen your learning?

1 Work out the surface area of this prism.

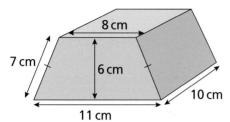

2 A shape is made by placing a cylinder on top of a cube, as shown.

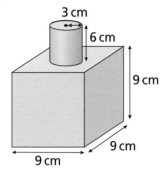

Calculate the total surface area of the solid shape.

3 An A4 piece of paper is rolled up to make a cylinder. The paper does not overlap.

Work out two possible values for the radius of the cylinder.

Reflect

The formula for the surface area of a cylinder uses the formulae for the area of a circle and for the circumference. Explain why this is and what mistakes someone might make when finding the surface area of a cylinder.

4.5 Volume

Small steps

- Find the volume of cubes and cuboids
- Find the volume of other 3-D shapes – prisms and cylinders
- Explore volumes of cones, pyramids and spheres (H)

Key words

Capacity – how much space a 3-D shape, or container, holds

Volume – the amount of space taken up by a 3-D shape

Cross-section – the shape you get when you slice a prism parallel to its base

Litre – 1000 cm³

Are you ready?

1 Work out the area of each shape.

a 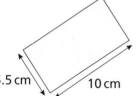 **b** 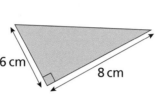 **c**   **d**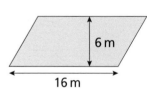

2 Round each of these numbers to 1 decimal place.

a 2.255 **b** 28.965 **c** 105.9702 **d** 0.9512

3 Given that $a = 6$ and $b = 2$, work out the value of

a $2a + b$ **b** $2b^3$ **c** $\frac{4a}{3b}$ **d** $a(b + 4)$ **e** $\frac{2}{3}a^2b$

4 Solve these equations. Give your answers to 1 decimal place.

a $2x = 2.8$ **b** $x^2 = 9$ **c** $x^2 = 12$ **d** $x^3 = 10$ **e** $2x^2 = 42$

Models and representations

Interlocking cubes

3-D shapes

In this chapter, you will learn how to find the **volume** of a variety of shapes.

You can find the volume of a shape by counting how many 1 cm³ cubes it is made up of.

This shape is made up of 7 cubes.

Each cube measures 1 cm by 1 cm by 1 cm ○——| This is the same as 1 cubic centimetre or 1 cm³

So the volume of this shape is 7 cm³

Volume is the amount of space a 3-D shape takes up. It is often measured in cm³ or m³

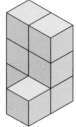

Example 1

Work out the volume of this cuboid.

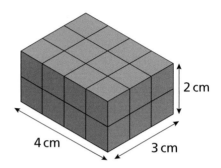

2 cm

4 cm   3 cm

You could count the cubes.

However, this would be difficult with a really large shape.

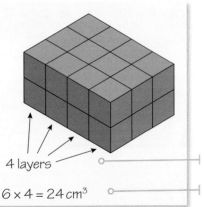

Instead, divide the cuboid into layers. This layer is made up of 2 × 3 cubes, so it has a volume of 6 cm³

2 × 3 = 6

There are a total of four layers, each with volume 6 cm³

4 layers ○——————

6 × 4 = 24 cm³ ○——————| So the total volume is 6 × 4 = 24 cm³

Volume = 24 cm³ ○——| 2 × 3 × 4 = 24 cm³

You could do this in one calculation:

length × width × depth

Example 2

Work out the volume of this shape.

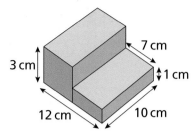

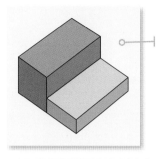 You can think of this shape as two cuboids.

Pink cuboid

Width = 12 cm – 7 cm = 5 cm

You need to work out the width of the pink cuboid. The overall width of the shape is 12 cm and the width of the blue cuboid is 7 cm

Volume = 3 × 5 × 10 = 150 cm³

Volume of a cube = length × width × height

Blue cuboid

Volume = 1 × 7 × 10 = 70 cm³

The total volume is 150 + 70 = 220 cm³

Finally, add up the volume of the two cuboids.

Practice 4.5A

1. Work out the volume of each shape. Each cube measures 1cm by 1cm by 1cm.

    a   **b**   **c**

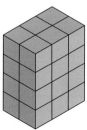
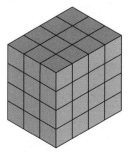
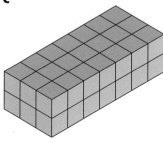

2. Work out the volume of each shape.

   a   **b**   **c**

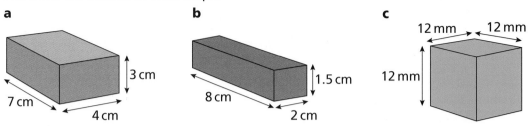

3. Both of these cuboids have the same volume.

   Work out the value of x

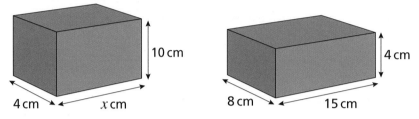

4. Here is the net of a cuboid.

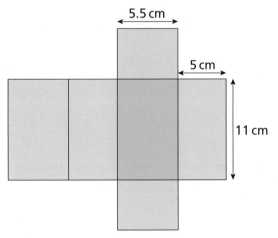

   Work out the volume of the cuboid.

5. A fish tank measures 80 cm by 60 cm by 50 cm

   Given that 1 litre = 1000 cm³, find, in litres, the volume of the fish tank.

6. Find the value of y in each of these cubes and cuboids.

   a   **b**   **c**   **d**

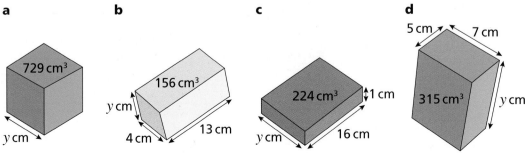

7 Here are three students' methods for working out the volume of this cuboid.

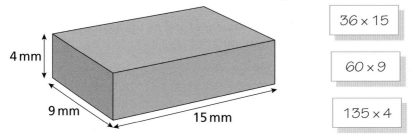

4 mm

9 mm

15 mm

36 × 15

60 × 9

135 × 4

Explain why all of these methods will give the correct answer.

8 Boxes measuring 1 m by 1 m by 0.5 m are packed into a shipping container.

The container measures 14 m by 12 m by 6 m. How many boxes will fit into the container?

9 A cube measures 5 cm by 5 cm by 5 cm

a Work out the volume of the cube.

The length of all the sides is increased by 10%

b Calculate the percentage increase in the volume of the cube.

c Explain why the answer to part **b** is not 10%

What do you think?

1 A cube has surface area 384 cm²

Work out the volume of the cube.

2 A swimming pool is in the shape of a cuboid measuring 20 m by 12 m by 2 m

The pool is filled with water at a rate of 750 litres per minute.

Given that 1 m³ = 1000 litres, work out how long it will take to fill the swimming pool with water.

Give your answer in hours and minutes.

3 A carton of juice measures 10 cm by 10 cm by 25 cm

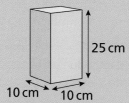

25 cm

10 cm   10 cm

The depth of the juice in the carton is 8 cm

Mario flips the carton so it now sits like this.

> The **capacity** of the carton is given by the calculation 10 × 10 × 25
> The question is about the volume of juice in the carton. Note that capacity and volume are slightly different.

Work out the depth of the juice in the carton now.

In this section, you are going to find the volume of some different prisms and cylinders.

Remember: a prism is a 3-D shape with a uniform **cross-section**.

You can find the volume of any prism using the same method as for a cuboid: find the number of cubes in one layer and then multiply by the number of layers.

The number of cubes in one layer represents the cross-sectional area.

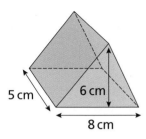

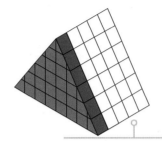

There are 24 cubes in the first layer.

The area of cross-section is $\frac{1}{2} b \times h = \frac{1}{2} \times 8 \times 6 = 24 \text{ cm}^2$

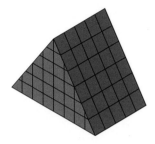

There are 5 layers in this prism.

So the total numbers of cubes is $24 \times 5 = 120 \text{ cm}^3$

Volume = area of cross-section × length

Example 3

Work out the volume of this cylinder.

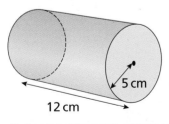

Although a cylinder is not a prism, you can find the volume in the same way.

Volume = area of circle × length ○—| The cross-section of a cylinder is a circle.

volume = area of cross-section × length

Volume = $\pi \times 5^2 \times 12 = 942.4777\ldots$ ○—| The area of a circle is $\pi \times r^2$

Volume = 942.5 cm^3 ○—| Round sensibly and remember to include the units.

Example 4

Work out the volume of this prism.

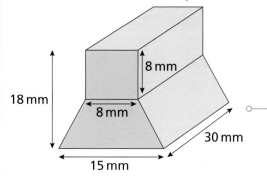

This shape is a prism so you can use the formula

volume = area of cross-section × length

The cross-section is made up of a square and a trapezium.

Area of square = 8 × 8 = 64 mm²

Area of trapezium = $\frac{1}{2}$ (8 + 15) × 10 = 115 mm²

The height of the trapezium is 18 – 8 = 10 mm

Total area of cross-section = 115 + 64 = 179 cm²

Add up the separate areas to get the total area of the cross-section.

Volume = 179 × 30 = 5370 mm³

Now you can use the formula for the volume of a prism.

Practice 4.5B

1 Work out the volume of each prism.

a

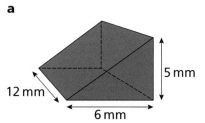

b

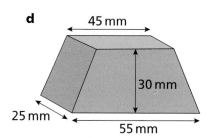

c

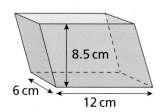

d

2 Work out the volume of each cylinder. Give your answers in terms of π

a

b
16 mm
35 mm

2 cm
10 cm

c
4.5 cm
3.2 cm

d
18 cm
14 cm

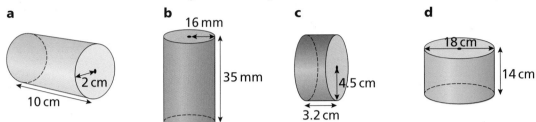

3 The cube and triangular prism have the same volume.

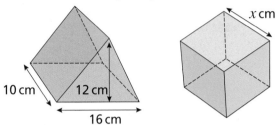

x cm

10 cm 12 cm
16 cm

Work out the value of x

4 A cylindrical tin of beans has radius 4 cm and a height of 10 cm

The can contains 400 cm³ of beans. How much empty space is there inside the tin?

5 Samira has a 2-litre bottle of cola. She pours the cola into cylindrical glasses with radius 3 cm and a height of 11 cm. How many full glasses can be filled from the bottle of cola?

6 Jakub fills an empty flowerpot with soil. The flowerpot is a prism. Its cross-section is a trapezium. Jakub has 2.5 litres of soil.

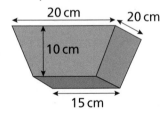

20 cm   20 cm
10 cm
15 cm

Does Jakub have enough soil to fill the flowerpot? Justify your answer.

7 Work out the value of y in each of these prisms.

a   **b**   **c**
Volume = 1344 mm³   Volume = 434 cm³   Volume = 2513 cm³
8 cm

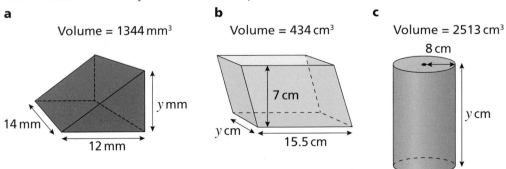

y mm
14 mm
12 mm

7 cm
y cm
15.5 cm

y cm

8 Work out the volume of this semicylinder.

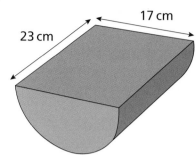

17 cm

23 cm

What do you think? 💭

1 A washing-up bowl is in the shape of an open cuboid, as shown.

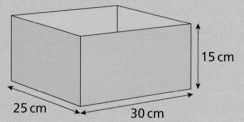

15 cm

25 cm   30 cm

It is $\frac{1}{3}$ full. Huda pours another 4 litres of water into the bowl.
Work out the new depth of the water.

2 Eight cylindrical cans of height 12 cm are placed into a box, as shown.

The box measures 42 cm by 21 cm by 12 cm.

Work out the percentage of the volume of the box that is empty space.

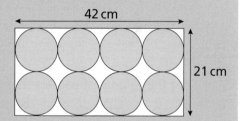

42 cm

21 cm

3 This cylinder is made of metal.

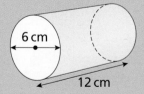

6 cm

12 cm

It is melted down and the metal is used to make a cube.
Work out the length of each side of the cube.

In this section, you will find the volume of cones, spheres and pyramids.

The volume of a sphere is given by

$V = \frac{4}{3}\pi r^3$, where r is the radius.

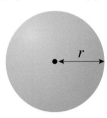

The volume of a cone is given by $V = \frac{1}{3}\pi r^2 h$, where r is the radius and h is the perpendicular height.

This should look familiar. It is exactly one third of the volume of a cylinder with the same radius and height. A cone takes up $\frac{1}{3}$ of the space of the cylinder.

The volume of a pyramid is given by $V = \frac{1}{3} \times$ area of base $\times h$, where h is the perpendicular height.

 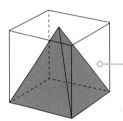

This means that the volume of a pyramid is $\frac{1}{3}$ of the volume of the prism with the same base area and the same height.

Example 5

Work out the volume of each of these shapes.

a   **b**   **c**

 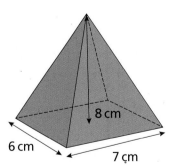 

a   Volume $= \frac{4}{3}\pi r^3$ Use the formula for the volume of a sphere.

Volume $= \frac{4}{3} \times \pi \times 6^3 = 904.7786\ldots$ Substitute the value of the radius, which is 6 cm

Volume $= 905\,\text{cm}^3$ to 3 s.f.

b $\text{Volume} = \frac{1}{3} \times \text{area base} \times \text{height}$ ○———⊣ Use the formula for the volume of a pyramid.

$\text{Volume} = \frac{1}{3} \times 42 \times 8 = 112\,\text{cm}^3$ ○———⊣ In this pyramid, the base is a rectangle measuring 6 cm by 7 cm
The area of the base is 6 × 7 = 42 cm²

c $\text{Volume} = \frac{1}{3}\pi r^2 h$ ○———⊣ Use the formula for the volume of a cone.

$\text{Volume} = \frac{1}{3} \times \pi \times 9.5^2 \times 22 = 2079.21\ldots$ ○———⊣ The diameter is 19 mm so the radius is 19 ÷ 2 = 9.5 mm and the height of the cone is 22 mm

$\text{Volume} = 2079\,\text{mm}^2$

Example 6

This shape is made up of a cube and a pyramid. Work out the volume of the shape.

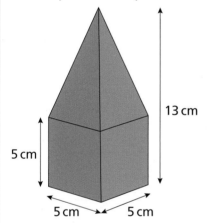

13 cm

5 cm

5 cm   5 cm

You can work out the volume of each part separately and then add them together.

Volume of cube = 5 × 5 × 5 = 125 cm³ ○———⊣ The cube has side length 5 cm

Volume of pyramid = $\frac{1}{3} \times$ area of base × height ○———⊣ Use the formula for the volume of a pyramid.

$\text{Volume} = \frac{1}{3} \times 25 \times 8 = 66.666\ldots = 66.7\,\text{cm}^3$ ○———⊣ The base of this pyramid is a square measuring 5 cm by 5 cm
Total volume = 125 + 66.7
The area of the base is 25 cm² and the height of the pyramid is 13 − 5 = 8 cm
= 191.7 cm³ ○———⊣ Finally, find the total volume.

Practice 4.5C

1 Work out the volume of each shape.

a

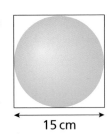

16 cm

b

7.5 cm

7.5 cm 8 cm

c

2.4 m

1.1 m

d

3 mm

2 A wooden toy is made from a cone placed on top of a cylinder. The cone and the cylinder have the same radius. Work out the total volume of the toy.

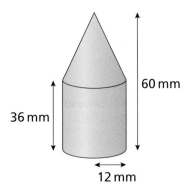

60 mm

36 mm

12 mm

3

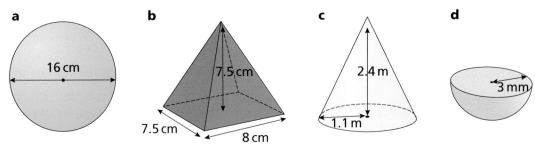

The cone and pyramid have the same dimensions so they will have the same volume.

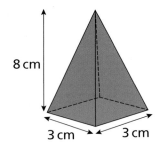

8 cm

3 cm 3 cm

8 cm

3 cm

Is Chloe correct? Justify your answer.

4 A sphere is placed tightly inside a cube so that it just touches each side. Work out the proportion of the cube that is empty space.

15 cm

5 Work out the value of x in each of these shapes.

a

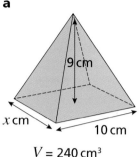

$V = 240 \text{ cm}^3$

b

$V = 10\,603 \text{ mm}^3$

c

$V = 1150 \text{ cm}^3$

What do you think? 💭

1 A metal sphere of radius 6 cm is melted down. The metal is then used to make three identical spheres. What is the radius of each of the new spheres?

2 A sphere has radius r. A cylinder has radius r and a height of 4 cm.

The sphere and cylinder have the same volume. Work out the value of r

Consolidate – do you need more?

1 Work out the volume of each of these cuboids.

a

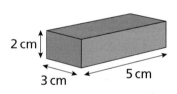

b

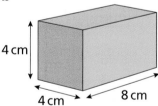

c

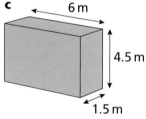

2 Work out the volume of these prisms.

a

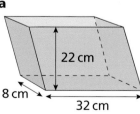

b

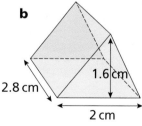

c

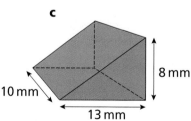

3 Work out the volume of these cylinders. Give your answers in terms of π

a

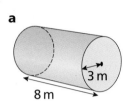

3 m

8 m

b 13 mm

30 mm

c

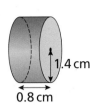

1.4 cm

0.8 cm

4 Find the missing length in each of these shapes.

a

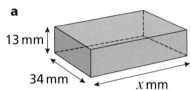

13 mm

34 mm

x mm

$V = 22\,100\text{ mm}^3$

b

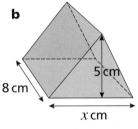

8 cm

5 cm

x cm

$V = 180\text{ cm}^3$

c

x cm

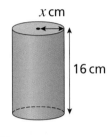

16 cm

$V = 615.8\text{ cm}^3$

5 Marta fills the cylindrical container with water. She then pours the water into the cuboid-shaped container. Will all of the water fit into the cuboid? Why or why not?

10 cm

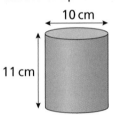

11 cm

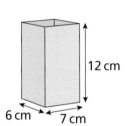

12 cm

6 cm   7 cm

6 The volume of the cuboid is double the volume of the cube.

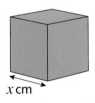

x cm

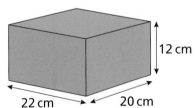

12 cm

22 cm   20 cm

Work out the value of x

7 Express the volume of the cylinder as a percentage of the volume of the cube.

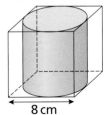

8 cm

8 Sort these shapes in order of volume, from least to greatest.

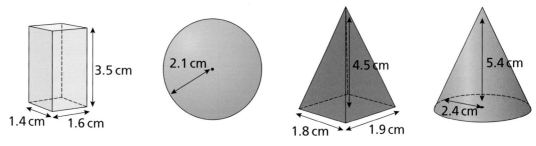

Stretch – can you deepen your learning?

1 A cylindrical container has a capacity of 2.4 litres and a length of 16 cm

Work out its diameter.

2 The volume of this cuboid is double the volume of the cube. Work out the ratio of the surface area of the cuboid to the surface area of the cube.

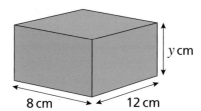

3 A cylindrical vase is filled with water to the height shown.

Sven pours in another 200 ml of water.

Calculate the new depth of the water.

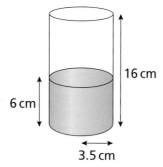

4 Faith puts pebbles and water into her fish tank so that it is $\frac{3}{4}$ full.

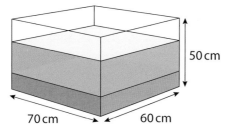

The ratio of the volume of pebbles to the volume of water is 2:3

Work out the volume of water in the tank. Give your answer in litres.

5

This cone is half full.

a Explain why Flo is wrong.

b Work out the fraction of the cone that is filled with water.

c How much more water would Flo need to add so that the cone will be half full?

Reflect

1 Volume can be measured in cm³, m³, ml and litres. Think of some real-life contexts when each one may be used and discuss these with a partner.

2 What is the difference between capacity and volume?

I have become **fluent** in...	I have developed my **reasoning** skills by...	I have been **problem-solving** through...
▦ identifying and naming different 2-D and 3-D shapes	▦ making connections between 3-D shapes and their nets	▦ working backwards to find lengths when given areas and volumes
▦ identifying prisms through their properties	▦ interpreting plans and elevations to identify 3-D shapes	▦ modelling more complex problems involving areas and volumes
▦ finding the areas of 2-D shapes	▦ making connections between the numbers of faces, edges and vertices in 3-D shapes	▦ setting up and using equations to solve problems.
▦ finding the surface areas of cubes, cuboids and prisms	▦ solving complex problems involving volume.	
▦ finding the volume of simple 3-D shapes.		

Check my understanding

1 How many faces, edges and vertices does a cuboid have?

2 Which of these shapes is a prism? How do you know?

A   B   C   D

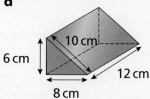

3 For each shape

 i state the mathematical name **ii** find the surface area **iii** find the volume.

a

3 mm

6 mm

7 mm

b

8 cm

c

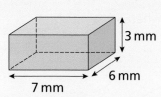
2.5 cm

9 cm

d

10 cm

6 cm

12 cm

8 cm

5 Constructions and congruency

In this block, I will learn...

how to construct and work with accurate drawings

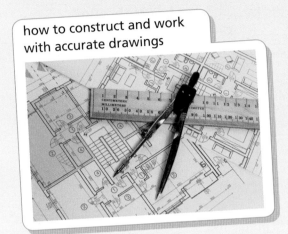

how to interpret and construct loci

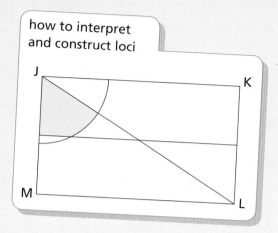

how to construct perpendiculars

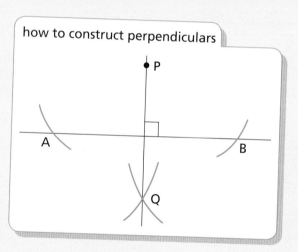

how to construct bisectors

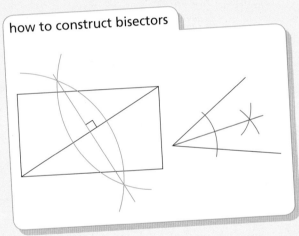

how to recognise congruent shapes

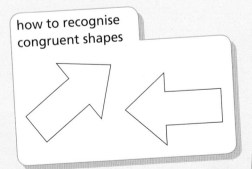

how to determine whether triangles are congruent

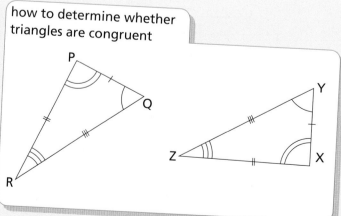

Small steps

- Draw and measure angles **R**
- Construct and interpret scale drawings **R**
- Construct triangles from given information **R**

> ### Key words
>
> **Construct** – draw accurately using a ruler and compasses
>
> **Sketch** – a rough drawing
>
> **Acute angle** – an angle less than 90°
>
> **Obtuse angle** – an angle more than 90° but less than 180°
>
> **Scale** – the ratio of the length in a drawing or a model to the actual object

Are you ready?

1 Use the information in the diagram to find the size of each of these angles.

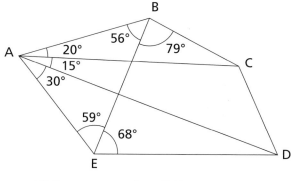

Diagram not drawn accurately

   a ∠EAD   **b** ∠EAB   **c** ∠EAC   **d** ∠DAC   **e** ∠DAB

2 Look again at the diagram in question **1**. Which other angles can you find

   a without doing any calculations   **b** by doing one addition?

3 Look at the diagram in question **1** again. Which of these angles are obtuse?

   ∠ABC   ∠ABE   ∠AED   ∠DAE   ∠DEA

4 How do you know, without using a protractor, that these diagrams are not accurately drawn?

   a

   b

65°

103°

Models and representations

Angles are measures of turn. This can be represented in many ways.

Geostrips

Card and paper fasteners

Pencils or rulers

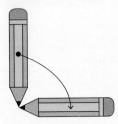

Parts of the body

Scale drawing

A scale drawing represents real objects so that you can see the relative size of each object.

This scale drawing is the floor plan of an apartment. You can see which rooms are larger and smaller than other rooms.

The **scale** is 1:100
Each cm on the scale represents 100 cm (or 1 m) in the real apartment.

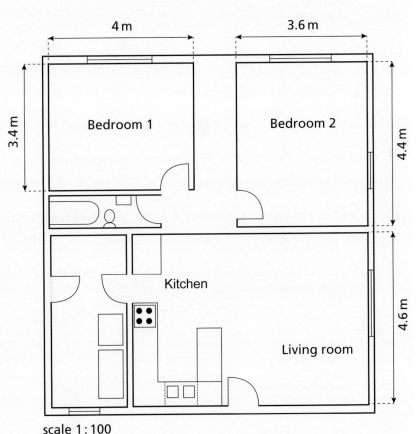

scale 1:100

You learned how to draw and measure angles in Book 1 and how to work with scale drawings in Book 2. Here is a reminder.

Example 1

Draw an angle of 130°

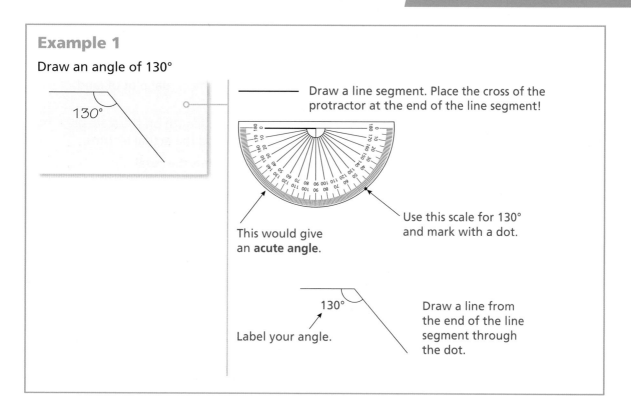

Draw a line segment. Place the cross of the protractor at the end of the line segment!

This would give an **acute angle**.

Use this scale for 130° and mark with a dot.

Label your angle.

Draw a line from the end of the line segment through the dot.

Example 2

Here is a scale drawing of a plot of land shaped like a trapezium.

The scale of the drawing is 1 : 2000

Find the perimeter of the plot of land.

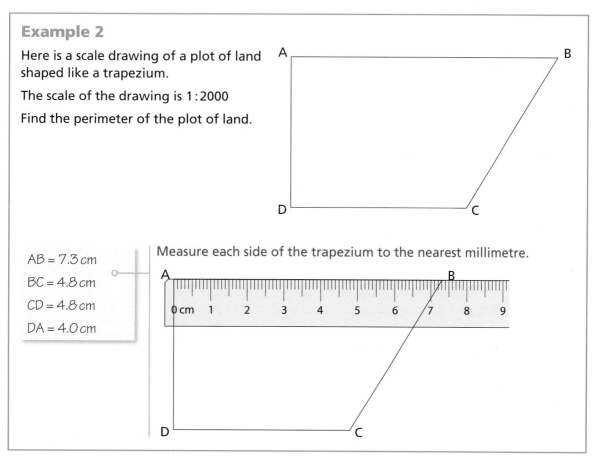

AB = 7.3 cm

BC = 4.8 cm

CD = 4.8 cm

DA = 4.0 cm

Measure each side of the trapezium to the nearest millimetre.

The actual lengths are

AB = 7.3 × 2000 = 14 600 cm = 146 m

BC = 4.8 × 2000 = 96 m

CD = 4.8 × 2000 = 96 m

DA = 4 × 2000 = 80 m

Perimeter = 146 + 96 + 96 + 80 = 418 m

A scale of 1 : 2000 means that each centimetre on the drawing represents 2000 cm on the actual plot of land.

Multiply each length on the scale drawing by 2000 to find the actual lengths.

The perimeter is the total of the four lengths.

Practice 5.1A

1 For each part of this question, you will need to start with a line segment AB of length 6 cm

A ——————————————— B

a Draw an angle of 50° from point A. Draw the angle **below** your line segment.

b Draw an angle of 50° from point B. Draw the angle **above** your line segment.

c Draw an angle of 125° from point B. Draw the angle **below** your line segment.

d Draw an angle of 135° from point A. Draw the angle **above** your line segment.

2 **a** Estimate the size of each of these angles.

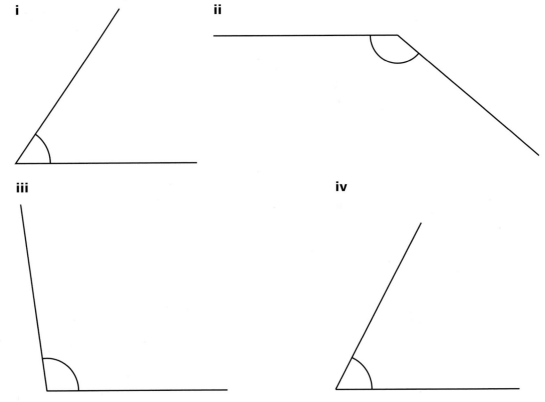

i

ii

iii

iv

b Use a protractor to measure each angle. Compare your answers with a partner's.

I draw an angle of 230° by adding 50° to a straight line.

Emily

I draw an angle of 230° by starting with an angle of 130°

Zach

a Explain why both Emily's and Zach's methods work.

b Use both methods to draw an angle of 230°

c Use your preferred method to draw these angles.

    i 300°   **ii** 195°   **iii** 340°   **iv** 211°

    Did you choose the same method each time? If not, how did you decide which method to use?

4 Parallelogram ABCD is drawn on dotty paper as shown.

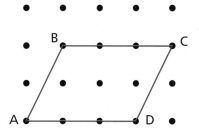

a Copy the diagram and measure the four angles in your parallelogram.

b Which angles are equal? Why?

c Which angles add up to 180°? Why?

5 Here is a scale drawing of a rectangle.

a Measure the sides of the rectangle in the scale drawing.

b Find the length of each side of the actual rectangle for each of these scales.

    i 1 cm represents 2 m

    ii 2 cm represents 1 m

    iii 1 : 500

    iv 10 : 1

What units would be appropriate for each answer?

6 This is a scale drawing of a field.

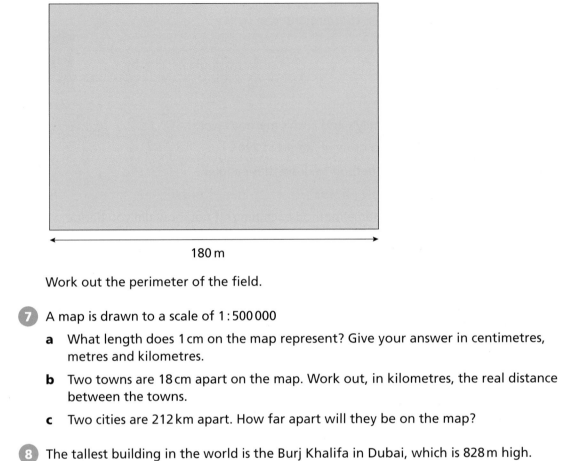

180 m

Work out the perimeter of the field.

7 A map is drawn to a scale of 1 : 500 000

a What length does 1 cm on the map represent? Give your answer in centimetres, metres and kilometres.

b Two towns are 18 cm apart on the map. Work out, in kilometres, the real distance between the towns.

c Two cities are 212 km apart. How far apart will they be on the map?

8 The tallest building in the world is the Burj Khalifa in Dubai, which is 828 m high.

a Calculate the height of a model of the Burj Khalifa made to each of these scales. Give your answers in centimetres.

  i 1 : 1000　　　　ii 1 : 10 000　　　　iii 1 : 500　　　　iv 2 : 500

b Suggest a good scale to use so that the height of the model is about

  i 1 metre　　　　ii 50 cm　　　　iii 40 cm

9 Here is some information about some of the world's tallest statues.

Statue of Unity	Spring Temple Buddha	Ushiku Daibutsu	Statue of Liberty	The Motherland Calls
India	China	Japan	USA	Russia
182 m	153 m	120 m	93 m	85 m

Discuss the scales used to make the drawings of the statues.

What do you think?

1 Why are there two scales on a protractor? How do you know which one to use?

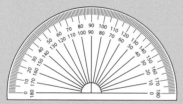

2 Suggest a sensible scale for drawing plans of these objects.

   a Your classroom   **b** Your school hall

   c Your desk   **d** Your school grounds

3 A rectangular field measures 200 m by 250 m

   a Calculate the area of the field.

   b A scale drawing of the field uses a scale of 1 : 1000. Calculate the area of the scale drawing.

   c Explain why your answer to part **b** is not one-thousandth of your answer to part **a**.

Here is a reminder of how to draw triangles accurately.

You learned this in Book 1

Example 3

a Construct triangle ABC with AB = 8 cm, ∠ABC = 65° and ∠BAC = 55°

b Measure ∠ACB. How does your measurement help you to check that you have drawn the triangle accurately?

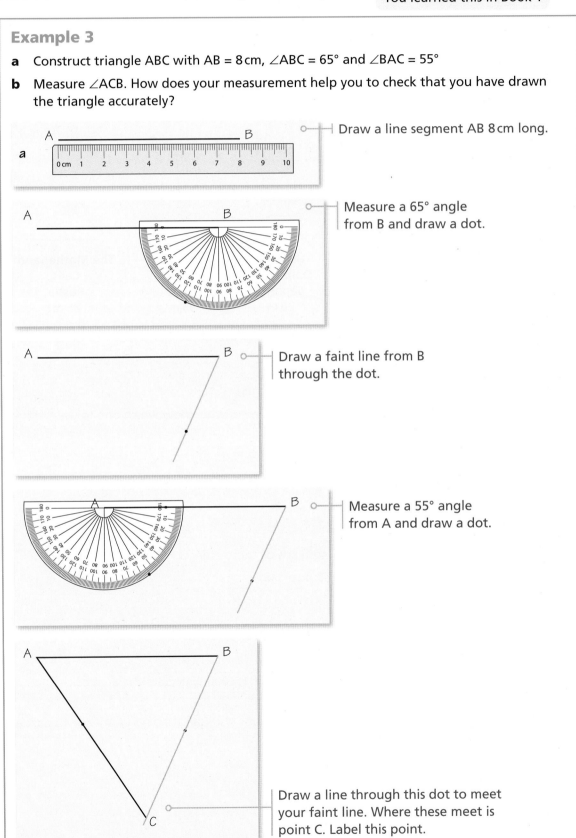

a

Draw a line segment AB 8 cm long.

Measure a 65° angle from B and draw a dot.

Draw a faint line from B through the dot.

Measure a 55° angle from A and draw a dot.

Draw a line through this dot to meet your faint line. Where these meet is point C. Label this point.

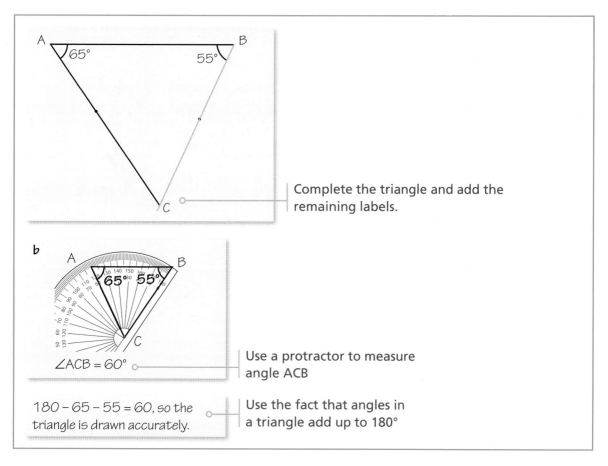

Complete the triangle and add the remaining labels.

b

Use a protractor to measure angle ACB

$\angle ACB = 60°$

$180 - 65 - 55 = 60$, so the triangle is drawn accurately.

Use the fact that angles in a triangle add up to 180°

Practice 5.1B

1 Use the method shown in Example 3 to construct accurate copies of these triangles.

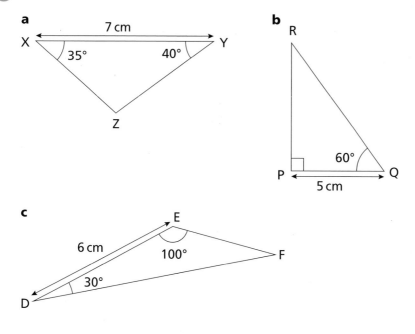

a

7 cm

X 35° 40° Y

Z

b

R

60°

P 5 cm Q

c

E

6 cm

100° F

30°

D

2 **a** Explain how the diagrams show how to construct a triangle with sides 7 cm, 6 cm and 5 cm

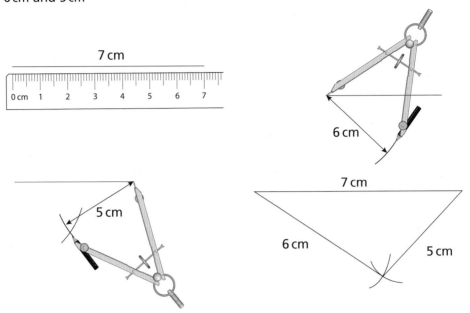

b Construct an accurate copy of the triangle.

c Use the same method to construct

   i an equilateral triangle with sides of length 8 cm

   ii an isosceles triangle with one side 8 cm and two sides 6 cm

   iii a scalene triangle with sides of length 9 cm, 7 cm and 6.5 cm

d Explain why it is not possible to construct a triangle with sides 9 cm, 3 cm and 5 cm

3 **a** Explain how the diagrams show how to construct triangle XYZ with XY = 5 cm, angle XYZ = 35° and YZ = 4 cm

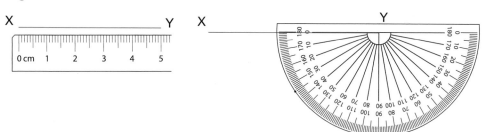

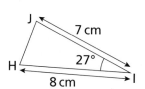

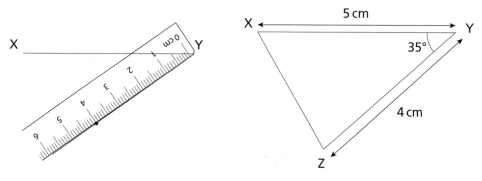

b Construct an accurate copy of triangle XYZ

c Use the same method to construct these triangles.

i

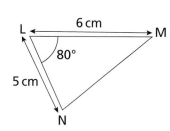

ii

iii

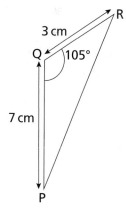

4 The three methods for constructing triangles are called AAS, SSS and SAS for short. Can you see why these letters are used?

5 Here is a sketch of a triangular field.

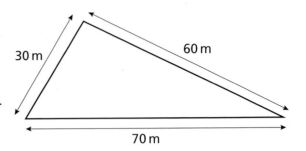

a Explain what is meant by the word "sketch".

b Use a scale of 1 : 500 to draw an accurate scale drawing of the field.

6 Construct each of these the triangles using the given information. You might find it useful to draw a sketch first.

a Triangle PQR such that PQ = 7 cm, QR = 5 cm and PR = 4.5 cm

b Triangle ABC such that AB = 6 cm, AC = 8 cm and ∠BAC = 45°

c Triangle XYZ such that XY = 5.6 cm, ∠XYZ = 65° and ∠YXZ = 35°

d Triangle DEF such that DE = EF = 8 cm and ∠DEF = 70°

e Use your knowledge of the angles in triangles to work out the unknown angles in triangles XYZ and DEF. Measure the angles to check that you have constructed the triangles accurately.

7 a The sides of a rhombus are 5 cm long. One of the angles in the rhombus is 55°
Construct the rhombus.

b The sides of a parallelogram measure 7 cm and 4 cm. One of the angles in the parallelogram is 140°. Construct the parallelogram.

What do you think?

1 a What is the minimum number of pieces of information you need to draw a triangle accurately?

b Explain why you need to know the length of at least one of the sides of a triangle if you are going to construct it.

2 The sides of a triangle are x cm, y cm and z cm. Write three inequalities in x, y and z

3 a Construct a regular hexagon ABCDEF with sides of length 4 cm

b Write down the size of the obtuse angle BXF

c What other angles can you find?

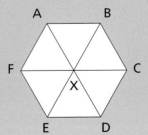

Consolidate – do you need more?

1 For each part of the question, you will need to start with a line segment AB of length 6 cm

A ————————————————— B

  a Draw an angle of 70° from point A. Draw the angle **below** your line segment.

  b Draw an angle of 75° from point B. Draw the angle **above** your line segment.

  c Draw an angle of 115° from point A. Draw the angle **below** your line segment.

  d Draw an angle of 95° from point B. Draw the angle **above** your line segment.

2 a Estimate the size of each of these angles.

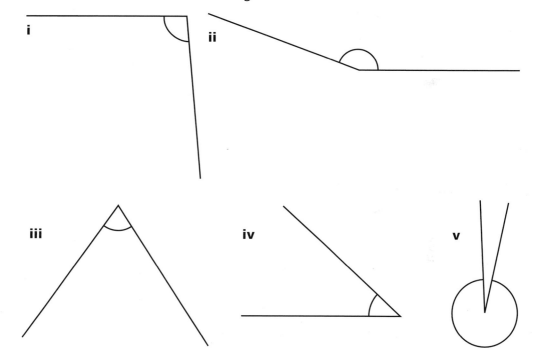

  b Use a protractor to measure each angle. Compare your answers with a partner's.

3 On a scale drawing, the length of a line is 6 cm

What real length does the line represent if each of these scales has been used?

  a 1 : 20 **b** 1 cm represents 20 m

  c 1 : 200 **d** 2 cm represents 10 m

  e 1 : 500 **f** 1 : 100 000

Give your answers using the most appropriate units.

4 On a map, a section of motorway measures 8 cm. The actual length of this section of motorway is 40 km

  a Find the scale of the map in the form "1 cm represents … km".

  b Give the scale of the map in the form 1 : n where n is an integer.

5 Construct accurate copies of these triangles.

a

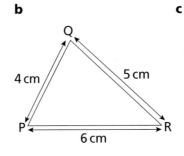

b

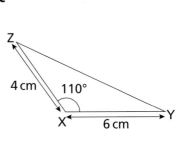

c

Stretch – can you deepen your learning?

1 A bearing is an angle in degrees measured clockwise from north. Bearings are usually written with three figures, so the bearing of B from O is written 090°

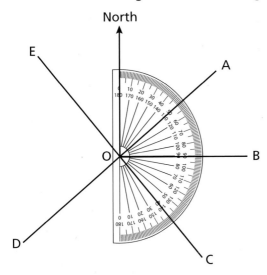

If the angle is less than 100°, then you need to add a zero before the angle.

a Explain how the diagram shows that the bearing of E from O is 320°

b Find the bearing from O to each of these points.

  i A   ii C   iii D

2 Mark a point P in the middle of a page of your exercise book.

Using a scale of 1:200, mark the following points.

a Q, which is 8 m from P on a bearing of 125°

b R, which is 12 m from P on a bearing of 240°

c S, which is 10.4 m from P on a bearing of 195°

d T, which is 11 cm from P on a bearing of 064°

e By drawing additional north lines, find

    i the bearing of P from Q   **ii** the bearing of P from R

    iii the bearing of P from S   **iv** the bearing of P from T

f What do you notice about the bearings you found in part **e** compared with the bearings you used to mark the points Q, R, S and T on your diagram?

g If the bearing of A from B is $x°$, find an expression for the bearing of B from A. Use your knowledge of parallel lines to explain your answer. (Hint: there are different expressions depending on the value of x)

3 The diagram shows a square ABCD. B is on a bearing of 220° from A

What other bearings can you calculate?

Investigate for different positions of the square ABCD, different regular shapes and different quadrilaterals. What patterns can you find?

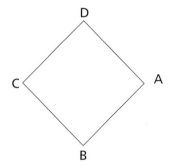

Reflect

1 When you look at an angle, how do you recognise what type of angle it is? How does this help you to measure the angle correctly?

2 What's the same and what's different about scales written in the form "1 cm represents…" or "2 cm represents…" and those using ratio notation? Which are easier to work with? Why?

3 Write a list of instructions for constructing a triangle given different types of information.

Small steps

- Locus of distance from a point
- Locus of distance from a straight line/shape

Key words

Locus (plural: **loci**) – a set of points that describe a property

Equidistant – at the same distance from

Are you ready?

1 Draw a circle of radius 5 cm

2 Draw a semicircle of diameter 12 cm

3 In this pattern, all five circles have radius 4 cm. The four outer circles pass through the centre of the circle in the middle of the pattern.

Make a copy of the pattern.

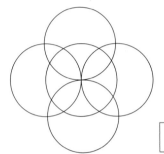

Not drawn to scale

Models and representations

You can use coins or counters to show points that are **equidistant** from other points or lines.

Coins

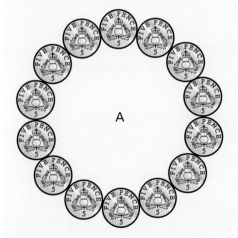

A

Counters

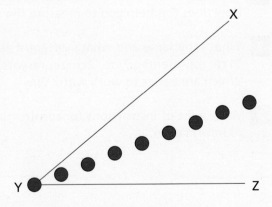

When you draw a circle of radius 5 cm from a point O, all the points on the circumference of the circle are exactly 5 cm from the centre, O

OX = OY = OZ = 5 cm

The circle is the **locus** of the points that are 5 cm from O

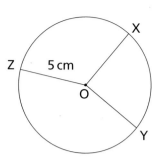

Example 1

The diagram shows a circle radius 2.5 cm inside a square of side 6 cm. The centre of the circle is at the centre of the square.

a Shade the locus of points inside the square that are more than 2.5 cm from O

b Draw the locus of points that are exactly 3 cm from O

c Shade the locus of points inside the square that are more than 2.5 cm from O but less than 3 cm from O

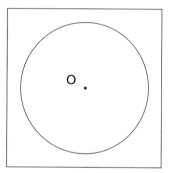

a

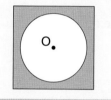

The circle is the locus of points that are exactly 2.5 cm from O. So you need to shade the region that is outside the circle, but inside the square.

b

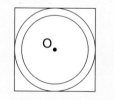

The locus of points that are exactly 3 cm from O is a circle, centre O, with radius 3 cm

c

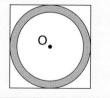

The locus of points inside the square that are more than 2.5 cm from O, but less than 3 cm from O, is all the points that are outside the smaller circle, but inside the larger circle.

For example, point P, such that OP is 2.8 cm, is one point in the locus.

Practice 5.2A

 1 Put a coin at the centre of your desk. Place counters so they are 10 cm from the coin, as shown, to complete the pattern.

Check that the locus of all the points 10 cm from the coin forms a circle of radius 10 cm with the coin at its centre.

How many counters can you fit? Can you make a complete circle?

2 **a** Mark a point O and draw the locus of points that are exactly 5 cm from O

b Mark a point X and draw the locus of points that are exactly 3 cm from X

c Mark a point Y and shade the locus of points that are less than 4 cm from Y

3 Mark two points, A and B, that are 5 cm apart.

a Draw the locus of points that are exactly 4 cm from A

b Draw the locus of points that are exactly 4 cm from B

c Shade the region where your two circles overlap.

d Which of these statements describes the points that are in the region that you have shaded?

A | More than 4 cm from A and more than 4 cm from B

B | More than 4 cm from A and less than 4 cm from B

C | Less than 4 cm from A and more than 4 cm from B

D | Less than 4 cm from A and less than 4 cm from B

4 Mark two points, X and Y, that are 6 cm apart.

a Draw the locus of points that are exactly 5 cm from X

b Draw the locus of points that are exactly 4 cm from Y

c Shade the locus of points that are more than 5 cm from X and less than 4 cm from Y

5 Construct the locus of points that are between 4 cm and 6 cm from a point O

6 AB is a line segment 5 cm long.

P is a point exactly 4 cm from A and 3 cm from B

a Show by construction there are exactly two possible locations of point P

b Verify that ∠APB is a right angle in both cases.

7 ABCD is a rectangle that measures 8 cm by 4 cm. M is the midpoint of CD

The diagram shows the locus of points that are less than or equal to 4 cm from M

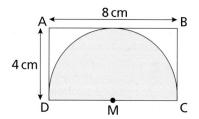

a Make four copies of the empty rectangle ABCD. In each rectangle, draw in M, the midpoint of CD. On your diagrams, show each of these loci.

Points inside the rectangle that are

i exactly 3 cm from A

ii less than 5 cm from B

iii less than 4 cm from C and less than 4 cm from M

iv less than 3 cm from M and more than 4 cm from A

b Challenge a partner to create and describe other regions in the rectangle ABCD. Use any vertices or midpoints that you like.

What do you think?

1 A rectangular lawn measures 5 m by 6 m

A sprinkler waters the grass within a circle of radius of 3 m

The sprinkler is placed at the centre of the lawn.

a Explain why Diagram B is a better model of the region of the lawn that gets watered than Diagram A

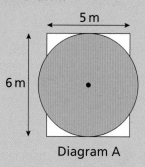

Diagram A

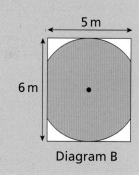

Diagram B

b Draw two scale drawings of the lawn using a scale of 1 cm represents 1 m. Label the corners of the lawn WXYZ

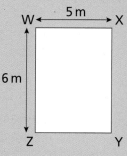

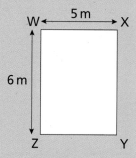

i On your first copy, shade the region watered if there are two sprinklers placed at W and Y

ii On your second copy, shade the region watered if there are two sprinklers placed at the midpoints of WZ and XY

2 Look again at question **1**. Investigate the best place to put the sprinklers so that as much of the lawn is watered as possible, without the regions overlapping. What is the best radius for the water sprinkler? What other questions can you ask?

Next you will explore the locus of points that are at a constant distance from a line.

Example 2

A square ABCD has sides of length 6 cm

a   On a copy of ABCD, draw the locus of points that are exactly 2 cm from AB

b   On a second copy of ABCD, draw the locus of points that are less than 4 cm from AB and less than 3 cm from point D

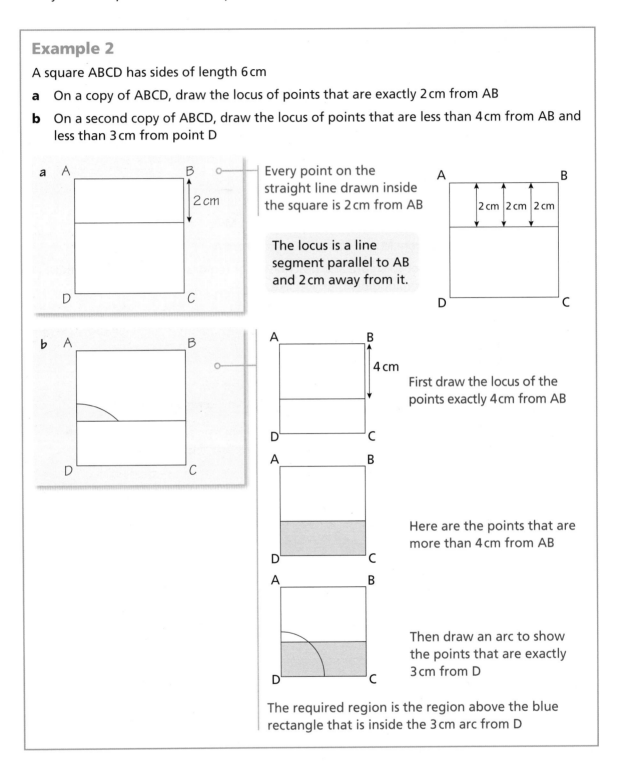

Every point on the straight line drawn inside the square is 2 cm from AB

The locus is a line segment parallel to AB and 2 cm away from it.

First draw the locus of the points exactly 4 cm from AB

Here are the points that are more than 4 cm from AB

Then draw an arc to show the points that are exactly 3 cm from D

The required region is the region above the blue rectangle that is inside the 3 cm arc from D

Practice 5.2B

 1 Place counters so that they are 15 cm from the bottom edge of your desk, as shown.

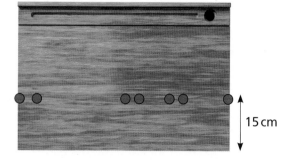

15 cm

Complete the pattern and verify that the locus of points that are 15 cm from the bottom edge of your desk is a straight line parallel to the edge of your desk and 15 cm away from it.

Explore other distances from other edges of your desk.

2 Here is a sketch of a classroom.

For this question, you will need several scale drawings of the classroom.

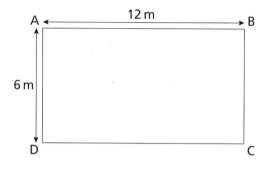

A ← 12 m → B

6 m

D   C

a Which is better, a scale of 1 cm represents 1 m or 1 cm represents 2 m?

b A teacher asks a class to stand in the following positions. Draw the locus of points where the students could stand in each case.

   i 2 m from wall AB   **ii** 4 m from wall AD

   iii Less than 4 m from the wall CD

   iv Less than 4 m from the wall CD and more than 4 m from the wall BC

   v 4 m from the corner D. Why is this part different from the other parts of the question?

3 PQ is a line segment 6 cm long. Seb and Chloe draw the locus of points that are exactly 2 cm from PQ

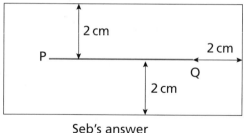

Seb's answer

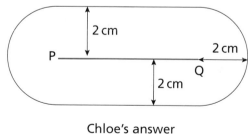

Chloe's answer

Who is correct? What mistake has the other student made?

4 Copy and complete the diagram showing the locus of points that are 3 cm from WXYZ on the outside of the rectangle.

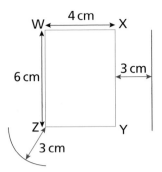

5 Construct an equilateral triangle, PQR, with sides of length 4 cm. Draw the locus of points that are 2 cm outside the triangle.

6 A garden measures 10 m by 6 m. A tree is to be planted in the garden. The tree must be at least 2 m from the edge of the garden. Make a scale drawing of the garden and shade the region where the tree could be planted.

What do you think? 💭

1 a A dog is tied to a wall by a 1.5 m-long lead. The lead is fixed at the midpoint of the wall. Draw a scale diagram to show the area in which the dog can go.

b Another dog is tied to a corner of a shed that measures 6 m long by 4 m. The length of the dog's lead is 5 m

The shed is in the middle of a large garden. Draw a scale drawing and shade the region of the garden that the dog can reach. Discuss with a partner a suitable scale to use.

2 How would your answer change if the shed in question **1b** were in a corner of the garden? Or in the middle of one side of the garden?

Investigate different positions of the shed, different lengths of the lead and different places where the lead is tied to the shed, for example the middle of a wall rather than a corner.

Consolidate – do you need more?

1　**a**　Mark a point O and draw the locus of points that are exactly 4 cm from O

　　b　Mark a point X and draw the locus of points that are exactly 4.5 cm from X

　　c　Mark a point Y and shade the locus of points that are less than 3 cm from Y

2　Mark two points, A and B, that are 4 cm apart.

　a　Draw the locus of points that are

　　i　exactly 3 cm from A

　　ii　exactly 3 cm from B

　b　Shade the locus of points that are more than 3 cm from A but less than 3 cm from B

3　For each part of this question, you need to draw a square, PQRS, with sides of length 5 cm

　a　Draw the locus of the points that are

　　i　exactly 4 cm from P
　　ii　more than 4 cm from P

　　iii　exactly 4 cm from PQ
　　iv　less than 4 cm from PQ

　　v　less than 3 cm from point R and less than 3 cm from side PS

　b　Find the point in the square that is 3 cm from PQ and 3 cm from point R

4　**a**　Draw a horizontal line segment, XY, 5 cm long. Draw the locus of points that are 3 cm from XY

　　b　Draw a vertical line segment, MN, 6 cm long. Draw the locus of points that are 2 cm from MN

5　Here is a rectangle, ABCD

　a　Draw a scale drawing of ABCD. Shade the locus of points outside rectangle ABCD that are less than 2 m from the sides of ABCD

　b　On another scale drawing of ABCD, draw the locus of points inside the rectangle that are more than 2 m from the sides of ABCD

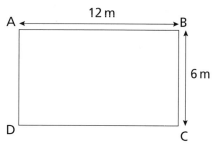

Stretch – can you deepen your learning?

1 Faith stands at point P behind a wall. The diagram shows the locus of points which she cannot see.

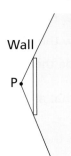

Investigate how the area that she cannot see changes for different positions of point P and different lengths of the wall.

2 Look again at question **1**. Investigate the areas that Faith cannot see for different positions of P between two walls.

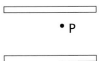

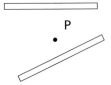

3 Imagine a square being rolled along a horizontal surface, as shown. What path would the point P trace out? Draw the locus of point P as the square moves. Investigate for other shapes.

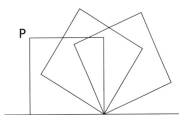

Reflect

1 What's the same and what's different about the locus of points that are a fixed distance from a point and the locus of points that are a fixed distance from a line?

2 How do you construct the locus of points that are a fixed distance from a shape, both on the inside and on the outside?

Small steps

- Locus equidistant from two points
- Construct a perpendicular bisector
- Construct a perpendicular from a point
- Construct a perpendicular to a point

Are you ready?

1 Which pair(s) of line segments are perpendicular?

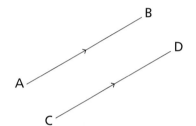
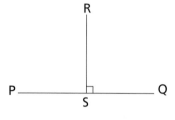
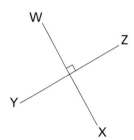

2 Mark a point O and draw the locus of the points that are 4.5 cm from O

3 Draw a line segment, AB, 6 cm long. Draw the locus of the points that are 3 cm from AB

Models and representations

You can use coins or counters to show points that are the same distance from other points or lines.

Coins

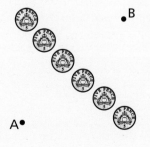

Counters

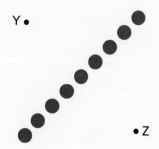

Paper folding

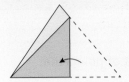

You can also make perpendicular lines by folding.

In the first part of this chapter, you will explore points that are the same distance from two fixed points. These points are **equidistant** from the fixed points.

Example 1

Mark two points, A and B, that are 6 cm apart.

How many points are exactly 4 cm from both A and B?

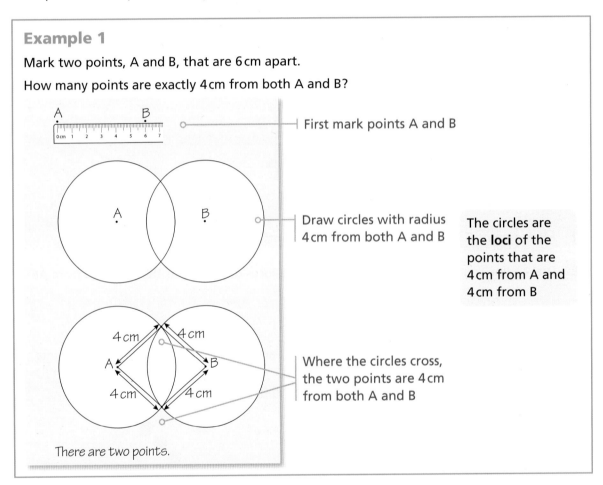

First mark points A and B

Draw circles with radius 4 cm from both A and B

The circles are the **loci** of the points that are 4 cm from A and 4 cm from B

Where the circles cross, the two points are 4 cm from both A and B

There are two points.

Practice 5.3A

 1 Put two coins on your desk and arrange a set of counters so that they are the same distance from both coins. Try this with your coins in different positions.

What do notice about your counters each time?

2 Abdullah solves Example 1 in a different way.

I can find two points that are equidistant from A and B without drawing circles.

A·   ·B

a Explain how Abdullah's method works.

b Draw points A and B 6 cm apart and either use circles or Abdullah's method to find points that are 4 cm, 5 cm and 6 cm from both A and B. What do you notice about your points?

3 **a** Follow these construction steps.

■ Draw a line segment, PQ, 5 cm long.

■ Use circles or arcs to find points that are the same distance from P and from Q.

■ Join the points that are the same distance from P and Q with a line segment that meets PQ at M.

■ Measure PM and QM. What do you notice?

■ Measure the angle PMQ.

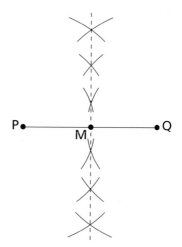

b Repeat the construction in part **a** with different line segments in different orientations. Do you get the same results?

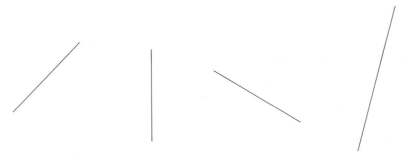

What do you think? 💭

Look again at question **2** from Practice 5.3A. How many points above and below a line segment do you need to draw so as to construct a line that passes through the midpoint of the line segment? What is the minimum number of points you need? Can they be on the same "side" of the line?

In question **2** of Practice 5.3A, you should have noticed that the locus of points which are equidistant from two fixed points from a straight line. This straight line cuts the line segment joining the two points in half and is at right angles to the line segment. This is called the perpendicular **bisector** of the line segment.

It is perpendicular because it is at right angles to the line segment and it is a bisector because it cuts the line segment into two equal parts.

In the next example, you will learn an efficient way to construct a perpendicular bisector.

You may have already seen this method in Book 2

Example 2

AB is a line segment with length 8 cm. Construct the perpendicular bisector of AB

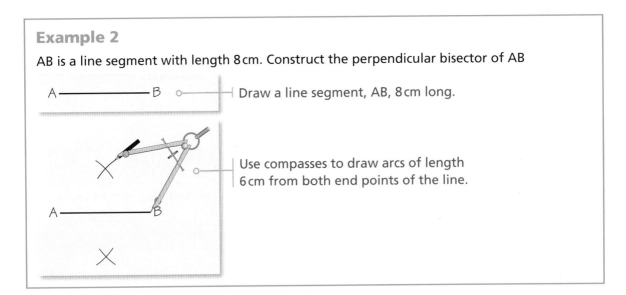

Draw a line segment, AB, 8 cm long.

Use compasses to draw arcs of length 6 cm from both end points of the line.

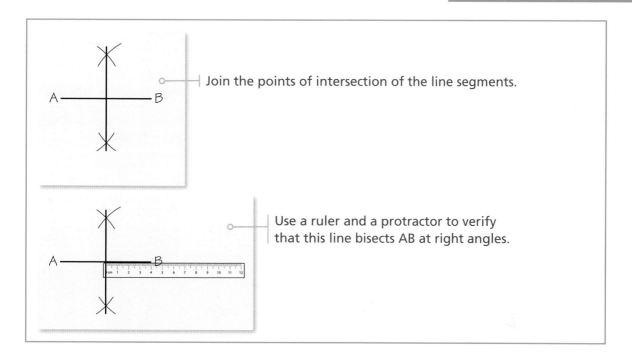

Join the points of intersection of the line segments.

Use a ruler and a protractor to verify that this line bisects AB at right angles.

Practice 5.3B

1. Repeat Example 2, but this time draw smaller arcs above and below AB instead of long arcs that go through AB.

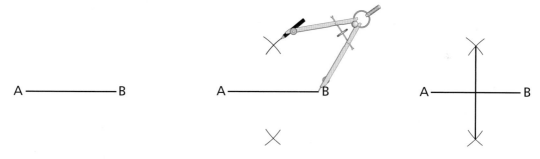

2. Repeat question **1** with

a AB as a vertical line

b AB at an angle of 30° to the horizontal.

3. Practise constructing perpendicular bisectors of line segments with different lengths and in different orientations. Use a ruler and a protractor to check how accurately you have drawn your perpendicular bisectors. Compare your constructions with a partner's.

4 PQ is a line segment 10 cm long. X is a point on the line 4 cm from Q

Here is how to construct a perpendicular to the line segment PQ that passes through point X

▦ Place the point of a pair of compasses on X and draw arcs on each side of X
  Label these points A and B

P ———————————— Q
            A   X   B

▦ Now draw the perpendicular bisector of AB

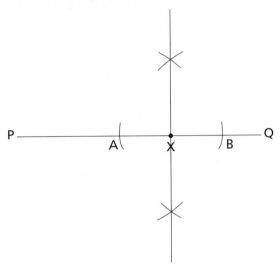

▦ This will pass through X and be perpendicular to PQ

a Make a copy of PQ and work through the steps described to draw a line perpendicular to PQ that passes through X

b Draw another copy of PQ and mark point Y, 3 cm from P. Construct the line perpendicular to PQ that passes through Y

5 Copy these line segments and draw perpendiculars through point X for each.

a   X   3 cm

6 cm

b   X   2 cm

7 cm

c   5 cm

X

9 cm

6 Benji

I can't draw a perpendicular through X because it's too close to the end of the line segment for me to draw two arcs on the line.

8 cm   1 cm

 Flo

You can extend the line segment.

X

8 cm   1 cm

a How does Flo's suggestion help?

b Copy the line segment and construct the perpendicular through point X, which is 1 cm from the end of the line.

c Construct this right-angled triangle on plain paper without using a protractor.

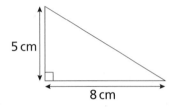

5 cm

8 cm

What do you think?

1 Look again at the method you used in Practice 5.3B question **4**. How can you adapt this to construct a perpendicular from a point such as P to a line?

 P

A ——————————————— B

Have a go on your own or with a partner before looking at the answer in question **3** of this section.

2 a Draw a right-angled triangle XYZ as shown. It does not matter what the lengths of the sides are.

Draw the perpendicular bisectors of XY and XZ and show that they meet at the midpoint of YZ. Try this with some other right-angled triangles.

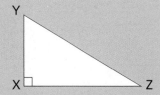

b Investigate where the perpendicular bisectors of the sides meet for

i acute-angled triangles   **ii** obtuse-angled triangles.

233

3 Here is the answer to question **1**

×P

A —————————————————————— B

Place the point of a pair of compasses on P and draw arcs that cut AB. Label the points of intersection X and Y

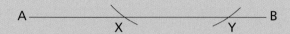

Now draw the perpendicular bisector of XY

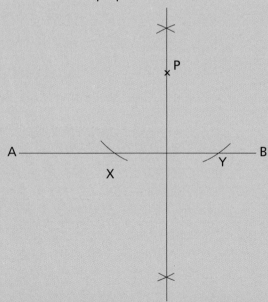

This will go through P and be perpendicular to AB

Copy these diagrams and draw perpendiculars from the points marked to the line segments. You may need to extend some of the segments.

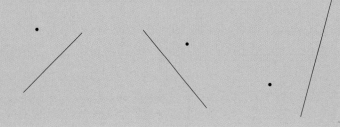

Consolidate – do you need more?

1 Copy the diagram and construct the perpendicular bisectors of AB and BC

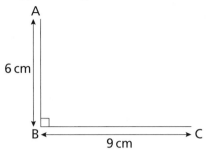

2 Draw line segments at three different angles and construct their perpendicular bisectors.

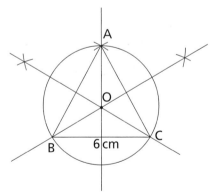

3 **a** Construct an equilateral triangle with sides 6 cm

b Show that the perpendicular bisectors of all three sides meet at a point in the triangle.

c Use this point to draw a circle that touches all three vertices of the triangle, as shown.

This is called the circumcircle of the triangle.

4 **a** Copy the diagram and construct a perpendicular to the line segment that passes through point X

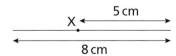

b Repeat part **a** with X positioned 5 cm from the left-hand end of the line segment.

5 **a** Copy the diagram and construct a perpendicular to the line segment from point P

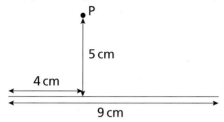

b Repeat part **a** with P positioned 2 cm above the line segment instead of 5 cm above it.

c Repeat part **a** with P positioned 5 cm below the line segment instead of 5 cm above it.

Stretch – can you deepen your learning?

1 Investigate how to draw perpendicular bisectors, perpendiculars through a point and perpendiculars from a point using dynamic geometry software. Include looking at perpendicular bisectors of the three sides of a triangle.

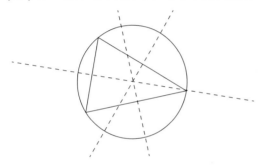

2 A point P moves so that it is always equidistant from two parallel lines l_1 and l_2

a Is the locus of P parallel to, or perpendicular to, l_1 and l_2?

b Describe the locus of P as fully as you can. You may use a diagram to help you.

3 A is the point with coordinates (−2, −5) and B is the point with coordinates (2, 3)

a Draw x- and y-axes numbered from −5 to 5. Plot points A and B and join them to form the line segment AB

b Work out the equation of the line that passes through A and B

c Find the coordinates of the midpoint of AB

d Draw the perpendicular bisector of AB and work out its equation.

e Investigate the relationship between gradients of straight lines and lines that are perpendicular to them.

Reflect

1 Explain how to construct the perpendicular bisector of a line segment.

2 What's the same and what's different about constructing a perpendicular from a point to a line and constructing a perpendicular through a point on a line? Use diagrams to help you explain.

Small steps

■ Locus of distance from two lines

■ Construct an angle bisector

Key words

Locus (plural: **loci**) – a set of points that describe a property

Equidistant – at the same distance from

Bisector – a line that divides something into two equal parts

Are you ready?

1 a Draw a horizontal line and construct its perpendicular bisector.

b Draw a vertical line and construct its perpendicular bisector.

2 Draw angles of

a 50°   **b** 140°   **c** 220°

3 Draw a square with sides of length 5 cm. Shade the locus of points inside the square that are more than 2 cm from any vertex of the square.

Models and representations

You can use coins or counters to show points that are the same distance from two points or two lines.

Coins

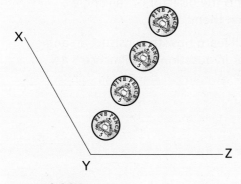

Counters

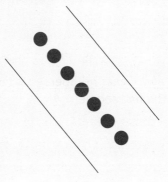

Paper folding

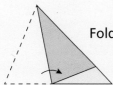

Folding an angle **bisector**

You can also find a line that is **equidistant** from two other lines by folding paper, by folding one line exactly onto the other.

In this chapter, you will explore the **locus** of points that are equidistant from two lines.

Example 1

Draw diagrams to show

a the locus of points that are equidistant from a pair of parallel lines

b the locus of points that are equidistant from a pair of line segments that meet at a point.

a

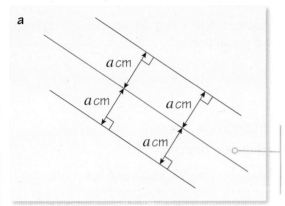

The red line is the locus of points that are equidistant from the pair of parallel lines.

It must be halfway between them and parallel to the two lines.

b

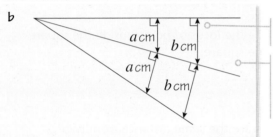

The shortest distance between two lines is the length of a perpendicular line between them.

The red line is the locus of points that are equidistant from the two line segments that meet at a point.

It bisects the angle where the lines meets.

> The locus of points that are equidistant from from two line segments that meet at a point is the bisector of the angle between them.
>
> You will learn how to construct an angle bisector in question **2** of Practice 5.4A

Practice 5.4A

 1 Place counters on your desk so that they are the same distance from the left-hand edge as from the bottom edge.

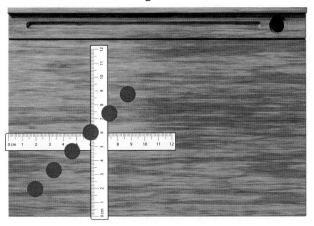

Use a protractor to check that the line of counters is at 45° to both edges of the desk.

2 Follow these instructions to construct an angle bisector.

Start by drawing an angle of 70°. Make the arms of the angle about 5 cm long.

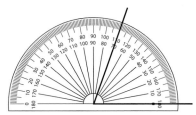

Put the point of a pair of compasses at the vertex of the angle. Draw an arc with radius 4 cm that cuts both arms of the angle. Label the points of intersection A and B, as shown.

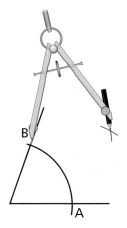

Next, draw arcs of the same radius from A and B, making sure that you draw them long enough so that they meet.

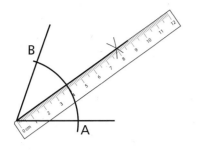

Draw a straight line from the vertex of the angle, through the point of intersection of the two arcs.

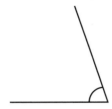

Use a protractor to verify that both angles formed measure 35°

You have constructed the bisector of the 70° angle.

> The bisector of an angle divides it into two equal parts.

3 Repeat the steps for question **2** with the angle facing in the opposite direction.

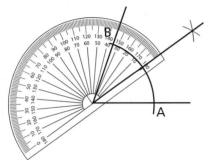

4 Copy these acute angles and construct their bisectors.

a

40°

b

48°

c

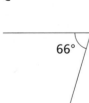

66°

d

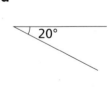

20°

5 Copy these obtuse angles and construct their bisectors.

a

120°

b

150°

c

112°

d

140°

6 a Construct an equilateral triangle with sides of length 8 cm

Look back at Chapter 5.1 for a reminder of how to construct triangles.

b Show that the angle bisectors of all three angles meet at the same point in the triangle.

c Use this point to draw a circle that touches all three edges of the triangle as shown.

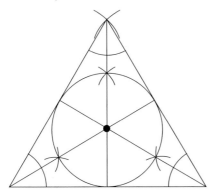

This is called the incircle of the triangle.

d Draw some other large acute-angled triangles and construct their incircles.

Is it necessary to bisect all three angles? Why or why not?

7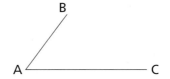

Is it possible to construct the bisector of a reflex angle?

Draw some reflex angles and see whether you can bisect them.

8 a Copy the diagram and use the method shown in question **2** to construct the locus of points that are equidistant from lines AB and AC

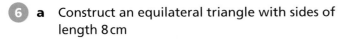

b Explain why the construction of the locus of points that are equidistant from points B and C would be different from the construction you have just drawn. Add this construction to your diagram.

What do you think?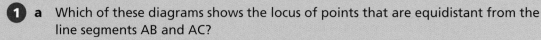

1 **a** Which of these diagrams shows the locus of points that are equidistant from the line segments AB and AC?

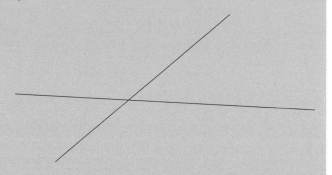

b Find the locus of points that are equidistant from a pair of lines that cross at a point.

2 Explain what is wrong with this attempt to draw an angle bisector.

3 Here is the plan view of a rectangular room. Make a scale drawing of the room using a scale of 1 cm to represent 2 m for each of parts **a** and **b**.

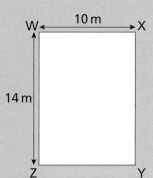

a Shade the region in the room that is closer to wall WZ than to wall WX

b Shade the region in the room that is closer to WX than to WY and closer to corner X than to corner W

c How would your regions change if WXYZ was a parallelogram? Would the shaded areas increase or decrease? How do the angles of the parallelogram change your answers?

Consolidate – do you need more?

1 Copy each diagram and construct the bisector of each angle.

a

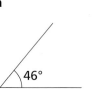
46°

b

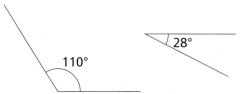
160°

c

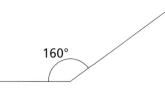
76°

d

144°

e

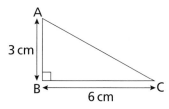
110°

f

28°

g

38°

h

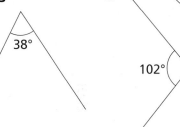
102°

2 **a** Draw a rectangle and construct the bisector of one of the angles. Does the bisector also bisect the opposite angle?

b Repeat part **a** for a

i parallelogram **ii** rhombus **iii** kite.

3 **a** Draw an accurate copy of triangle ABC

A
3 cm
B
6 cm
C

Find the point where the perpendicular bisector of AC meets the bisector of ∠ACB

b Draw an accurate copy of triangle XYZ. Find the point where the perpendicular bisector of YZ meets the bisector of ∠XYZ

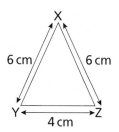
X
6 cm   6 cm
Y
4 cm
Z

c Draw an accurate copy of triangle PQR. Find the point where the perpendicular bisector of PQ meets the bisector of ∠PQR

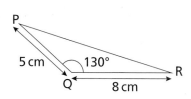
P
5 cm   130°
Q   8 cm   R

Stretch – can you deepen your learning?

1 **a** Construct an equilateral triangle with sides of length 4 cm

  b Construct an angle of 60° using the method for constructing an equilateral triangle but only drawing two sides of the triangle, as shown. Practise the construction using several different initial line segments in different orientations.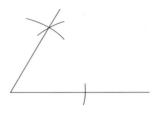

  c Use the construction in part **b** and your knowledge of angle bisectors to construct angles of 30° and 15°

  d Use your knowledge of constructing perpendiculars to construct an angle of 90° Bisect your angle to create an angle of 45°

  e Which of these angles can you construct without using a protractor? Compare your methods with a partner's and check how accurate your constructions are by measuring your angles with a protractor.

| 120° | 150° | 15° | 22.5° | 67.5° | 240° | 75° | 330° |

2 The angle bisector theorem states: "An angle bisector of a triangle divides the opposite side into two segments that are proportional to the other two sides of the triangle."

Investigate this theorem.

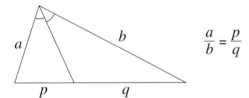

$$\frac{a}{b} = \frac{p}{q}$$

Reflect

1 Explain why the locus of points that are equidistant from two lines bisects the angle between the two lines.

2 Compare constructing the perpendicular bisector of a line to constructing the bisector of an angle.

5.5 Congruence

Small steps

- ■ Identify congruent figures
- ■ Explore congruent triangles
- ■ Identify congruent triangles

Key words

Congruent – exactly the same size and shape, but possibly in a different orientation

Orientation – the position of an object based on the direction it is facing

Hypotenuse – the side opposite the right angle in a right-angled triangle

Are you ready?

1 Construct each triangle accurately.

a

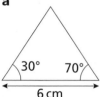

30°   70°
6 cm

b

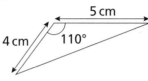

5 cm
4 cm   110°

2 Describe each triangle as fully as you can.

a   **b**   **c**   **d**   **e**

3 What is the sum of the angles in a

    a triangle   **b** quadrilateral?

Models and representations

Geoboards

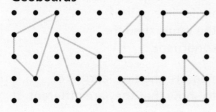

You can use geoboards (or geoboard apps) to make and compare shapes, or you can represent them on squared paper or dotty paper.

Pattern blocks

Squared paper or dotty paper

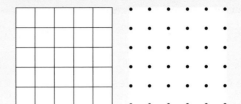

You can also use blocks to make and compare combinations of shapes.

Two shapes are **congruent** if they are exactly the same shape and size.

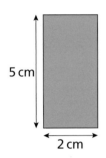

5 cm

2 cm

5 cm

2 cm

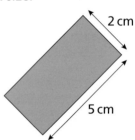

2 cm

5 cm

These three rectangles are congruent even though they are in different **orientations** (facing in different directions).

Example 1

a Are shapes F and H congruent? How do you know?

b Find all the pairs of congruent shapes shown in the diagram.

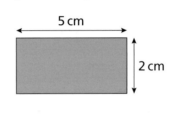

a No, because one of the angles in shape F is a right angle, but none of the angles in H are.

If the lengths of all the sides or the sizes of all the angles in a pair of shapes are not the same, then the shapes are not congruent.

You could also measure the side lengths of shapes F and H to show that they are different.

b A and D
C and G
B and E

You can trace the shapes and put the tracing of one on top of the other to show that they are congruent.

You would need to turn the tracing paper over for shapes C and G.

Practice 5.5A

1 Shapes X and Y are congruent. How many more shapes can you draw that are congruent to X and Y but are in different orientations?

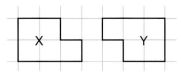

2 Draw two triangles congruent to each of these.

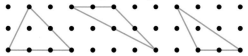

3 Which shape is not congruent to the other three?

A   B   C   D

4 Explain how you know that the quadrilaterals shown are congruent.

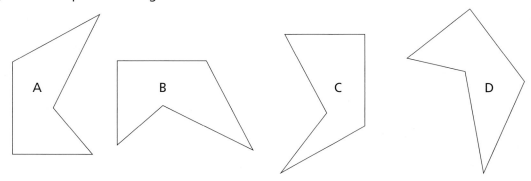

5 On dotty paper, draw at least three shapes that are congruent to each of A, B and C

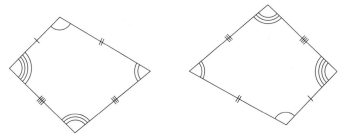

A   B   C

6

All rectangles with side lengths 6 cm by 3 cm are congruent.

Marta

All parallelograms with side lengths 6 cm by 3 cm are congruent.

Seb

Investigate Marta's and Seb's conjectures.

7. Here are two of the possible triangles that can be made on a 3 by 3 pinboard.

There are six more possibilities. Draw them. Make sure that you do not draw any triangles that are congruent.

8. These two quadrilaterals are congruent. Which unlabelled sides and angles can you work out?

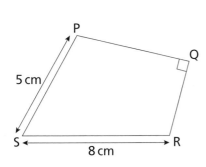

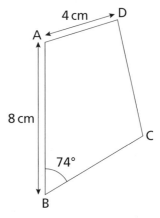

What do you think?

1. Here are some pairs of triangles. For each pair decide whether

   - they are definitely congruent
   - they are definitely not congruent
   - you need more information to decide whether they are congruent or not.

   a   **b**   **c**   **d**

2. Construct the triangles described below. In each case, compare your triangle with a partner's. Are your triangles congruent? Will they always be congruent?

   a ABC with AB = 7 cm, BC = 5 cm, AC = 4 cm

   b DEF with ∠DEF = 60°, ∠DFE = 75°, ∠EDF = 45°

   c JKL with JK = 8 cm, ∠LJK = 40°, ∠JKL = 35°

   d PQR with PQ = 6 cm, QR = 4 cm and ∠PQR = 55°

   e XYZ with XY = 7 cm, XZ = 6 cm and ∠XZY = 45°

Question **2** of the What do you think? exercise illustrates some of the following sets of conditions that you can use to determine whether two triangles are congruent.

Side (SSS) If three pairs of corresponding sides of the two triangles are equal, then the two triangles are congruent.

Side Angle Side (SAS) If two sides and the angle between them in one triangle are equal to the corresponding sides and angle of another triangle, then the two triangles are congruent.

Angle Angle Side (AAS) If two angles and a side of one triangle are equal to the corresponding angles and side of another triangle, then the triangles are congruent.

There is another condition for right-angled triangles that you will explore in the next What do you think? exercise.

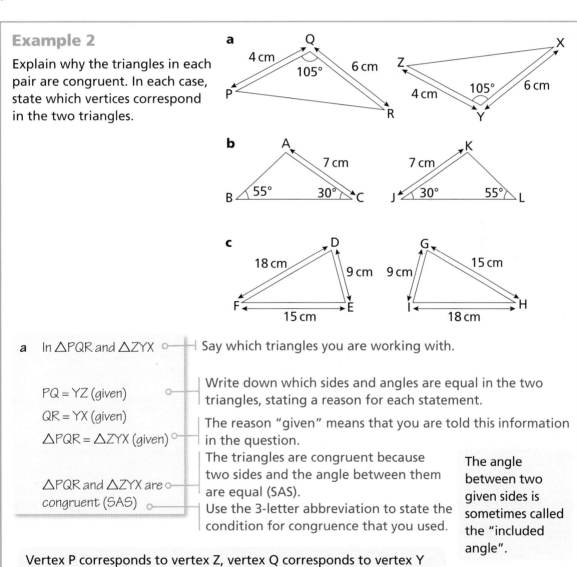

Example 2

Explain why the triangles in each pair are congruent. In each case, state which vertices correspond in the two triangles.

a In △PQR and △ZYX ○————┤ Say which triangles you are working with.

PQ = YZ *(given)* ○————┤ Write down which sides and angles are equal in the two triangles, stating a reason for each statement.

QR = YX *(given)*

△PQR = △ZYX *(given)* ○— The reason "given" means that you are told this information in the question.

The triangles are congruent because two sides and the angle between them are equal (SAS).

△PQR and △ZYX are ○— congruent (SAS) ○————┤ Use the 3-letter abbreviation to state the condition for congruence that you used.

The angle between two given sides is sometimes called the "included angle".

Vertex P corresponds to vertex Z, vertex Q corresponds to vertex Y and vertex R corresponds to vertex X: the triangles are named using the corresponding order.

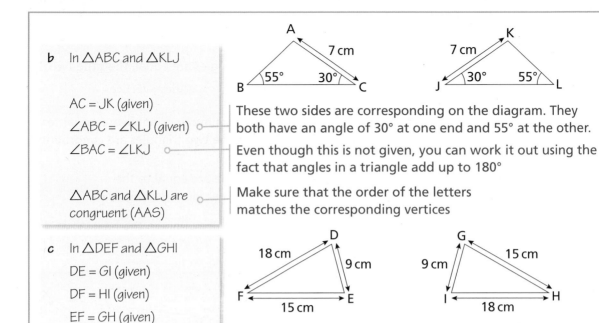

b In △ABC and △KLJ

AC = JK (given)

∠ABC = ∠KLJ (given) ○— These two sides are corresponding on the diagram. They both have an angle of 30° at one end and 55° at the other.

∠BAC = ∠LKJ ○— Even though this is not given, you can work it out using the fact that angles in a triangle add up to 180°

△ABC and △KLJ are ○— Make sure that the order of the letters matches the corresponding vertices
congruent (AAS)

c In △DEF and △GHI

DE = GI (given)

DF = HI (given)

EF = GH (given)

△DEF and △GIH are ○— The corresponding sides of the triangles are equal in length.
congruent (SSS)

Practice 5.5B

1 Sketch each pair of triangles.

For each pair, copy and complete this statement:

"△ _____ and △ _____ are congruent (SSS)."

Make sure that the order of the letters matches the corresponding vertices.

a

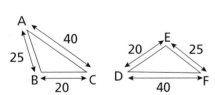

b

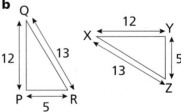

c

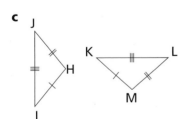

2 Sketch each pair of triangles.

For each pair, copy and complete this statement:

"△ _____ and △ _____ are congruent (SAS)."

Make sure that the order of the letters matches the corresponding vertices.

a

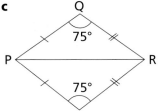

b

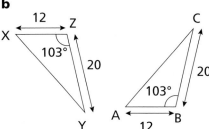

c

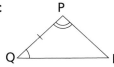

3 Sketch each pair of triangles.

For each pair, copy and complete this statement:

"△ _____ and △ _____ are congruent (AAS)."

Make sure that the order of the letters matches the corresponding vertices.

a

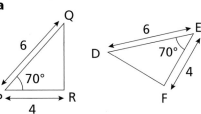

b

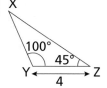

c

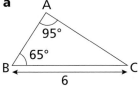

4 Prove that the triangles in each pair are congruent, stating the condition for congruence that you use.

Make sure that the order of the letters matches the corresponding vertices.

a

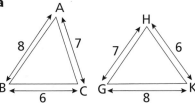

b

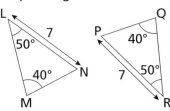

c

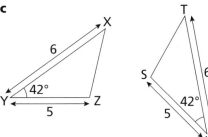

d

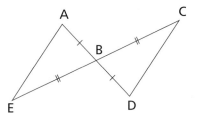

5 Line segments AD and CE meet at B, as shown.

a Explain why angles ABE and CBD are equal.

b Determine whether triangles ABE and BCD are congruent, giving a reason for each stage of your answer.

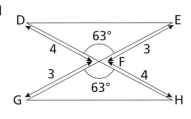

6 a PR is a diagonal of kite PQRS

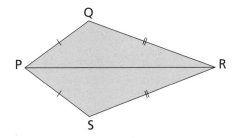

Copy and complete the proof that triangles PQR and PSR are congruent.

> In ΔPQR and ΔPSR
>
> PR = PR (common side)
>
> PQ = PS (given)

b ABCD is a rectangle. Prove that the diagonal AC splits ABCD into two congruent triangles.

c Do the diagonals of a rhombus split it into two congruent triangles? Explain how you know.

7 Explain how the condition AAS can be used to show that these two triangles are congruent.

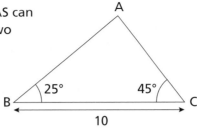

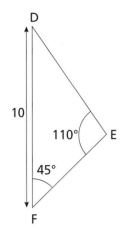

8 Explain why the diagram shows that "two equal sides and one equal angle" is not enough to show that two triangles are congruent.

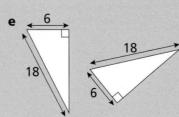

What do you think?

1 The hypotenuse of a right-angled triangle is the side opposite the right angle.

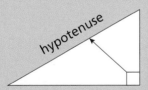

Right angle Hypotenuse Side (RHS) If the **hypotenuse** and one other side of a right-angled triangle are equal to the hypotenuse and one other side of another right-angled triangle, then the two triangles are congruent.

Draw some right-angled triangles to explore this set of conditions for showing that a pair of triangles are congruent.

2 State which condition you can use to show that the triangles in each pair are congruent.

a

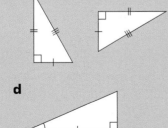

b

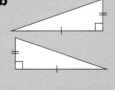

c

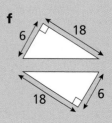

d

e

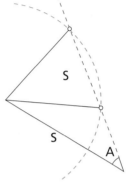

f

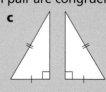

Consolidate – do you need more?

1 Which of these diagrams show a pair of congruent shapes? How can you tell?

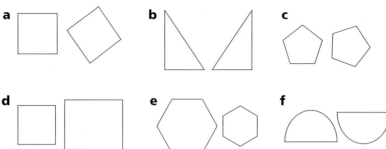

2 On dotty paper, draw at least three shapes that are congruent to each of A, B and C

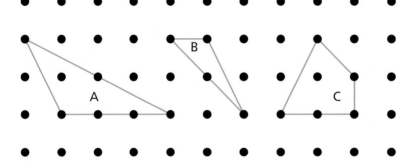

3 Shape A and shape B are congruent. State whether each of these statements is true or false.

   a Shape B has the same area as shape A

   b Shape A has the same perimeter as shape B

   c The smallest angle in shape A is equal to the smallest angle in shape B

   d The ratio of the longest side in shape A to the longest side in shape B is 1 : 1

4 The five rectangles in this shape are congruent.

Work out

   a the width of a single rectangle

   b the area of the shape.

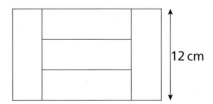

12 cm

5 Huda draws a line segment AB that is 5 cm long. She makes a triangle by drawing a 40° angle at point A and a 60° angle at point B

Darius draws a line segment XY that is 5 cm long. He makes a triangle by drawing a 40° angle at point X and a 60° angle at point Y

Will Huda's and Darius's triangles be congruent? How do you know?

6 Name a pair of congruent triangles in each diagram.

a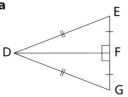

b

c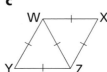

d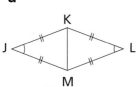

Stretch – can you deepen your learning?

1 **a** A triomino is produced by joining three congruent squares edge to edge. Faith says that there are two possible triominoes, but Abdullah says that there are five. Who is correct?

b A tetromino is produced by joining four congruent squares edge to edge. Draw all the possible tetronimoes.

2 **a** Prove that a diagonal of a square splits the shape into two congruent triangles. Prove this in four different ways that each use one of the conditions for congruence.

b How many sets of congruent triangles are produced when the diagonals of a parallelogram are drawn?

3 Prove that the perpendicular bisector of the base of an isosceles triangles splits the triangle into two congruent triangles. In how many ways can you do this by using different conditions for congruence?

4 In how many ways can you prove that triangles ABE and BCD are congruent?

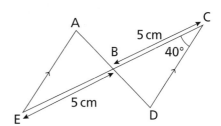

Reflect

1 Explain the meaning of the word congruent.

2 Draw sketches to show the four sets of conditions that you can use to show that two triangles are congruent.

I have become fluent in...

- constructing shapes
- recognising congruent shapes
- knowing and using the criteria for congruence of triangles
- constructing loci
- constructing perpendiculars to and from points
- constructing bisectors of lines and angles.

I have developed my reasoning skills by...

- reasoning deductively in geometry using geometrical constructions
- interpreting complex diagrams
- explaining why pairs of triangles are or are not congruent
- using properties of shapes to justify answers about congruence.

I have been problem-solving through...

- breaking down problems into smaller parts
- looking at non-routine problems
- using diagrams to represent situations
- applying construction skills in context.

Check my understanding

1 **a** Construct an accurate scale drawing of the trapezium. Use a scale of 1 cm to represent 5 m

   b Measure the obtuse angle in your scale drawing.

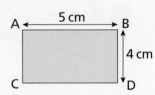

2 Mark a point P and draw the locus of points that are 3.5 cm from P

3 **a** Make a copy of rectangle ABCD. Shade the locus of points inside ABCD that are more than 3 cm from A and closer to AB than to AC

   b Make a second copy of rectangle ABCD. Draw the locus of points outside ABCD that are exactly 2 cm from ABCD

   c Make a third copy of rectangle ABCD. Construct the perpendicular bisector of BC

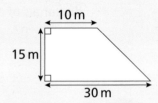

4 Draw an angle of 74° and construct the bisector of the angle.

5 On squared paper, draw three shapes that are congruent to shape A but in different orientations.

6 Are these triangles congruent? Explain your answer.

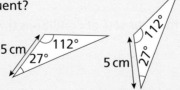

6 Numbers

In this block, I will learn...

about types of number

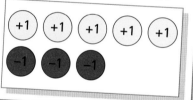

rational numbers

irrational numbers

integers

real numbers

how to round to
1 significant figure to estimate
answers to calculations

$$\frac{4.27 \times 18.01}{0.209} \approx \frac{4 \times 20}{0.2}$$

how to solve problems
involving directed numbers

+1 +1 +1 +1 +1

−1 −1 −1

how to use Venn diagrams to find the
LCM and HCF of two numbers

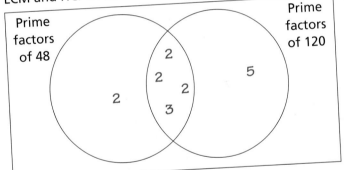

Prime factors of 48

Prime factors of 120

2

2   2

2

3

5

how to solve problems in a
variety of contexts

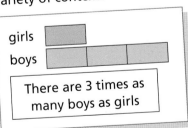

girls

boys

There are 3 times as
many boys as girls

how to solve problems
involving fractions

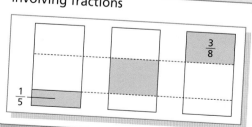

$\frac{3}{8}$

$\frac{1}{5}$

how to write numbers in standard form

Standard form is written in the form
$A \times 10^n$, where $1 \leqslant A < 10$ and n is an integer.

Small steps

- Integers, real and rational numbers
- HCF and LCM **R**
- Understand and use surds **H**

Key words

Integer – a whole number

Real number – all positive and negative numbers including decimals and fractions

Rational number – a number that can be written in the form $\frac{a}{b}$ where a and b are integers

Irrational number – a number that cannot be written in the form $\frac{a}{b}$ where a and b are integers

Factor – a positive integer that divides exactly into another positive integer

Multiple – the result of multiplying a number by a positive integer

Surd – a root that cannot be written as an integer

Are you ready?

1 List the first five multiples of each number.

   a 20   **b** 16   **c** 32   **d** 48   **e** 17

2 List the factors of each number.

   a 20   **b** 16   **c** 32   **d** 48   **e** 17

3 List all the numbers on the grid that are multiplies of both 2 and 3

Venn diagrams

21	22	23	24	25	26	27	28	29	30
31	32	33	34	35	36	37	38	39	40
41	42	43	44	45	46	47	48	49	50

Models and representations

Venn diagrams

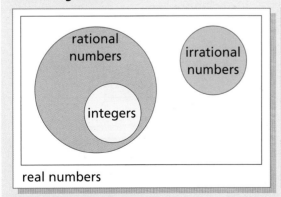

This Venn diagram shows where each type of number sits within the number system.

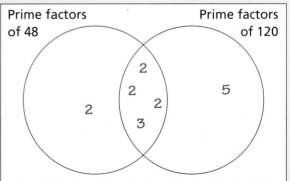

This Venn diagram can be used to find the HCF and LCM of two numbers.

Integers are whole numbers, like 4 and –8

Rational numbers can be written as fractions, for example $\frac{3}{4}$. They are called rational numbers as you can compare the numerator to the denominator like you compare two numbers in a ratio. Mixed numbers are rational because you can rewrite them as improper fractions, for example $3\frac{1}{2} = \frac{7}{2}$. Integers are rational too as you can write, for example, 8 as $\frac{8}{1}$ or $\frac{16}{2}$

Numbers that cannot be written as fractions are called **irrational**. $\sqrt{2}$ and π are examples.

The set of **real numbers** includes all positive and negative numbers, rational and irrational.

Example 1

Show that 1.739 is a rational number.

1.739 is 1 whole and 739 thousandths.

$1.739 = 1 + \frac{739}{1000} = \frac{1000}{1000} + \frac{739}{1000} = \frac{1739}{1000}$

1739 and 1000 are integers, so 1.739 is rational.

A rational number can be written in the form $\frac{a}{b}$ where a and b are integers.

Practice 6.1A

1 Which of the following numbers are integers?

10   9.6 $\frac{3}{4}$ –3   117

2 List

  a four integers between 5 and 15

  b four non-integer real numbers between 2 and 10

  c four real numbers between 3 and 4

3 Which of these numbers are rational?

$\frac{7}{2}$ $\sqrt{3}$ 1.5   11

4 Look at these numbers.

$\sqrt{2}$ $\sqrt{3}$ $\sqrt{4}$ $\sqrt{5}$ $3\sqrt{7}$ $2\sqrt{100}$ $-\sqrt{3}$ $-\sqrt{25}$

Which numbers are

  a irrational   **b** real?

Discuss your answers with a partner.

5 **a** Jakub says that 0.7 is a rational number. Explain why Jakub is correct.

  b Flo says that 0.7777… is irrational as you cannot write it as a fraction.

    Is Flo correct?

6 Are either of these numbers irrational? Explain your answer.

1.76 503 $0.\dot{6}$

What do you think? 💭

1 **a** Emily says that $\frac{10}{2}$ is not an integer.

Seb says that it is.

Who is correct? Explain your answer.

b Faith says that π is an irrational number as you cannot write it as a fraction.

Darius says, "π is rational as you can approximate it using $\frac{22}{7}$"

Who is correct? Explain your answer.

c Beca says that $\sqrt{25}$ is an integer.

Zach says that it is not an integer.

Who is correct? Explain your answer.

2 Investigate this conjecture.

$2\sqrt{9}$ is irrational as it contains a square root.

Example 2

a Write these numbers as products of their prime factors.

i 24   **ii** 100

b Represent the prime factors of 24 and 100 in a Venn diagram.

c Use the Venn diagram to find

i the highest common factor (HCF) of 24 and 100

ii the lowest common multiple (LCM) of 24 and 100

a **i**

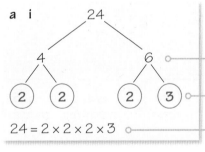

$24 = 2 \times 2 \times 2 \times 3$

You could start with any pair of **factors** of 24

Branches end when the end numbers are prime.

You will get the same answer whichever pair of factors you start with.

ii

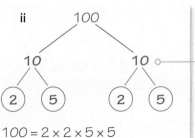

$100 = 2 \times 2 \times 5 \times 5$

Alternatively, you could have started with 2 and 50, 4 and 25 or 5 and 20

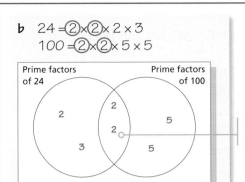

b $24 = 2 \times 2 \times 2 \times 3$
$100 = 2 \times 2 \times 5 \times 5$

Both 24 and 100 have two 2s as prime factors so these go in the intersection.

c i $2 \times 2 = 4$ so the HCF of 24 and 100 is 4

Multiply the common factors to find the HCF.

ii $2 \times 3 \times 2 \times 2 \times 5 \times 5 = 600$ so the LCM of 24 and 100 is 600

To find the LCM, you find the product of all the numbers in the Venn diagram.

Practice 6.1B

1 a List all the factors of 20   b List all the factors of 50
c List the common factors of 20 and 50
d Find the highest common factor of 20 and 50

2 a List the first 10 multiplies of 4   b List the first 10 multiples of 6
c Identify the common multiples of 4 and 6 from your lists.
d State the lowest common multiple of 4 and 6

3 a Write these numbers as products of their prime factors.
i 45   ii 120
b Sort the prime factors of 45 and 120 into the Venn diagram.

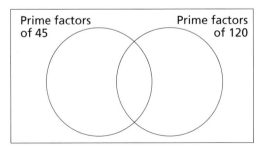

c Use your Venn diagram to find
i the highest common factor of 45 and 120
ii the lowest common multiple of 45 and 120

4 Find the highest common factor of each pair of numbers.

    a   60 and 150   **b**   42 and 840   **c**   140 and 80   **d**   240 and 144

5 Find the lowest common multiple of each pair of numbers.

    a   40 and 50   **b**   24 and 36   **c**   75 and 100   **d**   64 and 132

6 The Venn diagram shows the prime factors of two numbers, A and B

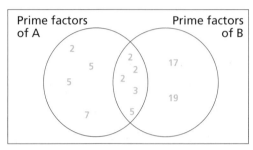

    a   Identify the highest common factor of A and B

    b   Find the lowest common multiple of A and B

    c   Which number is greater, A or B? Explain your answer.

7 $P = 2^3 \times 3 \times 5^2 \times 11$ $Q = 2 \times 3^2 \times 5^3 \times 7$

  Find

    a   the highest common factor of P and Q

    b   the lowest common multiple of P and Q

8 The lowest common multiple of 12 and M is 36

    a   Explain why

        i   M cannot be 6   **ii**   M could be 18

    b   Is there another possible value for M?

9 Trains to Bradford leave Leeds station every 18 minutes.

  Trains to Harrogate leave Leeds station every 24 minutes.

  A train to Bradford and a train to Harrogate both leave Leeds station at 10 a.m.

  When will trains for Bradford and Harrogate next leave Leeds station at the same time?

  Show your working.

What do you think?

1

1 is a common factor of any two numbers.

Beca

1 can never be the highest common factor of two numbers.

Abdullah

Who do you agree with? Explain your answer.

2 Here are four pairs of numbers.

7 and 11   5 and 11   3 and 11   11 and 15

a What do you notice about the highest common factor of each pair of numbers? Why does this happen?

b The highest common factor of 11 and X is 1. What does this tell you about X?

c The highest common factor of 11 and Y is not 1. What does this tell you about Y?

3 The highest common factor of two numbers is 18

What could the numbers be?

How many possible pairs of numbers can you find? Explain your method.

Example 3

Write $\sqrt{175}$ in the form $a\sqrt{b}$ where a and b are integers.

The factor pairs of 175 are | List the factor pairs of 175
| Find any factors that are square numbers.

1×175

5×35

7×25 | 25 is the highest square factor.

$\sqrt{175} = \sqrt{25} \times \sqrt{7}$
$\sqrt{25} \times \sqrt{7} = 5 \times \sqrt{7}$ so
$\sqrt{175} = 5\sqrt{7}$

25 is a square number so its square root is an integer: $\sqrt{25} = 5$
$\sqrt{7}$ is a **surd** because it is a root that cannot be written as an integer.

Practice 6.1C

1 Sort the cards into two headings "surd" and "not a surd".

| $\sqrt{25}$ | $\sqrt{24}$ | $\sqrt{2}$ | $\sqrt{1}$ | $\sqrt{169}$ | $\sqrt{8}$ | $\sqrt[3]{8}$ | $\sqrt{130}$ |

Explain your reasoning.

2 **a** Write down the value of

   i $\sqrt{4}$ **ii** $\sqrt{100}$

  b Hence or otherwise, show that

   i $\sqrt{4} \times \sqrt{100} = \sqrt{400}$ **ii** $\sqrt{100} \div \sqrt{4} = \sqrt{25}$

3 **a** Write down the value of

   i $\sqrt{9}$ **ii** $\sqrt{16}$

  b Hence or otherwise, show that $\sqrt{9} + \sqrt{16} \neq \sqrt{25}$ and $\sqrt{16} - \sqrt{9} \neq \sqrt{7}$

4 Use the fact that $\sqrt{a} \times \sqrt{b} = \sqrt{ab}$ to write each expression as a single surd.

  a $\sqrt{5} \times \sqrt{3}$ **b** $\sqrt{7} \times \sqrt{11}$ **c** $\sqrt{51} \times \sqrt{2}$ **d** $\sqrt{15} \times \sqrt{21}$

5 Use the fact that $\sqrt{a} \div \sqrt{b} = \sqrt{\dfrac{a}{b}}$ to write each expression as a single surd.

  a $\sqrt{21} \div \sqrt{3}$ **b** $\sqrt{26} \div \sqrt{2}$ **c** $\sqrt{51} \div \sqrt{17}$ **d** $\sqrt{3} \div \sqrt{21}$

6 Write each surd in the form $\sqrt{a} \times \sqrt{b}$ where a is a square number greater than 1 and b is an integer.

  a $\sqrt{8}$ **b** $\sqrt{50}$ **c** $\sqrt{27}$ **d** $\sqrt{1100}$

  e $\sqrt{490}$ **f** $\sqrt{48}$

Is there more than one answer for each number?

7 Write each surd in the form $c\sqrt{b}$ where b and c are integers.

You can use your answers to question **6** to help you.

  a $\sqrt{8}$ **b** $\sqrt{50}$ **c** $\sqrt{27}$ **d** $\sqrt{1100}$

  e $\sqrt{490}$ **f** $\sqrt{48}$

Is there more than one answer for each number?

What do you think? 💭

1 Samira has simplified $\sqrt{200}$

Here is her working.

$$\sqrt{200} = \sqrt{4} \times \sqrt{50} = 2 \times \sqrt{50} = 2\sqrt{50}$$

a Identify one thing that Samira has done well.

b Explain how Samira can improve her answer.

2 Show that $\dfrac{\sqrt{18y}}{\sqrt{2y}}$ is an integer.

3 **a** Show that $9\sqrt{3} = \sqrt{243}$

b Write $6\sqrt{5}$ as a single surd.

Consolidate – do you need more?

1 List all the factors of these numbers.

a 12   **b** 30   **c** 100

2 Write the common factors of each pair of numbers.

a 12 and 30   **b** 12 and 100   **c** 30 and 100

3 **a** Write these numbers as products of their prime factors.

    i 80   **ii** 56

b Find

    i the lowest common multiple of 80 and 56

    ii the highest common factor of 80 and 56

Stretch – can you deepen your learning?

1 Decide whether each statement is always true, sometimes true or never true.
Explain your answers.

a A number can be both rational and irrational.

b Recurring decimals are real numbers.

c Recurring decimals are rational numbers.

d A number with a square root is irrational.

e A real number is an integer.

f An integer is a real number.

2 Ed has drawn a Venn diagram to find the highest common factor and lowest common multiple of two integers, X and Y

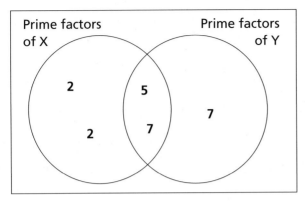

a Calculate the highest common factor and lowest common multiple of X and Y

b Ed says, "20 is a factor of X but not of Y"

Explain why Ed is correct.

c Work out the value of X and the value of Y

3 Abdullah has formed a conjecture.

> Any number of the form $\sqrt{x}$ can be written in the form $a\sqrt{b}$ where a is an integer greater than 1

Give an example that

a supports Abdullah's conjecture

b disproves Abdullah's conjecture.

4 Work out each value of x

a $\sqrt{150} = 5\sqrt{x}$ **b** $\sqrt{300} = x\sqrt{3}$ **c** $\sqrt{x} = 7\sqrt{13}$

Reflect

1 Give three examples of each type of number.

a integer   **b** real number   **c** rational number

2 Explain what is meant by

a the highest common factor   **b** the lowest common multiple.

6.2 Estimation

Small steps

- ■ Solve problems with integers
- ■ Solve problems with decimals

Key words

Integer – a whole number

Significant figure – the most important digits in a number that give you an idea of its size

Estimate – an approximate answer or to give an approximate answer

Error interval – the range of values a number could have taken before being rounded

Are you ready?

1 Round each number to 1 significant figure.

    a 563   **b** 14 637   **c** 0.0736

    d 7.105   **e** 135.48

2 Work out

    a 300 + 700   **b** 5 × 600   **c** 20 000 ÷ 50

    d 30 ÷ 0.5   **e** 50 000 – 7000

3 Which of these numbers round to 1000 to 1 significant figure?

    500   1200   950   1100   1500   750

Models and representations

Number lines

These can be used to show rounding and **error intervals**.

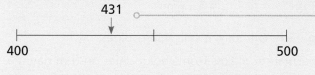

The first **significant figure** of 431 is in the hundreds column, so 431 rounds to 400 to 1 significant figure.

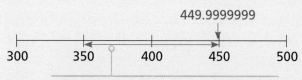

This shows all the values that round to 400 to 1 significant figure.

Example 1

Estimate the answer to $\dfrac{4.27 \times 18.01}{0.209}$

$\dfrac{4.27 \times 18.01}{0.209} \approx \dfrac{4 \times 20}{0.2}$ — Round each number to 1 significant figure.

The symbol $\approx$ means "approximately equal to".

$\dfrac{4 \times 20}{0.2} = \dfrac{80}{0.2} = \dfrac{800}{2} = 400$ — You can multiply the numerator and denominator of the fraction by 10 and not change its value.

$\dfrac{4.27 \times 18.01}{0.209} \approx 400$ — The answer to the calculation is approximately 400

Practice 6.2A

1. **a** Round each number to 1 significant figure
   i 916   **ii** 289

   b Use your answers to parts **a i** and **ii** to estimate the answer to each calculation.
   i $916 + 289$ **ii** $916 - 289$ **iii** 916×289 **iv** $916 \div 289$

2. By rounding to 1 significant figure, estimate the answer to each calculation.
   a 325×16 **b** $746 + 8351$ **c** $971 - 250$ **d** $28380 \div 299$
   e $476 - 312 + 987$ **f** $0.82 + 3.55$ **g** $72.15 - 8.9$ **h** 0.036×5.81
   i $0.971 \div 0.249$ **j** $10.4987 - 9.5$

3. Estimate the answer to each calculation.
   a $17 + 325 \times 16$ **b** $\dfrac{746 + 8351}{18}$ **c** $3.2 \times (971 - 250)$
   d $\dfrac{28380}{299} - 23.71$ **e** $\dfrac{0.82 + 3.55}{0.0128}$ **f** $\dfrac{15 + 13.687}{3.2 - 0.99}$

4. A supermarket sells cookies for 89p each. One day, it sells 438 cookies.
   Estimate how much money the supermarket makes from cookie sales on that day.

5 Estimate the area of each shape.

a

24 m

7 m

b

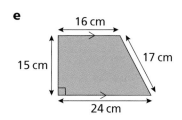

43 mm

16 mm

c

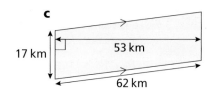

17 km   53 km

62 km

d

27 cm

12 cm

29 cm

e

16 cm

15 cm   17 cm

24 cm

f

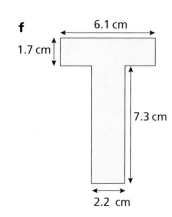

6.1 cm

1.7 cm

7.3 cm

2.2 cm

6 £1 is worth 1.83 Australian dollars (AUD).

Estimate the equivalent value of these amounts of money in pounds.

a 100 AUD   **b** 2000 AUD   **c** 50 000 AUD

7 Estimate the value of each expression when $x = 265.43$ and $y = 14.8399$

a $x + y$ **b** $x - y$ **c** xy

d $\dfrac{x}{y}$ **e** $x^2 + y^2$ **f** $(x + y)^2$

8 Estimate the mean mass of these parcels.

4.6 kg   3.2 kg   10.1 kg   4.4 kg   1.8 kg   8.3 kg

What do you think?

1 a Estimate the answer to $\dfrac{3.79 + 2.961}{0.534}$

b Is your estimate in part **a** an overestimate or an underestimate?
Explain how you know.

2 Abdullah is working out $\dfrac{300 \times 50}{2}$ to estimate the answer to a calculation.

What could the actual calculation be if Abdullah's estimate is an

a overestimate   **b** underestimate?

3 Chloe has estimated the answer to $\dfrac{8.74 + 16.391}{0.107}$

Here is her working.

$$\frac{8.74 + 16.391}{0.107} \approx \frac{9 + 20}{0.1} \approx \frac{29}{0.1} \approx \frac{30}{0.1} \approx 300$$

a Explain what Chloe has done well.

b Identify where Chloe has gone wrong.

4 Marta and Ed are estimating the value of $\sqrt{119}$

119 rounds to 100 to 1 significant figure so the answer is approximately 10

Marta

119 is closer to 121 than 100 so the answer is approximately 11

Ed

Who do you agree with? Explain your answer.

Example 2

A stick is 17 cm long, measured to the nearest centimetre.

a Which of these measurements could be the length of the stick?

16.36 cm   16.52 cm   16.601 cm   17.01 cm   17.49 cm   17.5 cm

b Write the error interval for the length, l, of the stick in cm

a   16.52 cm, 16.601 cm, 17.01 cm, 17.49 cm ○—| All of these numbers round to 17 to the nearest **integer**.
Notice that 16.36 would round to 16 and 17.5 would round to 18

b $16.5 \leq l < 17.5$ ○—| You read this as "l is greater than or equal to 16.5 and less than 17.5".

17.5 is not included as 17.5 rounds to 18 to the nearest integer.

16.5 is included in the error interval as 16.5 rounds to 17 to the nearest integer.

Practice 6.2B

1 **a** Which of these numbers round to 4 to the nearest integer?

| 4.4 | 4.403 | 3.91 | 4.51 | 3.51 | 3.3999 | 4.06 |

b Which is the correct error interval for a number x that rounds to 4 to the nearest integer?

| $3.5 < x < 4.5$ | $3.5 < x \leq 4.5$ | $3.5 \leq x < 4.5$ | $3.5 \leq x \leq 4.5$ |

2 The size of an angle a is given by the error interval $87.5° \leqslant a < 88.5°$

    a Write three possible values of a correct to 1 decimal place.

    b What is the size of angle a correct to the nearest integer?

3 **a** A number, y, is given as 12.4 to 1 decimal place. Copy and complete the error interval for y.

      $12.35 \leqslant y <$ ☐

    b Write error intervals for each of these numbers.

      i $p = 60$ to the nearest integer

      ii $q = 24.7$ to 1 decimal place

      iii $r = 5.61$ to 2 decimal places

    c Compare the error intervals for 85, correct to the nearest integer, and 85.0, correct to 1 decimal place.

4 The length, x, of a field is given as 90 m. Write the error interval for x if this measurement is correct to

    a the nearest metre   **b** 1 significant figure.

5 The length of a swimming pool is 40 m, correct to the nearest metre.

    a What is the shortest possible length of the pool?

    b Write the error interval for the length, l, of the pool.

6 Copy and complete the error interval for each statement.

$a = 5000$ to the nearest integer   ☐ $\leqslant a <$ ☐

$b = 5000$ to the nearest ten   ☐ $\leqslant b <$ ☐

$c = 5000$ to the nearest hundred   ☐ $\leqslant c <$ ☐

$d = 5000$ to the nearest thousand   ☐ $\leqslant d <$ ☐

7 A number is 10 000 to 1 significant figure. What's the same and what's different about the possible values of the number if it represents the area of a field in m² or the number of people at concert?

8 A number n when rounded is 300. Write the error interval for n if it has been rounded to

    a 1 significant figure

    b 2 significant figures

    c 3 significant figures.

What do you think? 💭

1 Points A, B and C lie on a straight line.

AB is 38 cm, correct to the nearest centimetre.

BC is 43 cm, correct to the nearest centimetre.

Work out

   a the greatest possible length of AC

   b the least possible length of AC

2 Write an error interval for each of these numbers.

   a $p = 50$ correct to the nearest 10

   b $q = 50$ correct to the nearest 5

   c $r = 50$ correct to the nearest 2

   d $s = 50$ correct to the nearest 0.5

   e $t = 50$ correct to the nearest x

Consolidate – do you need more?

1 Round these numbers to 1 significant figure.

   a 233   **b** 14   **c** 58.7

   d 0.000 766   **e** 40 319

2 One apple costs 49p.

Estimate the cost of 51 apples.

3 Work out an estimate for each calculation.

   a 324 + 712   **b** 2744 − 287   **c** 18.09 + 32.9

   d 3950 × 19.9   **e** 488 ÷ 5.2   **f** 488 ÷ 0.52

4 A theatre has 57 rows. Each row has 38 seats.

Estimate the number of people in the theatre when every seat is taken.

Is your estimate an underestimate or an overestimate?

5 Estimate the answer to each calculation.

   a $\dfrac{21 + 8.4}{3.9}$ **b** $\dfrac{398 \times 95}{12}$ **c** $\dfrac{4.3^2}{0.446}$

Stretch – can you deepen your learning?

1 Look at this diagram.

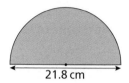

21.8 cm

    a Estimate the area of the semicircle, giving your answer in terms of π

    b Use $\pi = 3$ to estimate the area of the semicircle.

    c Which estimate is more accurate? How do you know?

2 The length of a square is given as 20 cm $\pm$ 0.5 cm

    a What do you think $\pm$ 0.5 cm means?

    b Write an error interval for

        i the length, l, of the square

        ii the perimeter, p, of the square

        iii the area, a, of the square.

3 Mario runs 100 m in 11.8 seconds. Find an error interval for Mario's speed in metres per second if

    a the distance is exact and the time is correct to 1 decimal place

    b the distance is correct to the nearest metre and the time is exact

    c the distance is correct to the nearest metre and the time is correct to 1 decimal place.

    Give the endpoints of your error intervals correct to 2 decimal places where necessary.

4 **a** Estimate the answer to 17.45 × 42.55

    b Without calculation, explain why your answer in part **a** is an overestimate.

5 Investigate this conjecture.

    If the error interval of a number n, is given by $a \leqslant n < b$, then $\boxed{n - a = b - n}$

    Explain your findings.

Reflect

1 Explain how to estimate the answer to a calculation.

2 Explain why error intervals are important.

Small steps

■ Work with directed numbers ®

■ Solve problems with integers

■ Solve problems with decimals

Key words

Decimal – a number with digits to the right of the decimal point

Negative number – a number less than 0

Directed number – a number that can be negative or positive

Are you ready?

1 Work these out without using a calculator.

   a $179 + 318$ **b** $2094 + 762$ **c** $915 - 172$ **d** $26\,731 - 3240$

   e 26×16 **f** 38×24 **g** 175×23 **h** 72×1800

   i $72 \div 2$ **j** $765 \div 5$ **k** $2394 \div 3$ **l** $2394 \div 9$

2 Work out

   a $10 - 7$ **b** $7 - 10$ **c** $-3 + 8$ **d** $-9 + 2$

3 Write three calculations that have an answer of -4

Models and representations

Double-sided counters

These can be used to represent **directed numbers**.

  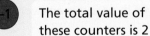   The total value of these counters is 2

Bar models

These are useful when solving problems.

girls [▯]
boys [▯ | ▯ | ▯]

There are 3 times as many boys as girls

Example 1

A toy car has a mass of 321 g

The mass of a toy train is twice the mass of the toy car.

A toy motorbike is 109 g lighter than the toy car.

Work out the total mass of the three toys. Give your answer in kilograms.

The mass of the toy train is $2 \times 321\,g = 642\,g$ ○─┤ You are told that the mass of the train is twice that of the toy car, so you need to double the mass of the car.

The mass of the toy motorbike is
$321g - 109g = 212g$

You're told that the motorbike is 109g lighter so you need to subtract 109g from the mass of the car.

The total mass is $321g + 642g + 212g = 1175g$

$= 1.175kg$

Remember: there are 1000g in 1kg

Practice 6.3A

1 Filipo has £36

Ali has five times as much money as Filipo.

How much money do they have altogether?

2 There are 365 children going to a theme park by coach. A coach can hold 45 children.

   a How many coaches are needed to transport the children to the theme park?

The coach company also has different-sized coaches. They have 3 coaches that can hold 70 children and 2 coaches that can hold 20 children.

   b What is the smallest number of coaches that the school can hire?

3 Each week, Leon puts some money into his bank account.

In week 1, he puts £120 into his account.

Each week, from week 2, he puts £15 more into his account than the week before.

How many weeks will it be before he has £1000 in the bank?

Show all of your working.

4 A theatre has 12 rows of 25 seats. Tickets for these seats cost £35 per seat.

There are also 20 rows of 29 seats. Tickets for these seats cost £14 per seat.

How much money does the theatre receive from ticket sales if the theatre is full?

5 Mr Singh buys a 5-litre carton of juice.

He pours all of the juice into glasses that each hold 300ml

He fills each glass before moving onto the next one.

How much juice does he pour into the final glass?

6 Biscuits are sold in two packet sizes, large and small.

A large packet contains 48 biscuits.

Benji buys 5 large packets and 2 small packets of biscuits.

He buys 302 biscuits in total.

How many biscuits are in each small pack?

What do you think? 💭

1 Here are some calculation cards.

48 + 76	1000 − 320	11 × 15	700 × 6000	360 ÷ 3

£29 × 5	26 × 2000	702 − 697	116 000 + 231 000

Work out the value of each card mentally.

What methods did you use? Did you use the same method as your partner?

2 Here are two problems.

A scooter costs £48

A bike costs £25 more than the scooter.

How much do the scooter and bike cost altogether?

A skateboard costs £28 more than a football.

The total cost of the football and skateboard is £66

How much does each item cost?

a What's the same and what's different about these problems?

b Work out the answer to each problem.

Discuss and compare the methods you use.

3 The table shows the cost of renting a holiday home by the sea.

Monday to Thursday	£55 per day per person
Friday to Sunday	£68 per day per person

A couple want to rent the holiday home for as many days as they can afford.

They have £700 to spend.

What is the maximum number of days for which they can rent the holiday home?

Example 2

28 × 41 = 1148

Use this to find the value of

a   2.8 × 4.1   **b**   0.28 × 0.041   **c**   1148 ÷ 2.8

a $2.8 \times 4.1 = 11.48$ ○———

2.8 = 28 ÷ 10 and 4.1 = 41 ÷ 10

Therefore 2.8 × 4.1 = 1148 ÷ 100 = 11.48

Each of the numbers in the question has been divided by 10, so the answer is 1148 divided by 100

b $0.28 \times 0.041 = 0.011\,48$ ○—

0.28 = 28 ÷ 100 and 0.041 = 41 ÷ 1000

So you need to divide 1148 by 100 000 (100 × 1000)

c	$1148 \div 2.8 = 410$	As there are 41 lots of 28 in 1148, there must be ten times as many 2.8s, because there are ten 2.8s in each 28

Practice 6.3B

1 Charlie is adding 2.17 and 11.9

```
    2   1   7
+   1   1   9
_____
    3   3   6
```

   a What mistake has Charlie made?   **b** Work out the correct answer.

2 Work these out without using a calculator.

   a $2.35 + 1.49$ **b** $16.2\,kg + 9.4\,kg$ **c** £1.89 − £1.45   **d** $10 − 1.3$

3 Rhys is trying to work out 1.7×3.9

His friend tells him to start by multiplying 17×39

×	10	7
30	300	210
9	90	63

```
    5   1   0
+   1   5   3
_____
    6   6   3
```

Rhys is not sure what to do next.

Explain to Rhys what he should do next and why this method works.

4 Work these out without using a calculator.

   a 45×1.6 **b** 3.7×8.2 **c** 2.35×14 **d** 0.175×13

Discuss your methods with a partner.

5 Work out the missing numbers.

   a $1.8 \div 2 = \boxed{}$ **b** $1.8 \div 3 = \boxed{}$ **c** $1.8 \div \boxed{} = 0.2$

   d $1.8 \div \boxed{} = 0.3$ **e** $\boxed{} \div 3 = 0.8$ **f** $\boxed{} \div 3 = 0.08$

6 $36 \times 14 = 504$

Use this to write down the answers to the following calculations.

a 3.6×14 **b** 1.4×3.6 **c** 36×28 **d** 36×2.8

e 0.14×36 **f** 0.14×3.6 **g** $504 \div 14$ **h** $504 \div 1.4$

i $504 \div 3.6$ **j** $50.4 \div 36$

Check your answers by estimation and by using a calculator.

 Which questions did you find the easiest? Which did you find the most challenging?

7 Junaid has £3.80. His friend Sven has £9.50 more than him.

a How much money do Junaid and Sven have altogether?

Sven gives some of his money to Junaid. They now have the same amount.

b How much money did Sven give Junaid?

8 Mr Hassan pays £1.72 for a pack of 5 rulers.

How much does it cost to buy 40 of these rulers?

£1.72

9 Flour is sold in two different-sized bags.

Beth buys 3 large and 2 small bags of flour.
She buys a total of 56.7 kg of flour.
What is the mass of flour in one small bag?

Flour

12.5 kg

Flour

10 A piece of ribbon is 16 metres long. The ribbon needs to go around the edges of two rectangular tables.

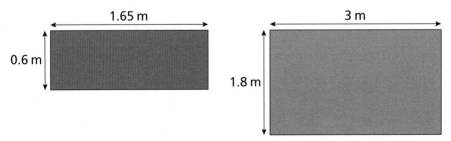

1.65 m

0.6 m

3 m

1.8 m

Is there enough ribbon to go around both tables? Show working to support your answer.

What do you think?

1 Work these out.

    a $360 \div 2$ **b** $36 \div 2$ **c** $3.6 \div 2$ **d** $0.36 \div 2$ **e** $0.036 \div 2$

What patterns do you notice?

2   **a**   Show that $48 \times 26 = 1248$

Here are the answers to five calculations that are related to part **a**.

    b   What numbers could go in the boxes?

    $\boxed{} \times \boxed{} = 124.8$ $\boxed{} \times \boxed{} = 12.48$ $\boxed{} \times \boxed{} = 1.248$

    $\boxed{} \div \boxed{} = 4.8$ $\boxed{} \div \boxed{} = 0.26$

Discuss and compare methods with a partner.

3 For each of these calculations, explain a method that can be used to work out the answer mentally.

    a   The total cost of 5 books that are £3.99 each.

    b   The change from a £10 note if the cost of shopping is £6.30

    c   The difference in height between two children who are 1.72 m and 1.69 m tall.

    d   The total mass of two objects that have masses of 2.8 kg and 3.5 kg

---

Example 3

Work out

a $7 + -4$ **b** 7×-4

a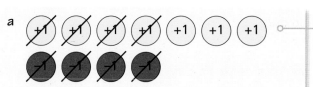

$7 + -4 = 3$

Remember: one +1 and one −1 counter make a zero pair because $1 + (-1) = 0$

b

$7 \times -4 = -28$

Each row contains seven −1 counters.

There are four rows in total.

There are twenty-eight −1 counters.

Practice 6.3C

 1 Faith is working out the answer to −5 − 3

 The answer is + 8 because two negatives make a positive.

a What mistake has Faith made?

b What is the correct answer?

2 Work out

a −7 − 3 **b** −7 + 4 **c** −6 − 5 **d** −60 − 50

 3 Rob uses some zero pairs to help him work out 5 + −3

Explain Rob's method.   5 + −3 = 2

 **4** Work out

a 6 + −2 **b** 6 − +2 **c** 3 − +7

d 3 + −7 **e** 4 − −2 **f** −7 − −4

5 Work out the missing numbers.

a −4 + ☐ = 3 **b** 6 + ☐ = 1 **c** −5 − ☐ = −8 **d** ☐ − 3 = −9

    −4 − ☐ = 3   6 − ☐ = 1   −5 + ☐ = −8   ☐ + 3 = −9

6 Copy and complete the multiplication grid.

×	−2	−3	−5	4	5	8
−2	4					
−3						
−5						
4						
5						
8						

What patterns do you notice?

7 Work out the missing numbers in these calculations.

a $-7 \times -3 = \boxed{}$ **b** $-20 \div 2 = \boxed{}$ **c** $20 \div -2 = \boxed{}$ **d** $-20 \div -2 = \boxed{}$

e $4 \times \boxed{} = -24$ **f** $15 \div \boxed{} = -3$ **g** $18 \div \boxed{} = 2$ **h** $\boxed{} \div -5 = -6$

i $\boxed{} \div 5 = -6$ **j** $6 \times \boxed{} \times 3 = -18$ **k** $-30 \div \boxed{} = 15$ **l** $8 \times \boxed{} = -4$

What do you think? 💡

1 Lydia is working out the square of −5 using a calculator.

$$-5^2$$
$$-25$$

a How do you know that Lydia's answer is wrong without doing a calculation?

b Explain the mistake Lydia made. How can she avoid making this mistake?

c What's the same and what's different about these cards?

$$\boxed{-6^2} \quad \boxed{(-6)^2}$$

2 Decide whether the value of each card is positive or negative. You do not need to work out the answers.

Explain your reasoning.

a $\boxed{-716 + 550}$ **b** $\boxed{835 - 1917}$ **c** $\boxed{-76 - +85}$

d $\boxed{30 - -100.6}$ **e** $\boxed{77 \times -30 \times -59 \times -8.8}$ **f** $\boxed{(-5)^6}$

3 Find as many pairs of numbers as you can to complete each calculation.

$\boxed{} + \boxed{} = -50$ $\boxed{} - \boxed{} = -50$

$\boxed{} \times \boxed{} = -50$ $\boxed{} \div \boxed{} = -50$

Consolidate – do you need more?

1 At a school there are

- 237 pupils in Year 7
- 241 pupils in Year 8
- 202 pupils in Year 9

a How many pupils are there altogether in Years 7, 8 and 9?

There are 12 more pupils in Year 10 than there are in Year 7

There are 17 fewer pupils in Year 11 than there are in Year 8

b Find the total number of pupils in Years 10 and 11

2

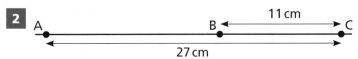

Points A, B and C lie on a straight line.

The distance between points A and C is 27 cm

The distance between points B and C is 11 cm

Find the distance between points A and B

3 Find the product of 17 and 63

4 How many hours are there in March?

5 Bobbie's birthday is in 273 days. How many weeks is this?

6 a Find the total value of each set of counters.

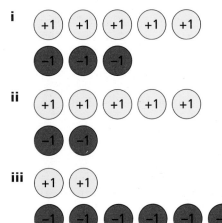

b Work out

i 5 + −3   **ii** 5 + −2   **iii** 2 + −7

7 Work out

a 9 − −2 b −9 − 2 c −9 + −2 d 9 + −2

e 9 × −2 f −9 × −2 g 9 ÷ −2 h −9 ÷ −2

Stretch – can you deepen your learning?

1 36 trees are planted in a straight line with the same space between each tree.

The distance between the first and ninth tree is 32 metres.

What is the distance between the first and last tree?

2 The total height of four children is 620 cm

a Work out the mean height of these children.

Two children have the same height.

The range of heights is 31 cm

The shortest child has a height of 135 cm

b Find the heights of the four children. Is there more than one answer?

3 A supermarket has a special offer.

What is the maximum number of fish fingers that can you buy with £21?

Discuss your method.

4 A barrel contains 32.8 litres of water. The water is poured into 4 buckets.

5.7 litres is poured into each of the first 2 buckets.

1.5 more litres are poured into the third bucket than the fourth.

How much water is poured into the fourth bucket?

5 1 kg of apples costs £1.12 more than 1 kg of pears.

1 kg of apples costs £2.35 less than 1 kg of bananas.

Jackson buys 3 kg of apples, 3 kg of pears and 2 kg of bananas. The total cost is £25.02

How much does 1 kg of apples cost?

6 The charges for parking a car in a car park are shown in the table.

First hour	£2.40
For each 30 mins (or part of) thereafter	40p
No change given	

Kate parks her car at 11:07 a.m.

She puts £5 into the parking machine.

By what time must she leave to avoid overstaying?

7 Here are some number cards

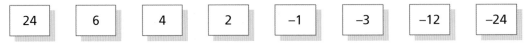

24   6   4   2   −1   −3   −12   −24

a Choose cards to make these statements correct. Find at least two different ways of completing each statement.

$\boxed{} \times \boxed{} = 24$ $\boxed{} \div \boxed{} = -6$ $\boxed{} \times \boxed{} < 100$ $\boxed{} \div \boxed{} < 1$

b What cards could complete this statement?

$-1 < \boxed{} \div \boxed{} < 1$

Compare your answers with a partner.

8 In a quiz, contestants score 5 points for each correct answer and lose 3 points for each incorrect answer.

a Explain how a contestant could get a negative score.

Abdullah answers an even number of questions.

b Is it possible for Abdullah to get an odd score?

Reflect

1 Which words in a question can help you to select the correct calculations to perform? What do each of these words mean?

2 Explain how to calculate with directed numbers.

6.4 Fractions

Small steps

■ Add and subtract fractions ®
■ Multiply and divide fractions ®
■ Solve problems with fractions

Key words

Numerator – the top number in a fraction that shows the number of parts

Denominator – the bottom number in a fraction; it shows how many equal parts one whole has been divided into

Reciprocal – the result of dividing 1 by a given number. The product of a number and its reciprocal is always 1

Are you ready?

1 Write these fractions in their simplest form.

  a $\frac{6}{8}$ **b** $\frac{20}{35}$ **c** $\frac{21}{70}$ **d** $\frac{150}{300}$

2 Write the mixed number and the improper fraction represented by the diagram.

3 Write these mixed numbers as improper fractions.

  a $1\frac{1}{2}$ **b** $1\frac{4}{5}$ **c** $3\frac{2}{3}$

4 Write these improper factions as mixed numbers.

  a $\frac{11}{6}$ **b** $\frac{15}{4}$ **c** $\frac{25}{3}$

5 Work out

  a $\frac{1}{5}+\frac{4}{5}$ **b** $\frac{9}{7}-\frac{2}{7}$ **c** $1-\frac{1}{6}$

Models and representations

Bar models

These bar models show that $\frac{1}{3}=\frac{3}{9}$ and hence $\frac{1}{3}+\frac{2}{9}=\frac{5}{9}$

Area models

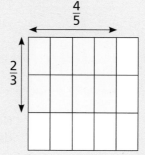

This area model shows that $\frac{2}{3}\times\frac{4}{5}=\frac{8}{15}$

Fraction walls

This fraction wall shows how fractions with different denominators relate to each other.

1											
$\frac{1}{2}$						$\frac{1}{2}$					
$\frac{1}{3}$				$\frac{1}{3}$				$\frac{1}{3}$			
$\frac{1}{4}$			$\frac{1}{4}$			$\frac{1}{4}$			$\frac{1}{4}$		
$\frac{1}{5}$		$\frac{1}{5}$		$\frac{1}{5}$		$\frac{1}{5}$		$\frac{1}{5}$			
$\frac{1}{6}$		$\frac{1}{6}$		$\frac{1}{6}$		$\frac{1}{6}$		$\frac{1}{6}$		$\frac{1}{6}$	
$\frac{1}{7}$	$\frac{1}{7}$		$\frac{1}{7}$		$\frac{1}{7}$		$\frac{1}{7}$		$\frac{1}{7}$		$\frac{1}{7}$
$\frac{1}{8}$		$\frac{1}{8}$		$\frac{1}{8}$		$\frac{1}{8}$		$\frac{1}{8}$		$\frac{1}{8}$	$\frac{1}{8}$
$\frac{1}{9}$	$\frac{1}{9}$	$\frac{1}{9}$	$\frac{1}{9}$	$\frac{1}{9}$	$\frac{1}{9}$	$\frac{1}{9}$	$\frac{1}{9}$	$\frac{1}{9}$			
$\frac{1}{10}$	$\frac{1}{10}$	$\frac{1}{10}$	$\frac{1}{10}$	$\frac{1}{10}$	$\frac{1}{10}$	$\frac{1}{10}$	$\frac{1}{10}$	$\frac{1}{10}$	$\frac{1}{10}$		
$\frac{1}{12}$	$\frac{1}{12}$	$\frac{1}{12}$	$\frac{1}{12}$	$\frac{1}{12}$	$\frac{1}{12}$	$\frac{1}{12}$	$\frac{1}{12}$	$\frac{1}{12}$	$\frac{1}{12}$	$\frac{1}{12}$	$\frac{1}{12}$

In this chapter, you will review what you learned about fractions in Books 1 and 2 and apply your knowledge to solving problems.

Example 1

Work out $\frac{1}{6} + \frac{3}{4}$

Give your answer in its simplest form.

Method A

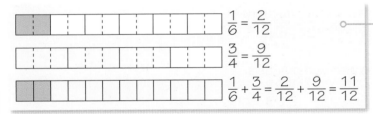

$\frac{1}{6} = \frac{2}{12}$

$\frac{3}{4} = \frac{9}{12}$

$\frac{1}{6} + \frac{3}{4} = \frac{2}{12} + \frac{9}{12} = \frac{11}{12}$

Start by finding a common **denominator**. Each bar is split into 12 equal parts because 12 is the lowest common multiple of 6 and 4

Method B

Multiples of 6: 6, 12, 18, 24, 30 …

Multiples of 4: 4, 8, 12, 16, 20, 24 …

LCM of 4 and 6 = 12

Start by finding the lowest common multiple of the denominators.

If you use a different common multiple, you will need to simplify your answer.

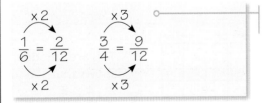

Then rewrite the fractions as equivalent fractions with the same denominator.

$\frac{1}{6} + \frac{3}{4} = \frac{2}{12} + \frac{9}{12} = \frac{11}{12}$

Now that the fractions have the same denominator you can just add the **numerators**.

Practice 6.4A

1 Here are three sets of fraction additions and subtractions.

Set A	Set B	Set C
$\frac{1}{3} + \frac{1}{4}$	$\frac{2}{5} + \frac{3}{10}$	$\frac{1}{4} + \frac{1}{6}$
$\frac{2}{5} + \frac{1}{3}$	$\frac{7}{12} - \frac{1}{4}$	$\frac{5}{6} + \frac{1}{9}$
$\frac{5}{9} - \frac{1}{2}$	$\frac{5}{9} - \frac{1}{3}$	$\frac{7}{10} - \frac{3}{15}$

a Work out the answers to the questions in each set.

b Why you do think the questions have been put into these sets?

c Write another question that fits into each of the sets.

Compare your question to your partner's.

2 Work these out. Give your answers in their simplest form.

a $\frac{1}{5} + \frac{2}{3}$ **b** $\frac{3}{5} - \frac{1}{4}$ **c** $\frac{3}{7} + \frac{1}{2}$ **d** $\frac{3}{10} + \frac{2}{5}$

e $\frac{8}{15} - \frac{1}{3}$ **f** $\frac{7}{10} - \frac{3}{20}$ **g** $\frac{5}{4} + \frac{5}{6}$ **h** $\frac{11}{18} + \frac{7}{12}$

i $\frac{7}{16} - \frac{3}{8}$ **j** $\frac{4}{9} + \frac{5}{6}$ **k** $\frac{19}{24} - \frac{35}{48}$ **l** $\frac{17}{10} + \frac{49}{100}$

3 Use the bar models to show that $2\frac{1}{3} + 1\frac{5}{6} = 4\frac{1}{6}$

Explain each step fully.

4 Work these out. Give your answers as mixed numbers in their simplest form.

a $1\frac{2}{5} + 1\frac{1}{2}$ **b** $2\frac{5}{9} - 1\frac{1}{3}$ **c** $1\frac{7}{15} + 3\frac{1}{5}$ **d** $5\frac{3}{10} - \frac{3}{5}$

e $2\frac{3}{4} + 3\frac{5}{6}$ **f** $7\frac{4}{5} - 2\frac{2}{3}$ **g** $7\frac{11}{20} + \frac{21}{4}$ **h** $16\frac{1}{3} - 7\frac{1}{3}$

i $145\frac{3}{4} + 72\frac{5}{6}$ **j** $19\frac{17}{18} - 4\frac{5}{12}$ **k** $7\frac{5}{6} + 1\frac{1}{4}$ **l** $9\frac{4}{15} + 5\frac{7}{10}$

5 Work these out. Give your answers as mixed numbers in their simplest form.

a $\frac{1}{2} + \frac{1}{4} + \frac{3}{12}$ **b** $\frac{5}{6} + \frac{2}{9} - \frac{1}{2}$ **c** $1\frac{1}{10} + 2\frac{2}{5} + \frac{3}{4}$

d $\frac{2}{3} + 1\frac{4}{5} - \frac{7}{10}$ **e** $7\frac{3}{8} - 3\frac{1}{6} + 1\frac{2}{3}$ **f** $3\frac{4}{5} - 1\frac{1}{3} - \frac{13}{15}$

What do you think?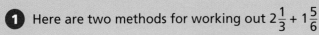

1 Here are two methods for working out $2\frac{1}{3} + 1\frac{5}{6}$

Method 1

Add the whole ones:

$2 + 1 = 3$

Add the fraction parts:

$\frac{1}{3} + \frac{5}{6} = \frac{2}{6} + \frac{5}{6} = \frac{7}{6} = 1\frac{1}{6}$

Add the whole ones and fraction parts:

$3 + 1\frac{1}{6} = 4\frac{1}{6}$

Method 2

Convert each fraction to a mixed number:

$2\frac{1}{3} = \frac{7}{3}$ $\qquad$ $1\frac{5}{6} = \frac{11}{6}$

$2\frac{1}{3} + 1\frac{5}{6} = \frac{7}{3} + \frac{11}{6}$

Now write the fractions over a common denominator and add them:

$\frac{7}{3} + \frac{11}{6} = \frac{14}{6} + \frac{11}{6} = \frac{25}{6} = 4\frac{1}{6}$

a Which method do you prefer? Why?

b Which method is more efficient for calculating $112\frac{11}{15} + 67\frac{3}{5}$?

Explain your reasoning.

2 Zach is subtracting these fractions.

$\frac{31}{36} - \frac{7}{18}$

I'm going to multiply the two denominators together $36 \times 18 = 648$. This is my common denominator.

a Explain the advantages and disadvantages of Zach's method.

b What would you advise him to do?

Example 2

Work out $\frac{3}{8} \div \frac{1}{4}$

Method A

$\frac{1}{8}$	$\frac{1}{8}$	$\frac{1}{8}$	$\frac{1}{8}$	$\frac{1}{8}$	$\frac{1}{8}$	$\frac{1}{8}$	$\frac{1}{8}$
$\frac{1}{4}$		$\frac{1}{4}$		$\frac{1}{4}$		$\frac{1}{4}$	

So $\frac{3}{8} \div \frac{1}{4} = 1\frac{1}{2}$

You want to know how many lots of $\frac{1}{4}$ there are in $\frac{3}{8}$

The fraction wall shows that one and a half lots of $\frac{1}{4}$ make up $\frac{3}{8}$

Method B

$$\frac{3}{8} \div \frac{1}{4} = \frac{3}{8} \times 4 = \frac{12}{8}$$

$$\frac{12}{8} = \frac{3}{2} = 1\frac{1}{2}$$

Dividing by a fraction is equivalent to multiplying by its **reciprocal**.

The reciprocal of $\frac{1}{4}$ is 4 because $\frac{1}{4} \times 4 = 1$

Practice 6.4B

1 Work out

a **i** $1 \times \frac{1}{5}$ **ii** $2 \times \frac{1}{5}$ **iii** $3 \times \frac{1}{5}$ **iv** $4 \times \frac{1}{5}$ **v** $5 \times \frac{1}{5}$ **vi** $6 \times \frac{1}{5}$ **vii** $7 \times \frac{1}{5}$

b **i** $2 \times \frac{2}{3}$ **ii** $2 \times \frac{3}{4}$ **iii** $2 \times \frac{5}{9}$ **iv** $2 \times \frac{2}{9}$ **v** $2 \times \frac{3}{7}$ **vi** $2 \times \frac{9}{10}$ **vii** $2 \times \frac{11}{20}$

What do you notice?

2 This area model represents the calculation $\frac{2}{3} \times \frac{3}{4}$

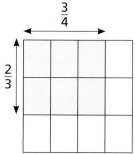

a Explain how the model shows that $\frac{3}{4} \times \frac{2}{3} = \frac{1}{2}$

b Draw diagrams to show that

i $\frac{1}{3} \times \frac{2}{3} = \frac{2}{9}$

ii $\frac{2}{3} \times 1\frac{1}{4} = \frac{10}{12}$

3 Work these out. Give your answers in their simplest form.

a $\frac{1}{5} \times \frac{2}{3}$ **b** $\frac{3}{5} \times \frac{1}{4}$ **c** $\frac{3}{7} \times \frac{1}{2}$ **d** $\frac{3}{10} \times \frac{2}{5}$

e $\frac{8}{15} \times \frac{4}{3}$ **f** $\frac{6}{7} \times 1\frac{1}{2}$ **g** $\frac{3}{10} \times 2\frac{1}{3}$ **h** $\frac{11}{15} \times 2\frac{1}{2}$

i $1\frac{1}{3} \times 2\frac{1}{4}$ **j** $1\frac{6}{11} \times 5$ **k** $3\frac{2}{7} \times 2\frac{1}{6}$ **l** $\frac{17}{5} \times \frac{5}{17}$

4 Work out

a $\frac{1}{2} \times \frac{1}{3} \times \frac{1}{4}$ **b** $\frac{2}{3} \times \frac{3}{4} \times \frac{7}{10}$ **c** $3 \times \frac{1}{2} \times \frac{2}{3}$

5 Find the missing number in each calculation.

a $\frac{1}{2} \times \boxed{} = \frac{3}{2}$ **b** $\frac{2}{3} \times \boxed{} = \frac{10}{3}$ **c** $\frac{3}{4} \times \boxed{} = 3\frac{3}{4}$

d $\frac{7}{10} \times \dfrac{\boxed{}}{\boxed{}} = \frac{7}{20}$ **e** $\dfrac{5}{\boxed{}} \times \dfrac{\boxed{}}{3} = \frac{20}{27}$ **f** $\frac{3}{1729} \times \boxed{} = \frac{15}{1729}$

6 Work out

a $\frac{1}{2} \div \frac{1}{3}$ **b** $\frac{3}{4} \div \frac{1}{7}$ **c** $\frac{2}{5} \div \frac{3}{7}$ **d** $\frac{9}{10} \div \frac{5}{4}$ **e** $\frac{6}{7} \div \frac{9}{10}$

7 **a** Here are two calculation cards.

$\boxed{2 \div \frac{1}{5}}$ $\boxed{\frac{1}{5} \div 2}$

What's the same? What's different?

Discuss with a partner the method you would use for each calculation.

b Work out

i $3 \div \frac{1}{2}$ **ii** $5 \div \frac{2}{3}$ **iii** $10 \div \frac{3}{4}$

iv $\frac{1}{2} \div 3$ **v** $\frac{2}{3} \div 5$ **vi** $\frac{3}{4} \div 10$

8 Work out

a $1\frac{1}{4} \div \frac{1}{2}$ **b** $2\frac{3}{5} \div \frac{1}{4}$ **c** $2\frac{2}{5} \div 1\frac{2}{3}$

d $5\frac{2}{3} \div 1\frac{1}{2}$ **e** $3 \div 2\frac{2}{5}$ **f** $5\frac{1}{2} \div 3$

What do you think?

1 Some of the values are missing from these calculation cards.

a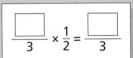
$\dfrac{\boxed{}}{3} \times \frac{1}{2} = \dfrac{\boxed{}}{3}$

b
$\frac{7}{9} \times \boxed{} = 3\frac{8}{9}$

c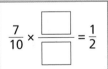
$\frac{7}{10} \times \dfrac{\boxed{}}{\boxed{}} = \frac{1}{2}$

d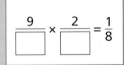
$\dfrac{9}{\boxed{}} \times \dfrac{2}{\boxed{}} = \frac{1}{8}$

For each calculation, decide whether there is

- one solution

- several possible solutions

- no solution.

Explain your reasoning.

2 Find an efficient method for calculating $\frac{121}{180} \times \frac{36}{55}$

Explain your choice of method to a partner.

Example 3

Evaluate $\frac{x^2 + y}{z}$ when $x = \frac{2}{5}$, $y = \frac{1}{4}$ and $z = \frac{3}{8}$

When $x = \frac{2}{5}$, $x^2 = \left(\frac{2}{5}\right)^2 = \frac{2}{5} \times \frac{2}{5}$ Use brackets to remind you that $\left(\frac{2}{5}\right)^2$ means $\frac{2}{5} \times \frac{2}{5}$

$= \frac{4}{25}$ Multiply the numerators and multiply the denominators.

$x^2 + y = \frac{4}{25} + \frac{1}{4}$

$= \frac{16}{100} + \frac{25}{100}$ The lowest common multiple of 4 and 25 is 100

$= \frac{41}{100}$

$\frac{x^2 + y}{z} = \frac{41}{100} \div \frac{3}{8} = \frac{41}{100} \times \frac{8}{3}$ Multiply $\frac{41}{100}$ by the reciprocal of $\frac{3}{8}$

$= \frac{328}{300}$

    Simplify the fraction and then convert the improper fraction to a mixed number.

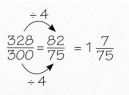

$\frac{328}{300} = \frac{82}{75} = 1\frac{7}{75}$

Practice 6.4C

1 Work out the perimeter of each shape.

a

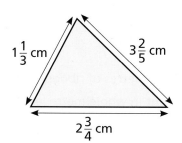

$1\frac{1}{3}$ cm $3\frac{2}{5}$ cm

$2\frac{3}{4}$ cm

b

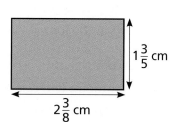

$1\frac{3}{5}$ cm

$2\frac{3}{8}$ cm

2 In this fraction pyramid, each number is the sum of the numbers in the two boxes below it.

Copy and complete the fraction pyramid.

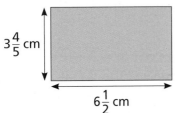

3 A rectangular piece of paper has length $6\frac{1}{2}$ cm and width $3\frac{4}{5}$ cm

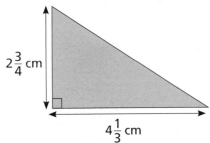

$3\frac{4}{5}$ cm

$6\frac{1}{2}$ cm

a Calculate the area of the piece of paper.

A square of side length 2 cm is cut out of the piece of paper.

b What is the area of paper remaining?

4 Find the area of the triangle.

$2\frac{3}{4}$ cm

$4\frac{1}{3}$ cm

5 A jug contains $2\frac{1}{2}$ litres of juice.

Marta pours herself $\frac{3}{4}$ of a litre of juice from the jug.

Ed pours himself $\frac{2}{5}$ of a litre of juice from the jug.

How much juice is left in the jug?

6 Ribbon is used to decorate skirts. Each skirt needs $1\frac{1}{5}$ metres of ribbon.

Ribbon costs £2.55 per metre and can only be bought in whole metre lengths.

Work out of the total cost of the ribbon needed for 7 skirts.

7 Parts of three congruent rectangles are shaded. What fraction of the middle rectangle is shaded?

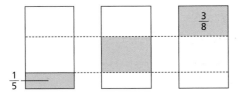

$\frac{1}{5}$

$\frac{3}{8}$

What do you think?

1 The symbol ⭐ means "double the first number and then add the second number".

a Work out

    i $\frac{2}{5}$ ⭐ $\frac{3}{10}$ **ii** $(1\frac{3}{4}$ ⭐ $2\frac{1}{3})$ ⭐ $\frac{5}{6}$

b Solve $(1\frac{4}{5}$ ⭐ $x) = 1\frac{1}{4}$

2 Work out the area of the trapezium.

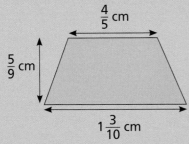

$\frac{4}{5}$ cm

$\frac{5}{9}$ cm

$1\frac{3}{10}$ cm

Compare your method with a partner. Did you work it out in the same way?

3 The area of the triangle is greater than 12 cm² but less than 20 cm²

The height of the triangle, h, is an integer.

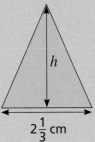

h

$2\frac{1}{3}$ cm

What could be the value of h? How many different answers can you find?

Consolidate – do you need more?

1 **a** What is the lowest common multiple of 3 and 8?

   b Work out $\frac{2}{3} + \frac{1}{8}$

2 Show how the bar model can be used to work out $\frac{1}{2} + \frac{1}{5}$

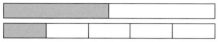

3 Work these out. Give your answers in their simplest form.

   a $\frac{1}{4} + \frac{1}{12}$ **b** $\frac{2}{5} + \frac{1}{6}$ **c** $\frac{3}{8} + \frac{1}{6}$ **d** $\frac{5}{7} - \frac{2}{3}$

   e $\frac{8}{9} - \frac{5}{6}$ **f** $\frac{1}{6} + \frac{3}{4} - \frac{1}{3}$

4 Ed reads $\frac{5}{36}$ of a book on Monday. He reads $\frac{1}{9}$ of the book on Tuesday. He reads $\frac{1}{6}$ of the book on Wednesday. He reads the rest at the weekend.

   What fraction of the book does he read at the weekend?

5 Work out

   a $\frac{1}{9} \times 8$ **b** $5 \times \frac{1}{7}$ **c** $\frac{2}{9} \times 4$ **d** $\frac{2}{7} \times 2$

6 Work these out. Give your answers in their simplest form.

   a $\frac{2}{9} \times \frac{3}{5}$ **b** $\frac{3}{8} \times \frac{2}{3}$ **c** $\frac{1}{2} \times \frac{21}{40}$ **d** $\frac{1}{6} \times \frac{3}{7}$

7 Work these out. Give your answers in their simplest form.

   a $\frac{7}{9} \times -\frac{2}{3}$ **b** $-\frac{4}{5} \times -\frac{5}{8}$ **c** $(-\frac{1}{15})^2$

8 Work these out. Give your answers in their simplest form.

   a $4 \div \frac{1}{5}$ **b** $6 \div \frac{2}{3}$ **c** $\frac{4}{5} \div \frac{1}{3}$

   d $\frac{3}{4} \div \frac{1}{8}$ **e** $\frac{5}{7} \div \frac{1}{10}$

9 Samira mixes $\frac{2}{3}$ of a litre of orange juice with $\frac{3}{4}$ of a litre of apple juice.

   She pours the same amount of juice into 5 glasses.

   How much juice is in each glass?

Stretch – can you deepen your learning?

1 **a** **i** Write 0.8 as a fraction.

   ii Write 1.3 as a fraction.

   iii Show that $0.8 \times 1.3 = \dfrac{26}{25}$

b Work these out. Give your answers as fractions in their simplest form.

   i 0.7×0.5 **ii** 0.31×0.03 **iii** $0.75 \times 0.999\,999\ldots$

2 Work out

$$\frac{\left(\frac{2}{3} + \frac{1}{9}\right)^2}{\left(1\frac{3}{4} - \frac{9}{10}\right)}$$

Show all the steps of your working.

3 Here are some fraction cards.

$2\frac{1}{4}$ $1\frac{3}{10}$ $1\frac{7}{8}$ $2\frac{3}{20}$

Identify which two fractions have

a the greatest total

b the smallest total

c the greatest difference

d the smallest difference.

Discuss your strategy with a partner.

4 Rectangles A and B have the same perimeter.

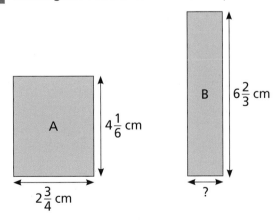

A $4\frac{1}{6}$ cm

$2\frac{3}{4}$ cm

B $6\frac{2}{3}$ cm

?

Find the width of rectangle B

5 Work these out. Give your answers in their simplest form.

   a $\frac{16}{25} \times \frac{15}{32}$ **b** $\frac{24}{35} \times \frac{15}{36}$ **c** $\frac{63}{75} \times \frac{22}{27} \times \frac{15}{44}$ **d** $\frac{48}{49} \div \frac{60}{77}$

6 A parallelogram has a base length of $\frac{3x}{20}$ mm, a slant height of $\frac{7x}{10}$ mm and a perpendicular height of $\frac{2x}{5y}$ mm

   Form expressions for

   a the area of the parallelogram

   b the perimeter of the parallelogram.

7 The point $(a, \frac{3p}{5})$ lies on the line with equation $7y = 4x - 3$

   Form an expression for a in terms of p

8 Show that the line $y = 1 - \frac{2gx}{9}$ is perpendicular to the line $\frac{9}{gx} + 15 = 2y$

Reflect

1 What's the same and what's different about adding fractions and subtracting fractions?

2 What's the same and what's different about multiplying fractions and dividing fractions?

3 Which do you find easier? Why?

   ▦ adding and subtracting fractions

   ▦ multiplying fractions

   ▦ dividing fractions

Small steps

■ Numbers in standard form ®

Key words

Power (or **exponent**) – this is written as a small number to the right and above the base number, indicating how many times to use the number in a multiplication. For example, the 5 in 2^5

Standard form – a number written in the form $A \times 10^n$ where A is at least 1 and less than 10, and n is an integer

Are you ready?

1 Work out

   a $400 + 50\,000$ b $700\,000 - 30\,000$ c $23\,000\,000 + 5800$ d $10\,000 - 400$

   e $0.4 + 0.005$ f $0.7 - 0.003$ g $2.3 + 0.58$ h $1 - 0.0004$

2 Work out

   a i 5×200 ii 50×200 iii 500×200 iv $50\,000 \times 200$

   b i $100\,000 \div 2$ ii $100\,000 \div 20$ iii $100\,000 \div 20\,000$ iv $100\,000 \div 0.2$

3 If $1 \leqslant t < 10$, which of these numbers are possible values of t?

   3.7   4   0.99   9.999   1   10

4 Find the value of

   a 10^2 b 10^3 c 10^4 d 10^6 e 10^0 f 10^{-1}

Models and representations

Place value chart

Place value charts are useful for comparing large or small numbers.

Millions			Thousands			Ones		
H	T	O	H	T	O	H	T	O
	2	6	9	1	1	1	0	3

This shows 26 millions, 911 thousands and 103 ones.

A calculator can be used to work with numbers in **standard form**.

There is often a "$\boxed{\times 10^x}$" button on a calculator that can be used to work with standard form.

You know from previous learning that $3 \times 10^9 \times 7 = 21 \times 10^9$

You can use a calculator to give this answer in correct standard form.

If you type in $\boxed{2}\boxed{1}\boxed{\times 10^x}\boxed{9}\boxed{=}$ your calculator will rewrite this in standard form.

$21\times_{10}9$

$2.1\times_{10}{}^{10}$

You learned about standard form in Book 2

In this chapter, you will practise converting numbers to and from standard form and solve problems involving standard form.

Remember: a number in standard form is written as $A \times 10^n$, where $1 \leqslant A < 10$ and n is an integer.

Example 1

a Write these numbers in standard form.

   i 602 700 000   **ii** 0.000 81

b Write these numbers in ordinary form.

   i 5.4×10^6 **ii** 5.4×10^{-6}

a **i** $602\,700\,000 = 6.027 \times 10^8$

    ii $0.000\,81 = 8.1 \times 10^{-4}$

Remember that the **power** will be negative when the number is between 0 and 1

b **i** $5.4 \times 10^6 = 5\,400\,000$

    ii $5.4 \times 10^{-6} = 0.000\,005\,4$

$10^6 = 1\,000\,000$, so you need to multiply 5.4 by 1 000 000

The power is negative so the number is less than 1

$10^{-6} = 0.000\,001$ or $\frac{1}{1\,000\,000}$, so you need to divide by 1 000 000

Example 2

Complete the calculations. Give your answers in standard form.

a $(1.6 \times 10^4) + (7 \times 10^2)$

b $(1.6 \times 10^4) \times (7 \times 10^2)$

Method A

$1.6 \times 10^4 = 16\,000$

$7 \times 10^2 = 700$

$10^4 = 10\,000$ and $1.6 \times 10\,000 = 16\,000$

$10^2 = 100$ and $7 \times 100 = 700$

a $16\,000 + 700 = 16\,700$

   $16\,700 = 1.67 \times 10^4$

b $16\,000 \times 700 = 16 \times 7 \times 1000 \times 100$

               $= 112 \times 100\,000$

               $= 11\,200\,000 = 1.12 \times 10^7$

Multiply the non-zero digits and then work out the power of 10

Method B

a $\quad 1.6 \times 10^4 + 7 \times 10^2 = 1.6 \times 10^4 + 0.07 \times 10^4$ ○——| Adjust the calculation so that the powers are the same.

$\qquad\qquad\qquad = 1.67 \times 10^4$

b $\quad 1.6 \times 10^4 \times 7 \times 10^2 = 1.6 \times 7 \times 10^4 \times 10^2$

$\qquad\qquad\qquad\qquad = 11.2 \times 10^6$

$\qquad\qquad\qquad\qquad = 1.12 \times 10 \times 10^6$ ○——| $11.2 = 1.12 \times 10$

$\qquad\qquad\qquad\qquad = 1.12 \times 10^7$

Practice 6.5A

1 Write these in standard form.

a $\quad 300\,000\,000$ b $\quad 370\,000\,000$ c $\quad 377\,000\,000$

What's the same and what's different about your answers?

2 Write these in standard form.

a $\quad 0.000\,000\,8$ b $\quad 0.000\,008$ c $\quad 0.000\,000\,81$

What's the same and what's different about your answers?

3 Write these in standard form.

a $\quad 61\,000$ b $\quad 74\,100\,000$ c $\quad 90\,300\,000$ d $\quad 5\,904\,000\,000\,000$

e $\quad$ 3 million   f $\quad$ four hundred and fifty-eight thousand

g $\quad$ 5.7 billion

4 Write these in standard form.

a $\quad 0.000\,000\,08$ b $\quad 0.000\,002\,6$ c $\quad 0.000\,050\,7$ d $\quad 0.05$

e $\quad 0.000\,400\,9$ f $\quad 0.000\,000\,007\,133$ g $\quad$ 4 hundredths   h $\quad$ 7 thousandths

5 Round each of these numbers to 1 significant figure and then write them in standard form.

The mass of Earth is $5\,972\,000\,000\,000\,000\,000\,000\,000\,000$ kg	On average, YouTube is watched for a total of 3.25 billion hours each month.

6 Write <, > or = to compare each pair of numbers.

a $\quad 7.5 \times 10^3 \bigcirc 7.2 \times 10^3$ b $\quad 4 \times 10^4 \bigcirc 25\,000$

c $\quad 3.03 \times 10^{-2} \bigcirc 3.5 \times 10^{-3}$ d $\quad 8.03 \times 10^{-2} \bigcirc 8.5 \times 10^{-1}$

7 None of these numbers are written in standard form. Adjust each one so that it is written in correct standard form.

a 12×10^3 **b** 34×10^9 **c** 0.7×10^5

d 0.0032×10^8 **e** 735×10^{11} **f** 7020×10^2

8 Work out

a $(3 \times 10^5) + (5 \times 10^4)$ **b** $(1.6 \times 10^5) + (7 \times 10^3)$

c $(2.8 \times 10^3) + (3.5 \times 10^2)$ **d** $(1.892 \times 10^5) - (1.7 \times 10^5)$

e $(7.45 \times 10^5) + (3.6 \times 10^3)$ **f** $(3.855 \times 10^7) - (1.31 \times 10^5)$

g $(3 \times 10^{-4}) + (4.1 \times 10^{-3})$ **h** $(6.8 \times 10^{-3}) + (3.3 \times 10^{-1})$

i $(6.64 \times 10^2) + (3.5 \times 10^3) + (1.8 \times 10^4)$

Check your answers using a calculator.

9 Work out

a $(5 \times 10^3) \times (3 \times 10^4)$ **b** $(3 \times 10^2) \times (5 \times 10^6)$

c $(7 \times 10^5) \times (1.2 \times 10^{-2})$ **d** $(3 \times 10^{-4}) \times (3.9 \times 10^7)$

e $(1.86 \times 10^8) \times (3 \times 10^4)$ **f** $(4.95 \times 10^5) \times (4 \times 10^{-1})$

10 Work out

a $(8 \times 10^7) \div (4 \times 10^4)$ **b** $(8 \times 10^5) \div (2 \times 10^3)$

c $(2.4 \times 10^6) \div (2 \times 10^4)$ **d** $(7.5 \times 10^5) \div (5 \times 10^2)$

e $(6.6 \times 10^7) \div (6 \times 10^6)$ **f** $(3.6 \times 10^4) \div (6 \times 10^2)$

11 An average adult has about 20 000 000 000 000 red blood cells. Each red blood cell has a mass of about 0.000 000 000 1 gram.

Work out the total mass of the red blood cells in an average adult. Give your answer in standard form.

12 Evaluate $\dfrac{T}{B}$ when $T = 4.2 \times 10^5$ and $B = 7 \times 10^{-3}$ Write your answer in words.

13 When the Sun, Earth and Mars are aligned, the distance from Earth to Mars is 5.46×10^7 km and the distance from the Sun to Mars is 2.06×10^8 km

What is the distance from the Sun to Earth?

What do you think? 🌐

1 A grain of salt has a mass of approximately 6×10^{-5} g

Approximately how many grains of salt are in a 750 g packet?

2 The International Space Station orbits Earth at a speed of 2.8×10^4 km per hour.

How far does it travel in

a 1 day **b** 1 year?

3 The United Kingdom is made up of England, Scotland, Wales and Northern Ireland.

The approximate population of each country is shown in the table.

England	Scotland	Wales	Northern Ireland
5.6×10^7	5.4×10^6	3.1×10^6	1 880 000

What is the approximate total population of the United Kingdom?

Consolidate – do you need more?

1 Here are some number cards.

29×10^2 0.12×10^2 $3.2 \times 10^{2.5}$ 3×10^{12} 27×10^2

Which number is written in correct standard form?

2 Match the cards that are equal in value.

a 2.9×10^4 2 900 000

b 2.09×10^3 29 000

c 2.9×10^6 2090

d 2.09×10^4 20 900

Explain how you know.

3 Write these numbers in standard form.

a 2000 **b** 2100 **c** 71 000 **d** 80 400

e 320 000 **f** 5 600 000 **g** 0.000 04 **h** 0.0001

i 0.000 087 **j** 0.000 000 943 **k** 0.000 070 2

4 Work these out. Give your answers in standard form.

a $(4 \times 10^6) + (3 \times 10^6)$

b $(5 \times 10^8) + (3.1 \times 10^8)$

c $(3 \times 10^{-3}) + (5 \times 10^{-3})$

d $(8 \times 10^6) + (7 \times 10^6)$

e $(3.5 \times 10^5) + (2.7 \times 10^4)$

f $(3.5 \times 10^5) + (2.7 \times 10^6)$

g $(3.4 \times 10^{-3}) + (2.1 \times 10^{-2})$

h $(4.241 \times 10^{-2}) + (5.2 \times 10^{-4})$

5 Work these out. Give your answers in standard form.

a $(6 \times 10^5) - (4 \times 10^4)$

b $(5 \times 10^7) - (2.7 \times 10^6)$

c $(3.4 \times 10^{-3}) - (2.1 \times 10^{-4})$

d $(4.241 \times 10^{-3}) - (5.2 \times 10^{-5})$

e $(5.71 \times 10^5) - (2 \times 10^2)$

f $(2.73 \times 10^7) - (2.5 \times 10^4)$

6 Work these out. Give your answers in standard form.

a $4 \times (2 \times 10^4)$

b $(3 \times 10^{-3}) \times 3$

c $(9.24 \times 10^6) \times 10$

d $(4 \times 10^4) \times (2 \times 10^2)$

e $(3 \times 10^{-1}) \times (3 \times 10^{-3})$

f $(3.4 \times 10^5) \times (2 \times 10^3)$

7 Work these out. Give your answers in standard form.

a $(8 \times 10^4) \div 2$

b $(6 \times 10^{-3}) \div 3$

c $(9.24 \times 10^6) \div 10^2$

d $(4 \times 10^6) \div (2 \times 10^2)$

e $(6 \times 10^4) \div (1.5 \times 10^3)$

f $(3.4 \times 10^5) \div (2 \times 10^3)$

Stretch – can you deepen your learning?

1 The table shows information about the approximate area of Earth's surface, in km²

Water	Land	Total
3.6×10^8		5.1×10^8

What percentage of Earth's surface is covered by land? Give your answer to 1 significant figure.

2 $A = 3 \times 10^5$, $B = 2.5 \times 10^4$ and $C = 3.4 \times 10^7$

Find the value of

a AB

b B^2

c $AB - C$

d ABC

e A^2C

3 A ship is carrying containers of two different sizes, standard and large.

Each standard container has the same mass and each large container has the same mass.

The total mass of 4 large and 3 small containers is 2.22×10^7 grams.

The total mass of 2 large and 1 small container is 1.02×10^7 grams.

Find the total mass of 1 large and 1 small container.

4 $p = 4 \times 10^{2x}$

Find an expression in terms of x for $\sqrt{p}$

5 $w \times 10^{-3} + x \times 10^{-5} = y \times 10^{-3}$

Form an expression for y in terms of w and x

6 A and B are numbers in standard form such that $A = x \times 10^a$ and $B = y \times 10^b$

  a Find an expression for AB in standard form if

    i $1 \leqslant xy < 10$ **ii** $10 \leqslant xy < 100$

  b Explain why xy cannot take any other value than those listed in part **a**

7 Find the value of each letter.

  a p and q are integers.

    $(p \times 10^7) + (q \times 10^2) = 50\,000\,400$

  b n is a positive integer less than 10

    $(n \times 10^8) - (n \times 10^5) = 399\,600\,000$

  c x and y are integers.

    $x - y = 1$

    $(y \times 10^{-2}) - (6 \times 10^{-3}) = 0.044$

8 The length and width of a rectangle are in the ratio $5:2$

The perimeter of the rectangle is 1.467×10^{20} mm

35% of the rectangle is shaded.

Calculate the area of the rectangle that is shaded.

Reflect

A number written as $A \times 10^n$ is in standard form. What do you know about A?
What do you know about n? Which of A and n can be positive, negative, zero, decimals or fractions?

6 Numbers
Chapters 6.1–6.5

White Rose Maths

I have become **fluent** in...	I have developed my **reasoning** skills by...	I have been **problem-solving** through...
▨ identifying integers, real and rational numbers	▨ explaining where a number fits in the number system	▨ answering challenging questions involving integers, decimals and fractions
▨ calculating with directed numbers	▨ interpreting the HCF or LCM of two or more numbers in context	▨ using prior knowledge to solve multi-step problems
▨ calculating with integers, decimals and fractions	▨ identifying common misconceptions	▨ exploring multiple methods to answer calculations.
▨ working with numbers in standard form.	▨ interpreting answers to calculations in context.	

Check my understanding

1 Complete the calculations.

    a $15 - 31$ **b** $-18 + 12$ **c** 16×-5 **d** $-121 \div -11$

2 **a** Write 145 000 000 in standard form.

    b Write 0.000 006 07 in standard form.

    c Write 4.7×10^{-5} in ordinary form.

3 The tables shows the cost of flights and accommodation for a holiday.

Return flight	
Adult	£210
Child	£90

Accommodation	
Room A (sleeps 3)	£45 per night
Room B (sleeps 4)	£55 per night

Work out the total cost of a 7-night holiday for 2 adults and 2 children.

4 The cost per ticket for a school play is £4.95

On the first night, the school sells 190 tickets.

On the second night, the school sells 162 tickets.

How much more money does the school make from ticket sales on night 1 than night 2?

5 Timer A beeps every 12 minutes. Timer B beeps every 8 minutes. They both beep at 6:15 a.m.

At what time will they next both beep at the same time?

6 Evaluate $\dfrac{a + bc}{d}$ for $a = \dfrac{1}{2}$, $b = \dfrac{3}{5}$, $c = \dfrac{3}{4}$ and $d = \dfrac{5}{8}$

7 Write $\sqrt{50}$ in the form $a\sqrt{b}$ where a and b are integers. Ⓗ

7 Percentages

In this block, I will learn...

how to convert between fractions, percentages and decimals

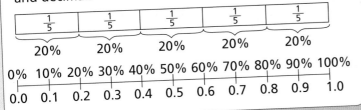

how to increase and decrease by a given percentage

increase £30 by 12%

decrease £30 by 12%

how to work out a percentage change

Find the percentage profit made when an item bought for £80 is sold for £100

profit = selling price – cost price

percentage profit = $\dfrac{\text{profit}}{\text{original amount}}$

how to find the original number given the result of a percentage change

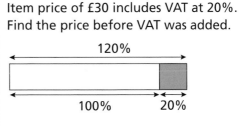

Item price of £30 includes VAT at 20%. Find the price before VAT was added.

how to apply repeated percentage change Ⓗ

Original price £180

Sale: 20% off

One day only: extra 10% off

Find the sale price

100% – 20% = 80%

80% as a decimal is 0.8

£180 × 0.8 = £144

100% – 10% = 90%

90% as a decimal is 0.9

£144 × 0.9 = £129.60

7.1 Percentage basics

Small steps

- Use the equivalence of fractions, decimals and percentages ®
- Calculate percentage increase and decrease ®
- Express a change as a percentage ®

Key words

Equivalent – numbers or expressions that are written differently but are always equal in value

Multiplier – a number you multiply by

Profit – if you buy something and then sell it for a higher amount, profit = amount received – amount paid

Loss – if you buy something and then sell it for a smaller amount, loss = amount paid – amount received

Are you ready?

1 Write 0.3 as

    a a percentage   **b** a fraction.

2 Write 55% as

    a a decimal   **b** a fraction.

3 Round 45.576 to

    a the nearest integer   **b** one decimal place   **c** two decimal places.

4 Find 30% of 150

5 Find $\frac{2}{5}$ of £240

Models and representations

Hundred square

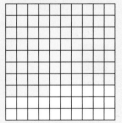

Bar model

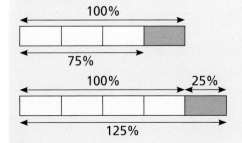

Fraction wall

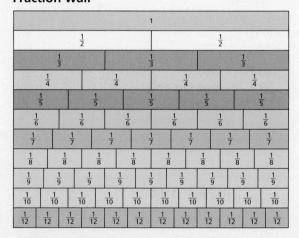

Number line

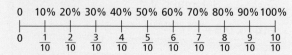

In this chapter, you will be reviewing your work on fractions, decimals and percentages from Books 1 and 2

Example 1

Express these fractions as percentages.

a $\dfrac{2}{5}$ **b** $\dfrac{5}{9}$ **c** $1\dfrac{1}{4}$

a

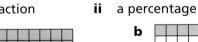

$\dfrac{2}{5} = \dfrac{4}{10} = \dfrac{40}{100} = 40\%$

Per cent means "out of 100" so you need to change $\dfrac{2}{5}$ to an **equivalent** fraction with 100 as the denominator.

$\dfrac{2}{5}$ is equivalent to $\dfrac{40}{100}$, which means 40 out of 100, or 40%

b $\dfrac{5}{9} = 5 \div 9 = 0.555\,555\dots$

$0.555\,555\dots \times 100 = 55.5555\dots\%$

$= 55.6\%$ to 3 s.f.

You can convert $\dfrac{5}{9}$ to a decimal by calculating $5 \div 9$ and then multiplying by 100

Round your answer to a suitable degree of accuracy.

9 is not a factor of 100 so you can't use the same method as in part **a**

c $1\dfrac{1}{4} = 1.25$

$1.25 \times 100 = 125\%$

$1\dfrac{1}{4}$ is the same as 1.25

To convert from a decimal to a percentage, you multiply by 100

Practice 7.1A

1 For each grid, write down the proportion of the total area that is shaded as

    i a fraction   **ii** a percentage   **iii** a decimal.

a
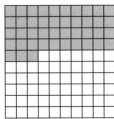

b
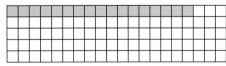

c

d

2 **a** Convert each of these percentages to a fraction and a decimal.

  i 20%   **ii** 42%   **iii** 17.5%   **iv** 140%   **v** 104%

  b Convert each of these fractions to a decimal and a percentage.

  i $\dfrac{3}{10}$ **ii** $\dfrac{4}{5}$ **iii** $\dfrac{6}{7}$ **iv** $1\dfrac{1}{8}$ **v** $\dfrac{1}{3}$

  c Convert each of these decimals to a fraction and a percentage.

  i 0.7   **ii** 0.63   **iii** 0.05   **iv** 1.32   **v** 1.456

3 Beca scored $\dfrac{35}{40}$ in a Maths test and $\dfrac{52}{60}$ in a Science test. Chloe says that Beca did better in Science because 52 is greater than 35. Is Chloe right?

Show working to explain your answer.

4 Determine whether each statement is true or false.

  a $0.25 = \dfrac{1}{4}$ **b** $0.04 = \dfrac{4}{10}$ **c** $2\% = 0.2$ **d** $18\% = \dfrac{9}{50}$

5 Write these in order of size, starting with the smallest.

  a 0.05 $\dfrac{1}{5}$ $\dfrac{1}{50}$ 15%   1.55 $\dfrac{15}{10}$

  b 1.6   16%   106%   0.106   0.06 $\dfrac{1}{6}$

6 Copy and complete the following, writing <, > or = in the circles to make each statement correct.

  a 0.4 ◯ 40%   **b** $\dfrac{3}{5}$ ◯ 35%   **c** 0.6 ◯ $\dfrac{5}{8}$ **d** $\dfrac{1}{50}$ ◯ 2%

7 If 50% of a number is 35, find

  a 10% of the number

  b 15% of the number

  c 44% of the number

  d 120% of the number.

8 There are some red, blue and yellow marbles in a bag. The probability of picking a red marble is 0.24. The probability of picking a blue marble is $\dfrac{3}{8}$

Calculate the probability of picking a yellow marble. Give your answer as a percentage.

What do you think? 💭

1. There are some pink, orange and purple sweets in a box. $\frac{2}{5}$ of the sweets are pink. $\frac{6}{25}$ of the sweets are orange. What percentage of the sweets are purple?

2. Zach puts some cubes into a box. The probability of selecting a red cube is 15% There are 36 red cubes in the box. How many cubes are there altogether in the box?

3. Marta makes 60 cupcakes for the school cake sale. She sells 20% of them at break time. She sells a further 30 cupcakes at lunch time. What percentage of the cupcakes remain?

4.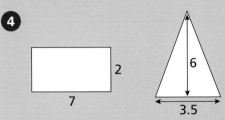

   Work out the area of the triangle as a percentage of the area of the rectangle.

5. Express $3x$ as a percentage of $15x$

Example 2

A coat costs £65. In a sale, the price is reduced by 15%
Work out the cost of the coat in the sale.

Method A

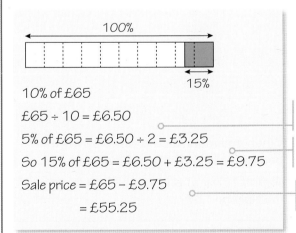

10% of £65

£65 ÷ 10 = £6.50

5% of £65 = £6.50 ÷ 2 = £3.25

So 15% of £65 = £6.50 + £3.25 = £9.75

Sale price = £65 – £9.75

= £55.25

You can find 15% of £65 by finding 10% and 5% of £65 and then adding them together.

Or you can use your calculator and enter 0.15 × 65

To find find 10% of a number you can divide by 10

To find find 5% of a number you can divide 10% by 2

Find the sale price by subtracting the discount from the original price of the coat.

Method B

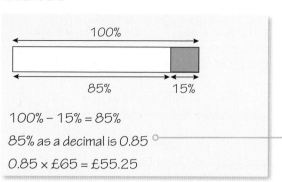

When you reduce an amount by 15%, there will be 85% of the original amount remaining.

Reducing by 15% is the same as finding 85% of an amount.

$100\% - 15\% = 85\%$

85% as a decimal is 0.85

$0.85 \times £65 = £55.25$

The **multiplier** for 85% is 0.85. So, you can multiply by 0.85 to reduce an amount by 15%

Example 3

Emily earns £34 000. This year, her salary will increase by 7%. Work out Emily's new salary.

Method A

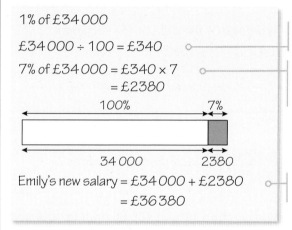

1% of $£34\,000$

$£34\,000 \div 100 = £340$

You can find 1% of any amount by dividing it by 100

7% of $£34\,000 = £340 \times 7$

$\qquad\qquad\qquad = £2380$

To find 7% of an amount you can find 1% and then multiply this by 7. Or you can use your calculator and enter 0.07 × 34 000

Emily's new salary $= £34\,000 + £2380$

$\qquad\qquad\qquad = £36\,380$

Find Emily's new salary by adding the increase of £2380 to her original salary.

Method B

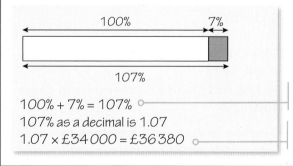

$100\% + 7\% = 107\%$

107% as a decimal is 1.07

$1.07 \times £34\,000 = £36\,380$

When an amount is increased by 7% it becomes 107% of the original amount.

To find 107% of an amount, you can multiply by 1.07

Practice 7.1B

1 A pair of shoes normally cost £60. In a sale, the price is reduced by 25%

a What percentage of the original price is the sale price of the shoes?

b Write the decimal equivalent of your answer to part **a**

c Use your answer to part **b** to find the sale price of the shoes.

2 A train company increases all fares by 3%. Lydia's old travel card cost £245 per month.

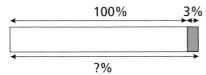

a What percentage of the original prices are the new fares?

b Write the decimal equivalent of your answer to part **a**

c Use your answer to part **b** to work out the price of Lydia's new travel card.

3 Write down the multiplier you would use to decrease an amount by the following percentages.

   a 20%   **b** 35%   **c** 5%   **d** 12.5%   **e** 1.5%

4 Write down the multiplier you would use to increase an amount by the following percentages.

   a 15%   **b** 28%   **c** 3%   **d** 17.5%   **e** 2.4%

5 What is the effect of each of these decimal multipliers?

   a 0.98   **b** 1.05   **c** 1.33

   d 0.34   **e** 0.765   **f** 1.235

6 A laptop costs £840 plus VAT at 20%. Work out the total cost of the laptop.

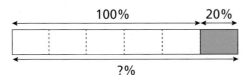

7 In 2001, the population of Manchester was 414 900. By 2011, the population had increased by 21%. Work out the population of Manchester in 2011

8 A car costs £22 000. Each year its value decreases by 12%

    a Work out the value of the car after one year.

    b Explain Flo's mistake and work out the correct value of the car after two years.

> To work out the value after two years, you can do 12% + 12% = 24%
> The multiplier for a 24% decrease is 0.76, so you can calculate
> £22 000 × 0.76 = £16 720

9 A rectangle has length 7 cm and width 3 cm

    a What is the area of the rectangle?

    b The length of the rectangle is increased by 10% and the width is decreased by 20% Work out the new area of the rectangle.

10 Emily wants to buy two boxes of chocolate. Which of these deals will be the better value for money?

A

£5 each with 10% discount.

B

£5.50 each. Buy one get one half price.

What do you think? 💡

1 In a sale, a shop reduces the prices of all products by 20%. For a special promotion, it offers an extra 10% off all sale prices. A candle cost £14 before the sale.

    a Calculate the cost of the candle in the special promotion.

    b Explain why a 20% decrease followed by a 10% decrease is not the same as a 30% decrease.

2 A rectangle has length 8 cm and width 10 cm. Its length is increased by 25% and its width is increased by 15%

    a Calculate the percentage increase in the area of the rectangle.

    b Is there more than one way of doing this?

3 Which of the following calculations can be used to find 24% of 63?

    A 2.4×63 B $2.4 \div 63 \times 100$ C 0.24×63

    D $100 \div 24 \times 63$ E $63 \div 100 \times 24$

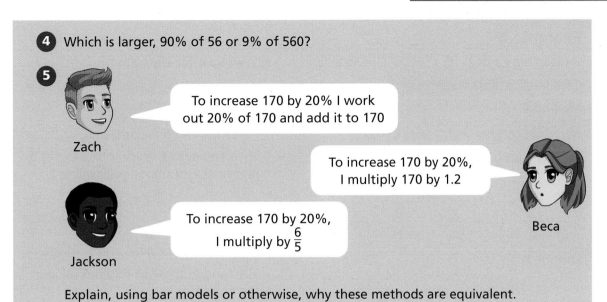

4 Which is larger, 90% of 56 or 9% of 560?

5

Zach: To increase 170 by 20% I work out 20% of 170 and add it to 170

Beca: To increase 170 by 20%, I multiply 170 by 1.2

Jackson: To increase 170 by 20%, I multiply by $\frac{6}{5}$

Explain, using bar models or otherwise, why these methods are equivalent.

In this section, you will learn how to express a change in value as a percentage, reviewing prior knowledge from Book 2

Example 4

A shop buys calculators for £8 each and sells them for £15 each. Calculate the percentage profit.

Method A

profit = £15.00 − £8.00 = £7.00 ○——— First work out the actual **profit**.

$$\text{percentage profit} = \frac{\text{profit}}{\text{original amount}} \times 100$$

Then you need to express the profit as a percentage of the original value.

$$= \frac{7}{8} \times 100$$

$$= 87.5\%$$

Use the formula for percentage profit that you learned in Book 2

Method B

$$8 \times \text{multiplier} = 15$$
$$\div 8 \qquad \qquad \div 8$$
$$\text{multiplier} = 1.875$$

The multiplier is 1.875

This is the same as 187.5% ○———

The percentage profit is 187.5% − 100% = 87.5%

You know the original and final values. You can use these to find the multiplier and then consider the effect of that multiplier.

Finding the multiplier is like solving an equation:

8 × ? = 15

This is the original 100% plus an increase of 87.5%

Example 5

Jackson buys a boat for £50 000. One year later he sells the boat for £36 000

Calculate his percentage loss.

Method A

loss = £50 000 – £36 000 = £14 000 ○————┤ First work out the actual **loss**.

$$\text{percentage loss} = \frac{\text{loss}}{\text{original amount}} \times 100$$ ○————┤ Then you need to express the loss as a percentage of the original value.

$$= \frac{14\,000}{50\,000} \times 100$$

$$= 28\%$$

Method B

$$50\,000 \times \text{multiplier} = 36\,000$$
$$\div 50\,000 \left(\right) \div 50\,000$$
$$\text{multiplier} = 0.72$$

The multiplier is 0.72

This is the same as 72%

The percentage loss is 100% – 72% = 28% ○————┤ This is the original 100% minus 28%

You know the original and final values. You can use these to find the value of the multiplier.

Finding the multiplier is like solving an equation:

50 000 × ? = 36 000

Practice 7.1C

Where appropriate, round your answers to 1 decimal place.

1 Abdullah buys a scooter for £75. He later sells it for £60

    a What loss does Abdullah make on the scooter?

    b What percentage loss does he make?

2 A shop buys printers for £80 and sells them for £95

    a How much profit does the shop make for each printer?

    b What percentage profit does the shop make for each printer?

3 Emily buys 6 mugs for £17.10. She sells each mug for £3.55. Work out her percentage profit.

4 Beth buys a car for £8000. Two years later, she sells the car for £5500. Work out her percentage loss.

5 During 2014, the population of a village decreased from 675 to 540. Work out the percentage decrease in population.

6 Copy and complete the table.

Original value	Multiplier	Final value	Effect
20	1.2		20% increase
45	0.9		10% decrease
120		150	
120		96	

7 Rob's monthly rent increased from £800 per month to £1000 per month.

a Find the percentage increase in his monthly rent.

At the same time, his wages increased from £2200 to £2600 per month.

b Find the percentage increase in Rob's monthly wages.

c Comment on your answers.

8 My test mark went up from 55% to 75%

My test mark went up from 42% to 62%

Seb

Ed

Who has made the most improvement?

What do you think?

1 At the start of January, a kitten weighed 0.8 kg. During January its weight increased by 35%. At the start of March, the kitten weighed 1.35 kg. What was the kitten's percentage gain in weight during the month of February?

2 Chloe buys 200 bracelets for £120. At the summer fete, she sells one-quarter of the bracelets for £2. She then reduces the price of the remaining bracelets by 25%. At the end of the day, she has 20 bracelets left. Work out:

a the total number of bracelets that she sold at the fete

b Chloe's total income

c the percentage profit that she made.

3 £400 to £800 is an increase of 100%. This means that £800 to £400 is a decrease of 100%

a Is Abdullah correct? Explain why or why not.

b Is it possible to increase a value by 250%? If so, what would the multiplier be? If not, why not?

c Is it possible to decrease a value by 250%? If so, what would the multiplier be? If not, why not?

Consolidate – do you need more?

1 Change each of these percentages to a decimal and a fraction.

   a 20%   **b** 50%   **c** 7.5%   **d** 12.5%   **e** 140%

2 Change each of these fractions to a decimal and a percentage.

   a $\frac{17}{50}$ **b** $\frac{3}{5}$ **c** $\frac{13}{20}$ **d** $\frac{25}{200}$ **e** $\frac{1}{3}$

3 Work out

   a 12 minutes as a fraction of 1 hour

   b 42 cm as a percentage of 1 m

   c 242 g as a percentage of 1 kg

4 Write these in order of size, starting with the smallest.

   a $\frac{1}{2}$ 0.22   0.52   53%   0.502 $\frac{2}{5}$

   b $\frac{3}{10}$ $\frac{1}{3}$ 33%   31%   0.35   3.3

5 Ed achieved these results in his end of year exams.

Maths	English	Science
84%	$\frac{44}{50}$	$\frac{17}{20}$

In which subject did he do best? Show working to explain your answer.

6 A box of cereal has a mass of 375 g. A promotional box contains 20% extra free. What is the mass of the promotional box?

7 A dress costs £42. It is reduced by 15% in a sale. What is the sale price of the dress?

8 Zach buys scarves for £15 and sells them for £21. What is his percentage profit?

9 Benji buys a bike for £120 and sells it for £85. What is his percentage loss?

10 Faith buys 12 cupcakes for £9. She sells each cupcake for £1.25. Calculate her percentage profit.

Stretch – can you deepen your learning?

1 Seb is investigating the areas of various rectangles.

   a The length of a rectangle is increased by 60% but its area remains the same. By what percentage is the width decreased?

   b The length of another rectangle is increased by 25% and its area remains the same. By what percentage is the width decreased?

   c By using your answers to parts **a** and **b** and writing them as fractions, can you generalise your findings?

2 A ball is dropped from a height. After each bounce, it reaches a height that is 80% of the previous maximum height. After how many bounces would it reach a height that is less than 10% of the original height?

Reflect

1 Explain how you can use both bar models and multipliers to answer percentage profit and loss questions.

2 Explain why you can't add percentages when working out repeated percentage change. For example, why isn't a 10% decrease followed by a 15% decrease the same as a 25% decrease?

Small steps

■ Solve reverse percentages with and without a calculator

Key words

Original value – a value before a change takes place

Reverse percentage – a problem where you work out the original value

Profit – if you buy something and then sell it for a higher amount, profit = amount received – amount paid

Are you ready?

1   Write each of these percentages as a decimal.

   a   30%   **b**   56%   **c**   17.4%   **d**   120%

2   Write down the decimal multiplier you would use to

   a   increase a quantity by 25%   **b**   decrease a quantity by 35%

   c   decrease a quantity by 42%   **d**   increase a quantity by 4.5%

   e   increase a quantity by 123%   **f**   decrease a quantity by 4%

3   Solve the following equations.

   a $1.2x = 48$ **b** $0.8x = 75$ **c** $1.45x = 210$ **d** $0.65x = 14.5$

Models and representations

When a value has been increased by 20%, the new value is 120% of the **original value**.

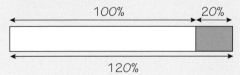

When a value has been decreased by 20%, the new value is 80% of the original value.

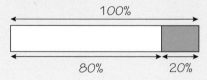

In this chapter, you will learn to solve problems involving finding the original amount prior to a percentage change.

These problems are often called **reverse percentage** problems. They can be approached using calculator and non-calculator methods.

Example 1

The price of a calculator is reduced by 20% to £6.40. Find the original price of the calculator.

Method A

100% – 20% = 80%

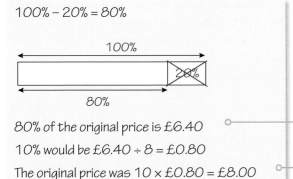

80% of the original price is £6.40

10% would be £6.40 ÷ 8 = £0.80

The original price was 10 × £0.80 = £8.00

Work out what percentage of the original price is left after the reduction.

80% = £6.40 so 10% = 80% ÷ 8

The original price is 100%, which is 10 × 10%

Method B

100% – 20% = 80%

80% = 0.8

Let x be the original amount.

$$x \times 0.8 = £6.40$$
$$÷0.8 \Big(\qquad \Big) ÷0.8$$
$$x = £8.00$$

So the original price was £8.00

Work out what percentage of the original price is left after the reduction.

Write this as the equivalent decimal multiplier.

For a 20% decrease, the multiplier is 0.8

Set up an equation to show the change that has taken place.

Solve the equation by balancing.

State the final answer.

Example 2

Beca's rent increases by 30%. After the increase, her rent is £780 per month.
Find the value of her rent before the increase.

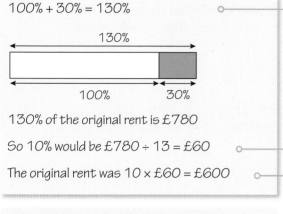

100% + 30% = 130% ○———— Work out what percentage of the original
rent the increased rent is.

130% of the original rent is £780

So 10% would be £780 ÷ 13 = £60 ○———— 130% = £780 so 10% = 130% ÷ 13

The original rent was 10 × £60 = £600 ○———— The original rent is 100%, which is
10 × 10%

State the final answer.

You could have worked this out by
finding 1%

130% = £780
÷ 130 ⟨   ⟩ ÷ 130
1% = £6
× 100 ⟨   ⟩ × 100
100% = £600

Example 3

Flo earns 3% interest on her savings each year. At the end of the first year, she has a total of
£1648 in her account. Work out how much she deposited before the interest was added.

100% + 3% = 103% ○———— Work out the new percentage after the
increase of 3%

103% = 1.03 ○———— Write this as the equivalent decimal multiplier.

The decimal multiplier for a 3% increase is 1.03

Let x be the original amount she deposited.

$x \times 1.03 = £1648$ ○———— Set up an equation to show the change that
has taken place.
÷ 1.03 ⟨   ⟩ ÷ 1.03

$x = £1600$ ○———— Solve the equation by balancing.

So the original amount was £1600 ○———— State the final answer.

Practice 7.2A

Answer questions 1–6 without a calculator.

1 40% of a number is 100

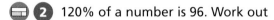

20%	20%	20%	20%	20%

100

Work out

  a 10% of the number   **b** 5% of the number

  c 60% of the number   **d** 100% of the number.

2 120% of a number is 96. Work out

  a 10% of the number   **b** 20% of the number

  c 35% of the number   **d** 100% of the number.

3 Abdullah saves 20% of his monthly salary. Each month he saves £160. What is his monthly salary?

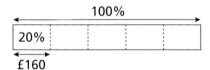

4 30% of the students in Year 9 have blue eyes. 36 students have blue eyes.

  a How many students are there in Year 9?

  b How many students do not have blue eyes?

5 In a sale, a shop reduces the price of all DVDs by 10%. Seb buys a DVD for £18. Work out the original cost of the DVD.

6 After a 20% pay rise, Ed earns £1680 per month. Work out his monthly salary before the pay rise.

7 The price of a computer is reduced by 40% to £564. What was the original price of the computer?

8 A promotional box of cereal contains 15% more than a normal box. The promotional box contains 690 g. How much does the normal box contain?

9 The price of a television set is £450 including VAT at 20%. Work out the price of the television set before VAT was added.

10 The price of a train ticket after a 5% fare increase is £65.10. Find the price of the ticket before the fare increase.

11 Zach is working on this problem.

> After a 15% pay rise, Benji earns £1380 per month. What did Benji earn before the pay increase?

> That's easy! 15% of £1380 is £207. So you do £1380 − £207 = £1173

a Explain why Zach is wrong.

b Work out the correct answer.

12 Copy and complete the table.

Original amount	Percentage change	New amount
45	increase by 20%	
20		16
	increase by 12%	134.4
	decrease by 45%	49.5
	increase by 7.5%	86

13 Beca sells her bike for £99. She makes a 10% loss on the original cost of the bike. How much did Beca pay for the bike?

14 Jakub invests money in a savings account. At the end of each year, he receives 4% interest on his savings. At the end of the first year, he has £1643.20 in the account. How much interest did he earn over the year?

15 Faith sells her car for £2760, making a 15% profit on the amount she originally paid for it. Work out how much profit Faith made.

What do you think?

1 Ed sells his bike to Seb and makes a 20% profit. Seb then sells the bike to Benji for £144. Seb makes a 20% loss. How much did Ed originally pay for the bike?

2 Flo sells a laptop to Chloe. Chloe sells the laptop to Jackson for £503.36. Both Flo and Chloe each make a 12% loss. How much did Flo pay for the laptop?

3

> If a value increases by 20% and then decreases by 20% it will be back where it started.

Try this out with some starting values of your choice. Is Marta correct? Explain your answer.

Consolidate – do you need more?

Answer questions 1–4 without a calculator.

1 30% of a number is 36

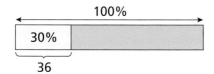

Work out

a 10% of the number   **b** 40% of the number

c 55% of the number   **d** 100% of the number.

2 100 ml of milk contains 124 mg of calcium. This is roughly 15% of an adult's recommended daily intake of calcium.

Using the above information, what is the recommended daily intake of calcium for an adult?

3 The prices of these items include VAT at 20%. How much did each item cost before VAT was added?

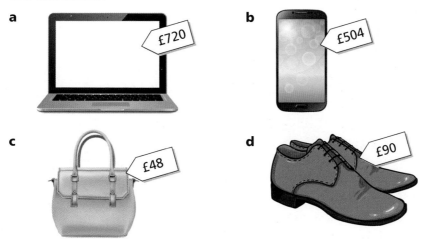

4 A bottle of shampoo is labelled "25% extra free". The bottle contains 250 ml of shampoo. Work out how much shampoo a normal-sized bottle contains.

5 Abdullah's restaurant bill comes to £38.50. This includes a 10% service charge. Calculate the cost of the meal before the service charge was added.

6 A sunflower is 180 cm tall. This is an increase of 5% on its height a week earlier. What was the height of the sunflower the previous week? Give your answer to 1 decimal place.

7 Emily sells her car for £3500. She makes a 12% profit on the price she paid. How much did Emily pay for the car?

8 After an 8% pay rise, Faith earns £91.80 a week. How much was Faith paid before the pay rise?

9 The cost of a theatre ticket is reduced by 5% to £42.75. What was the cost before the decrease?

10 This week, Chloe went to the gym for 40 minutes each day.

a If this is an increase of 25% per day compared with the previous week, how many minutes each day did she spend at the gym last week?

b Next week Chloe plans to go to the gym for 48 minutes per day. Calculate the percentage increase from 40 minutes to 48 minutes.

Stretch – can you deepen your learning?

1 Zach and Beca both get a pay rise of the same percentage. Zach's salary increases from £22 500 to £23 400

a Work out the percentage increase in Zach's salary.

b Beca's salary increases to £23 920. Use your answer to part **a** to work out Beca's salary before her pay increase.

2

Ed

54 × 0.8 represents finding 80% of 54

But 54 × 0.8 represents decreasing 54 by 20%

Faith

Explain why both Ed and Faith are correct.

3 A rectangle is enlarged so that the area increases by 25%. The enlarged rectangle measures 5 cm by 6 cm

a Calculate the area of the original rectangle.

b Given that the original rectangle also has length 6 cm, find the original width.

c Find the percentage increase in the width of the rectangle. Comment on your answer.

4   There are 3000 people at a football match. Benji says that this is exactly 30% more people than at the last match. Explain why Benji is wrong.

5   Emily buys a games console and a mobile phone. The games console costs £200. She sells both items for a total of £720. She made a 50% profit on the price of the games console and a 20% profit on the price of the mobile phone. Work out the original cost of the mobile phone.

6   Zach is 10% taller than he was last year and Jackson is 20% taller than he was last year. Zach is now 132 cm tall. The ratio of Zach's height to Jackson's height is now 11 : 10

   a   Work out Jackson's height last year.

   b   Work out the ratio of Zach's height to Jackson's height last year.

Reflect

1   By giving an example, explain how you can use bar models and equations to find the original amount when you know the result of a percentage change.

2   Write down any key words or phrases that are clues that you are working on a reverse percentage problem.

Small steps

- Recognise and solve percentage problems (non-calculator)
- Recognise and solve percentage problems (calculator)

Key words

Multiplier – a number you multiply by

Original value – a value before a change takes place

Reverse percentage – a problem where you work out the original value

Are you ready?

1 Round each of these numbers to the given degree of accuracy.

    a 2.354 (2 d.p.)   **b** 1424 (2 s.f.)   **c** 0.0431 (2 d.p.)   **d** 56.784 (1 d.p.)

2 Copy and complete this table.

Original price	Selling price	Profit or loss
£6	£8	£2 profit
£5	£4	
£19		£3 loss
	£18	£7 profit

3 Work these out without using a calculator.

    a 30% of 240   **b** 15% of 60   **c** 65% of 180   **d** 12% of 45

4 Work these out using a calculator, giving your answers to 1 decimal place.

    a 15% of 76   **b** 23% of 121   **c** 9% of 63   **d** 11.5% of 3353

5 Explain the effect of each of these multipliers as either an increase or a decrease.

    a 1.1   **b** 1.34   **c** 0.98   **d** 0.6   **e** 1.045   **f** 0.975

Models and representations

Bar model

Number line

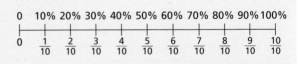

In this chapter, you will solve a mixture of percentage problems using methods from Books 1 and 2 and from earlier in this block.

The first section will focus on developing non-calculator skills.

Example 1

Without using a calculator

a   express 14 as a percentage of 40

b   work out 14% of 40

a

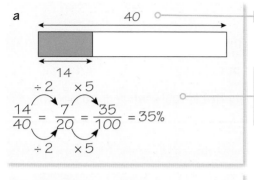

40 is the whole and 14 is the part.

You can write $\frac{14}{40}$ as a fraction and then use equivalent fractions to convert the denominator to 100

Remember: "per cent" means out of 100

b

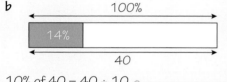

10% of 40 = 40 ÷ 10
          = 4

1% of 40 = 40 ÷ 100
         = 0.4

4% = 0.4 × 4
   = 1.6

14% = 10% + 4%
    = 4 + 1.6
    = 5.6

10% is 100% ÷ 10, so to find 10% you divide by 10

1% is 100% ÷ 100, so to find 1% you divide by 100

4% = 4 × 1%

Find 10% of 40 and 4% of 40, and then add them to find 14% of 40

Example 2

In a sale, a shop reduces the prices of all items by 20%. Without using a calculator, work out

a   the sale price of a coat originally costing £60

b   the original price of a skirt that costs £36 in the sale.

a

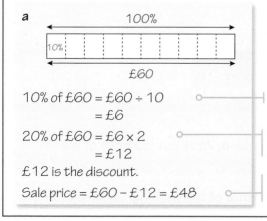

10% of £60 = £60 ÷ 10
           = £6

20% of £60 = £6 × 2
           = £12

£12 is the discount.

Sale price = £60 − £12 = £48

You can find the discount by finding 20% of 60

10% = 100% ÷ 10, so to find 10% you divide by 10

20% is double 10%, so to find 20% you multiply £6 by 2

Then you can subtract the discount from the **original** price.

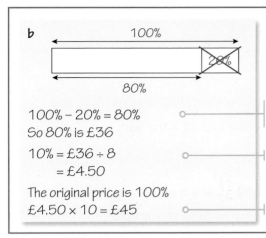

b

100% − 20% = 80%
So 80% is £36

10% = £36 ÷ 8
= £4.50

The original price is 100%
£4.50 × 10 = £45

The word "original" is a clue that this is a **reverse percentage** question.

After a 20% discount, 80% of the original price is left.

80% ÷ 8 = 10%

The original price is 100% which is 10% × 10

Practice 7.3A

Answer all these questions without a calculator.

1 **a** Express 60 as a percentage of 240

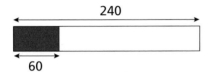

b Work out 60% of 240

2 What percentage of each shape is shaded?

a

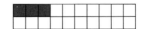

b

c

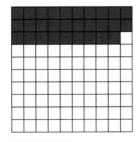

d

e

3 A computer costs £800 before VAT at 20% is added. Work out the price of the computer after VAT is added.

4 A washing machine costs £450. Work out the price of the washing machine after a 20% discount.

5 Beca scores 68 out of 80 in a Science test. Express her score as a percentage.

6 Marta scored 84 marks in a test. As a percentage, Marta scored 70%. Work out the total number of marks in the test.

 7 In a school, 15% of the students wear glasses. If 63 students wear glasses, how many students are in the school?

8 Jackson buys a scooter. He pays £300 including VAT at 20%. Work out the price of the scooter before VAT is added.

9 Here is some information about the students in a class.

	Left-handed	Right-handed
Boys	3	12
Girls	1	9

   a What percentage of boys are right-handed?

   b What percentage of left-handed students are girls?

   c What percentage of the students are girls?

10 Faith and Flo each earn a basic salary of £140 per week. Last week Faith was given a 15% bonus and Flo was given a £40 bonus. Work out the difference between the total amounts that Flo and Faith were paid last week.

11 In a restaurant, a tip jar containing £24 is shared between three waiting staff and two cooks. 30% of the tips are shared equally between the cooks. The rest is shared equally between the waiting staff. How much more money does each of the waiting staff receive than each of the cooks?

What do you think? 💭

1 Zach and Beca work for a company that sells cartons of orange juice. The company is planning a special offer.

I think we should increase the carton size by 20% and keep the price the same

Zach

I think we should reduce the price by 20% and keep the size of the carton the same

Beca

Which special offer would be better value for money for the customers?

2 In a school, 40% of the students are boys. 20% of the girls walk to school and 10% of the boys walk to school. What percentage of all the students walk to school?

In this section, you will solve a variety of problems involving percentages. The focus will be on using your calculator efficiently.

Example 3

Ed gets a 3% pay rise. His new salary is £24720. Work out Ed's salary before he was given the pay rise.

$100\% + 3\% = 103\%$ ○————————— Work out the total percentage after an increase of 3%

$103\% = 1.03$ ○————————— Write this as the equivalent decimal **multiplier**.

Let x be Ed's original salary:

> The decimal multiplier for a 3% increase is 1.03

$x \times 1.03 = 24720$ ○

$\div 1.03$ (   )$\div 1.03$ ○————— Set up an equation to show the change that has taken place.

$x = 24000$

Solve the equation by balancing.

Ed's original salary was £24000   ○————— End by stating the answer to the question that was asked.

Example 4

Last year, Marta paid £122 for pet insurance. This year, she has to pay £185 for pet insurance. Work out the percentage increase in the cost of her pet insurance.

increase $= 185 - 122 = £63$ ○————— First, work out the actual increase.

percentage increase $= \dfrac{change}{original\ amount} \times 100$ ○————— Then you need to express the increase as a percentage of the original amount.

$\dfrac{63}{122} \times 100 = 51.639\ldots$

$= 51.6\%\ (3\ s.f.)$ ○————— Round your answer to a sensible degree of accuracy.

Practice 7.3B

1 A necklace costs £23. In a sale, the price is reduced by 12%. What is the sale price of the necklace?

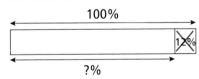

2 In a survey last year, the population of a village was 4752. One year later, the population has increased by 3%. Work out the new population of the village. Give your answer to the nearest integer.

3 Flo scored 56 out of 80 in a Maths test. Ed scored 64 out of 92 in a different test.

I did better than Flo because 64 is greater than 56

Is Ed right? Explain your answer.

4 Seb buys a bike for £245 and sells it for £380. Work out

   a Seb's profit   **b** Seb's percentage profit.

5 In a sale, the price of a dress is reduced from £26 to £20.80. Work out the percentage reduction in the price of the dress.

6 After a 5% fare increase, a travel card costs £13.23. Work out the original price of the travel card.

7 A juice carton measures 8 cm by 8 cm by 15 cm. It is 80% full. Abdullah uses the juice to fill some cups. Each cup holds 230 ml. How many cups of juice will Abdullah be able to fill?

15 cm

8 cm   8 cm

8 A ball is dropped from a height of 150 cm. After each bounce, it reaches a height 15% lower than its previous height. What height does it reach after the third bounce?

9 Which of these calculations represents increasing 55 by 12%?

  A 55×0.12 B 55×1.12 C $0.12 \times 55 + 55$ D $\frac{55}{100} \times 112$ E $55 \div 0.88$

10 In a sale, the price of a laptop is reduced by 25%. A week later, the price is reduced by a further 15%. The final price of the laptop is £586.50. Work out the original price of the laptop before the sale.

11 A   B   C

The area of circle B is 10% greater than the area of circle A. The area of circle C is 20% greater than the area of circle B. By what percentage is the area of circle C greater than the area of circle A?

12 Ali buys a car for £18 000. The value of the car depreciates by 16% in the first year and 12% each year after the first. Work out the value of the car at the end of 3 years.

What do you think? 💭

1 Because of a 10% increase, the price of Chloe's train ticket increases **by** £2.30. Work out the price of Chloe's ticket before the increase.

2 Benji buys a car for £4000 plus VAT at 20%. He pays a deposit for the car and pays the remainder of the cost in 12 equal monthly payments of £350

  a Work out the cost of the car after VAT is added.

  b Work out how much Benji pays as a deposit.

  c Find the ratio of Benji's deposit to the total monthly payments. Give your answer in its simplest form.

3 In an office, the ratio of men to women is 3 : 4. 12 men, which is 20% of the number of men in the office, wear glasses.

  a How many men work in the office?

  b How many women work in the office?

  c 15% of the women wear glasses. Work out the total percentage of employees who wear glasses.

Consolidate – do you need more?

1 Express 23 as a percentage of 50

2 Zach has 360 marbles. He gives 5% of his marbles to Jackson.

  a How many marbles does Jackson receive?

  b What fraction of his marbles does Zach have left?

3 **a** Work out 90% of 60

  b Work out 60% of 90

  c What do you notice?

4 A watch costs £48. Its price is reduced by 12% in a sale. Work out the sale price of the watch.

5 Flo spends 30 minutes on her homework each night. She plans on increasing this by 16% next week. Work out how long Flo plans to spend on her homework each night next week. Give your answer to the nearest minute.

6 Abdullah earns £1520 per month. He spends £532 on rent. What percentage of his salary does Abdullah spend on rent?

7 In a sale, all prices are reduced by 35%

    a Work out the sale price of a kettle that cost £36 before the sale.

    b Work out the original price of a sofa that costs £442 in the sale.

8 Whose test score is best?

I scored 53 out of 60

Seb

I got $\frac{2}{3}$ of the questions right.

Ed

I scored 79%

Jackson

9

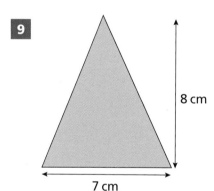

8 cm

7 cm

A triangle has base 7 cm and height 8 cm. Its base is increased by 10% and its height is decreased by 10%. Work out the new area of the triangle.

Stretch – can you deepen your learning?

1 Zach works 4 days a week. Each working day, he works from 9 a.m. to 4 p.m. He is paid £8 per hour. At the end of the week, he is given a 20% bonus on his earnings. Work out how much Zach earns each week.

2 In a sale, the price of a fridge is reduced by $\frac{2}{5}$. The fridge now costs £279. Work out the original price of the fridge.

3 Flo buys 12 cupcakes for £5.64. She sells the cupcakes for 60 pence each. Work out Flo's percentage profit.

4 Ed, Seb and Faith share £770 between them. The ratio of Ed's share to Seb's share is 2:5. Seb gets £210 more than Ed.

    a How much money do they each get?

    b What percentage of the £770 does Faith get?

5 Marta invests £4000 in a savings account. At the end of the year, interest is added to her account. Marta spends 15% of the total amount of money in her account. She has £3468 left. Work out the percentage interest rate on the account.

Reflect

Identify any key words that can help you to decide which type of percentage question you are working on.

⊕ 7.4 Repeated percentage change

Small steps

■ Solve problems with repeated percentage change ⊕

Key words

Repeated percentage change – when an amount is changed by one percentage followed by another

Depreciate – reduce or decrease in value

Are you ready?

1 Write the decimal multiplier that will increase an amount by each of these percentages.

     a 10%   **b** 15%   **c** 28.5%   **d** 3%   **e** 2.5%

2 Write the decimal multiplier that will decrease an amount by each of these percentages.

     a 10%   **b** 4%   **c** 49%   **d** 12.5%   **e** 5.5%

3 Work out

     a 10% of 80   **b** 25% of 120   **c** 34% of 75   **d** 19% of 63   **e** 8.5% of 90

4 Round each of these numbers to the given degree of accuracy.

     a 4.335 (2 d.p.)   **b** 426.78 (1 d.p.)

     c 54.82 (3 s.f.)   **d** 53 014 (2 s.f.)

     e 15.699 (2 d.p.)

Models and representations

Bar models

When a value has been increased by 20%, the new value is 120% of the original value. The equivalent decimal multiplier is 1.2

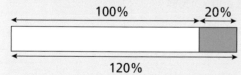

When a value has been decreased by 20%, the new value is 80% of the original value. The equivalent decimal multiplier is 0.8

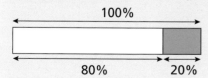

In this chapter, you will be exploring the effects of **repeated percentage changes**; this occurs when a value is increased or decreased by a given percentage more than once.

Example 1

The number of bacteria cells in a test tube increases by 5% each hour.

If there are originally 400 cells, how many cells will there be after

a 1 hour

b 4 hours?

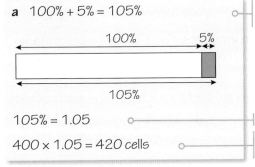

a $100\% + 5\% = 105\%$

First, work out the decimal multiplier for a 5% increase.

$105\% = 1.05$

$400 \times 1.05 = 420$ cells

The multiplier for a 5% increase is 1.05
Then use the multiplier to calculate the number cells after the increase.

b

After 1 hour: $400 \times 1.05 = 420$

After 2 hours: $420 \times 1.05 = 441$

After 3 hours: $441 \times 1.05 = 463.05$

After 4 hours: $463.05 \times 1.05 = 486.202\ldots$

After 4 hours there will be 486 cells.

Use the multiplier to work out the number of cells after each hour for 4 hours.

Round your answer sensibly.

On my calculator I have entered $400 \times 1.05 \times 1.05 \times 1.05 \times 1.05$.
This is the same as 400×1.05^4

Ed

Example 2

Jakub buys a car for £20 000. The value of the car **depreciates** by 8% per year.

Find the value of Jakub's car after 3 years.

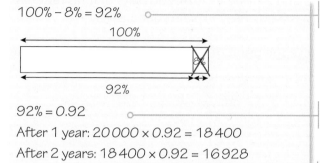

100% – 8% = 92% ○———— First, work out the decimal multiplier for an 8% decrease.

100%

92%

92% = 0.92 ○———— The multiplier for an 8% decrease is 0.92

After 1 year: 20 000 × 0.92 = 18 400

After 2 years: 18 400 × 0.92 = 16 928

After 3 years: 16 928 × 0.92 = 15 573.76 ○———— Then use the multiplier to find the value of the car each year for 3 years.

Value of car after 3 years: £15 574

How could you have used Ed's method to work out the answer more efficiently?

Practice 7.4A

1 **a** Increase £200 by 50%

b Now decrease your answer to part **a** by 50%

c What do you notice?

2 In a sale, the prices of all items in a shop are reduced by 10%. For a promotion, the shop is offering an extra 20% off sale prices. Work out the price of a handbag that cost £60 before the sale.

3 A ball is dropped from a height of 250 cm. After each bounce, it loses 50% of its height. What height will it reach on its

a 1st bounce **b** 4th bounce?

4 There are 300 cells in a Petri dish. The number of cells increases by 6% every hour. Work out the number of cells in the dish after

a 1 hour **b** 5 hours (give your answer to the nearest integer).

5 The value of a car depreciates by 12% per year. Faith's car cost £15 000 when new. Work out the value of the car after

a 1 year **b** 4 years **c** 10 years

6 Abdullah invested £2000 in shares. In March, the value of his shares increased by 4%. In April, the value of his shares decreased by 6%. Work out the value of Abdullah's shares at the start of May.

7 There are 244 people living in a village. The population of the village is growing at a rate of 8% per year. Work out the population of the village in 5 years' time if this rate of growth continues. Give your answer to the nearest integer.

8 Zach's rent increased by 10% in 2019 and by 15% in 2020. Zach says that his rent has increased by 25% in these two years. Is he correct? Explain your answer.

9 Which of these calculations would give the result of a 15% decrease, repeated 4 times?

A $\times 1.5^4$ B $\times 0.15^4$ C $\times 0.85^4$

10 A cube has sides of length 6 cm.

   a   Calculate the volume of the cube.

   Each side of the cube is increased by 10%

   b   Find the new volume of the cube.

   c   Work out the percentage increase in the volume of the cube.

   d   Would your answer to part **c** change if the original side lengths were different? Explain your answer.

What do you think?

1 The value of a car depreciates by 10% each year.

> 10 × 10% = 100%, so after 10 years, the car will be worth £0

Is Chloe correct? Explain your answer.

2 A sunflower is 50 cm tall. Each week its height increases by 8% of its value at the start of that week. How many weeks will it take for the sunflower to double in height?

3 In January, the prices of all items in a shop are reduced by 10%. The following month they are all increased by 10%.

> This means that the prices are now the same as they were originally.

Is Jakub right? Explain your answer.

4 Calculate the overall percentage change if

   a   a 20% decrease is followed by a 30% decrease

   b   a 30% increase is followed by a 15% increase

Consolidate – do you need more?

1 Faith buys a laptop for £450. Faith sells it to Benji, making a 10% loss. Benji sells the laptop to Abdullah. Benji also makes a 10% loss. How much did Abdullah pay for the laptop?

2 Two years ago Seb was 154 cm tall. In the first year his height increased by 2% and in the second year it increased by 1%. What is Seb's height now?

3 There are 400 cells in a test tube. The number of cells increases by 4% every hour. Work out the number of cells in the test tube after

   a 1 hour   **b** 3 hours   **c** 8 hours

4 The value of a car decreases by 8% each year. Ed paid £24 000 for a new car. Work out the value of the car after

   a 1 year   **b** 4 years   **c** 12 years

5 Chloe bought a house for £300 000. In the first year, the value of the house increased by 2%. In the second year, the value of the house increased by 3%. In the third year, the value of the house decreased by 4%. Work out the value of the house after 3 years.

6 Explain why a 20% discount followed by another 20% discount is not the same as a 40% discount.

7 In 2019, the population of London was 8.9 million and the population of New York City was 8.4 million. The population of London is growing at a rate of 1% per year and the population of New York City is growing at a rate of 2% per year. Which city will have the bigger population in 2025?

Stretch – can you deepen your learning?

1 A golf club has 250 members. In January, the membership increases by $\frac{1}{5}$. In February, the membership decreases by 10%. How many members are there at the end of February?

2 The population of a town increased by $\frac{2}{5}$ in 1990 and then decreased by 10% in 1991. Calculate the overall percentage change over the two years. Did the population increase or decrease?

3 In a sale, all items are reduced by 20%. As a special promotion, sale items are reduced by a further 25%

   a Work out the sale price of a cooker that cost £350 before the sale.

   b What is the overall percentage reduction in cost of the combined offers?

4 What is the overall percentage change if an amount is increased by 20% a total of three times?

5 Benji bought a new motorbike. Each year the value of the motorbike decreases by 15%. Work out the number of years it will take for the motorbike to halve in value.

Reflect

1 Explain how to use multipliers to find the effect of more than one percentage change.

2 Explain how using powers can be a more efficient way of calculating the result of repeated percentage changes.

I have become **fluent in...**

■ converting between fractions, decimals and percentages

■ working out percentages of amounts

■ working out percentage increases and decreases

■ working out the original amount given the result of a percentage change

■ working with percentages with and without a calculator.

I have developed my **reasoning** skills by...

■ interpreting the structure of a mathematical problem

■ interpreting relationships multiplicatively

■ understanding percentages greater than 100%

■ understanding and working with percentage changes.

I have been **problem-solving** through...

■ interpreting and solving mathematics problems in a variety of contexts

■ choosing appropriate methods to solve percentage problems

■ solving multi-step problems

■ solving problems involving repeated percentage change. Ⓗ

Check my understanding

1 Convert $\frac{2}{5}$ to

   a a decimal   **b** a percentage.

2 Convert 35% to

   a a fraction   **b** a decimal.

3 Which of these calculations give the same answer?

   A $\frac{1}{3}$ of 60   B 40% of 120   C 0.4 × 5   D 40% of 50

4 The price of a £26 dress is reduced by 15%. Work out the sale price of the dress.

5 A shop raises the prices of all items by 2%. Work out the new price of a desk that cost £125 before the price rise.

6 There are 280 students in a school. 35 of the students attend Science club. What percentage of the students attend Science club?

7 A shop buys picture frames for £5.20 and sells them for £7.28. Work out the percentage profit.

8 A laptop costs £420 including VAT at 20%. Work out the price of the laptop before VAT is added.

9 Marta buys a bike for £210. She sells the bike to Ed at a 10% loss. Ed sells the bike to Jakub for a 15% profit. How much does Jakub pay for the bike? Ⓗ

341

8 Maths and money

In this block, I will learn...

how to analyse bank statements and utility bills

Gas bill

Previous meter reading	2	3	6	4	Price per unit 14p
Current meter reading	2	5	4	6	Fixed charge £19

How many units have been used?

How much does the gas used cost?

How can you find the total charged?

how to calculate simple interest

£1500 is invested in an account paying 4% simple interest per annum. How much money will be in the account after 5 years?

how to solve problems involving wages and salaries

Wages: £9 per hour for a 35-hour week. Overtime paid at time and a half.

Mario works for 42 hours one week. How much does he earn?

How much overtime has Mario done?

What is the hourly rate for overtime?

How can you find his total pay?

how to calculate compound interest

£3400 is invested in an account paying 5% compound interest per annum. How much money will be in the account after 3 years?

how to calculate value-added tax

A laptop costs £450 plus VAT at 20%. What is the total cost of the laptop?

100%	20%
£450	

about exchange rates and how to solve problems involving foreign currency

£1 = €1.15
Change £240 into euros.

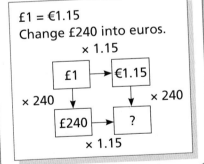

how to identify best value for money

Which is better value for money?

4 tins for £1.85:
1 tin costs £1.85 ÷ 4

6 tins for £2.99:
1 tin costs £2.99 ÷ 6

4 tins for £1.85

6 tins for £2.99

Small steps

■ Solve problems with bills and bank statements

Key words

Bill – shows how much money is owed for goods or services

Balance – an amount of money in an account

Credit – an amount of money paid into an account

Debit – an amount of money taken out of an account

Are you ready?

1 Work out

a £2.90 + £3.15 **b** £4.25 + £0.85 **c** £8 – £6.47 **d** £4.50 – 75p

2 Round each number to 2 decimal places.

a 3.643 **b** 12.4789 **c** 4.0243 **d** 0.4398

3 Work out

a 15 – 45 **b** –24 + 46 **c** –13 – 6 **d** 17 + –10

4

MENU	
tea £1.45	muffin £2.75
coffee.................£3.50	Danish pastry £2.50
orange juice £2.80	croissant £1.85
bottled water £0.90	flapjack £0.85

Filipo buys two cups of tea, a muffin and a croissant.

a How much does Filipo spend?

b How much change will Filipo receive if he pays with a £20 note?

Models and representations

Bank statement

MyBank

ACCOUNT NUMBER: 10045321   STATEMENT: 17
SORT CODE: 90-99-19   PAGE: 1 of 1

TRANSACTION DETAILS

DATE	DESCRIPTION	DEBITS		CREDITS		BALANCE	
Balance brought forward						35	00
6 May	CD07 High St	10	00			25	00
8 May	DC07 Pet care	3	00			22	00
11 May	BACS Regular Times			46	50	68	50
19 May	CH07			2	70	65	80

Utility bill

Reading last time	Reading this time	Tariff C – Customer reading E – Estimated reading No code – Company reading	Units	Price of each unit in pence	Amount
Meter number(s)		K96J747920 Q1 Standard			
64695 E	65295 E	–Standard energy	600	5.72	39.47
		Standing charge at 17.530p for 91 day(s)			15.96
		CCL on 0 units at 0.5090p			0.00
		VAT at 5.00% on charges of 55.42			2.77
		Total this invoice			58.19
		Balance from previous bill			64.52
		Payment received 29 May 2021			64.52 CR
		PAYMENT NOW DUE			58.19

S 03 801 201
S 13 0000 9526 306

Receipt from shops/online shops

```
**********
LOCAL SHOP
**********
MUSSELBURGH
-------------------
SANDWICH          4.00
CHOC BAR*         0.25
CHOC BAR*         0.25
BEETROOT          0.20
SWEETS            0.35
SWEETS            0.35
POTATOES          0.18
NEWSPAPER         0.35
BANANAS           0.79
TOTAL             6.72
COUPON            0.00
CASH              7.00
CHANGE DUE        0.28
===================
```

In this chapter, you will look at household **bills** such as gas, electricity and water bills and how they are calculated using meter readings.

You will also look at bank statements and learn to interpret the different features in order to solve problems.

Gives details about the transaction

Credits are money added to the account

Debits are money taken from the account

This is the **balance** at the start of the statement period

Balance is a running total Credits are added to the total and debits are subtracted from the total

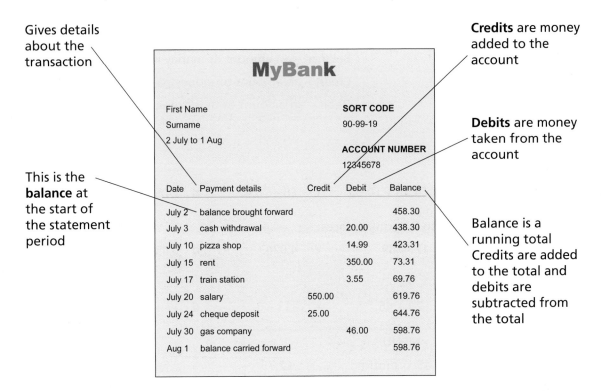

MyBank

			SORT CODE	
First Name			90-99-19	
Surname				
2 July to 1 Aug			ACCOUNT NUMBER	
			12345678	

Date	Payment details	Credit	Debit	Balance
July 2	balance brought forward			458.30
July 3	cash withdrawal		20.00	438.30
July 10	pizza shop		14.99	423.31
July 15	rent		350.00	73.31
July 17	train station		3.55	69.76
July 20	salary	550.00		619.76
July 24	cheque deposit	25.00		644.76
July 30	gas company		46.00	598.76
Aug 1	balance carried forward			598.76

Example 1

Sven has £55 in his bank account.

He spends £32.50 in the supermarket and buys two coffees, each costing £3.40

Later, he pays his gas bill costing £46

a How much money does Sven spend altogether?

b Work out Sven's bank balance after these transactions.

a   £3.40 × 2 = £6.80

Total amount spent:
£32.50 + £6.80 + £46 = £85.30

Sven buys two coffees, so you need to multiply £3.40 by 2

You can use a calculator, which gives the answer 85.3

Remember that working with money means using 2 decimal places and 85.3 is the same as £85.30

b   £55 − £85.30 = − £30.30

So he will be £30.30 overdrawn.

Sven's balance will be what he originally had in his account, subtract what he spent.

A negative balance means Sven owes the bank £30.30

Example 2

Here is part of Zach's gas bill. Work out how much Zach has to pay for the units of gas he has used.

Gas bill				
Old meter reading	8	6	4	2
New meter reading	9	3	1	5

Price per unit 44p
Fixed charge £12

$9315 - 8642 = 673$ units used ○─┤ First, you need to work out how many units of gas Zach has used.

$673 \times 0.44 = £296.12$ ○─┤ Multiply the number of units used by the cost of each unit. Remember to convert 44p to £0.44

$£296.12 + £12 = £308.12$
Zach owes £308.12 ○─┤ Then add on the fixed charge of £12

Practice 8.1A

1 Here is part of Lydia's shopping receipt showing everything she bought.

a How many oranges did Lydia buy?

b Altogether, how much did Lydia spend on drinks?

c What was the total cost of Lydia's shopping?

d Lydia paid with a £20 note. How much change did she receive?

```
**********
LOCAL SHOP
**********
Date 12-09-21
Time    12:19

Oranges
6 @ £0.45        £2.70
Apples
3 @ £0.35        £1.05
Grapes (500g)    £2.00
Orange juice
2 @ £1.50        £3.00
Cola
(6 bottles)      £5.25
Cereal (375g)    £2.19
Jam (300g)       £1.09
```

2 Here is a café menu.

CAFÉ MENU	
bottled water £0.80	muffin £3.55
tea £2.55	Danish pastry £3.00
coffee £4.45	croissant £2.55
orange juice £3.00	fruit bread £1.95
apple juice £4.10	pain au chocolat £2.85

For breakfast, Emily buys a coffee, a croissant and a bottle of water.

Jackson buys a tea, a Danish pastry and a muffin.

Who spends more money on breakfast?

3 At the greengrocer's, Mario buys 500 g of apples and 1 kg of pears.

Apples cost 35p per 100 g and pears cost 55p per 100 g

How much does Mario spend?

4 Seb's water bill shows that he has used 65 cubic metres of water since his last bill.
He is charged 112p per cubic metre plus a fixed charge of £8.50

What is the total cost of Seb's water bill?

5 The diagram shows the reading on Abdullah's gas meter.

Gas bill				
Old meter reading	2	3	5	9
New meter reading	2	5	0	4

Price per unit 22.8p
Fixed charge £9.50

Work out the total cost of Abdullah's gas bill.

6 Benji uses 46 units of electricity in one month.
He is charged 106p per unit.

46 × 106 = 4876p
So my bill is
£487.60

Explain Benji's error.

7 Marta's phone company charges her 3p per minute for calls plus 5p for each text message. She also pays a fixed cost of £23 each month.

In August, Marta uses 134 minutes of call time and sends a total of 45 text messages.

Work out the total cost of her phone bill in August.

8 Explain what it means for a bank account to be overdrawn.

9 Amina has £135 in her bank account. She makes payments of £27.99 and £65.45. She then withdraws £20 in cash.

a How much money will Amina have left in her account?

b Amina then pays a £40 parking fine using money from her account. By how much will she be overdrawn?

10 Work out the correct amount of money for each box, **a–h**, in this bank statement.

MyBank

First Name		SORT CODE		
Surname		11-11-11		
5 Sep to 4 Oct		**ACCOUNT NUMBER**		
		12345678		

Date	Payment details	Credit	Debit	Balance
5 Sep	balance brought forward			£215.56
7 Sep	supermarket		£23.55	a
10 Sep	gym membership		£30.00	£162.01
12 Sep	cash withdrawal		b	£142.01
15 Sep	cash deposit	£50.00		c
18 Sep	water bill		£45.00	d
21 Sep	salary paid in	e		£597.01
28 Sep	coffee shop		£10.19	f
3 Oct	supermarket		£27.66	g
4 Oct	balance carried forward			h

What do you think? 💭

1 Beth has £46 in her bank account. Her bank charges £25 if she becomes overdrawn.

Beth buys some groceries for £25.66 and withdraws £30 in cash.

How much will Beth now owe her bank?

2 Rob buys 600 g of blueberries and 400 g of strawberries. He pays £3.99 in total.

Blueberries cost £3.85 per kilogram. Find the cost of one kilogram of strawberries.

3 Faith received an electricity bill for £39.21. Her bill includes a fixed charge of £14.50 and she is charged 17.4p per unit of electricity.

Calculate, to the nearest whole number, the number of units of electricity Faith has used.

Consolidate – do you need more?

1 Look at the prices of these items.

£2.99

£3.15

£0.65

£5.59

£0.35

£3.99

a Work out the total cost of two notebooks, three pens and a ruler.

b Rhys buys colouring pencils and a notebook. He pays with a £10 note.
How much change will he receive?

2 Faith buys 1.5 kg of rice at a cost of £3.80 per kilogram.

a How much does Faith spend?

b Faith has £10 in her purse. Will she have enough money to buy a box of chocolates costing £4.50 as well as the rice? Show your working to explain how you know.

3 Here is Huda's electricity bill.

Electricity bill				
June 30	2	3	8	8
September 30	2	6	0	8

Price per unit 38p
Fixed charge £18.15
Admin fee £2.00

Work out the total cost of Huda's bill.

4 Kate uses these meter readings to calculate her gas bill.

Gas bill				
Latest meter reading	3	4	7	0
Previous meter reading	3	2	4	1

Cost: 22.5p per unit

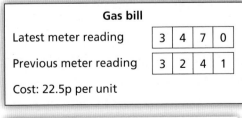

3241 + 3470 = 6711 units used.

6711 × 22.5p = £1509.98

Explain Kate's error and calculate the correct amount she owes.

5 Bobbie has £45.90 in her bank account.

On Monday, she buys some books for £18.98 and a coffee that costs £3.35

On Tuesday, she spends £12 at the cinema and £35 at the pet shop.

Work out the final balance of her bank account after these transactions.

6 Beca travels by car to meetings. Her company pays her 46p for each mile travelled.

The diagram shows the mileage of Beca's car at the beginning and the end of the week.

a How many miles did Beca travel this week?

b Calculate how much Beca's company will pay her for travel. (Assume Beca did not use her car for any other travel.)

Stretch – can you deepen your learning?

1 Emily is working out the cost of the electricity that she used in June.

She has used 146 units of electricity.

She pays 22.5p for the first 100 units used and 10.5p for all other units used.

She is then charged a fixed cost of £7.50

Finally, a 5% administration fee is added.

Work out the total cost of Emily's bill in June.

2 Ali compares two mobile phone tariffs.

Blue **network**	*Red* **network**
£10 per month plus 2p per minute	No fixed cost and only 6.5p per minute

a Ali uses his phone for approximately 150 minutes per month. Which tariff should he choose? Show how you decide.

b How many minutes would he need to be using each month for the other tariff to become the cheaper option?

Reflect

Most bills have a variety of different charges, fixed costs, costs per unit, interest, VAT and administrative fees. Why is it important to make sure that you understand the different tariffs or plans offered by utility companies?

Small steps

- ■ Calculate simple interest
- ■ Calculate compound interest

Key words

Interest – a percentage fee paid when borrowing money or a percentage earned when you deposit money into a savings account

Deposit – amount of money paid into a bank account

Per annum – means "per year"

Annual – covers a period of one year

Principal – an initial amount invested

Are you ready?

1 Work out

   a 30% of 70   **b** 45% of 60   **c** 75% of 180   **d** 3% of 720

2 Write down the multiplier equivalent to a

   a 10% increase   **b** 3% increase   **c** 20% decrease   **d** 4.5% increase.

3 Increase 550 by

   a 10%   **b** 14%   **c** 22%   **d** 3%

4 Decrease 340 by

   a 35%   **b** 22%   **c** 1%   **d** 17%

5 Work these out. Round your answers to 3 decimal places.

   a 1.1^3 **b** 1.05^4 **c** 1.54^6 **d** 1.13^2

Models and representations

Bar models

Bar models can be used to represent percentage increase.

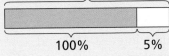

Part-whole bar models

Part-whole models can help you to work backwards to find an original amount.

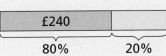

Money invested in a savings account *earns* **interest**, which can be paid monthly or **annually**.

You *pay* interest on money borrowed from a bank or other source of finance. The rate of interest is often given **per annum**, which means every year.

In this chapter, you will learn about two different types of interest: simple and compound.

Simple interest is paid on savings but does not earn interest itself. For example, if you had £100 and earned £5 interest in one year, you would now have £105. If you did not **deposit** or withdraw any money from the account, in the following year, you would earn interest only on the £100, not the extra £5.

Example 1

Find the simple interest earned on £600 invested for 3 years in an account earning 4% simple interest per year.

The initial amount invested is called the **principal**.

£600 × 0.04 = £24.00 in one year ○——— First you find 4% of £600. Remember: to find 4% of an amount you can multiply by 0.04

£24.00 × 3 = £72.00 ○———

The interest earned would be £72.00

The amount of interest earned does not change (unless money is added/removed), so multiply by 3 because the money is invested for three years.

Practice 8.2A

1 Flo invests £300 in an account paying simple interest at 5% per annum.

  a Work out 5% of £300

  b Work out how much money Flo will have in her account after

    i 1 year   **ii** 3 years.

2 Junaid invests £6000 in an account earning simple interest at 4% per annum.

  a Work out 4% of £6000

  b Work out the total amount of money in Junaid's account after 4 years.

3 Amina invests £2500 at 6% simple interest per annum. Work out the value of her savings after 6 years.

4 A bank pays 3% simple interest per annum. Work out how much interest each person will earn.

  a Bobbie invests £2000 for 3 years.

  b Rhys invests £2500 for 2 years.

  c Beth invests £1500 for 5 years.

5 Ed buys a washing machine costing £650. He pays a 10% deposit.

   a How much does Ed still owe?

   Ed pays 5% interest on the remaining amount.

   b How much does Ed owe after interest is added?

   c Ed pays the amount he owes in 12 equal monthly instalments. Work out how much Ed pays each month.

6 Marta buys a computer costing £780

   She pays a 15% deposit and then pays the remainder of the cost over 12 months at 8% interest.

   How much will Marta pay for the computer in total?

 7

I've invested £2000 at 4% simple interest per year.

Chloe

I've invested £4000 at 2% simple interest per year.

Jackson

Who will have earned the most interest in one year? Show working to explain how you decide.

What do you think?

1 Marta invests £600 at 4% simple interest per annum. After n years, she has a total of £720 in her bank account. Work out the value of n.

2 Faith invests £850 at x% simple interest per year. After 3 years, she has £1054 in her account. Work out the value of x.

3 Emily invests £n at x% simple interest per year. After 3 years, she has £440 in her account, and after 5 years, she has £520 in her account. Work out the values of n and x.

In this section, you will learn about compound interest and how to calculate the interest earned over several years.

Compound interest is added to your savings each year and you will then earn interest on the total amount in future years.

For example, if you have £100 and earned £5 in interest in one year, you would now have £105. The following year, you would earn interest on the full £105 in your account.

Example 2

Mario invests £1500 in a savings account earning 4% compound interest per year.

How much money will he have in his account after 3 years?

100% + 4% = 104%	First work out the required decimal multiplier. Remember: the multiplier for a 4% increase is 1.04
After 1 year: £1500 × 1.04 = £1560 After 2 years: £1560 × 1.04 = £1622.40 After 3 years: £1622.40 × 1.04 = £1687.296	Then work out the total value of Mario's savings after each year for 3 years. Multiply the total amount in the account at the end of each year by 1.04
This rounds to £1687.30	Round your answer to 2 decimal places for money.

So, on my calculator I enter 1500 × 1.04 × 1.04 × 1.04
This is the same as 1500 × 1.04³

Ed

Practice 8.2B

1 Huda invests £500 at 2% compound interest per annum.

 a Work out 2% of £500

 b How much money will Huda have in her account after 1 year?

 c How much interest will she earn in the second year?

 d How much money will Huda have in her account after 2 years?

2 Flo invests £2540 at 5% compound interest per annum.

 a How much money will Flo have in her account after 3 years?

 b How much interest will Flo earn in 5 years?

3 Beca invests £250 at 6% compound interest per annum for 3 years.

 Chloe invests £350 at 6% simple interest per annum for 3 years.

 a Who will have earned more interest?

 b How much more interest will they have earned?

4 Benji saves for his retirement. He invests £1000 into an account paying 7% compound interest per annum. Assuming he makes no further withdrawals or deposits, how much money will he have after 40 years?

5 Which earns more interest?

  A £600 invested for 5 years at 7% compound interest per annum

  B £500 invested for 6 years at 7% compound interest per annum

6 Lydia invests £550 in an account earning 6% compound interest per annum. How long will it take for her investment to double in value?

7 Look at this advert.

MyBank

Savings account pays 6% interest per annum, compound.

Double your money in 12 years!

  a Choose some starting investment values to check the bank's claim.

  b Does the initial investment value matter?

8 Which calculation represents the total amount of money after £3000 is invested at 2% compound interest for 6 years?

A | 3000×1.2^6

B | 3000×1.06^2

C | 3000×1.02^6

D | $3000 \div 1.02^6$

What do you think?

1. Zach takes out a loan of £1500

   He is charged interest at a rate of 23% per annum. Interest is calculated on the outstanding loan amount at the start of each year.

   a How much will Zach owe immediately after the loan is taken out?

   Zach makes annual payments of £400

   b How much will he owe at the start of the fourth year?

   c Comment on the length of time it will take Zach to pay off the loan.

2. Samira deposits £100 in a savings account earning 7.5% compound interest per annum.

   a Show that her savings will double in value after 10 years.

   Faith

   > If I deposit £300 in the same account, it will take much longer to double.

   b Is Faith correct? Explain your answer.

Consolidate – do you need more?

1. Kate invests £7500 in an account paying 4% simple interest per annum.
   How much interest will she earn in 5 years?

2. Ali and Charlie each have £3450 to invest.

   Ali chooses an account paying simple interest at 7% per annum and invests for 4 years.

   Charlie chooses an account paying compound interest at 4% per annum and invests for 7 years.

   Who will have earned more interest? Show working to explain how you know.

3. Rob invests £2400 in a savings account paying compound interest at 8% per annum.

   a How much will he have in his account after 4 years?

   b After how many years will his balance exceed £4000?

   c How would your answers to parts **a** and **b** change if the interest paid was simple interest?

4. Copy and complete the table.

Principal value	Interest rate	Interest type	After 2 years	After 5 years
£5000	3%	compound		
£200		simple	£208.00	
£1700	2.5%		£1786.06	£1923.39
£23 000	8%	simple		

5 Filipo saves £8000 at 5% compound interest per annum. He needs £14 000 for a deposit to purchase a house. Will he have enough in his account after 3 years?

6 Look at these adverts.

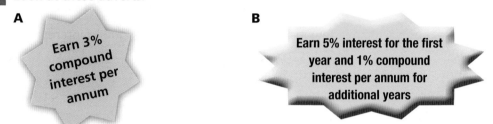

A Earn 3% compound interest per annum

B Earn 5% interest for the first year and 1% compound interest per annum for additional years

Abdullah has £10 000 to invest. He wants to earn the maximum amount of interest in 3 years. Which account should he choose?

Stretch – can you deepen your learning?

1 Jakub invests £2000 at 5% compound interest per annum.

Ali invests £2500 at 2% compound interest per annum.

In which year will Jakub's investment exceed Ali's?

2 Mario invests some money in an account paying 7% compound interest per year.

After how many years will his investment triple?

3 Kate invests £7000 at x% compound interest per annum. After 3 years, she has £9065.20 in her account. Work out the value of x

4 Samira invests £8000 at 2.5% compound interest per annum. After n years, she has £9051.27 in her account. Work out the value of n

5 Marta invests £n at x% compound interest per annum for 3 years. After 4 years, she has £5035.26 and after 5 years she has £5639.49 in her account. Work out the values of n and x

Reflect

1 Explain the difference between simple and compound interest. You can use examples to help.

2 Explain the dangers of using loans and credit cards with high interest rates when paying for goods. Why can it be a problem if you only make small repayments?

8.3 Taxes

Small steps

■ Solve problems with value-added tax

■ Calculate wages and taxes

Key words

Income tax – a tax that is payable on personal income such as wages and salary

VAT (value-added tax) – a tax that is added to the price of some goods and services

Overtime – the time worked in addition to a person's normal working hours

Tax allowance – the amount that you can earn before being taxed

Are you ready?

1   Work out

   a   20% of 45   **b**   15% of 423   **c**   12% of 485   **d**   8% of 32

2   Calculate

   a   £12.03 + £23.66   **b**   £1530 − £299   **c**   0.4 × £35.65   **d**   1.5 × £18.55

3   State the effect of each of these multipliers.

   a   1.3   **b**   0.8   **c**   1.12   **d**   0.975

4   A taxi driver charges 45p per mile travelled. Work out the cost of a journey for each of these distances.

   a   14 miles   **b**   23 miles   **c**   8 miles   **d**   49 miles

Models and representations

Bar models

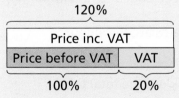

Bar models are useful when working out amounts before and after a percentage change.

Some goods and services are subject to a tax called **VAT (value-added tax)**.

VAT is calculated as a percentage of the price paid for the goods or services.

In the UK, VAT is set at 20% for most items and 5% for some items, such as gas and electricity. Some items are exempt from VAT.

Example 1

a Chloe buys a jacket costing £65 plus VAT at 20%. Work out the total cost of the jacket.

b Amina buys a pair of boots for £72. The price includes VAT at 20%. Work out the price of the boots before VAT was added.

a

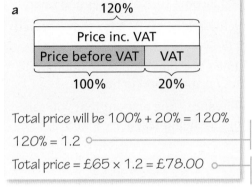

Total price will be 100% + 20% = 120%

120% = 1.2 ○————————————————

Total price = £65 × 1.2 = £78.00 ○————

First work out the multiplier you need to use.
The multiplier for a 20% increase is 1.2
Use the multiplier to work out the price after the increase.

b

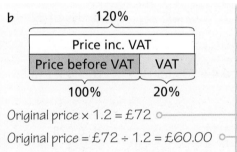

Original price × 1.2 = £72 ○————

Original price = £72 ÷ 1.2 = £60.00 ○——

The multiplier is still 1.2, but this time you have been given the final amount.
To find the original price, you need to divide by 1.2

Practice 8.3A

1 Filipo's gas bill is £186 plus VAT at 5%

    a Work out 5% of £186

    b Work out how much Filipo has to pay after VAT has been added to his bill.

2 Huda receives an electricity bill for £84 plus VAT at 5%. Calculate the amount of VAT she is charged.

3 A video game costs £38 before VAT at 20% is added. Work out the total price of the game.

4 A car costs £21 000 plus VAT at 20%. Work out the total price of the car.

5 VAT on goods and services used to be 17.5% and is now 20%. Work out the increase in the price of a phone that cost £125 before VAT was added.

6 Look at this sign.

Sven hires a van for three days and drives a total of 75 miles. VAT at 20% is added to his charges.

Work out the total amount that Sven has to pay for hiring the van.

£55 per day plus 40p per mile

7 A laptop costs £550. The price includes VAT at 20%. Work out the cost of the laptop before VAT was added.

8 A coat costs €44 before VAT is added.

VAT in Germany is 19% and in Greece, it is 24%

Work out the difference in the cost of the coat in Germany and Greece.

9 Look at this cinema advert.

Marta buys 2 adults' and 3 children's tickets. She pays 20% VAT on top of the total cost.

How much does she pay in total?

Cinema tickets –
Adults £14, Children £8
Buy four or more tickets and save 10%

What do you think?

1 A decorator charges £450 plus a 5% administration fee, plus VAT at 20%

So I would work out £450 plus 25%

a Explain why Ed is wrong.

b Work out the total cost of the decorator's bill.

2 Mario buys a dining table and a set of four chairs. The dining table is originally priced at £340
Mario spends a total of £615.60

The shop gave a discount of 5% and VAT at 20% was added to the total.

Work out the original price of one chair.

3 Rhys pays £231.34 for a carpet. This includes VAT at 20%, an administrative fee of 5% and a 2.5% surcharge for paying by credit card.

a Work out the cost of the carpet before the extra charges were added.

b Does the order in which the charges are added to the cost matter? Explain why or why not.

In this section, you will learn more about income and taxes.

Some people are paid by the hour; they will have an hourly rate of pay, for example, £12 an hour. They may also earn **overtime** for any extra hours worked (especially on weekends and in the evenings).

Some people have an annual salary paid to them in monthly instalments.

Most people pay **income tax** on their earnings. In many countries, everyone is allowed to earn a certain amount that is not taxed (this is called your **tax allowance**). The remaining income is taxed. Income tax is worked out as a percentage of the taxable income.

Example 2

Mario is paid £12.50 an hour for a 35-hour week.

Overtime is paid at "time and a half".

One week, Mario works for 41 hours.

How much does he earn?

"Time and a half" means 1.5 × the normal hourly rate.

35 hours at £12.50

$35 \times £12.50 = £437.50$

For the first 35 hours, Mario earns £12.50 per hour.
You can work out his basic pay by multiplying 35 by his hourly rate.

$41 - 35 = 6$

This is how many extra hours he worked.

Overtime rate = $1.5 \times £12.50 = £18.75$

His overtime rate of pay is 1.5 × his basic rate of pay.

So, 6 hours at £18.75
$6 \times £18.75 = £112.50$

Multiply the overtime rate of pay by the extra number of hours he works.

Total = $£437.50 + £112.50 = £550.00$

Finally, add the basic pay and overtime pay together.

Example 3

The table shows the income tax rates in the UK (at the time of writing).

Flo earns £43 500 per year.

How much income tax will she pay?

	Income	Tax rate
Tax allowance	£12 500	0%
Basic rate	£12 501–£50 000	20%
Higher rate	£50 001–£150 000	40%
Additional rate	Over £150 000	45%

$£43 500 - £12 500 = £31 000$

The first £12 500 of income is not taxed.

Flo's salary falls into the "basic rate" band so she will pay 20% on the difference between what she earns and £12 500

$£31 000 \times 0.2 = £6200$

Flo pays £6200 income tax.

The multiplier for working out 20% is 0.2

Practice 8.3B

1 Huda earns £9.50 per hour. She works a 37-hour week. Work out her weekly earnings.

2 A waiter is paid £11.50 per hour for a 35-hour week.

    a Work out his weekly earnings.

    He is paid time and a half for overtime. One week, he works a total of 50 hours.

    b Work out his total pay for the week.

3 Ed earns £395.20 per week. He is paid £10.40 an hour. How many hours a week does he work?

4 Abdullah is paid overtime at time and a half. He is paid £16.20 per hour for overtime. What is his basic hourly rate of pay?

5 Chloe earns £18 450 per year. Her tax allowance is £12 500. She pays 20% income tax on the remaining income.

    a How much income tax does Chloe pay?

    b Chloe is paid monthly. Work out how much she earns each month after tax is deducted.

6 Emily earns £62 050 per year. Her tax allowance is £12 500

    a What is Emily's taxable income?

    She pays 20% income tax on the first £37 500 of her taxable income and 40% on the remainder.

    b Work out how much income tax Emily pays each year.

7 Zach earns £440 per week. He has a tax allowance of £12 500 and pays income tax at 20% on the remaining income.

    a Calculate Zach's annual earnings (assume that he works for 52 weeks a year).

    b Calculate the amount of income tax Zach pays each year.

    c Zach pays his income tax monthly. Calculate his monthly take-home pay.

    d Work out the percentage of Zach's total salary that he pays in income tax.

What do you think? 💭

1 Kate earns £11.80 per hour. She has an annual tax allowance of £12 500

Work out the maximum number of hours per week that Kate can work so that she does not pay any income tax.

2 The tables show income tax rates for Italy and France.

Italy

€0–€15 000	23%
€15 001–€28 000	27%
€28 001–€55 000	38%

France

€0–€10 084	0%
€10 085–€25 710	11%
€25 711–€73 516	30%

Faith has been offered a job in Italy paying €46 500 per year and a job in France paying €38 000 per year.

Faith will choose the job with the highest pay after tax has been deducted. Which job should she accept?

Consolidate – do you need more?

1 A computer costs £620 plus VAT at 20%

    a Work out the total amount of VAT charged.

    b Work out the total cost of the computer.

2 Junaid receives a bill for £168 plus VAT at 5%. How much is the total bill?

3 Samira earns £8.40 per hour. She works for 37.5 hours per week. Calculate her weekly wages.

4 A cook is paid £13 per hour for a 38-hour week. He is paid overtime at double time.

    a What is his rate of overtime pay?

> Double time means the normal hourly rate is multiplied by 2

One week, the cook works 45 hours.

    b Calculate his total income for that week.

5 A fridge costs £340 after VAT is added. Given that the rate of VAT is 20%, work out the price of the fridge before VAT was added.

6 Beth is charged £67.50 for her water bill. This includes VAT at 5%. Work out the cost of the bill before VAT was added.

7 Bobbie earns £28 250 per year. Her tax allowance is £12 500. She pays 20% income tax on the remaining income.

    a How much income tax does Bobbie pay each year?

    b Bobbie is paid monthly. Work out how much she earns per month after tax is deducted.

Stretch – can you deepen your learning?

1 The table shows the income tax rates for the UK.

	Income	Tax rate
Tax allowance	£12 500	0%
Basic rate	£12 501–£50 000	20%
Higher rate	£50 001–£150 000	40%
Additional rate	Over £150 000	45%

Jackson earns £56 000 per year before income tax is deducted. He is offered a job in Egypt earning £48 000 tax-free. Should he accept the job offer? Justify your answer.

2 VAT in the UK is charged at 20%. In Switzerland, it is 7.7%

The exchange rate is £1 = 1.2 Swiss francs.

A laptop costs £640 plus VAT in the UK.

The same laptop costs 750 Swiss francs plus VAT in Switzerland.

In which country would it be cheaper to buy the laptop and by how much?

3 A camera costs £212.50 plus VAT at 20% in the UK.

Junaid sees the same camera on sale in Germany costing €210 plus 19% VAT.

Junaid has to pay 10% import duties and buy a compatible charger for £24.99

The exchange rate is £1 = €1.15

If Junaid buys the camera from Germany, will it cost him more than buying it in the UK? What is the difference in cost?

Reflect

1 Life can be expensive. Imagine that you earn £30 000 per year before income tax is deducted. Work out your post-tax salary and then try to budget your monthly spending. Be sure to include things like rent/mortgage, food costs, travel costs, gas, electricity and water bills, phone bills, broadband and TV.

2 In reality, as well as income tax, most people pay National Insurance contributions. Find out more about rates of payment and what National Insurance contributions are used for.

Small steps

■ Solve problems with exchange rates

■ Solve unit pricing problems

Key words

Currency – the system of money used in a particular country

Convert – change from one form to another, for example a percentage to a decimal

Exchange rate – the value of a currency compared to another

Unit cost/price – the cost or price of 1 item or 1 unit of an item

Are you ready?

1 Work these out and round your answers sensibly.

    a £2.55 × 5   **b** £39.50 ÷ 7   **c** £25.45 ÷ 1.45   **d** £647.33 × 1.05

2 Work out the cost of 1 pencil if

    a 3 pencils cost 21p   **b** 8 pencils cost 184p

    c 6 pencils cost £1.32   **d** 10 pencils cost £2

3 What is the missing number in each ratio?

    a $1:5 = 3:\boxed{}$ **b** $3:5 = \boxed{}:20$ **c** $4:\boxed{} = 16:28$ **d** $1:\boxed{} = 2.5:10$

Models and representations

Proportion diagrams

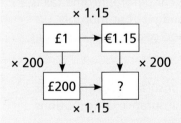

Double number lines

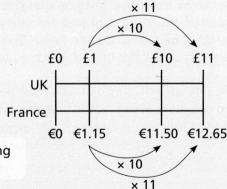

Both of these diagrams are useful for showing the multiplicative nature of **exchange rates**.

Bar models

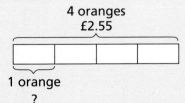

Bar models can help when working out best buys and for finding **unit costs**.

In this section, you will learn about different currencies and solve problems which involve exchange rates.

Exchange rates tell you how much of one **currency** can be exchanged for another.

For example, if £1 = €1.15, it means that each British pound can buy 1.15 euros.

Some sellers charge commission, which is an extra fee for completing the transaction. This is usually a percentage of the total sale price.

Example 1

Exchange rates	
£1 = €1.15 (euro)	£1 = 102 rubles (Russia)
£1 = $1.37 (USA)	£1 = 100 rupee (India)
£1 = 145 yen (Japan)	£1 = 20.4 rand (South Africa)

a Convert £240 into Russian rubles.

b Convert 215 South African rand into British pounds.

c Convert $440 into Indian rupees.

a

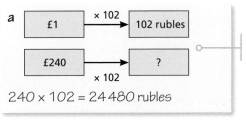

If £1 buys you 102 rubles, then £240 would buy you 240 times as many rubles

$240 \times 102 = 24\,480$ rubles

b

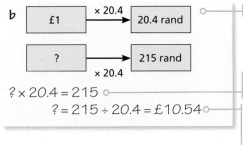

$? \times 20.4 = 215$

$? = 215 \div 20.4 = £10.54$

£1 is equal to 20.4 rand

To **convert** from rand to pounds, you need see how many lots of 20.4 rand you have in 215 rand So this time, you need to divide by the exchange rate.

Make sure you round your answer sensibly. For money, 2 decimal places is correct.

c $? \times 1.37 = 440

$? = 440 \div 1.37 = £321.17$

£321.17 $\times 100 = 32\,117$ rupees

First you need to convert the US dollars into pounds. Use the exchange rate £1 = $1.37

Then change the pounds into rupees £1 = 100 rupee

Practice 8.4A

1 £1 = 27.78 Mexican pesos. Use this information to complete the table.

British pounds	Mexican pesos
£2	
£100	
	277.8
	555.6

2 Look at these exchange rates.

Exchange rates	
£1 = €1.15 (euro)	£1 = 102 rubles (Russia)
£1 = $1.37 (USA)	£1 = 100 rupee (India)
£1 = 145 yen (Japan)	£1 = 20.4 rand (South Africa)

Convert the following amounts of money.

a £300 into euros   **b** £150 into yen

c £950 into rand   **d** £55.50 into dollars

e 425 rand into pounds   **f** 53 000 yen into pounds

g 2 million rubles into pounds   **h** 355 rupees into pounds

3 £1 = 145 Japanese yen

Rhys changes £350 into Japanese yen. How many yen does he get?

4 £1 = €1.15

Amina returns to the UK from France with €135. She exchanges this into British pounds. How many pounds will she get?

5 $1 = £0.73

A phone costs £350 in the UK and $420 in the USA. In which country is the phone cheaper?

6 £1 = 5.26 Saudi Arabian riyals   £1 = 1.79 Australian dollars

Huda has 150 riyals and Flo has 50 Australian dollars. Who has more money?

7 In which country would the same handbag be cheapest?

Exchange rates		
£1 = €1.15 (euro)	£1 = $1.37 (USA)	£1 = 145 yen (Japan)

UK – £135   USA – $160   FRANCE – €150   JAPAN – 20 000 yen

8 £1 = 145 Japanese yen

Seb has 3500 yen left from his holiday in Japan. The bank charges him 3% commission to convert the yen back into British pounds. How many pounds will Seb receive?

9 Use the conversion graph to convert the following amounts of money.

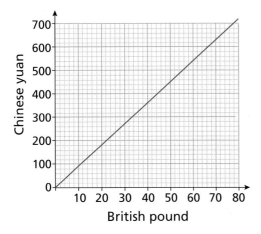

You studied graphs like this in Book 2 and you will learn more about them in Block 13

a Convert £70 into Chinese yuan.

b Convert 500 yuan into British pounds.

What do you think?

1 On the plane back from Germany, Beth buys some snacks.

Snack Prices	
Crisps £1.25	Pastry £2.10
Chocolate bar £1.25	Soft drink £2.50
** Pay in pounds or euros**	
£1 = €1.15	

Beth buys two bars of chocolate, a bag of crisps and two cans of cola.

She pays part of the cost with a €5 note and the rest in pounds.

How much does Beth pay in pounds?

2 In a London bank, the exchange rate is £1 = €1.15

In a Madrid bank, the exchange rate is €1 = £0.85

Rob wants to change £250 into euros. In which city would he get the better deal?

3 £1 = €1.15

Darius buys a motorbike in Italy for €2540. He pays €590 to have the motorbike delivered to the UK.

Once in the UK, he pays £350 in import taxes. He then sells the motorbike for a 10% profit.

Work out the selling price of the motorbike, giving your answer to the nearest pound.

Shops often have lots of offers such as "two-for-one", "buy one get one half price" or "20% off".

In this section, you will learn how to work out which offer gives best value for money.

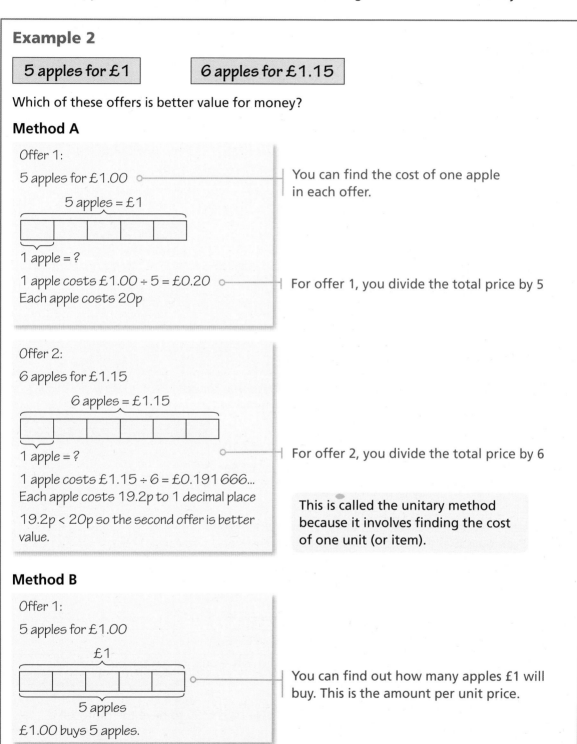

Example 2

| 5 apples for £1 | 6 apples for £1.15 |

Which of these offers is better value for money?

Method A

Offer 1:

5 apples for £1.00 ○——————— You can find the cost of one apple in each offer.

5 apples = £1

1 apple = ?

1 apple costs £1.00 ÷ 5 = £0.20 ○——— For offer 1, you divide the total price by 5
Each apple costs 20p

Offer 2:

6 apples for £1.15

6 apples = £1.15

1 apple = ? ○——— For offer 2, you divide the total price by 6

1 apple costs £1.15 ÷ 6 = £0.191 666...
Each apple costs 19.2p to 1 decimal place

19.2p < 20p so the second offer is better value.

This is called the unitary method because it involves finding the cost of one unit (or item).

Method B

Offer 1:

5 apples for £1.00

£1

5 apples

£1.00 buys 5 apples.

○——— You can find out how many apples £1 will buy. This is the amount per unit price.

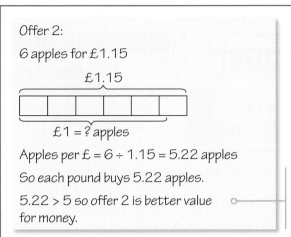

Offer 2:

6 apples for £1.15

£1.15

£1 = ? apples

Apples per £ = 6 ÷ 1.15 = 5.22 apples

So each pound buys 5.22 apples.

5.22 > 5 so offer 2 is better value for money.

You get more apples per £1 spent on offer 2. In this case, you want to get more apples for each £1 so the larger number tells you which is better value for money.

Practice 8.4B

1 Look at these offers.

Offer A: 9 oranges for £3.20

Offer B: 4 oranges for £1.35

a What is the cost of one orange in

i offer A **ii** offer B?

b Which offer is better value for money?

2 At a wholesaler, onions cost £8.50 for an 8 kg bag.

In the supermarket, a 2 kg bag costs £2.10

a What is the cost of 1 kg of onions at

i the wholesaler **ii** the supermarket?

b In which shop are the onions better value for money?

3 The same tea is available in two different sizes.

Which size box is better value for money? Show working to explain how you decide.

4 Three packs of cola are available at a supermarket.

A 12 cans for £7.50

B 8 cans for £4.50

C 6 cans for £3.50

Which pack offers best value for money?

5 Rob wants to buy six DVDs. He sees these offers on three different websites.

A  B  C DVDs 3 for £35 or £16 each

Rob wants to buy the DVDs as cheaply as possible. From which website should Rob buy them?

6 Calculators cost £5 each. A school wants to buy 30 calculators.

Which offer should the school choose if they want to pay as little as possible?

Edu Supplies – 15% off all orders   School Stuff – Buy 4 and get 1 free

7 Mario is doing his Maths homework. He needs to decide which is better value: 200 g of pasta for 80p or 500 g of pasta for £1.95

Here is his working.

$$200 \div 80 = 2.5 \qquad 500 \div 195 = 2.564$$

2.5 is smaller, so 80p for 200 g is better value.

Is Mario right? Explain your answer.

8 Look at these offers.

Ed

I'm going to find the cost of 1 g of biscuits in each offer.

I'm going to find the cost of 50 g of biscuits in each offer.

Tommy

Chloe

I'm going to find out how many biscuits you can get for £1

Compare these methods.

What do you think? 💭

1 Two offers are available on boxes of cereal at the supermarket.

A
| 350g plus 20% extra free £3.80 |

B
| 350g with 15% off the normal price £3.80 |

Which offer is better value for money?

2 Kate wants to buy at least 2 kg of cat food. Which of these offers would give best value for money?

A
| 420g tins at 80p each |

B
| Box of 6 × 100g pouches for £1.75 |

C
| 1 kg of dry food for £2.49 |

3 Socks are sold in three different packs containing 3 pairs, 6 pairs or 10 pairs.

The pack of 3 pairs costs £5.90. The pack of 10 pairs costs £18.50

The pack of 6 pairs is better value than buying 3 pairs but is not as good value as buying 10 pairs.

Work out the range of possible prices for the pack of 6 socks.

Consolidate – do you need more?

1 £1 = $1.75 Canadian dollars

  a Sven changes £150 into Canadian dollars. How many dollars will he get?

  b Emily changes $375 into British pounds. How many pounds will she get?

2 £1 = 102 rubles

At the end of her holiday, Flo has 1430 rubles remaining. She spends 42 rubles at the airport.

She changes the remainder into British pounds. How many pounds will Flo get?

3 $1 = £0.73

A necklace costs £86.99 in Manchester and $120 in New York. In which city is the necklace cheaper and by how much?

4 Washing-up liquid is available in two sizes.

  A 500 ml for £2.55   **B** 850 ml for £5.55

Which size is better value for money?

5 Croissants come in packs of two sizes.

A
| 6 croissants for £3.00 |

B
| 4 croissants for £1.96 |

Which size pack is better value for money?

6 Shampoo is sold in bottles with three different sizes.

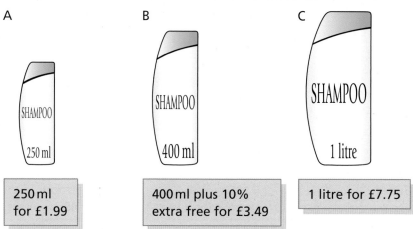

A   B   C

| 250 ml for £1.99 | 400 ml plus 10% extra free for £3.49 | 1 litre for £7.75 |

Which size bottle is best value for money?

Stretch – can you deepen your learning?

1 £1 = €1.15   US$1 = 73p

 a Convert €50 into US dollars.

 b Work out the exchange rate between euros and dollars.

2 Orange juice is sold in three different sizes.

 A   A 1 litre bottle costs £2.55

 B   A carton measuring 8 cm by 8 cm by 18 cm costs £2.75

 C   A pack of 6 × 200 ml bottles costs £3

 Which size offers best value for money?

3 Filipo wants to buy a games console and two games. He wants to pay as little as possible. Which of these options should he choose?

Option A	Option B	Option C
Console £275	Console + 1 game £300	Console £280
Games £18.99	Games £15.99	Games 2 for £34
Free delivery	10% discount	8% discount
	Delivery £4.99	Delivery £2.99

4 Emily wants to buy four books, each costing £10

  a Which of these offers gives best value for money?

  A
  | Buy one, get one half price |

  B
  | Buy two, get one free |

  C
  | 25% off each book |

  b Would your answer change if

  i the books cost £8 each

  ii she wanted to buy five books?

Reflect

1 Are there times when you might not choose to buy the item that is best value for money? Why might this be?

2 You can exchange foreign currency in banks, post offices, travel agents and at airports. Investigate who offers the best rates. Why might this be?

I have become fluent in...

- carrying out calculations involving money
- calculating interest as a percentage of an amount
- finding the cost after VAT has been added
- working out the cost of one unit of an item.

I have developed my reasoning skills by...

- recognising the difference between simple and compound interest
- solving problems involving different tax brackets
- solving problems involving one or more exchange rates
- comparing different offers to identify the best value.

I have been problem-solving through...

- solving backwards problems involving percentages
- using algebraic notation to model problems
- solving multi-step problems involving two different currencies
- comparing and contrasting earnings in different countries.

Check my understanding

1. Darius buys two notebooks costing £2.99 each and a pen costing £1.49. He pays with a £10 note. How much change will he get?

2. Mario invests £3000 in an account earning simple interest at 5% per annum.
   a. How much interest will he earn in a year?
   b. How much money will he have in his account after 4 years?

3. Abdullah invests £1750 in an account earning compound interest at 3.5% per annum.
   a. How much money will he have in his account after 4 years?
   b. How much interest will he earn in 6 years?

4. A bike costs £219 plus VAT at 20%. Work out the total price of the bike.

5. A laptop costs £455 after VAT at 20% is added. Work out the price of the laptop before VAT was added.

6. Samira earns £11 per hour for a 35-hour week. Her overtime rate of pay is double the normal rate. One week, she works for a total of 45 hours. Calculate her earnings that week.

7. £1 = $1.37
   a. Convert £240 to dollars.
   b. Convert $1400 to pounds.

8. A shop sells coffee in jars with two different sizes.

   A   200 g for £2.45   B   500 g for £5.99

   Which size is better value for money?

9 Deduction

In this block, I will learn...

how to find missing angles in a variety of shapes

Find the size of each angle labelled with a letter.

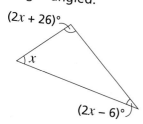

135°

35°

d c

65° 88°

e

---

how to find missing angles on parallel lines

Find the size of each angle labelled with a letter.

105°

a d c

b

84°

---

how to solve complex angle problems

Find the size of each angle labelled with a letter.

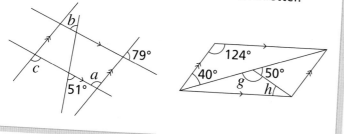

b

79°

c

a

51°

124°

40° 50°

g h

---

how to use algebra to solve angle problems

Find the value of x

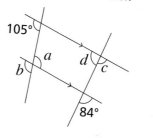

$(2x + 4)°$

$(3x - 11)°$

---

how to prove geometrical facts about shapes and angles

Prove that this triangle is right-angled.

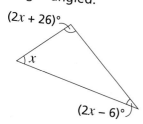

$(2x + 26)°$

x

$(2x - 6)°$

Prove that $a + b = 50°$

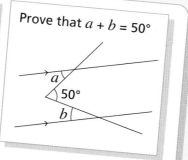

a

50°

b

---

to link constructions to angle and shape facts

Explain how this angle bisector can be used to construct a kite.

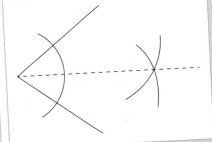

375

Small steps

■ Angles in parallel lines ®

Key words

Transversal – a line that crosses at least two other lines

Corresponding angles – a pair of angles in matching positions compared with a transversal

Alternate angles – a pair of angles between a pair of parallel lines on opposite sides of a transversal

Co-interior angles – a pair of angles between a pair of parallel lines on the same side of a transversal

Give a reason – state the mathematical rule(s) you have used, not just the calculations you have done

Are you ready?

1 Work out

    a 360 – 44   **b** 180 – 124   **c** 360 – 249   **d** 180 – 39

2 ABCD is a trapezium.

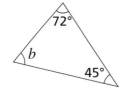

Determine whether each statement is true or false.

    a AB is parallel to DC   **b** AD is parallel to BC

    c DBA is an acute angle   **d** The angles all add up to 360°

3 Work out the size of each angle labelled with a letter.

    a   **b**   **c**

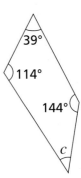

    d   **e**

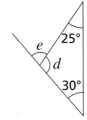

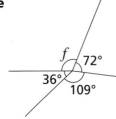

Models and representations

Geometry software

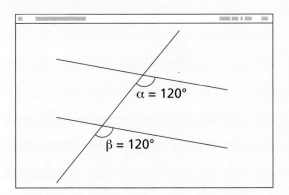

$\alpha = 120°$

$\beta = 120°$

Straws

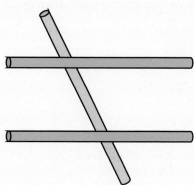

In this chapter, you will review your work on angles from Book 1 and Book 2 as you will need to use the angle facts that you have already met.

Summary of angle facts

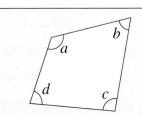

The angles in a triangle add up to 180°

$a + b + c = 180°$

The angles in a quadrilateral add up to 360°

$a + b + c + d = 360°$

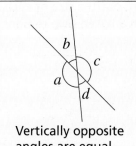

Vertically opposite angles are equal

$a = c$ and $b = d$

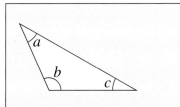

Alternate angles are equal

$a = b$

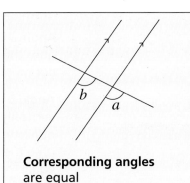

Corresponding angles are equal

$a = b$

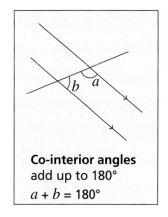

Co-interior angles add up to 180°

$a + b = 180°$

Example 1

Work out the size of each angle labelled with a letter. Give reasons for your answers.

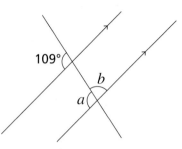

$a = 109°$

Reason: corresponding angles are equal.

Angle a corresponds to the 109° angle since they are in corresponding positions when compared to the **transversal**.

$b = 180° - 109° = 71°$

Reason: angles on a straight line add up to 180°

Angles a and b lie on a straight line, so they add up to 180°. Subtract 109° from 180° to get the size of b

Example 2

Work out the size of each angle labelled with a letter. Give reasons for your answers.

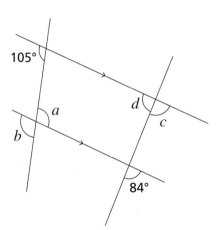

$a = 105°$

Reason: alternate angles are equal.

Angle a is alternate to the 105° angle since they are within the parallel lines but on opposite sides of the transversal.

$b = 105°$

Reason: vertically opposite angles are equal.

Angle b is vertically opposite angle a, so they are equal.

You could also use the fact that angle b is corresponding to the 105° angle, so they are equal.

$c = 84°$

Reason: corresponding angles are equal.

Angle c is corresponding to the given 84° angle, therefore they are equal.

$d = 180° - 84° = 96°$

Reason: angles on a straight line add up to 180°

Angles c and d lie on a straight line so they add up to 180° To find d, you can subtract 84 from 180

Practice 9.1A

1 Look at the diagram.

 a Name two pairs of corresponding angles.

 b Name two pairs of alternate angles.

 c Name two pairs of co-interior angles.

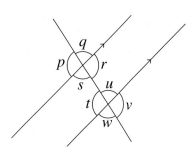

2 Work out the size of each angle labelled with a letter. Give reason for your answers.

 a 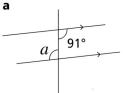 **b** 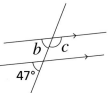 **c** 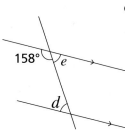 **d**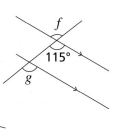

 e 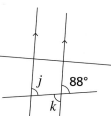 **f** 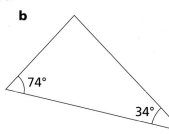 **g**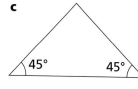

3 Identify each type of triangle.

 a 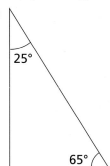 **b** 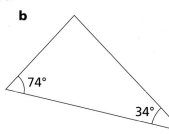 **c**

④ Work out the size of each angle labelled with a letter in these quadrilaterals.

a

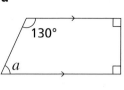

b

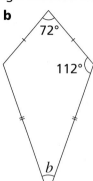

c

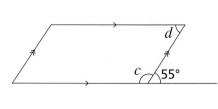

d

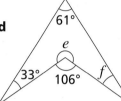

e

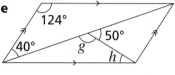

⑤ Seb is completing his homework. He has been asked to work out the sizes of angles
a and *b*

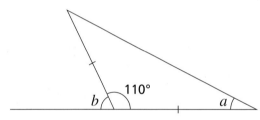

180 − 110 = 70 (angles in a triangle
add up to 180°)

It's an isosceles triangle so 70 ÷ 2 = 35

Angle *a* = 35°

Angles *a*, 110° and *b* are all on a
straight line and add up to 180°

So, 180 − 110 − 35 = 35°

Angle *b* = 35°

Identify Seb's error.

⑥ In each case, determine whether PQ is parallel to RS. Give reasons for your answers.

a

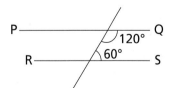

b

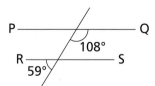

c

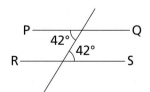

⑦ Based on the given angles, how many pairs of parallel sides does this shape have?

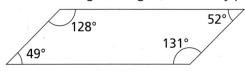

8 Work out the size of each angle labelled with a letter in these shapes.

a
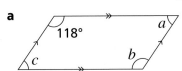
118°
a
c
b

b
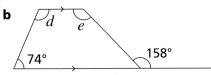
d   e
74°
158°

9 If I know just one angle in a parallelogram, I can work out all of the others.

Is Seb right? Explain your answer.

What do you think?

1 Work out the size of each angle labelled with a letter.

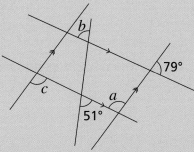

b
79°
c
a
51°

2 Which of the four lines, AB, CD, EF, GH, are parallel?

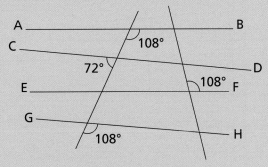

A _____ B
C _____ 108°
72°
D
E _____ 108° F
G _____ H
108°

3 Work out the values of x and y

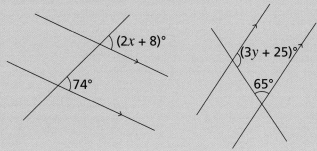

$(2x + 8)°$
74°

$(3y + 25)°$
65°

Consolidate – do you need more?

1 From the diagram, name an angle that is

   a   corresponding to a

   b   alternate to a

   c   co-interior to a

   d   vertically opposite to a

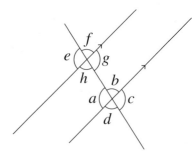

2 Work out the size of each angle labelled with a letter. Give reasons for your answers.

a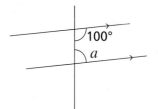

b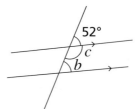

c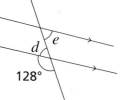

d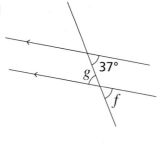

e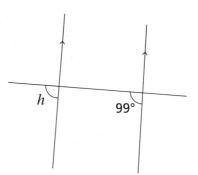

3 Is line AB parallel to line CD? How do you know?

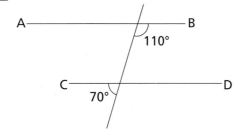

4 Find the size of each of these angles, giving reasons for your answers.

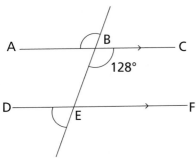

a ∠ABE **b** ∠FEB **c** ∠DEB

5 Find the size of each angle labelled with a letter.

a

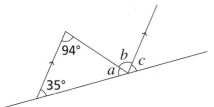

b

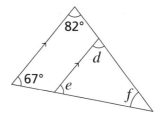

6 Look at this diagram.
a What type of triangle is triangle ABC?
b Find the size of each angle labelled with a letter.

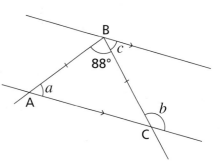

Stretch – can you deepen your learning?

1 Work out the size of each angle labelled with a letter.

a

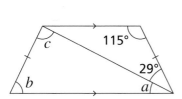

b

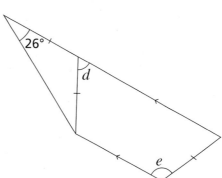

2 The diagram shows two overlapping squares. Work out the size of angles a and b

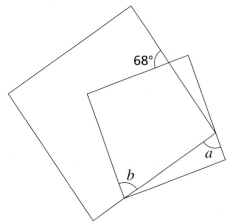

3 The diagram shows a quadrilateral inside an equilateral triangle. Find the size of angles a and b

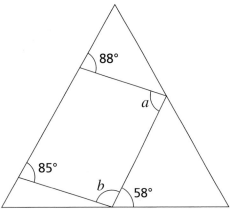

4 Work out the value of x

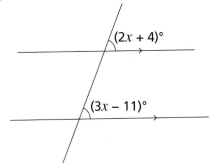

Reflect

Make revision cards or posters covering all the angle facts you know. Can you think of some useful strategies for learning them and how to recognise them?

Small steps

■ Solve angle problems using chains of reasoning

Key words

Corresponding angles – a pair of angles in matching positions compared with a transversal

Alternate angles – a pair of angles between a pair of parallel lines on opposite sides of a transversal

Co-interior angles – a pair of angles between a pair of parallel lines on the same side of a transversal

Give a reason – state the mathematical rule(s) you have used, not just the calculations you have done

Are you ready?

1 Work out

   a 360 – 45   **b** 180 – 125   **c** 360 – 77   **d** 180 – 99

2 Find the size of each angle labelled with a letter.

   a   **b**   **c**   **d**

3 Match each diagram to the correct mathematical relationship.

   a   **b**   **c**   **d**

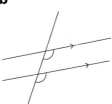  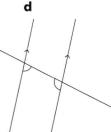

| corresponding angles | co-interior angles | alternate angles | vertically opposite angles |

Models and representations

Geometry software

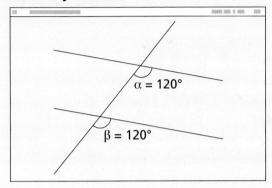

α = 120°

β = 120°

Straws

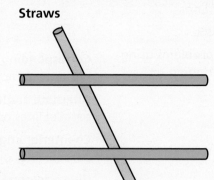

In this chapter, you will use your knowledge of angle facts to solve increasingly complex problems.

Example 1

Find the size of each missing angle.

Give a reason for each step in your working.

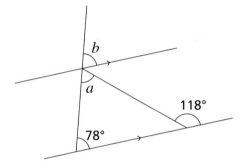

b

a

118°

78°

To find angle *a*, you need to find some other angles first.

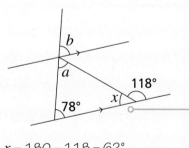

b

a

78° *x* 118°

$x = 180 - 118 = 62°$

Reason: angles on a straight line add up to 180°

$a = 180 - (78 + 62) = 40°$

Reason: angles in a triangle add up to 180°

$b = 78°$

Reason: corresponding angles are equal.

It is a helpful strategy to label other unknown angles with letters.

You can find the size of angle *x*, because it forms a straight line with the 118° angle.

$x + 118° = 180°$

Now you know two of the angles in the triangle, so you can find *a*

$a + x + 78° = 180°$

Angle *b* is **corresponding** to the 78° angle, so they are equal.

There are lots of way to find *b*; some are more efficient than others!

Example 2

ABCD is a rectangle.

PQR is a triangle.

Work out the size of

a ∠AQP   **b** ∠QPR   **c** ∠RPC

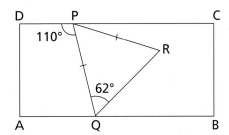

a

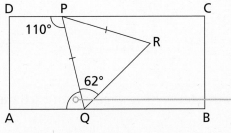

∠AQP = 180° − 110° = 70°

Co-interior angles add up to 180°

Marking the angle you want to find is a useful strategy.

Since ABCD is a rectangle, you know that AB and DC are parallel. This means that ∠DPQ and ∠AQP are **co-interior**.

b

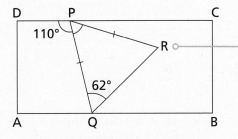

∠QPR = 180° − (62° + 62°) = 56°

Angles in a triangle add up to 180°

PQR is an isosceles triangle.

This means that ∠PRQ = ∠RQP = 62°

∠QPR + 62° + 62° = 180

c

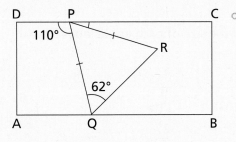

110° + 56° + ∠RPC = 180°

Angles on a straight line add up to 180°

∠RPC = 180° − 110° − 56° = 14°

DPC is a straight line.

For each part **give the mathematical reason** for your working.

Practice 9.2A

1 For each diagram, find the size of the required angle.

Show each step in your working and give reasons for your answers.

a ∠DEH

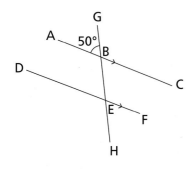

b ∠RSU

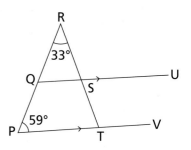

2 Work out the size of each angle labelled with a letter. Give a reason for each stage in your working.

a

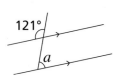

b

c

d

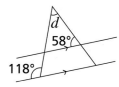

e

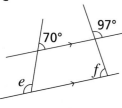

f

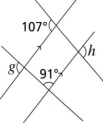

Hint: You may find it helpful to label some of the other angles with letters.

3 The diagrams all show parallelograms. Find the size of each angle labelled with a letter.

a

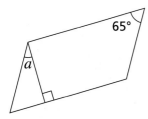

b

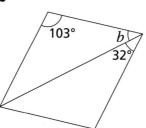

c

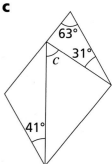

Hint: You may find it helpful to label some of the other angles with letters.

4 Ed is working on this question.

Work out the size of ∠HEF

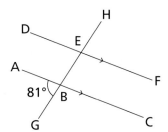

BEF = 81° because it is alternate to ABG, so they are equal.

So HEF = 180° – 81° = 99°, because angles on a straight line add up to 180°

Identify Ed's mistake and correct his solution.

5 Work out the size of angles a and b

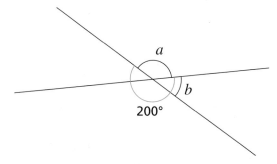

6 The diagram shows two triangles. Find the size of ∠BCD

Show your method and give reasons for each stage of your working.

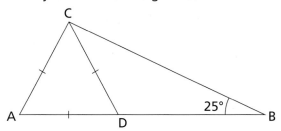

7 The diagram shows a triangle in a square. Find the size of ∠BQP

Show your method and a give reason for each stage of your working.

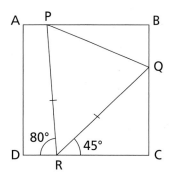

8 Look at this diagram.

a Using the definition of alternate angles, explain why a and b are known as alternate exterior angles.

b If $a = 115°$, show that b would also equal 115°

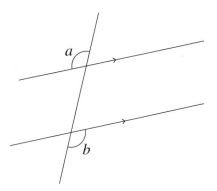

What do you think?

1 The diagram shows a pair of overlapping equilateral triangles.
Sketch the diagram and find the size of *all* the angles in the shape.

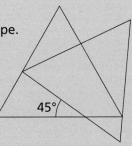

2 The diagram shows a pair of isosceles triangles.
Work out the size of angle y

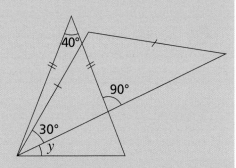

3 Junaid is exploring angles by folding rectangular strips of paper.

Work out the size of each angle marked x

a

b

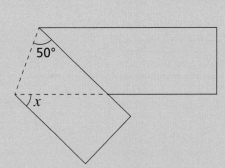

c Fold a rectangular piece of paper yourself and explore the relationships between the angles created.

Consolidate – do you need more?

1 Work out the size of each angle labelled with a letter.

Give reasons for each stage of your working.

a

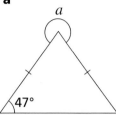

b

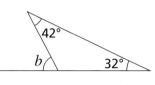

c

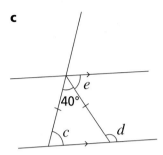

2 Work out the size of each angle labelled with a letter.

Give a reason for each stage of your working.

a

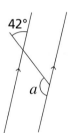

b

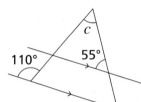

c

d

Hint: You may find it helpful to label some of the other angles.

3 All of these diagrams show rectangles. Work out the size of each angle marked with a letter.

a

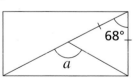

b

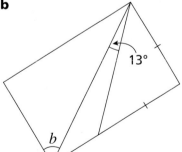

c

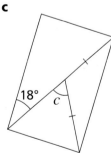

4 Work out the size of the required angle in each of these triangles. Show your method clearly and give reasons for your answers.

a ∠ABC

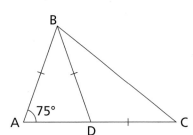

b ∠EFG

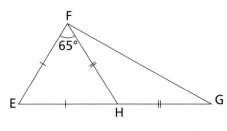

5 Do you agree with Flo? Give a reason for your answer.

Angle b = 65° because vertically opposite angles are equal.

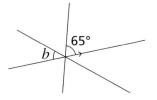

6 Work out the size of each angle labelled with a letter.

a

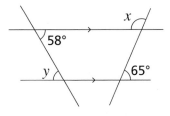

b

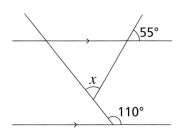

c

Stretch – can you deepen your learning?

1 Show that triangle ABC is equilateral. Give a reason for each stage of your working.

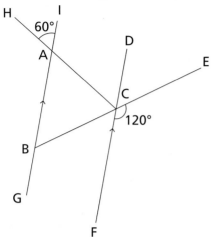

2 Work out the value of x

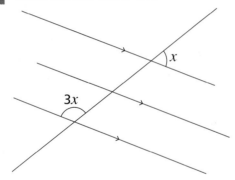

3 a A pair of parallel lines crosses another pair of parallel lines. How many angles are formed? How much information do you need in order to work out all of the angles formed?

b Investigate the number of angles formed and how much information you need to work them all out for different arrangements of four lines.

Reflect

1 What is the difference between showing your method and giving a reason for your answer? Use examples to explain.

2 Make up your own angle problem where it is necessary to find several other angles before the one you ask for. What is the minimum number of angles you need to give so that someone else can find all the other angles in your problem?

Small steps

■ Solve angle problems with algebra

Are you ready?

1 Copy and complete each statement using a number or a word.

   a Angles in a triangle add up to _____.

   b Angles in a quadrilateral add up to _____.

   c Vertically opposite angles are _____.

   d Exterior angles of a polygon add up to _____.

2 For each polygon, state the sum of the interior angles.

   a pentagon   **b** hexagon   **c** octagon

3 Simplify each expression.

   a $2x + 7 + 3x - 5$ **b** $2y + 2 + 3y - 15 + y$ **c** $5z + 13 - 2z + 18 - z$

4 Solve each equation.

   a $2x + 25 = 180$ **b** $3x + 18 = 360$

   c $9x - 9 = 540$ **d** $2x + 22 = 6x - 34$

Models and representations

Geoboards (or geoboard apps)

These are useful for reviewing angles in polygons.

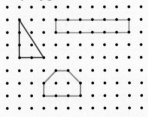

Bar models

These are useful for solving equations.

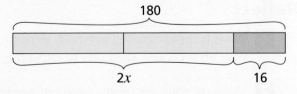

In this chapter, you will focus on solving algebraic problems involving all of the angle facts you have met so far.

As well as angles in parallel lines, you need to use your knowledge of angles in polygons from Book 2

Recall these facts about **interior** and **exterior angles** of polygons.

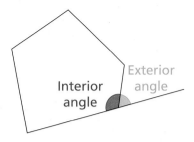

Interior angle

Exterior angle

The exterior angles of a polygon always add up to 360°

To find the sum of the interior angles of a polygon, you can split the shape into triangles or use the formula: sum = $(n - 2) \times 180°$, where n is the number of sides of the polygon.

The interior and exterior angles of a polygon lie on a straight line and so will always add up to 180°

Example 1

Work out the value of each letter.

a

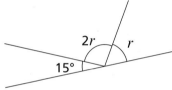

$2r$ r

15°

b

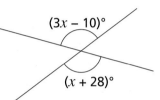

$(3x - 10)°$

$(x + 28)°$

a
$$15 + r + 2r = 180$$
$$15 + 3r = 180$$
$$3r = 180 - 15$$
$$3r = 165$$
$$r = 165 \div 3 = 55$$

You need to use the fact that angles on a straight line add up to 180°. Write an equation to show this. Simplify the equation.

Solve by balancing. You can use a bar model to help.

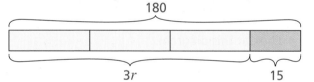

180

$3r$ 15

b

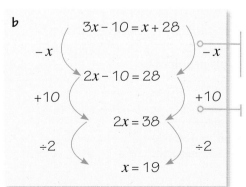

$$3x - 10 = x + 28$$
$-x$ $-x$
$$2x - 10 = 28$$
$+10$ $+10$
$$2x = 38$$
$\div 2$ $\div 2$
$$x = 19$$

You need to use the fact that vertically opposite angles are equal. Write down an equation to show that the two angles are equal.

Solve this equation by balancing.

Example 2

Work out the value of t

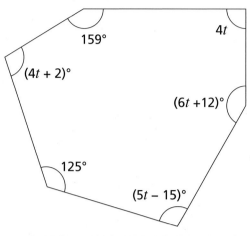

Sum of interior angles = $(n - 2) \times 180$ ○———— You can use this formula to work out the sum of all the interior angles.

$n = 6$, so $(6 - 2) \times 180 = 720°$ ○———— The shape is a hexagon, so $n = 6$

Now that you know that all the angles must add up to 720°, you can write down an equation.

$4t + 6t + 12 + 5t - 15 + 125 + 4t + 2 + 159 = 720$ ○————

$19t + 283 = 720$ ○———— Simplify the equation by collecting like terms.

$19t = 720 - 283$

$19t = 437$ ○———— Solve by balancing to find t

$t = 23$

Practice 9.3A

1 Look at this diagram.

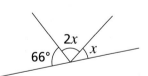

a Which mathematical fact is needed to solve this problem?

b Which equation correctly describes this fact?

| $3x = 66°$ | $3x + 180° = 66°$ | $3x + 66° = 180°$ |

c Solve this equation to find the value of x

2 Look at this diagram.

   a What do the angles in a pentagon add up to?

   b Write and solve an equation to find the value of y

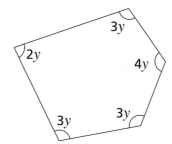

3 Work out the size of the angle represented by a letter in each of these diagrams.

   a

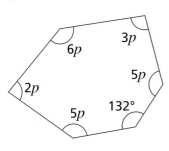

   b

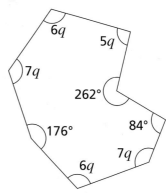

4 Work out the size of the angle represented by a letter in each of these diagrams.

   a

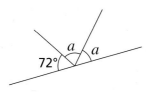

   b

   c

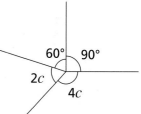

   d

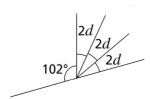

   e

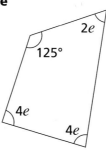

   f

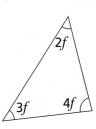

5 Work out the size of the angle represented by a letter in each of these diagrams.

a

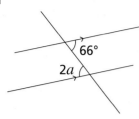

b

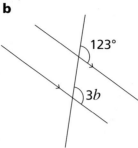

c

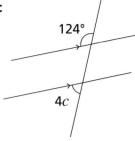

d

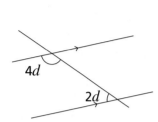

e

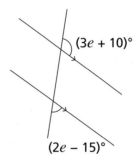

f

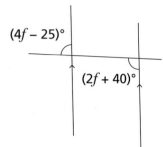

6 Work out the value of x and y in each of these diagrams.

a

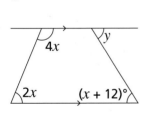

b

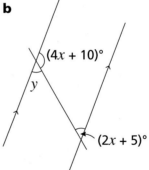

7 Work out the value of x

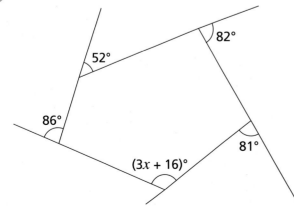

8 Work out the size of the angle represented by a letter in each of these diagrams.

a

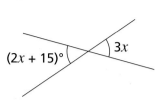

$(2x + 15)°$ $3x$

b

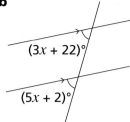

$(3x + 22)°$

$(5x + 2)°$

c

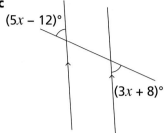

$(5x - 12)°$

$(3x + 8)°$

9 OB is the angle bisector of ∠AOC
Work out
   a the value of x
   b the size of ∠AOC

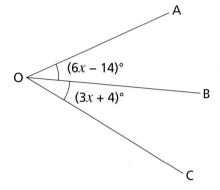

$(6x - 14)°$

$(3x + 4)°$

What do you think? 🌐

1 ABC is a straight line. DEF is a straight line.

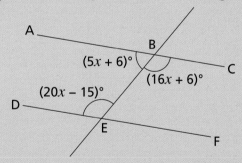

A

B

$(5x + 6)°$

$(16x + 6)°$

C

$(20x - 15)°$

D

E

F

Is ABC parallel to DEF? Show working to explain how you know.

2 In a regular polygon, each exterior angle is 160° less than each interior angle.
How many sides does the polygon have?

3 Is this shape a parallelogram?

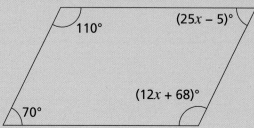

110°

$(25x - 5)°$

$(12x + 68)°$

70°

Consolidate – do you need more?

1 Work out the size of the angle represented by a letter in each of these diagrams.

a

b

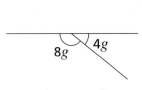

c

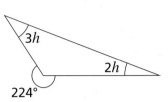

d

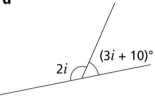

e

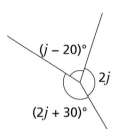

f

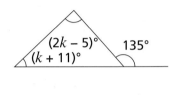

2 Work out the size of the angle represented by a letter in each of these diagrams.

a

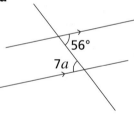

b

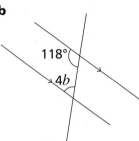

c

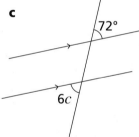

d

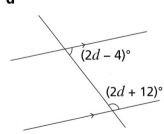

e

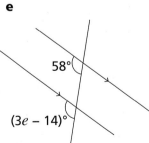

f

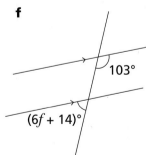

3 Work out the size of the angle represented by a letter in each of these diagrams.

a

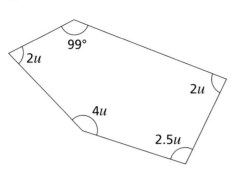

b

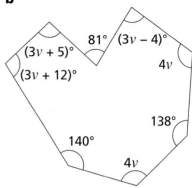

4 ABCDE is a regular pentagon. Work out the value of x

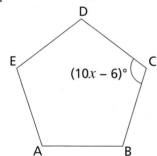

5 In each diagram, work out the size of each angle labelled with a letter.

a

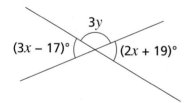

b

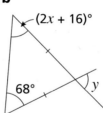

c

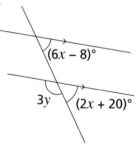

Stretch – can you deepen your learning?

1 Look at this triangle.

a If $x + y = z$, show that z must equal 90°

b If $x + y = 2z$, show that z must equal 60°

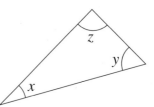

2 Show that this triangle is isosceles.

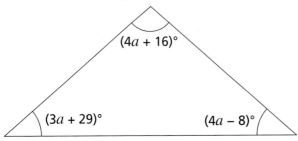

$(4a + 16)°$

$(3a + 29)°$ $(4a - 8)°$

3 If the angles in a triangle go up in equal steps, for example 50, 60, 70, the middle angle is always 60°

Investigate Chloe's conjecture. Can you prove this using algebra?

Reflect

Explain why using algebraic notation can be helpful when solving problems like those in this chapter. Are there other ways you could solve them, for example using bar models or other manipulatives?

9.4 Geometric conjectures

Small steps

- Conjectures with angles
- Conjectures with shapes
- Link constructions and geometrical reasoning (H)

Key words

Conjecture – a statement that might be true that has not yet been proved

Proof – an argument that shows that a statement is true

Counterexample – an example that disproves a statement

Are you ready?

1 Work out the size of each angle labelled with a letter.

a   **b**   **c**   **d**

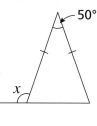
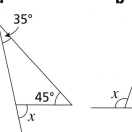
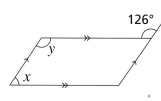

2 Work out the size of each angle labelled with a letter.

a   **b**   **c**   **d**

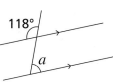

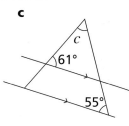
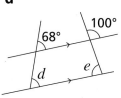

3 Identify the mathematical relationship shown in each diagram.

a   **b**   **c**   **d**

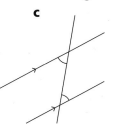
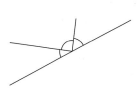

Models and representations

Bar models

These are useful for solving equations.

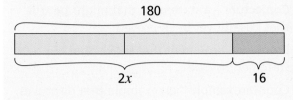

These are useful for showing equivalences.

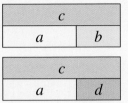

If $a + b = c$ and $a + d = c$, then $b = d$

In this chapter, you will develop your reasoning and justification skills by making and proving **conjectures**.

A **proof** is a formal way to explain your answer. Each statement you write needs to follow in a logical manner and be justified with a mathematical reason.

Example 1

Abdullah says that a hexagon cannot have a reflex angle.

Show, by giving a counterexample, that Abdullah is wrong.

You can investigate Abdullah's conjecture by using a geoboard or dynamic geometry software.

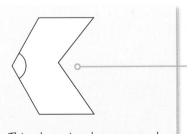

You only need to give one **counterexample** to disprove a conjecture. A hexagon has 6 sides. So you need to draw a 6-sided shape, with a reflex angle.

Remember: a reflex angle is between 180° and 360°

This shape is a hexagon and has a reflex angle. Therefore, Abdullah is wrong.

Example 2

Prove that triangle ABC is isosceles. Give a reason for each step in your working.

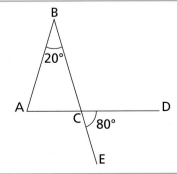

∠ACB = 80°, since vertically opposite angles are equal.

To prove that ABC is isosceles you need to show that two of the angles are equal in size.

∠BAC = 180° − (20° + 80°) = 80°

Now that you know two of the angles in the triangle, you can work out the size of the third angle.

Triangle ABC has angles of 20°, 80° and 80°

Since two angles are the same, it must be an isosceles triangle.

A proof should have a conclusion to summarise what you have just shown.

Example 3

Prove that $a + b = 90°$

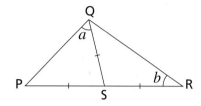

In triangle PQS,

∠QPS = a (since two angles in an isosceles triangle are equal).

You know triangles PQS and QRS are both isosceles because of the hatch marks on the diagram.

In triangle QRS,

∠SQR = b (since two angles in an isosceles triangle are equal).

In triangle PQR the angles are a, b and $(a + b)$

There is also a bigger triangle, PQR. You can start by finding the missing angles in triangle PQR

It is really helpful to sketch the diagram and mark on the angles you have found.

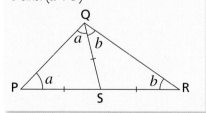

$a + b + a + b = 180°$ (angles in a triangle add up to 180°)

Use angle facts to set up an equation.

$2a + 2b = 180°$

$2(a + b) = 180°$

$a + b = 90°$

Simplify in order to show the required fact.

Practice 9.4A

1 Mario is investigating triangles. He makes the following three conjectures. Determine whether each one is true or false, justifying your answers.

    a A scalene triangle has one line of symmetry.

    b A right-angled triangle has no lines of symmetry.

    c An equilateral triangle has three lines of symmetry.

2 Beth is investigating angles and turns. She makes the following three conjectures. Determine whether each one is true or false, justifying your answers.

    a A quarter turn followed by a quarter turn is always the same as a half turn.

    b An acute angle plus an obtuse angle will always give a reflex angle.

    c An obtuse angle plus an obtuse angle will always give a reflex angle.

3 Marta believes that triangle ABC is equilateral.

    a Work out the size of angle BCA. Give a reason for your answer.

    b Work out the size of angle BAC. Give a reason for your answer.

    c Is Marta correct?

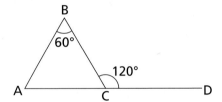

4 Show that triangle PQR is isosceles. Give a reason at each stage in your working.

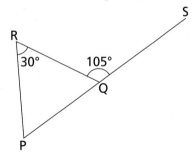

5 A quadrilateral can't have a reflex angle.

Show, by giving a counterexample, that Ed is wrong.

6

All quadrilaterals can be split into two congruent triangles.

   a   Test Chloe's conjecture by drawing quadrilaterals of your own.

   b   Comment on whether the conjecture is always true, sometimes true or never true.

7 Decide whether each statement is always true, sometimes true or never true.
Explain your answers.

   a   The diagonals of a rectangle cross at 90°

   b   If two rectangles have an area of 48 cm², they will also have the same perimeter.

   c   The longest side of a triangle is shorter than the sum of the other two sides.

   d   A parallelogram has two lines of reflection symmetry.

8 Show that angle BAD is equal to 80°. Give a reason for each stage in your working.

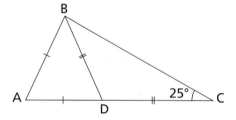

9 Show that this shape is a parallelogram. You must justify your answer clearly.

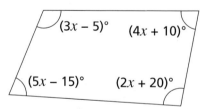

10 Show that angle y is half the size of angle x

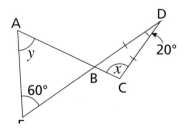

What do you think?

1 Show that $x + y = 60°$ Give a reason for each step of your working.

> Hint: Try drawing another line parallel to the ones already there.

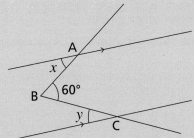

2 Prove that $3a + b = 180°$ Give a reason for each step of your working.

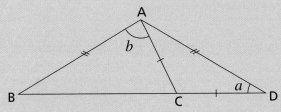

3 What is the relationship between angles x and y? Give a reason for each step of your working.

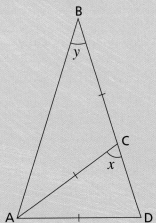

In the final section of this chapter, you will make connections between constructions and geometric reasoning and proof.

Example 4

Draw a line segment AB

Construct the perpendicular bisector of AB and label it CD

Label the point of intersection between the two lines as O

Join up the points ABC to create a triangle.

a What type of triangle is ABC? How do you know?

b Show that triangles AOC and BOC are congruent.

a

Draw line AB and use your compasses to construct the perpendicular bisector.

You learned how to draw perpendicular bisectors in Chapter 5.3 of this book.

Label CD and the point O

Then join up the points to form triangle ABC

ABC is an isosceles triangle. This is because AC and BC are the same length.

You know this because you did not change the width of the compasses, so both arcs have the same radius.

All points on the perpendicular bisector of AB are the same distance from A as from B

b AC = BC (shown in part **a**)

You need to show that one of the triangle congruence conditions hold. You learned these in Chapter 5.5 of this book.

Angle AOC and BOC are both 90° as their bisector is perpendicular.

Perpendicular means at 90°

AO = OB

This is because the bisector cuts AB exactly in half.

Explain why AO and AB are equal.

RHS condition is met, therefore triangles are congruent.

Both triangles have a right angle, both have the same length hypotenuse and both have a side length in common: R, H and S

Practice 9.4B

1 Draw a line segment AB. Construct the perpendicular bisector of AB. Label it CD. Join up points ABCD to create a quadrilateral.

Give the name of quadrilateral ABCD

Explain how you know.

2 Jackson constructs the angle bisector of angle AOB as shown.

He joins up points OACB to make a quadrilateral.

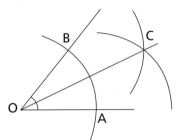

a Give the name of the quadrilateral Jackson has constructed. How do you know?

b Show that triangles OBC and OAC are congruent.

3 Amina constructs an angle of 60° as shown.

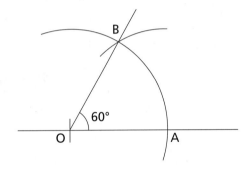

Explain why this method works.

4 Rob conjectures that none of the triangles below can be constructed.

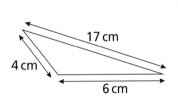 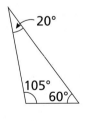 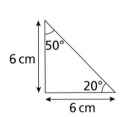 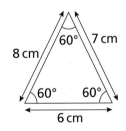

a Try to construct them yourself.

b In each case, explain why Rob is right.

5 Draw a circle using a pair of compasses. Draw two chords in your circle.

  a Construct the perpendicular bisector of each chord.

  b Extend the bisectors so that they meet. What do you notice?

Ali draws three points A, B and C. He joins them up as shown.

> Remember: a chord joins two points on the circumference of a circle.

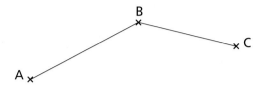

  c Copy Ali's drawing. Construct a circle which has AB and BC as chords.

What do you think?

1 Copy the diagram by constructing a circle. Then, from a point outside the circle, draw two tangents to the circle, as shown.

> A tangent touches the circumference of a circle.

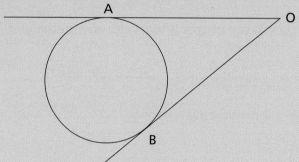

  a Construct a line perpendicular to OB, going through B

  b Construct a line perpendicular to OA going through A

  Label the point where the two perpendiculars meet as D

> You learned how to draw perpendiculars from a point in Chapter 5.3 of this book.

  c What do you notice about the point D?

  d Measure angles DBO and DAO – what do you notice?

  e Give the name of shape OADB

  f Will you always get the same answer for parts **c** to **e** if you repeat parts **a** and **b** with a different circle and a different point O? Test this with another example.

2 By constructing the perpendicular from point A to the line DB, explain why the areas of the two shaded regions are equal.

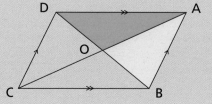

Consolidate – do you need more?

1 Decide whether each statement is true or false. Give reasons for your answers.

   a   Two acute angles will always add up to make an obtuse angle.

   b   The interior angle of a regular hexagon is obtuse.

   c   No regular polygon has interior angles less than 90°

2 Marta conjectures that all quadrilaterals have at least one line of symmetry.

   a   Investigate Marta's conjecture.

   b   What conclusion can you draw?

3 Filipo is investigating the properties of rectangles. He conjectures that as the perimeter of a rectangle becomes greater, so does the area.

   a   Investigate Filipo's conjecture using examples of your own.

   b   What conclusion can you draw?

4 Huda believes that a regular hexagon can be split into six equilateral triangles.

   a   Show that Huda is right.

   Huda conjectures that this means a regular pentagon can be split into five equilateral triangles.

   b   Explore Huda's conjecture and show that she is wrong.

5 Show that angle BCD = 27.5°

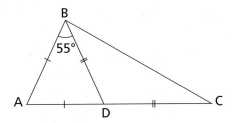

6 Show that this shape is a right-angled triangle.

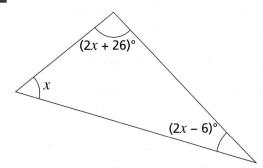

Stretch – can you deepen your learning?

1 Show that $p + q = r + s$

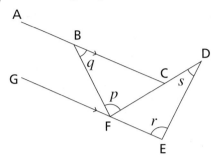

2 A triangle can have a maximum of one right angle.

A quadrilateral can have a maximum of four right angles.

Investigate the maximum number of right angles you can have in pentagons, hexagons and other shapes.

3 Jakub and Faith are constructing kites.

Jakub starts by drawing an angle and bisecting it.

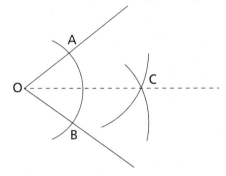

Faith starts by drawing the perpendicular from a point to a line.

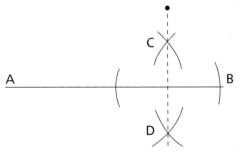

Explain why each method works and describe the next steps.

Reflect

1 What is the difference between a proof and a demonstration?

2 Make up an example of your own to show what is meant by a counterexample.

I have become **fluent in...**	I have developed my **reasoning** skills by...	I have been **problem-solving** through...
▧ finding missing angles on straight lines, around a point and in 2-D shapes	▧ justifying my calculations by giving reasons	▧ solving more complex problems involving several steps
▧ finding missing angles in problems involving parallel lines	▧ identifying which mathematical rule to use	▧ setting up and solving equations, including with unknowns on both sides
▧ using algebra to set up and solve simple equations	▧ forming chains of reasoning to find a missing angle	▧ proving geometric facts
▧ determining whether a statement about a shape is true or false.	▧ making and testing simple conjectures about shapes and angles.	▧ making connections between constructions and geometric facts. Ⓗ

Check my understanding

1 Work out the size of each angle labelled with a letter. Give reasons for your answers.

a

b

c

d

2 Work out the size of each angle labelled with a letter. Give reasons for your answers.

a

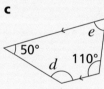

b

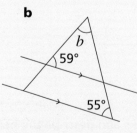

c

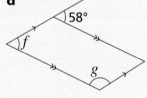

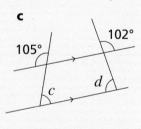

3 Determine whether each statement is always true, sometimes true or never true.

a The angles in a regular polygon are always greater than or equal to 90°

b Vertically opposite angles add up to 180°

c All triangles have at least one line of symmetry.

In this block, I will learn...

about rotational symmetry in real-life objects and with 2-D shapes

to compare rotational and line symmetry

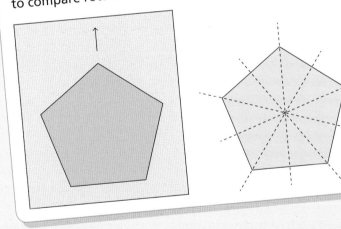

how to rotate a shape on a grid

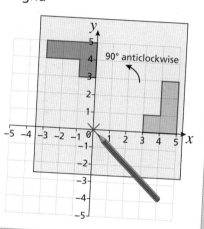

90° anticlockwise

how to perform a translation described using vector notation

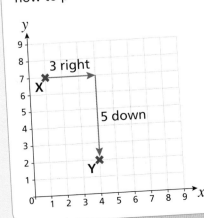

3 right

5 down

The vector describing the movement from X to Y is written like this:

$$\begin{pmatrix} 3 \\ -5 \end{pmatrix} \begin{array}{l} \rightarrow \text{ 3 right} \\ \rightarrow \text{ 5 down} \end{array}$$

Small steps

■ Identify the order of rotational symmetry of a shape

■ Compare and contrast rotational symmetry with line symmetry

Key words

Rotational symmetry – when a shape still looks the same after turning

Line of symmetry – a line that cuts a shape exactly in half

Irregular – a shape that has unequal sides and unequal angles

Regular – a shape that has equal sides and equal angles

Are you ready?

1 Match each diagram with the turn that it shows.

  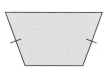

  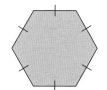 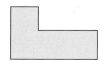

| full turn | half turn | quarter turn |

2 How do you know that this reflection hasn't been performed correctly?

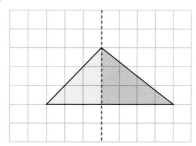

3 **a** Write down the mathematical name of each of these polygons.

i    **ii** 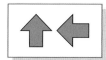   **iii**

iv   **v**   **vi**

b Which of the polygons are regular?

Models and representations

There are many examples of **rotational symmetry** in the real world.

When working out the order of rotational symmetry, it is a good idea to use tracing paper.

Place a piece of tracing paper over the shape and draw its outline. Draw an arrow pointing up.

 This will show you when you have completed a full turn.

Turn the tracing paper until you get back to the starting point. Count the number of times the shape exactly fits its outline in one rotation.

Example 1

State the order of rotational symmetry of each of these shapes.

a    b    c

a 5 ○—│ You can test this using tracing paper.

b 5 ○—│ You may have thought that the answer was 10 because there are 10 petals. However, the pattern only repeats 5 times.

c 1 ○—│ Be careful with images like this. If the traffic lights were not on the sign, the answer would have been 3

All shapes have at least order 1 rotational symmetry, this being the original position before any turn.

417

Practice 10.1A

1 State the order of rotational symmetry for each of the regular shapes.

a   b   c

What do you notice?

2

This rectangle will have rotational symmetry of order 4 because it has four sides.

Do you agree with Abdullah? Explain how you decided.

3 State the order of rotational symmetry for each shape.

a    b   c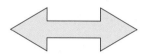

4 Copy and complete the sentence.

The order of rotational symmetry is equal to the number of sides if the shape is _____.

5 Draw a 2-D shape that has rotational symmetry of order 2

Compare your shape with a partner's. What's the same and what's different?

6 State the order of rotational symmetry for each of these road signs.

a   b   c

7 State the order of rotational symmetry for each of these car wheels.

a    b    c

What do you think? 🌐

1. Which 2-D shape would have an infinite order of rotational symmetry?

2. Is this statement always true, sometimes true or never true?

   When a regular shape is drawn inside another regular shape, the order of rotational symmetry is always that of the inner shape.

   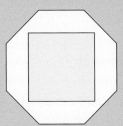

   Rotational symmetry of order 4

   Draw a diagram to explain your answer.

In this section, you will compare rotational symmetry with line symmetry.

You studied line symmetry in Book 2. Recall that a **regular** shape has the same number of lines of symmetry as sides.

For example

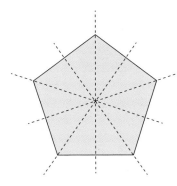

A regular pentagon has 5 lines of symmetry.

Example 2

For each shape, state the number of lines of symmetry and the order of rotational symmetry.

a 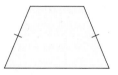 **b**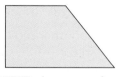

a lines of symmetry = 1	
order of rotational symmetry = 1	

 line of symmetry

b lines of symmetry = 0	The shape can't be folded in such a way that one
order of rotational symmetry = 1	half fits exactly on top of the other, so it has no lines of symmetry.

Practice 10.1B

1 Here are three regular shapes.

i   **ii**   **iii**

    a How many lines of symmetry does each shape have?

    b State the order of rotational symmetry of each shape.

2 Here are three irregular shapes.

i 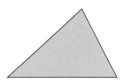   **ii**    **iii**

    a How many lines of symmetry does each shape have?

    b State the order of rotational symmetry of each shape.

3 What do you notice about your answers to questions **1** and **2**?

4

Irregular shapes have no lines of symmetry.

Is Zach's statement always true, sometimes true or never true? Use examples to explain your reasoning.

 5

This shape is irregular so it must have rotational symmetry of order 1

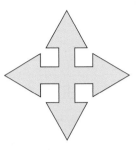

a Do you agree with Marta?

b How many lines of symmetry does the shape have?

What do you think?

1 **a** On a squared grid, shade a pattern that has 1 line of symmetry and rotational symmetry of order 1

Here is an example.

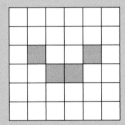

b How many different patterns can you make?

2 Beca has formed a conjecture.

If a shape has no lines of symmetry, it can only have rotational symmetry of order 1

Draw an example that

a supports Beca's conjecture

b disproves Beca's conjecture.

Consolidate – do you need more?

1 For each shape, state

  i the number of lines of symmetry

  ii the order of rotational symmetry.

a   **b**   **c**    **d**

2 For each image, state

  i the order of rotational symmetry

  ii the number of lines of symmetry.

a 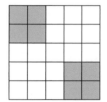   **b**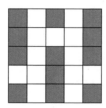

3 On a squared grid, shade squares to make a pattern that has

  a 2 lines of symmetry

  b rotational symmetry of order 4

Stretch – can you deepen your learning?

1 a What fraction of the upper case letters of the alphabet have an order of rotational symmetry greater than 1?

  b What percentage of the upper case letters of the alphabet have no lines of symmetry? Round your answer to the nearest integer.

2 The interior angles of a regular polygon sum to 540°. What is the order of rotational symmetry of the polygon?

3 One of the interior angles of a regular polygon is 135°. How many lines of symmetry does the polygon have?

Reflect

1 Explain the difference between rotational symmetry and line symmetry.

2 Imagine you are explaining to someone how to find the order of rotational symmetry of a shape. What tips would you give them?

10.2 Rotation

Small steps

■ Rotate a shape about a point on the shape

■ Rotate a shape about a point not on the shape

Key words

Rotation – turn a shape around a fixed point called the centre of rotation

Vertex (plural: **vertices**) – a point where two line segments meet; a corner of a shape

Congruent – exactly the same size and shape, but possibly in a different orientation

Are you ready?

1 Write down the coordinates of points A, B, C and D

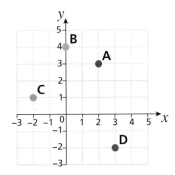

2 Write down the letters of two shapes that are congruent.

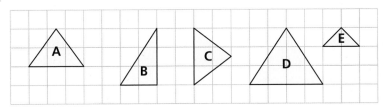

3 Match each turn to the correct number of degrees.

| quarter turn | half turn | three-quarter turn | full turn |

| 180° | 360° | 90° | 270° |

Models and representations

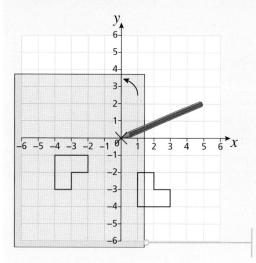

Tracing paper will help you to rotate shapes on a grid.

In this chapter, you will learn how to perform **rotations**, clockwise or anticlockwise, through 90°, 180° or 270°

You can see rotations all around you in real life, for example big wheels and wind turbines.

It is important to know the difference between "clockwise" and "anticlockwise". Consider the hands on a clock. As time moves on, the hands turn clockwise. The opposite direction is anticlockwise.

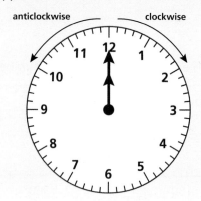

Example 1

Draw this shape after a 180° rotation.

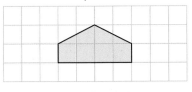

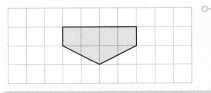

○─┤ A 180° rotation is the same as half a turn.

Example 2

Rotate this shape 90° anticlockwise about the origin.

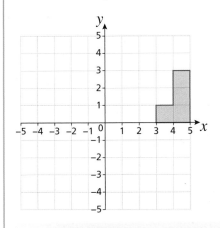

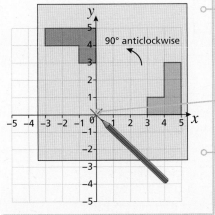

○─┤ Draw the outline of the shape on tracing paper.

Then put your pencil on the centre of rotation, which, in this case, is the origin.

─┤ The origin is the point (0, 0)

○─┤ Turn the tracing paper a quarter turn anticlockwise and draw the shape in its new position.

Practice 10.2A

1 Draw each shape after these rotations.

i 180°   **ii** 90° clockwise

iii 270° clockwise   **iv** 90° anticlockwise

a  **b** 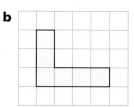 **c** 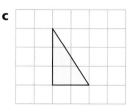 **d**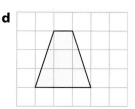

What do you notice about your answers?

2 Benji, Chloe and Seb have been asked to rotate triangle A 90° anticlockwise about the cross (✖). All three of them have made a mistake. Explain each mistake.

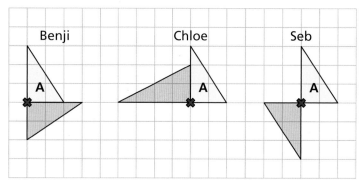

3 Copy each shape onto squared paper and rotate it 90° clockwise about the cross (✖)

a 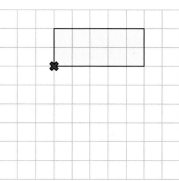 **b**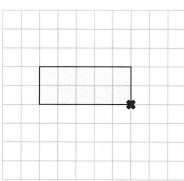

c 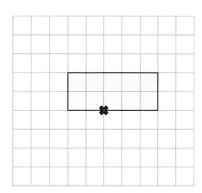 **d**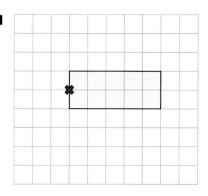

4 Copy each grid and rotate each shape 90° clockwise about the given centre.

a (3, 6)

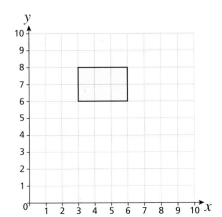

b (7, 3)

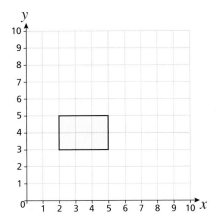

c (5, 4)

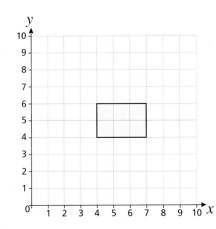

d (1, 6)

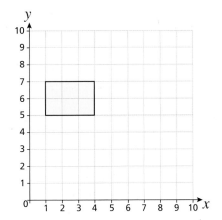

5 Copy each grid and rotate each shape 90° clockwise about the given centre.

a (2, 6)

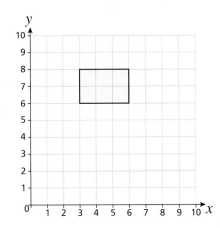

b (7, 2)

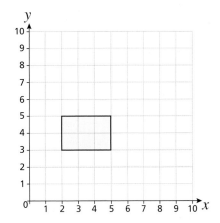

c (5, 7)

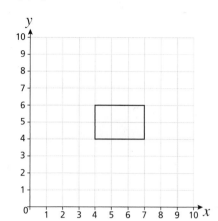

d (4, 3)

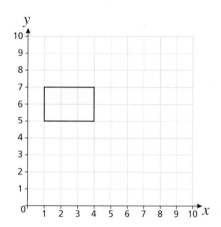

6 What's the same and what's different about questions **4** and **5**?

7 In each diagram, shape A has been rotated to give shape B. Describe fully each of these rotations. Comment on the direction, the angle turned through and the centre of rotation.

a

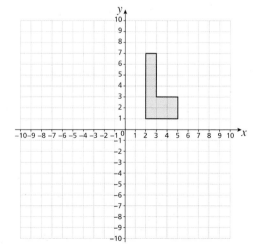

b

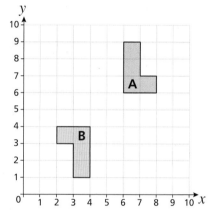

c Is there more than one possible answer for each diagram?

8 Copy each grid and complete the rotations.

a 180° about (2, 1)

b 90° clockwise about (–6, 1)

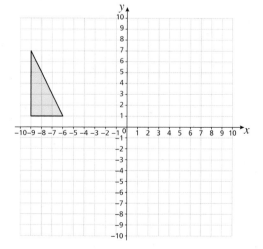

c 90° anticlockwise about (4, −2)

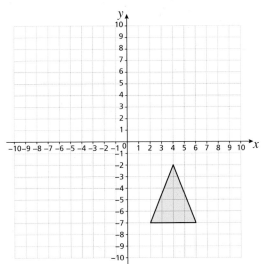

d 270° clockwise about (5, −5)

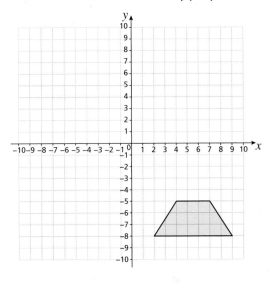

e 270° anticlockwise about the origin

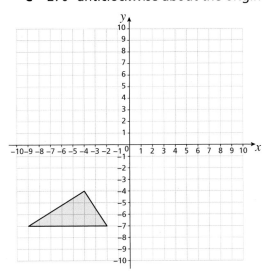

f 180° about (−2, −3)

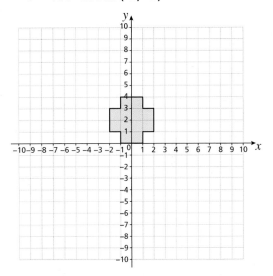

9 In each diagram, describe the single rotation that maps shape A onto shape B

a

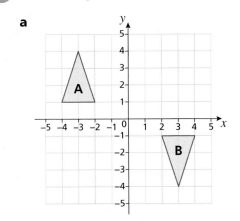

b

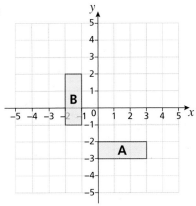

c

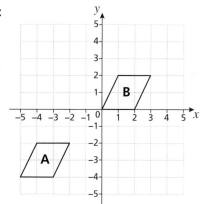

d

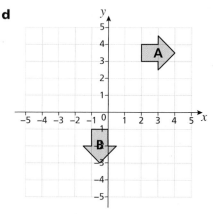

What do you think?

1. Explain why you don't need to give the direction when rotating a shape through 180°

2. What would happen if you rotated a shape through 360°?

3. Abdullah has rotated a shape through 630° anticlockwise. State two rotations that would give the same result as Abdullah's.

4. Is this statement always true, sometimes true or never true?

   The result of a rotation is congruent to the original shape.

Consolidate – do you need more?

1. Copy each diagram onto squared paper. Rotate each shape about point A as instructed.

   a 90° anticlockwise

   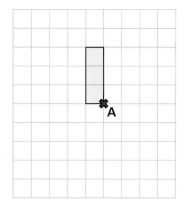

   b 270° clockwise

   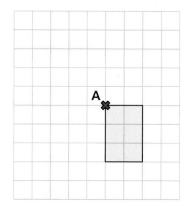

c 180°

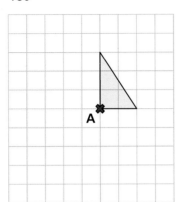

d 90° clockwise

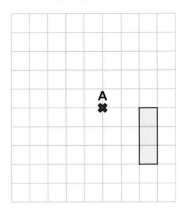

e 180°

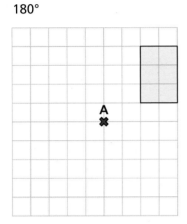

f 270° anticlockwise

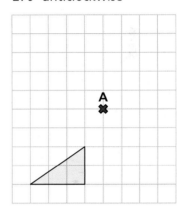

2 Copy each diagram onto squared paper. Rotate each shape as instructed using the origin as the centre of rotation.

a 180°

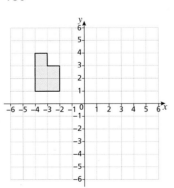

b 270° clockwise

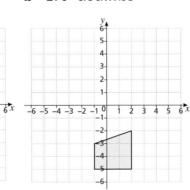

c 90° anticlockwise

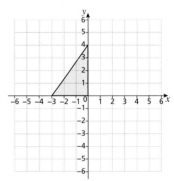

3 Beca has rotated shapes G, H and I 90° clockwise about the origin. Each time, she has made a mistake. Explain each mistake.

a

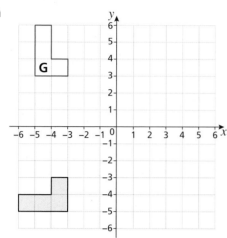

b

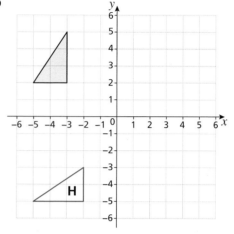

c

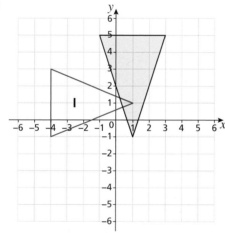

Stretch – can you deepen your learning?

1 A rectangle drawn on a coordinate grid has vertices at the points (2, 3), (2, 5), (5, 3) and (5, 5). After a rotation, the vertices are at the points (7, 5), (7, 8), (9, 5) and (9, 8). Describe fully the rotation of the rectangle.

2 ABCD is a rectangle drawn on a coordinate grid. The coordinates of the vertices are A (–4, 4), B (–3, 4), C (–3, 1) and D (–4, 1). Rectangle ABCD is rotated 180° about the point (–3, 0)

   a Chloe says, "After the rotation, line segment AB is between (–3, –1) and (–2, –1)."

       Explain the mistake that Chloe has made.

   b Write down the coordinates of each vertex after the rotation.

3 Points P (2, 1), Q (2, 5) and R (4, 2) are joined to form triangle PQR. Triangle PQR is rotated so that exactly one vertex does not move. How many possibilities are there for the centre of rotation?

4 Seb forms this conjecture.

If the centre of rotation isn't on the original shape, then it can't be on the rotated shape either.

Investigate Seb's conjecture and explain your findings.

Reflect

What information do you need to be given in order to perform a rotation?

Small steps

- Translate points and shapes by a given vector
- Compare rotation and reflection of shapes
- Find the result of a series of transformations (H)

Are you ready?

1 **a** Copy and complete the sentences to describe the movement of each point on the grid.

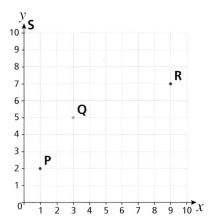

 i From point P to point Q is ☐ squares right and ☐ squares up.

 ii From point Q to point R is ☐ squares right and ☐ squares up.

 iii From point R to point S is ☐ squares left and ☐ squares up.

 iv From point S to point P is _____

 v From point Q to point P is _____

b What are the new coordinates of point R when moved 4 squares to the left and 6 squares down?

Models and representations

Vectors

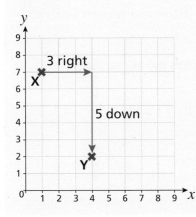

The vector describing the movement from X to Y is written like this:

$$\binom{3}{-5} \begin{array}{l} \longrightarrow \text{3 right} \\ \longrightarrow \text{5 down} \end{array}$$

In this chapter, you will learn how to describe **translations** using vector notation.

Vectors describe horizontal and vertical movements and are written in a similar way to coordinates.

For example, the coordinates (2, 4) represent the point 2 units right and 4 units up from (0, 0) and the coordinates (–3, –7) represent the point 3 units left and 7 units down from (0, 0)

Similarly, the vector $\binom{2}{4}$ represents a movement of 2 units right and 4 units up from any

starting point and $\binom{-3}{-7}$ represents a movement of 3 units left and 7 units down from any starting point.

Example 1

Translate shape A by vector $\binom{-5}{-2}$

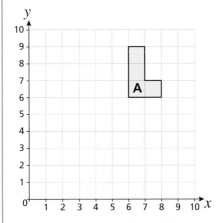

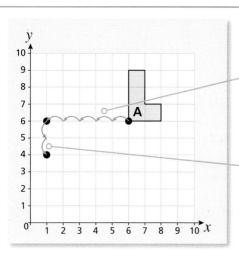

−5 in the vector $\begin{pmatrix} -5 \\ -2 \end{pmatrix}$ means move 5 squares to the left.

Choose a vertex of the shape and count 5 squares to the left…

−2 in the vector $\begin{pmatrix} -5 \\ -2 \end{pmatrix}$ means move 2 squares down.

…then move 2 squares down.

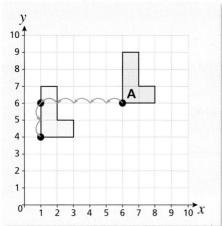

Make sure you start drawing your translated shape from the same vertex you counted from. Not doing this is a common mistake.

You can count from every vertex to make sure that your answer is correct.

The translated shape should be congruent to the original shape.

Remember from Block 5, "congruent" means it is exactly the same size.

Practice 10.3A

1 Match each column vector to its description.

$\begin{pmatrix} -6 \\ 1 \end{pmatrix}$

$\begin{pmatrix} 6 \\ 1 \end{pmatrix}$

$\begin{pmatrix} -6 \\ -1 \end{pmatrix}$

$\begin{pmatrix} 6 \\ -1 \end{pmatrix}$

$\begin{pmatrix} -1 \\ 6 \end{pmatrix}$

$\begin{pmatrix} 1 \\ 6 \end{pmatrix}$

1 right and 6 up

6 left and 1 down

1 left and 6 up

6 right and 1 up

6 right and 1 down

6 left and 1 up

2 Ed has translated rectangle A 2 squares left and 3 squares down.

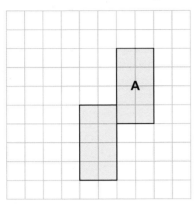

a Describe Ed's translation using vector notation.

b Beca has also tried to translate rectangle A 2 squares left and 3 squares down. Here is her answer.

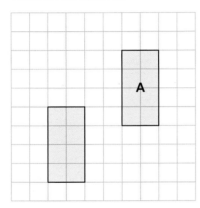

i Explain the mistake that Beca has made.

ii Use vector notation to describe Beca's translation.

3 Use vector notation to describe each translation from shape P to shape Q

a

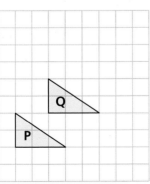

b

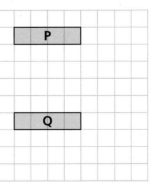

c

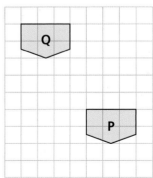

4 Copy each diagram onto squared paper and translate each shape by the given vector.

a $\begin{pmatrix} 3 \\ 2 \end{pmatrix}$

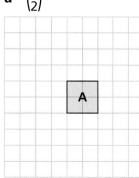

b $\begin{pmatrix} -4 \\ 0 \end{pmatrix}$

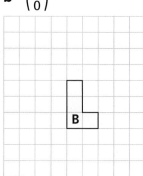

c $\begin{pmatrix} 0 \\ 4 \end{pmatrix}$

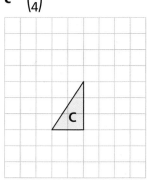

5 Zach thinks that triangle A has been translated by the vector $\begin{pmatrix} -5 \\ -3 \end{pmatrix}$ to give triangle B

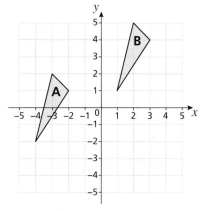

Do you agree with Zach? Explain your answer.

6 Copy each grid and translate each shape by the given vector.

a $\begin{pmatrix} -6 \\ -2 \end{pmatrix}$

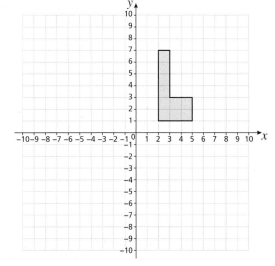

b $\begin{pmatrix} 10 \\ -5 \end{pmatrix}$

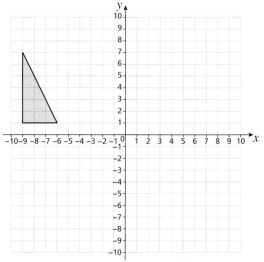

c $\begin{pmatrix} 0 \\ 8 \end{pmatrix}$

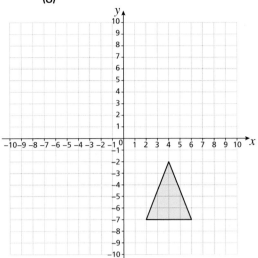

d $\begin{pmatrix} -9 \\ -1 \end{pmatrix}$

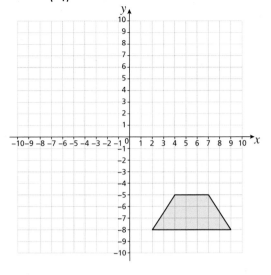

e $\begin{pmatrix} 4 \\ 11 \end{pmatrix}$

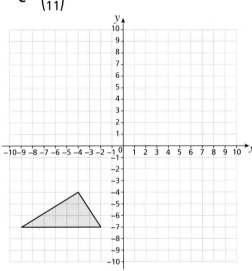

f $\begin{pmatrix} -3 \\ 4 \end{pmatrix}$

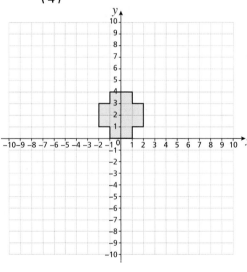

7 Use vector notation to describe each translation from shape A to shape B

a

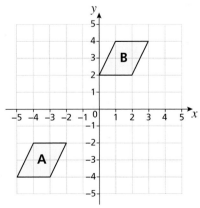

b

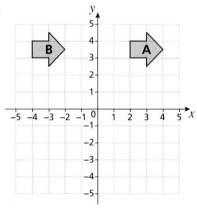

c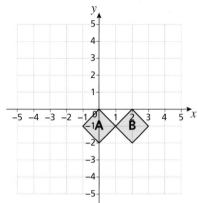

d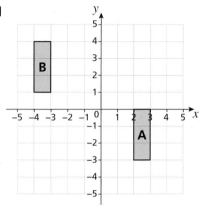

e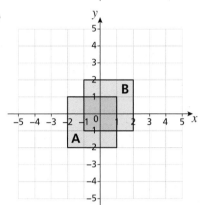

f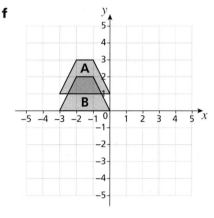

8 **a** Write the coordinates of points J, K, L and M

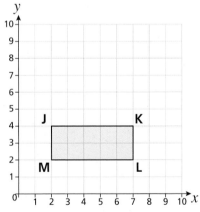

b On a copy of the grid, translate rectangle JKLM by vector $\begin{pmatrix} 2 \\ 4 \end{pmatrix}$ and label the vertices WXYZ

c Write the coordinates of points W, X, Y and Z

d What do you notice about the coordinates of J, K, L, M and W, X, Y, Z?

e A square has vertices with coordinates A (3, 5), B (3, 8), C (6, 5) and D (6, 8)

Write the coordinates of square ABCD after a translation by vector $\begin{pmatrix} 3 \\ 7 \end{pmatrix}$

What do you think?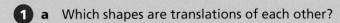

1 **a** Which shapes are translations of each other?

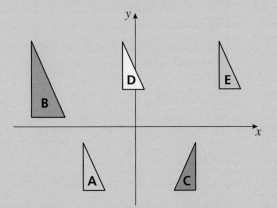

b How can you tell which are translations and which are not? Use mathematical language to explain your reasons.

2 Triangle ABC is drawn on a coordinate grid with vertices at points A (–4, –7), B (–4, 12) and C (–9, –7)

a What type of triangle is triangle ABC?

b Triangle ABC is translated by vector $\binom{4}{7}$ to give triangle DEF. Write the coordinates of triangle DEF

As well as single **transformations**, you can also look at a series of transformations that take place one after another.

Example 2

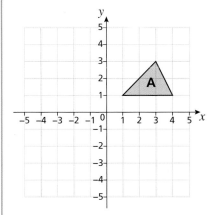

a Reflect triangle A in the line $x = 1$

b Rotate triangle A 90° clockwise about (0, 0)

c Translate triangle A by vector $\binom{-3}{-3}$

d What's the same and what's different about each transformation?

a

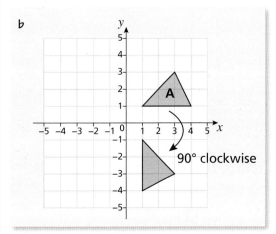

Draw the line $x = 1$

Remember: the x-value of all points on this line will be 1, for example (1, 5) and (1, –3)

Each vertex of the mirror image should be the same distance from the mirror line as the original object.

b

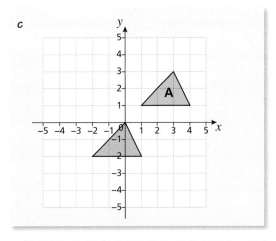

A 90° turn is a quarter turn.

Tracing paper can help you to rotate shapes; you saw this in Chapter 10.2

c

To translate the triangle by vector $\begin{pmatrix} -3 \\ -3 \end{pmatrix}$ move each vertex 3 squares left and 3 squares down.

d All the transformations give a congruent shape.

Reflection and rotation result in a congruent shape with a different orientation to the original, whereas a translation results in a congruent shape with the same orientation.

All of the triangles are the same size and shape as the original.

The triangles are facing a different direction after **reflection** and rotation.

Practice 10.3B

1 A triangle is drawn on the grid.

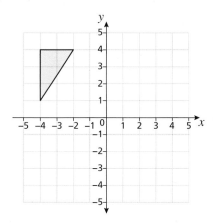

On a copy of the grid

a reflect the triangle in the y-axis and label the result A

b rotate the triangle 180° about the origin and label the result B

c translate the triangle by vector $\begin{pmatrix} 1 \\ -6 \end{pmatrix}$ and label the result C

d What's the same and what's different about your answers to parts **a**, **b** and **c**?

2 Which transformations do you think are the easiest to mix up? What mistakes might be made?

3

I'm going to reflect this shape in the x-axis and then rotate the new shape through 180° about the origin.

Chloe

I'm going to rotate this shape through 180° about the origin and then reflect the new shape in the x-axis.

Jackson

Will Chloe's shape and Jackson's shape end up in the same position? Perform their transformations to check.

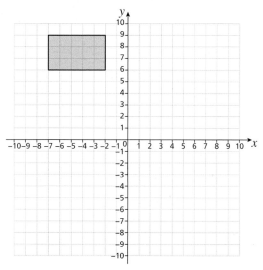

4 Shape P is drawn on the grid.

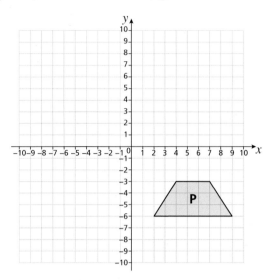

On a copy of the grid

a reflect shape P in the line $y = 1$ and label the result Q

b rotate shape Q through 90° clockwise about (3, 6) and label the result R

c translate shape R by vector $\begin{pmatrix} 0 \\ -7 \end{pmatrix}$ and label the result S

5 Shape A is drawn on the grid.

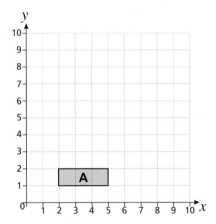

a On a copy of the grid, reflect shape A in the line $y = x$ and label the result B

b Describe a different transformation that would have resulted in shape B being in the same position.

6

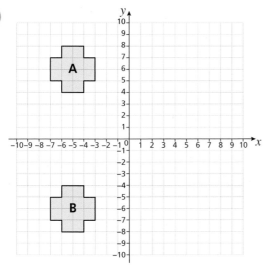

a Describe a single transformation that could map shape A onto shape B. Is there more than one answer?

b Describe a series of two transformations that could map shape A onto shape B

c Describe a series of three transformations that could map shape A onto shape B

What do you think? 💭

1 Emily and Faith are discussing this transformation.

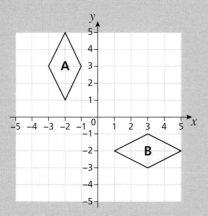

Emily says, "Shape B is a reflection of shape A."

Faith says, "Shape B is a rotation of shape A."

By describing fully the single transformations that can map shape A to shape B, show that Emily and Faith are both correct.

2 Are these statements true or false?

a The result of a reflection could also be the result of a rotation.

b The result of a reflection could also be the result of a translation.

Consolidate – do you need more?

1 Describe these movements using vector notation.

a 3 left and 4 up

b 5 right and 7 down

c 10 up

d 1 left and 17 down

e 15 left

2 Use vector notation to describe each translation from A to B

a
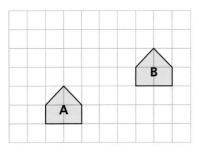

b
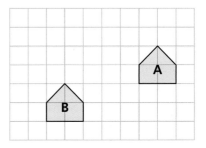

3 Use vector notation to describe each translation from X to Y

a
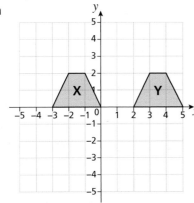

b
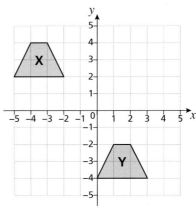

c
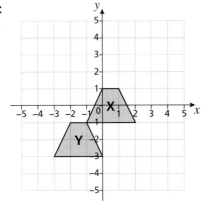

d
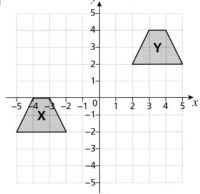

4 Triangle A is reflected in the line $x = -1$ and the result is reflected in the line $y = 1$ to give triangle B

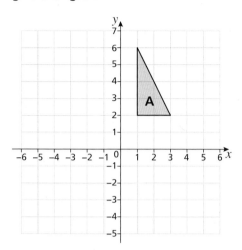

Describe fully the single transformation that maps triangle A onto triangle B

Stretch – can you deepen your learning?

1 A shape is translated on a grid. What can you say about

   a the area of the original shape and the area of the translated shape

   b the perimeter of the original shape and the perimeter of the translated shape

   c the orientation of the original shape and the orientation of the translated shape?

2 A shape is reflected on a grid. What can you say about

   a the area of the original shape and the area of the reflected shape

   b the perimeter of the original shape and the perimeter of the reflected shape

   c the orientation of the original shape and the orientation of the reflected shape?

3 The point with coordinates $(-5, 9)$ is translated by the vector $\begin{pmatrix} -8 \\ 4 \end{pmatrix}$ onto the point $(3a + 8, 5 - 2b)$. Work out the values of a and b

4 Point A has coordinates $(2, 3)$. It is reflected in the line $x + y = 7$ to give point A'. Find an equation of the straight line that passes through A and A'

Reflect

How can you tell whether a shape has been reflected, rotated or translated?

I have become **fluent in…**	I have developed my **reasoning** skills by…	I have been **problem-solving** through…
▦ identifying the order of rotational symmetry of a shape ▦ identifying line symmetry in a shape ▦ rotating a shape about a point ▦ translating points and shapes by a given vector.	▦ comparing and contrasting rotational symmetry with line symmetry ▦ comparing rotation and reflection of shapes ▦ identifying common misconceptions ▦ describing transformations.	▦ exploring different transformations that would give the same result ▦ using prior knowledge in the context of transformations ▦ finding the result of a series of transformations. Ⓗ

Check my understanding

1 For each shape, state

   i   the order of rotational symmetry

   ii the number of lines of symmetry.

a

b

2 **a** State the order of rotational symmetry of a regular nonagon.

   b How many lines of symmetry does a regular nonagon have?

3 Make three copies of this grid.

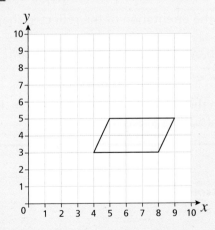

  a Rotate the parallelogram 90° anticlockwise about the point (4, 3)

  b Rotate the parallelogram 180° about the point (5, 6)

  c Translate the parallelogram by the vector $\begin{pmatrix} -3 \\ 4 \end{pmatrix}$

4 Rectangle A has vertices at the points with coordinates (3, 4), (8, 4), (3, 6) and (8, 6)

It is reflected in the y-axis then translated by the vector $\begin{pmatrix} 1 \\ -9 \end{pmatrix}$ to give rectangle B

Write the coordinates of the vertices of rectangle B Ⓗ

11 Pythagoras' theorem

In this block, I will learn...

how to identify the hypotenuse in any right-angled triangle

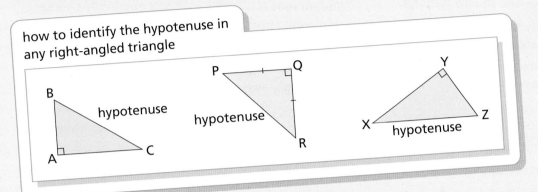

how to use Pythagoras' theorem to find the length of a line segment on a grid

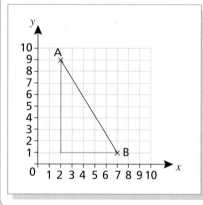

the formula for Pythagoras' theorem

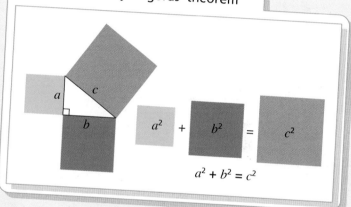

$$a^2 + b^2 = c^2$$

how to use Pythagoras' theorem to find unknown sides in a variety of contexts

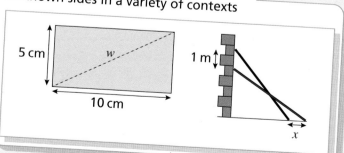

how to work with Pythagoras' theorem in 3-D **H**

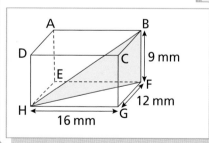

Small steps

- Squares and square roots Ⓡ
- Identify the hypotenuse of a right-angled triangle
- Determine whether a triangle is right-angled

Key words

Square root – a square root of a number is a value that, when multiplied by itself, gives the number

Hypotenuse – the side opposite the right angle in a right-angled triangle

Right angle – an angle of 90°

Are you ready?

1 Find the square of each of these numbers.

    a 6   **b** 8   **c** 1.5   **d** $\frac{2}{3}$ **e** −6

2 Find the square roots of each of these numbers.

    a 25   **b** 100   **c** 169   **d** 2.25   **e** $\frac{4}{9}$

3 Is this statement true or false?

    -4^2 is equal to $(-4)^2$

4 Work these out without using a calculator.

    a $3^2 + 4^2$ **b** $\sqrt{64 + 36}$

Models and representations

When you draw a square on each side of a right-angled triangle, the sum of the areas of the two smaller squares is equal to the area of the largest square.

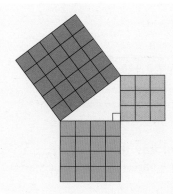

Example 1

a Identify the longest side in each of these right-angled triangles.

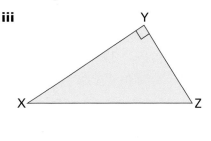

i B   **ii** P   Q   **iii**   Y

A   C   R   X   Z

b What do you notice?

a **i** BC

    ii PR

    iii XZ

You could measure the sides with a ruler to check.

b The longest side in a right-angled triangle is always opposite the right angle.

The longest side of a right-angled triangle is called the **hypotenuse**.

Practice 11.1A

1 Here is a sketch of a right-angled triangle.

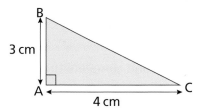

B

3 cm

A   4 cm   C

    a Make an accurate drawing of this triangle.

    b Measure the length of BC

    c Which side of the triangle is the longest?

    d Did you need to measure the length of the sides to know which is the longest?

2 Identify the hypotenuse in each of these right-angled triangles.

a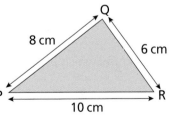

b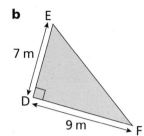

c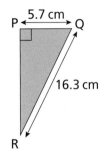

d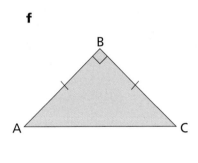

e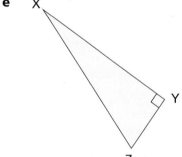

f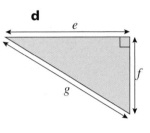

3 Look back at your answers to question **2** and then copy and complete this sentence.

The hypotenuse in a right-angled triangle is always _____

4 Here are three right-angled triangles.

a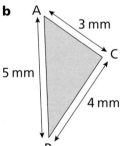

b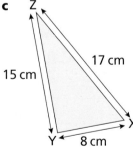

c

For each triangle, identify which angle is the right angle.

What do you think?

1 Do you agree with Chloe? Explain your answer.

> I can identify the hypotenuse of a right-angled triangle without measuring or knowing any of the lengths.

2 A right-angled triangle has vertices at (3, 5), (7, 5) and (3, –2)

Write the coordinates of the end points of the hypotenuse.

3 A ladder is leant against a wall so that the base of the ladder is on the floor and the top of the ladder is leaning on the wall.

Which statement describes the hypotenuse?

A   The horizontal distance between the base of the ladder and the wall

B   The vertical distance between the floor and the top of the ladder

C   The length of the ladder

Draw a diagram to support your answer.

Many thousands of years ago, mathematicians discovered an important fact about right-angled triangles that we still make use of today. This fact is often credited to a Greek mathematician called Pythagoras, but was known long before he lived.

The connection is: "In a right-angled triangle, the square on the hypotenuse is equal to the sum of the squares on the other two sides."

Visually, it looks like the diagram below.

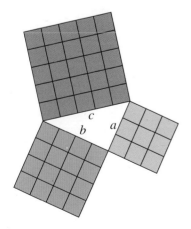

In this example, the sides of the triangle have length 3 units, 4 units and 5 units:

$3^2 + 4^2 = 5^2$

$9 + 16 = 25$

This can be generalised using a, b, c as the lengths of the sides, where c is the hypotenuse.

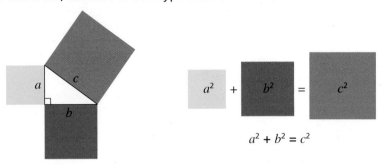

$$a^2 + b^2 = c^2$$

This formula is known as Pythagoras' theorem. It can also be written as $a^2 + b^2 = \text{hyp}^2$ to emphasise which side is the hypotenuse.

You can use Pythagoras' theorem to find missing lengths of sides in right-angled triangles and to prove that a triangle has a **right angle**.

Example 2

a Prove that this triangle is right-angled.

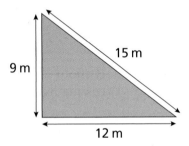

b Prove that this triangle is not right-angled.

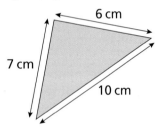

a
$$9^2 = 81$$
$$12^2 = 144$$
$$15^2 = 225$$
$$81 + 144 = 225$$

Therefore $9^2 + 12^2 = 15^2$ so the triangle is right-angled.

To prove that a triangle is right-angled you need to show that the square of the hypotenuse is equal to the sum of the squares of the two shorter sides.

b
$$6^2 = 36$$
$$7^2 = 49$$
$$10^2 = 100$$
$$36 + 49 \neq 100$$

Therefore $6^2 + 7^2 \neq 10^2$ so the triangle is not right-angled.

Practice 11.1B

1 Prove that this triangle is right-angled.

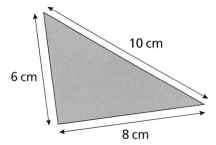

10 cm

6 cm

8 cm

2 Prove that this triangle is not right-angled.

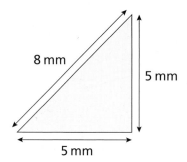

8 mm

5 mm

5 mm

3 Emily draws a sketch of a triangle.

Abdullah

This triangle is not right-angled. It doesn't have a right angle.

It's just a sketch. My triangle is right-angled.

Emily

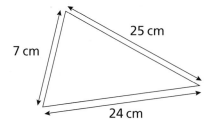

25 cm

7 cm

24 cm

How can you check whether the triangle is right-angled?

4 Show your working to decide whether each of these triangles is right-angled.

a

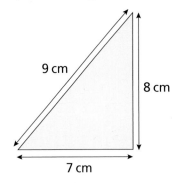

9 cm

8 cm

7 cm

b

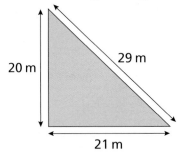

20 m

29 m

21 m

c

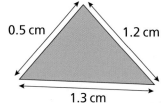

0.5 cm

1.2 cm

1.3 cm

d

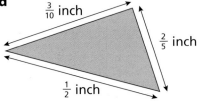

$\frac{3}{10}$ inch

$\frac{2}{5}$ inch

$\frac{1}{2}$ inch

💬 Discuss your reasoning with a partner.

5 Show that this triangle is right-angled.

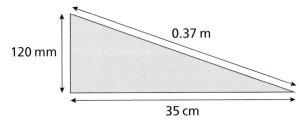

0.37 m

120 mm

35 cm

What do you think? 💭

1 Flo draws a right-angled isosceles triangle on a centimetre squared grid.

I know that the length of the longest side is 10 cm because I measured it.

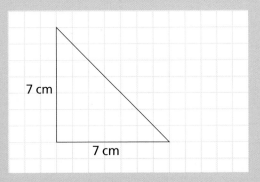

7 cm

7 cm

a Copy Flo's drawing.

b Measure the length of the hypotenuse.

c How do you know that the length of the longest side is not exactly 10 cm?

2 Here is a triangle.

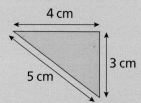

4 cm

3 cm

5 cm

a Prove that the triangle is right-angled.

b Here are some more right-angled triangles. Compare each of these triangles with triangle A. What's the same and what's different?

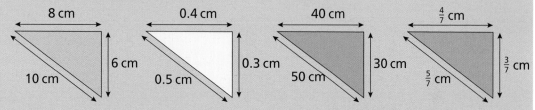

8 cm

6 cm

10 cm

0.4 cm

0.3 cm

0.5 cm

40 cm

30 cm

50 cm

$\frac{4}{7}$ cm

$\frac{3}{7}$ cm

$\frac{5}{7}$ cm

c Write down the dimensions of three more right-angled triangles that are based on the 3, 4, 5 triangle.

Discuss your method with a partner.

Consolidate – do you need more?

1 For each triangle, identify which side is the hypotenuse.

a

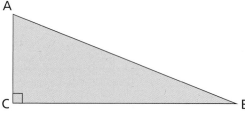

A

C   B

b

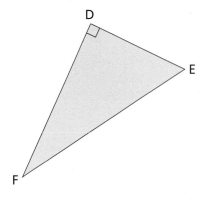

D

E

F

c

G   H

I

d

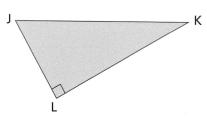

J   K

L

2 For each triangle, identify which angle is the right angle.

a

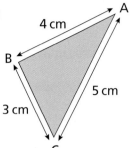

4 cm
B
5 cm
3 cm
A
C

b

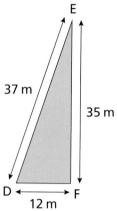

E
37 m
35 m
D
12 m
F

c

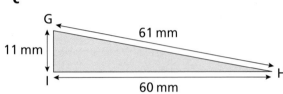

G
61 mm
11 mm
I
60 mm
H

d

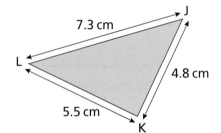

7.3 cm
J
L
4.8 cm
5.5 cm
K

3 Prove that these triangles are right-angled.

a

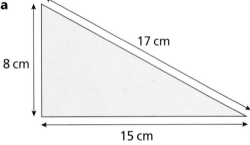

17 cm
8 cm
15 cm

b

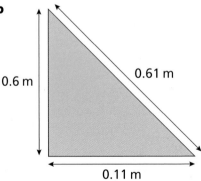

0.6 m
0.61 m
0.11 m

4 Prove that this triangle is not right-angled.

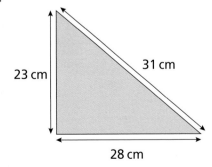

23 cm
31 cm
28 cm

Stretch – can you deepen your learning?

1 Triangle ABC is drawn on a coordinate grid.

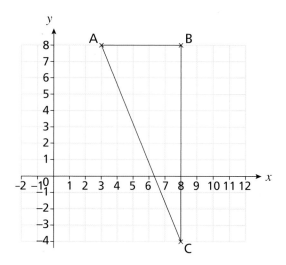

a Explain how you know that triangle ABC is right-angled.

b Show that the length of AC must be 13 units

2 Pythagoras' theorem can be applied to other shapes drawn on the sides of a right-angled triangle.

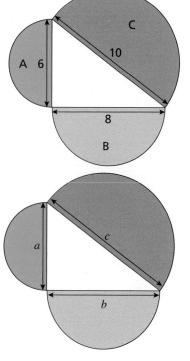

a Show that the sum of the area of the two smaller semi-circles is equal to the area of the largest semi-circle.

b Use Pythagoras' theorem to show that your findings from part **a** are always true.

3

3, 4, 5 and 5, 12, 13 are examples of Pythagorean triples.

a Why do you think these are called Pythagorean triples?

b 6, 8, 10 and 12, 16, 20 are also examples of Pythagorean triples.

I think these are based on the 3, 4, 5 triple.

Explain why Jackson might think this.

c Prove that any multiple of 3, 4 and 5 is a Pythagorean triple.

Reflect

1 Explain how you can identify which side in a right-angled triangle is the hypotenuse.

2 In your own words explain Pythagoras' theorem.

Small steps

- Calculate the hypotenuse of a right-angled triangle
- Calculate missing sides in right-angled triangles
- Explore proofs of Pythagoras' theorem

Are you ready?

1 Find the area of each square.

a

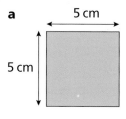

5 cm
5 cm

b

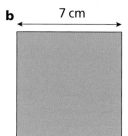

7 cm

c

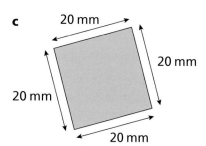

20 mm
20 mm
20 mm
20 mm

2 The area of each square is shown. Find the side length of each square.

a

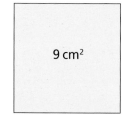

9 cm²

b

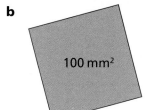

100 mm²

c

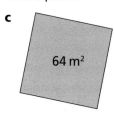

64 m²

3 Identify the hypotenuse in each of these triangles.

a

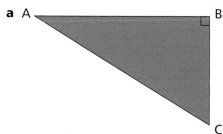

A   B
C

b

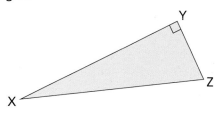

Y
X   Z

4 Show that this triangle is right-angled.

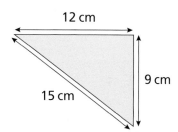

12 cm
9 cm
15 cm

Models and representations

Bar models

These show fact families and can help you to rearrange basic formulae.

3	7
10	

a^2	b^2
c^2	

$$3 + 7 = 10$$
$$7 + 3 = 10$$
$$10 - 3 = 7$$
$$10 - 7 = 3$$

$$a^2 + b^2 = c^2$$
$$b^2 + a^2 = c^2$$
$$c^2 - b^2 = a^2$$
$$c^2 - a^2 = b^2$$

In the last chapter, you learned about Pythagoras' theorem and saw how it can be used to prove that a triangle is right-angled.

In this chapter, you will learn how to find missing sides in right-angled triangles.

Pythagoras' theorem states: "In a right-angled triangle, the square of the **hypotenuse** is equal to the sum of the squares of the other two sides."

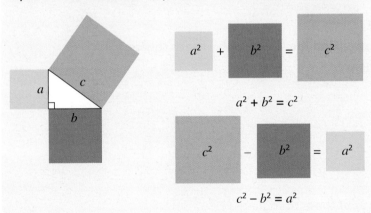

$$a^2 + b^2 = c^2$$

$$c^2 - b^2 = a^2$$

Example 1

Find the length of the side labelled x

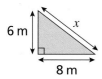

Method A

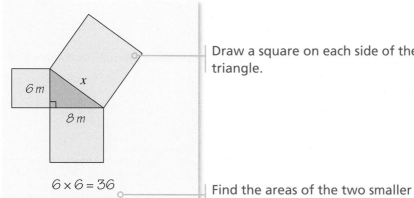

Draw a square on each side of the triangle.

$$6 \times 6 = 36$$
$$8 \times 8 = 64$$

Find the areas of the two smaller squares.

$$36 + 64 = 100$$

Add the areas.

$$\sqrt{100} = 10$$

Therefore $x = 10\,\text{m}$

To find the length of the hypotenuse, find the square root of 100

Method B

Label the sides a, b and c. Ensure that c is the hypotenuse.

Or you could use $a^2 + b^2 = \text{hyp}^2$

$$a^2 + b^2 = c^2$$
$$6^2 + 8^2 = x^2$$

Substitute the values you know into the formula for Pythagoras' theorem and solve for x

$$36 + 64 = x^2$$
$$100 = x^2$$
$$\sqrt{100} = x$$
$$10 = x$$

Therefore the length of the side labelled x is $10\,\text{m}$

Practice 11.2A

1 Copy this diagram onto squared paper.

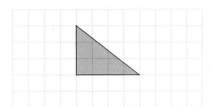

a Draw a square on each side of the triangle.

b Work out the area of each square.

c Show that the sum of the areas of the two smaller squares is equal to the area of the biggest square.

2 Lydia is finding the length of the hypotenuse in this right-angled triangle.

Here is Lydia's working.

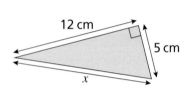

$$a^2 + b^2 = c^2$$
$$5^2 + 12^2 = x^2$$
$$25 + 144 = x^2$$
$$169 = x^2$$

The length of x is $169\,cm$

What mistake has Lydia made?

3 Find the length of the hypotenuse for each of these triangles.

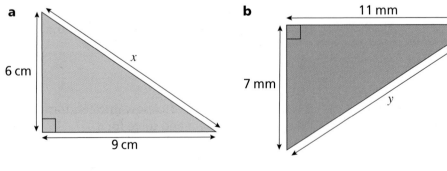

a 6 cm, 9 cm, x

b 11 mm, 7 mm, y

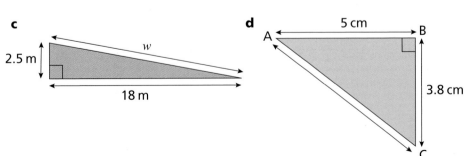

c 2.5 m, 18 m, w

d 5 cm, 3.8 cm, A, B, C

4 Find the length of the side labelled x. Give your answer to 3 significant figures.

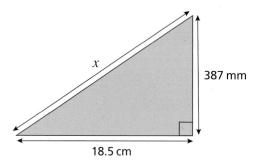

387 mm

x

18.5 cm

5 This shape is made up of two right-angled triangles.

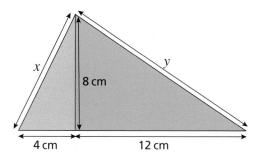

x

8 cm

y

4 cm

12 cm

Find the perimeter of the shape. Give your answer to 1 decimal place.

6 Find the length of the diagonal of this rectangle. Give your answer to 1 decimal place.

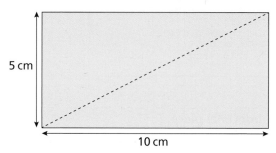

5 cm

10 cm

7 Find the length of the hypotenuse of this triangle. Do not use a calculator.

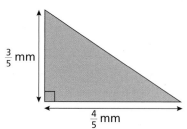

$\frac{3}{5}$ mm

$\frac{4}{5}$ mm

What do you think? 💭

1 Do you have enough information to find the lengths of all of the sides of these shapes?

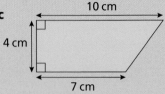

a

5 cm

7 cm

b

6.5 cm

c

10 cm

4 cm

7 cm

Explain your answers.

2 a Find the perimeter of this shape. Give your answer to 1 decimal place.

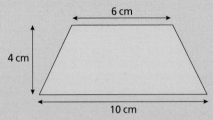

6 cm

4 cm

10 cm

b What assumption did you make?

c What if your assumption is incorrect? Would your answer to part **a** change?

3 A suitcase has dimensions 80 cm by 55 cm by 10 cm

Will a walking stick of length 97 cm fit flat inside the suitcase?

Use diagrams and calculations to support your answer.

Example 2

Find the length of XY. Give your answer to 1 decimal place.

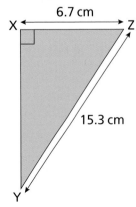

6.7 cm

X   Z

15.3 cm

Y

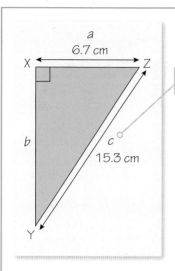

a
6.7 cm

X ⟶ Z

b

c
15.3 cm

Y

Label the triangle a, b, c where c is the longest side.
It doesn't matter which is a and which is b

$$c^2 - a^2 = b^2$$
$$15.3^2 - 6.7^2 = b^2$$
$$234.09 - 44.89 = b^2$$
$$189.2 = b^2$$
$$\sqrt{189.2} = b$$
$$13.7549990912 = b$$
$$13.8 = b$$

You know that $a^2 + b^2 = c^2$

a^2	b^2
c^2	

You could use $a^2 + b^2 = hyp^2$

Therefore $c^2 - b^2 = a^2$ and $c^2 - a^2 = b^2$

Write all the digits on your calculator display.

Round your answer to 1 decimal place.

Therefore XY = 13.8 cm
(to 1 d.p.)

Write the units with your answer.

Practice 11.2B

1 Bobbie is working out the length of AB in this triangle.

Here is her working.

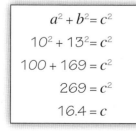

$$a^2 + b^2 = c^2$$
$$10^2 + 13^2 = c^2$$
$$100 + 169 = c^2$$
$$269 = c^2$$
$$16.4 = c$$

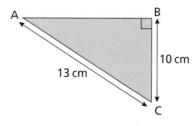

A

B

10 cm

13 cm

C

a What mistake has Bobbie made?

b Work out the length of AB

2 Find the length of the unknown side in each of these triangles. Give your answers to 1 decimal place, where appropriate.

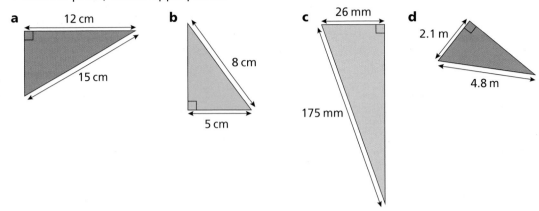

a   12 cm   15 cm

b   8 cm   5 cm

c   26 mm   175 mm

d   2.1 m   4.8 m

3 The size of a TV screen is determined by the length of the diagonal. Here is a 45-inch TV.

45 inches

The height of the TV screen is 30 inches.

Calculate the width of the TV screen. Give your answer to 1 decimal place.

4 a Find the perpendicular height, h, of this isosceles triangle.

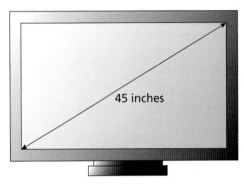

8 cm   8 cm h 6 cm

b Work out the area of the isosceles triangle.

Give your answers to parts **a** and **b** to 1 decimal place.

What do you think? 💡

1 Jakub says, "The length of the missing side is the same in both of these triangles."

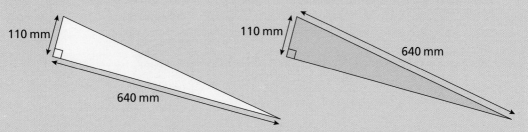

110 mm

640 mm

110 mm

640 mm

Do you agree with Jakub? Explain your answer.

2 A ladder rests on horizontal ground against a vertical wall. The ladder is 5 metres long and is placed 2.8 metres away from the base of the wall.

a Draw a diagram to show this situation.

b Will the ladder reach a window that is 3 metres up the wall? Show your working to explain how you decide.

3 Discuss how these diagrams illustrate Pythagoras' theorem. Could they be described as a proof? Why or why not?

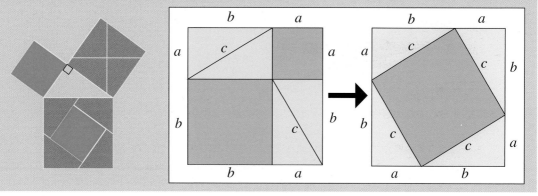

Consolidate – do you need more?

1 Which equation would help you to work out the length of the unknown side of this triangle?

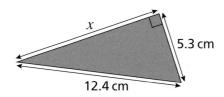

x

5.3 cm

12.4 cm

A $5.3^2 + 12.4^2 = x^2$ B $5.3^2 - 12.4^2 = x^2$ C $12.4^2 - 5.3^2 = x^2$

2 Work out the length of the unknown side in each of these right-angled triangles.
Give your answers to 1 decimal place.

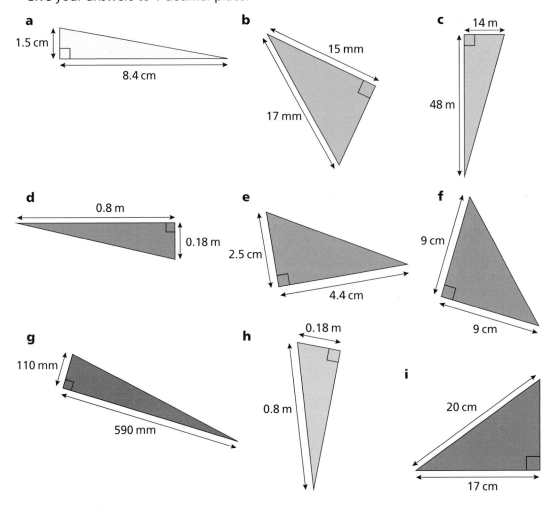

a 1.5 cm, 8.4 cm

b 15 mm, 17 mm

c 14 m, 48 m

d 0.8 m, 0.18 m

e 2.5 cm, 4.4 cm

f 9 cm, 9 cm

g 110 mm, 590 mm

h 0.18 m, 0.8 m

i 20 cm, 17 cm

Stretch – can you deepen your learning?

1 Here is a kite.

a Find the length labelled x

b Find the length of the side labelled y

c Find the perimeter of the kite.

d Find the area of the kite.

Give your answers to 1 decimal place.

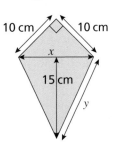

10 cm, 10 cm, x, 15 cm, y

2 A boat leaves a harbour.

It travels for 30 minutes due north at a speed of 30 mph

It then travels due east at the same speed for 15 minutes.

How far away is the boat from the harbour now? Give your answer to 1 decimal place.

3 A 7-metre ladder rests against a wall.

The ladder reaches 5.5 metres up the wall.

The ladder is moved so that it reaches 1 metre
less high than in its previous position, as shown in the diagram.

How far has the base of the ladder moved away from the
wall? Give your answer to 1 decimal place.

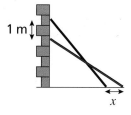

1 m

x

4 The length of the diagonal of a square is 15 cm

  a Find the length of each side of the square.

  b Hence find the area of the square.

Give your answers to 1 decimal place.

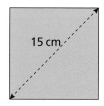

15 cm

5 The diagram shows three circles with centres A, B and C. The circles touch
without overlapping.

Show that triangle ABC is right-angled.

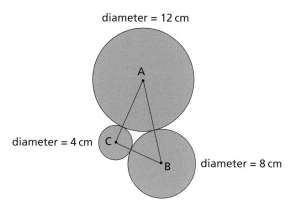

diameter = 12 cm

A

diameter = 4 cm C

B

diameter = 8 cm

6 Show how this diagram proves Pythagoras' theorem.

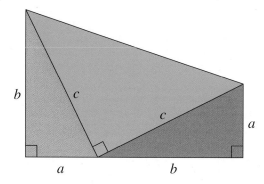

b

c

c

a

a

b

Reflect

1 Explain when you would use the formula $c^2 - b^2 = a^2$ instead of $a^2 + b^2 = c^2$

2 Does it matter which sides of a right-angled triangle you label a, b and c?

Small steps

- Use Pythagoras' theorem on coordinate axes
- Use Pythagoras' theorem in 3-D shapes **Ⓗ**

Key words

Line segment – part of a line that connects two points

Midpoint – the point halfway between two others

Dimensions – measurements such as the length, width and height of something

Are you ready?

1 Work out the length of the hypotenuse of each triangle. Give your answers to 1 decimal place, where appropriate.

a
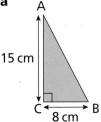
A
15 cm
C —— B
8 cm

b
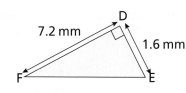
D
7.2 mm
1.6 mm
F
E

c
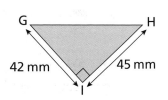
G   H
42 mm   45 mm
I

2 Work out the length of the unknown side of each triangle. Give your answers to 1 decimal place, where appropriate.

a
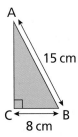
A
15 cm
C —— B
8 cm

b
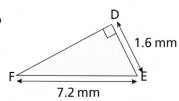
D
1.6 mm
F
E
7.2 mm

c
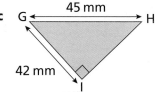
45 mm
G   H
42 mm
I

3 Find the length of each of these line segments.

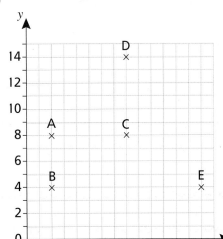

a AB

b AC

c CD

d BE

Models and representations

Coordinate grids

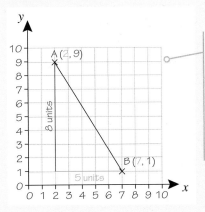

To calculate the length of a **line segment** joining two points, you can show the difference between the x-values and the difference between the y-values on a grid.

Example 1

Point A has coordinates (2, 9)

Point B has coordinates (7, 1)

Work out the length of line segment AB

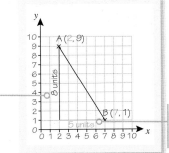

The line segment connecting the points is the hypotenuse of a right-angled triangle.

Vertically, going from A to B is from 9 to 1, which is 8 units.

Horizontally, going from A to B is from 2 to 7, which is 5 units.

$a^2 + b^2 = c^2$

$5^2 + 8^2 = AB^2$

$25 + 64 = AB^2$

$89 = AB^2$

so $AB = \sqrt{89}$

$= 9.4$ units

You know the two shorter sides of the right-angled triangle and want to find the hypotenuse.

473

Practice 11.3A

1 **a** Explain why PQ = 7 units

b Explain why QR = 5 units

c Use Pythagoras' theorem to show that PR = 8.6 units to 1 decimal place.

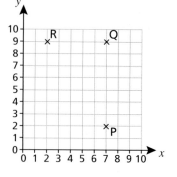

2 **a** Write the length of AB

b Write the length of BC

c Use Pythagoras' theorem to work out the length of AC.
Give your answer to 1 decimal place.

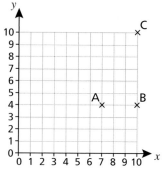

3 **a** Find the length of each of these line segments.

i AB   **ii** BD   **iii** CD   **iv** DE

b How far is point C from the origin?

Point F has coordinates (9, 14)

c Find the length of BF

Give your answers to 1 decimal place.

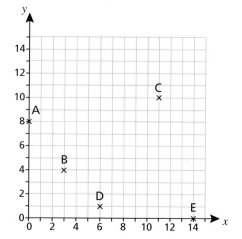

④ Work out the distance between each pair of points. Give your answers to 1 decimal place.

a (3, 7) and (5, 10)

b (1, 9) and (10, 3)

c (8, 2) and (3, 3)

d (4, 0) and (0, 4)

⑤ **a** Find the length of each of these line segments.

i AB   **ii** BC   **iii** CD   **iv** DE

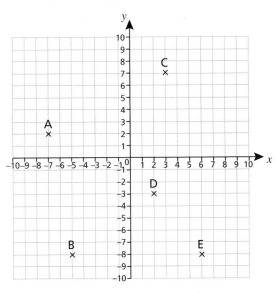

b How far is point E from the origin?

Point F has coordinates (−15, 13)

c Find the length of BF

Give your answers to 1 decimal place.

⑥ Work out the distance between each pair of points. Give your answers to 1 decimal place.

a (−3, 0) and (2, 6)

b (5, 1) and (−4, 6)

c (9, −8) and (8, −5)

d (−3, 7) and (2, −1)

⑦ A rectangle has vertices at the points with coordinates (−3, 5), (2, 5), (−3, −1) and (2, −1)

Find the length of the diagonal of the rectangle. Give your answer to 1 decimal place.

What do you think? 💬

1 Point X has coordinates (2, 8)

Point X is translated by the vector $\begin{pmatrix} 5 \\ -2 \end{pmatrix}$ to give the point X′

a Work out the distance between X and X′. Give your answer to 1 decimal place.

b Faith forms a conjecture. She says, "It doesn't matter what the coordinates of point X are, the distance between X and X′ will be the same."

Investigate Faith's conjecture and explain your findings.

2 Points A, B and C lie on a straight line.

Point A has coordinates (−4, 1)

Point B has coordinates (7, −5)

Point C is such that AB : BC = 4 : 7

Work out the length of line segment AC. Give your answer to 1 decimal place.

You will now explore finding right-angled triangles in 3-D shapes.

Example 2

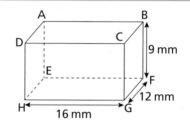

Here is a cuboid.

Work out the length of

a FH

b BH, giving your answer to 1 decimal place.

a

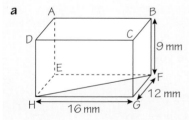

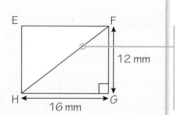

EFGH is the rectangular base of the cuboid.

This means that FGH is a right-angled triangle and FH is its hypotenuse.

Using Pythagoras' theorem

$GH^2 + FG^2 = FH^2$

so $FH^2 = 16^2 + 12^2 = 400$

$FH = \sqrt{400} = 20$ mm

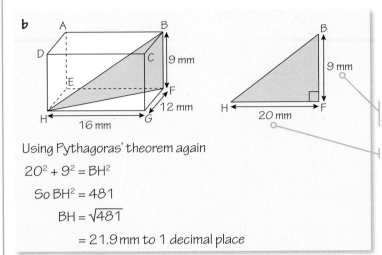

b

Using Pythagoras' theorem again

$20^2 + 9^2 = BH^2$

So $BH^2 = 481$

$BH = \sqrt{481}$

$= 21.9\,\text{mm}$ to 1 decimal place

BFH is a right-angled triangle with BH as the hypotenuse. Drawing this separately helps you to label the correct side lengths.

BF is 9mm from the original diagram.

FH is 20mm from part **a**

Practice 11.3B

For all the questions in this exercise, give your answers to 1 decimal place, where appropriate.

1 Here is a cuboid.

a i Sketch the rectangular face DCGH and label its dimensions.

ii Show that CH = 10 cm

b i Sketch the rectangular face BFGC and label its dimensions.

ii Work out the length of CF

c i Sketch the rectangular face EFGH and label its dimensions.

ii Work out the length of FH

d i Sketch the right-angled triangle BFH and label its dimensions.

ii Work out the length of BH

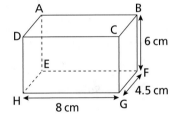

2 Here is a cuboid.

a Work out the length of these sides.

i EG   **ii** DE   **iii** BD

b By drawing right-angled triangle AEG, work out the length of AG

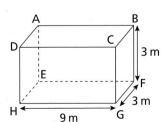

3 ABCDEFGH is a cuboid.

Work out the length of AG

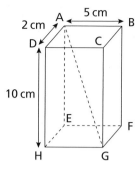

4 ABCDEFGH is a cube.

Work out the length of FD

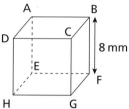

What do you think? 💭

1 Here is a square-based pyramid.

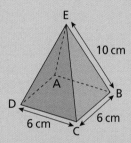

a Work out the length of BD

b M is the midpoint of BC

i Write the length of BM

ii Work out the length of EM

iii Work out the perpendicular height of the pyramid.

2 A pencil case is in the shape of a cuboid.

The cuboid has dimensions 15 cm by 4 cm by 6 cm

A pencil is 16 cm long.

Will the pencil fit inside the pencil case? Show your working to explain how you decide.

3 A cylindrical container has diameter 82 mm and height 200 mm

The container will be used to hold straws.

What is the maximum length of a straw that will fit inside the container?

Consolidate – do you need more?

1 Calculate the length of each line segment AB. Give your answers to 1 decimal place.

a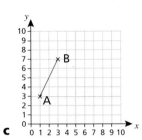

b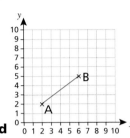

c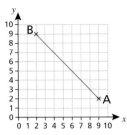

2 Calculate the length of each line segment AB. Give your answers to 1 decimal place.

a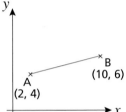

b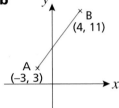

c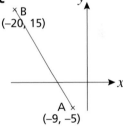

3 Here is a triangular prism.

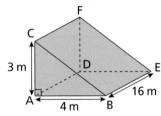

Calculate the length of

a BC **b** AE **c** AF

Give your answers to 1 decimal place, where appropriate.

4 ABCDEFGH is a cube with side length 3 cm

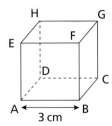

a Work out the length of AC

b Work out the length of AG

Give your answers to 1 decimal place.

Stretch – can you deepen your learning?

1 Point M has coordinates $(-3, 4)$

Point N has coordinates $(5, a)$

The length of line segment MN is 12 units.

Work out two possible values of a

2 Point A has coordinates $(4, -1)$

Point B has coordinates $(15, 3)$

Work out the length of line segment AB. Give your answer to 1 decimal place.

3 Write an expression for the distance between

a the point (x, y) and the origin

b the points (x_1, y_1) and (x_2, y_2)

4 The diagonal of a cube measures 20 m. Calculate

a the volume of the cube

b the surface area of the cube.

Give your answers to 1 decimal place.

5 A cuboid has length a cm, width b cm and height c cm

Write an expression for the length of the longest diagonal of the cuboid.

Reflect

1 Explain why Pythagoras' theorem can be used to work out the distance between any two pairs of points.

2 What's the same and what's different about using Pythagoras' theorem in two dimensions and three dimensions?

11 Pythagoras' theorem
Chapters 11.1–11.3

I have become **fluent in…**	I have developed my **reasoning** skills by…	I have been **problem-solving** through…
◾ calculating squares and square roots ◾ identifying the hypotenuse of a right-angled triangle ◾ calculating side lengths in right-angled triangles.	◾ being able to explain why a triangle is/is not right-angled ◾ identifying common misconceptions ◾ exploring proofs of Pythagoras' theorem.	◾ using Pythagoras' theorem on coordinate axes ◾ using Pythagoras' theorem in 3-D shapes ⒣ ◾ applying Pythagoras' theorem in other mathematical contexts.

Check my understanding

1 Work out the value of

   a 9^2 **b** $\sqrt{9}$ **c** $3^2 + 4^2$ **d** $\sqrt{100 - 64}$

2 Identify the hypotenuse of each triangle.

   a 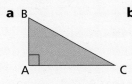   **b** 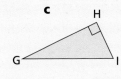   **c**

3 **a** Show that a triangle with side lengths 12 cm, 16 cm and 20 cm is right-angled.

     b Show that a triangle with side lengths 50 mm, 30 mm and 20 mm is not right-angled.

4 Work out the unknown length in each triangle. Give your answers to 1 decimal place.

   a 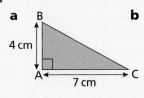   **b**   **c**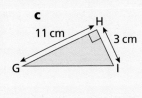

5 Calculate the distance between the points (4, 7) and (−1, 15)

   Give your answer to the nearest integer.

6 A cube has sides of length 15 mm

   Find the length of the longest diagonal of the cube. ⒣

12 Enlargement and similarity

In this block, I will learn...

about enlargement and similarity in real-life contexts

how to work out missing angles and sides in similar shapes

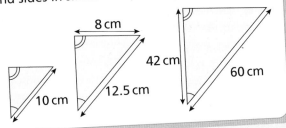

how to enlarge shapes on a grid

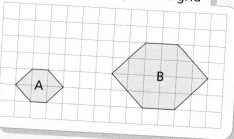

how to enlarge shapes from a given point

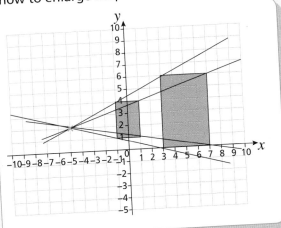

how to enlarge a shape by a negative scale factor ⓗ

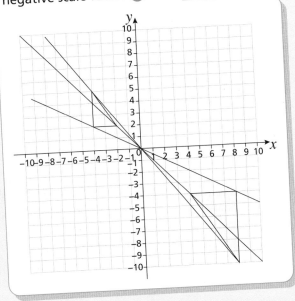

that an enlargement can make a shape bigger or smaller

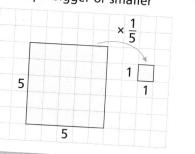

12.1 Calculating with enlargement

Small steps

- Recognise enlargement and similarity
- Work out missing sides and angles in a pair of given similar shapes

Key words

Enlargement – making a shape bigger, or smaller

Similar – two shapes are similar if their corresponding sides are in the same ratio

Scale factor – how much a shape has been enlarged by

Are you ready?

1 Work out

   a $30 \div 2$ **b** $30 \times \frac{1}{2}$ **c** $30 \div 5$ **d** $30 \times \frac{1}{5}$ **e** $30 \div 3$ **f** $30 \times \frac{1}{3}$

2 Work out

   a $\frac{3}{5}$ of 30   **b** $\frac{5}{3}$ of 30

3 Simplify these ratios.

   a $6:9$ **b** $20:12$ **c** $30:15$ **d** $18:24$ **e** $42:35$ **f** $80:120$

4 Simplify these fractions.

   a $\frac{6}{9}$ **b** $\frac{12}{20}$ **c** $\frac{15}{30}$ **d** $\frac{18}{24}$ **e** $\frac{35}{42}$ **f** $\frac{80}{120}$

Models and representations

Enlargement and similarity are all around you in real life. Here are some examples.

Example 1

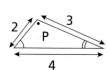 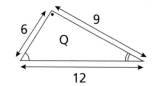 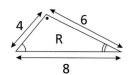

a Explain why triangles Q and R are similar to triangle P

b Identify the scale factor of enlargement from

    i   P to Q   **ii**   Q to P   **iii**   R to Q   **iv**   R to P

a   From P to Q, each of the side lengths has been multiplied by 3

     From P to R, each of the side lengths has been multiplied by 2

     Each of the side lengths of P has been multiplied by the same scale factor
to give the corresponding side length in triangles Q and R

     Q and R are enlargements of P so they are similar to P

> To find the **scale factor** you divide the length of a side in the
> image by the corresponding length in the original shape.

b   **i** $12 \div 4 = 3$

> You can use the marked angles to see which sides are corresponding.
>
> You could also have worked out $9 \div 3$ or $6 \div 2$
>
> As the shapes are **similar**, you will get the same answer.

     ii $4 \div 12 = \dfrac{1}{3}$

> It is usual to give scale factors less than 1 as fractions rather than
> decimals.

     iii $12 \div 8 = \dfrac{3}{2}$

> You can leave this as an improper fraction, or give the answer as 1.5

     iv $4 \div 8 = \dfrac{1}{2}$

> Be careful to divide in the correct order. The enlargement from P to R
> would have scale factor 2, but the enlargement from R to P has scale
> factor $\dfrac{1}{2}$

Example 2

Triangles ABC and PQR are similar.

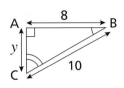

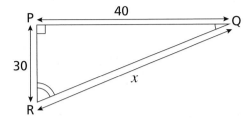

Work out the length of the sides labelled x and y

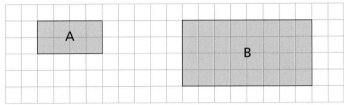

Scale factor = $40 \div 8 = 5$ — Use the corresponding side lengths AB and PQ

Therefore $x = 10 \times 5 = 50$ — You know that the triangles are similar so all the sides of ABC are enlarged by the same scale factor to give the corresponding sides of PQR

$y \times 5 = 30$ so $y = 6$

Practice 12.1A

1 Here are two rectangles, A and B, drawn on a squared grid.

a Explain why rectangle B is similar to rectangle A

Here is a third rectangle, C

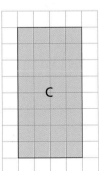

b Is rectangle C similar to rectangle A? Explain your answer.

2 Here are two triangles, X and Y

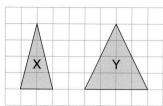

Explain why triangle X is not similar to triangle Y

3 Sketch two shapes that are similar to each of these. Label the dimensions on your diagrams.

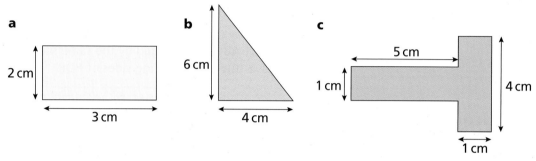

a

2 cm

3 cm

b

6 cm

4 cm

c

5 cm

1 cm

4 cm

1 cm

4 The following shapes are enlargements of rectangle A

Find the scale factor of enlargement from A to each shape.

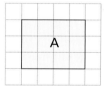

A

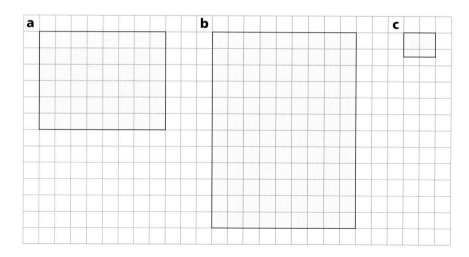

a

b

c

5 These two shapes are similar.

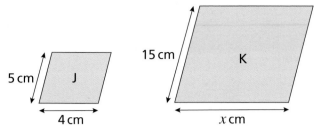

a What is the scale factor of the enlargement from J to K? How do you know?

b Work out the value of x

6 These two rectangles are similar.

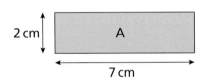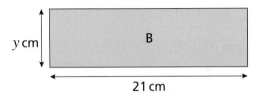

a What is the scale factor of the enlargement from A to B?

b Write the ratio of the width of rectangle A to the width of rectangle B

c Work out the value of y.

7 These two triangles are similar.

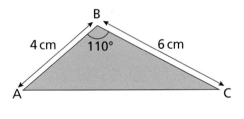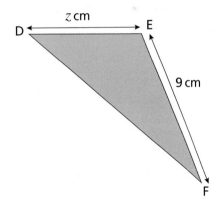

a What is the scale factor of the enlargement from ABC to DEF?

b Find the value of z

c What is the size of angle EDF?

8 Here are two similar right-angled triangles.

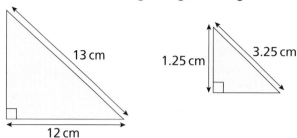

a Work out the lengths of the two missing sides.

b Find the perimeter of each triangle.

c Is the ratio of the perimeters the same as the ratio of the sides? Show working to explain how you decide.

9 Here are two similar triangles.

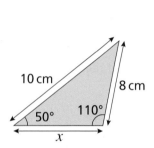
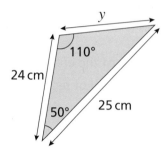

Find the length of the side labelled

a x b y

What do you think?

1 A rectangle has length 3.7 cm and width 2 cm. Work out the dimensions of this rectangle after it has been enlarged by each of the given scale factors.

a scale factor 10   b scale factor 100   c scale factor 1000

2 Here are three similar triangles.

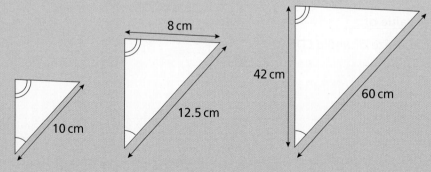

What information can you work out about the triangles?

Compare your findings and methods with a partner.

3 Here is a triangle.

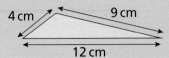

4 cm   9 cm

12 cm

The triangle is enlarged. One of the side lengths of the new triangle measures 6 cm

a Work out the possible scale factors for the enlargement.

b Sketch all the possible enlarged triangles.

Consolidate – do you need more?

1 Copy triangle P onto squared paper. Draw two triangles that are similar to triangle P

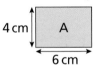

2 Here is rectangle A

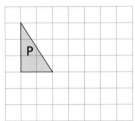

4 cm   A

6 cm

Which of these rectangles are similar to rectangle A?

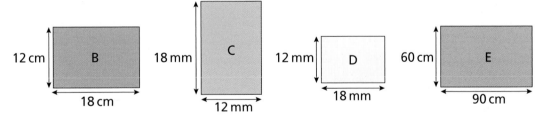

12 cm   B   18 cm

18 mm   C   12 mm

12 mm   D   18 mm

60 cm   E   90 cm

3 Work out the length of the side marked x in each pair of similar shapes.

a

9 cm | A | 9 cm

x | B | 27 cm

b

90 mm | A | 60 mm

45 mm | B | x

c

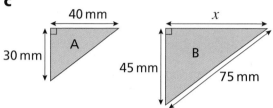

40 mm | A | 30 mm

x | B | 45 mm | 75 mm

d

 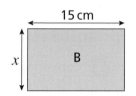

2.5 cm | 1.5 cm | A

15 cm | x | B

Stretch – can you deepen your learning?

1 Here is a triangle ADE

BC is parallel to DE

DE = 20 mm, BC = 8 mm, AB = 6 mm

Work out the length of AD

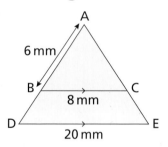

A

6 mm

B 8 mm C

D 20 mm E

2 These two triangles are similar.

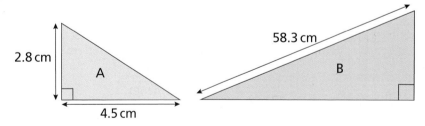

2.8 cm

A

4.5 cm

58.3 cm

B

Work out the perimeter of triangle B.

3

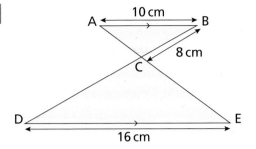

a Show that ABC and CDE are similar.

b Four students have calculated the missing lengths. Here are their answers.

| CD = 12.8 cm | CD = 20 cm | CE = 12.8 cm | CE = 20 cm |

 i Which answer is correct? Show working to explain your answer.

 ii For any incorrect answers, state and explain the mistake that has been made.

4 Here are two similar triangles.

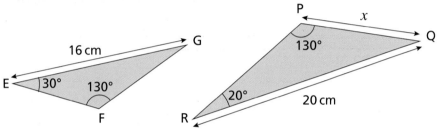

a Find an expression for the length of EF in terms of x

The perimeter of triangle EFG is 33.6 cm

b Find an expression for the length of FG in terms of x

The perimeter of PQR is 42 cm and PR = 13.05 cm

c Calculate the length of each unknown side of the triangles.

Reflect

1 Explain what it means for shapes to be similar.

2 Explain why an enlargement does not always make a shape larger.

Small steps

- Enlarge a shape by a positive integer scale factor
- Enlarge a shape by a positive integer scale factor from a point

Key words

Enlargement – making a shape bigger, or smaller

Scale factor – how much a shape has been enlarged by

Dimensions – measurements such as the length, width and height of something

Centre of enlargement – point from which an enlargement is made

Are you ready?

1   On a squared grid, draw a rectangle that has a length of 4 squares and a width of 3 squares.

2   A shape is drawn on a grid.

Draw a shape that is twice as long and twice as wide as this shape.

3   Copy and complete this sentence.

A shape in which each side has been tripled has been enlarged by scale factor _____.

Models and representations

Ray lines

Straight lines from the **centre of enlargement** to each vertex of a shape can help you to draw and describe **enlargements**.

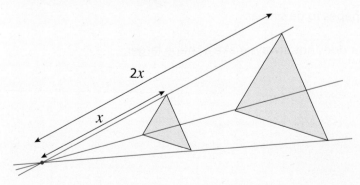

In this chapter, you will learn how to enlarge a shape by a given **scale factor**.

Example 1

Enlarge the triangle by each scale factor.

a scale factor 2

b scale factor 3

c scale factor 4

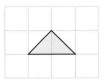

a

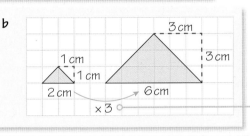

The diagonal sides make a triangle tricky to enlarge. It is easier to count the squares horizontally and vertically then multiply this by the scale factor.

To enlarge a scale by scale factor 2 you need to multiply all the **dimensions** of the shape by 2

b

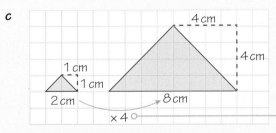

To enlarge a shape by scale factor 3 you need to multiply all the dimensions of the shape by 3

c

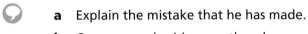

To enlarge a shape by scale factor 4 you need to multiply all the dimensions of the shape by 4

Practice 12.2A

1 Seb has tried to enlarge rectangle A by scale factor 3

Seb has made a mistake.

a Explain the mistake that he has made.

b On a squared grid, correctly enlarge rectangle A by scale factor 3

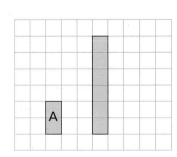

2 Copy each shape onto a squared grid and enlarge it by the given scale factor.

a scale factor 2

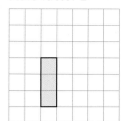

b scale factor 4

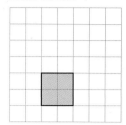

c scale factor 3

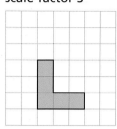

d scale factor 2

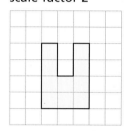

3 Abdullah has been asked to enlarge this triangle by scale factor 2

He has made a start but is not sure how to draw the rest of the triangle.

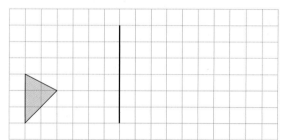

a What instructions would you give to Abdullah?

b Copy and complete Abdullah's enlargement.

4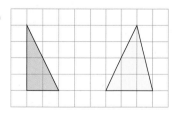

a Copy these triangles onto squared paper. Enlarge each triangle by scale factor 2

b Which enlargement was easier to draw. Why?

5 Copy each shape onto squared paper and enlarge it by the given scale factor.

a scale factor 3

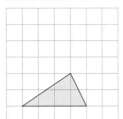

b scale factor 2

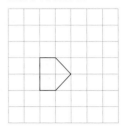

c scale factor 4

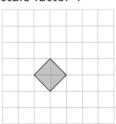

d scale factor 2

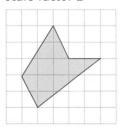

6 Work out the scale factor of each enlargement from A to B

a

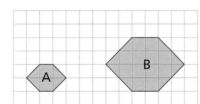

b

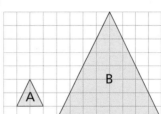

What do you think? 💭

1 Is it possible to enlarge a shape by scale factor 1?

2 Is enlarging a shape by scale factor 2 then by scale factor 3 the same as enlarging the shape by scale factor 6?

Use drawings to support your reasoning.

So far you have looked at enlargements without a centre of enlargement. This means they can be drawn anywhere. Enlargements can also be drawn using a centre of enlargement. This means that the enlarged shape must be drawn in a specific place.

Example 2

Enlarge shape P by scale factor 3, centre (−7, −4)

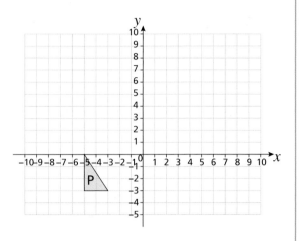

Method A

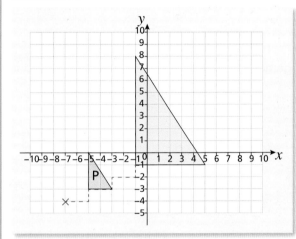

First plot the point (−7, −4) on the grid. This is the centre of enlargement.

Next count how far along and up it is from the centre of enlargement to one vertex of the shape.

As you are enlarging this shape by scale factor 3 you need to do this 3 times.

You could repeat this for every vertex to check.

Method B

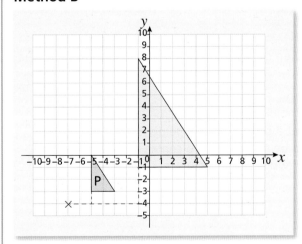

First plot the point (−7, −4) on the grid. This is the centre of enlargement.

Count how far along and up it is from the centre of enlargement to one vertex of the shape. In this case it is 2 squares right and 1 square up. You can write this as the vector $\begin{pmatrix} 2 \\ 1 \end{pmatrix}$

You need to enlarge this shape by scale factor 3, so you can multiply this vector by 3:

$\begin{pmatrix} 2 \\ 1 \end{pmatrix} \times 3 = \begin{pmatrix} 6 \\ 3 \end{pmatrix}$

Use the vector $\begin{pmatrix} 6 \\ 3 \end{pmatrix}$ from the centre of enlargement to decide where to draw the first vertex of the new shape.

Example 3

Describe fully the single transformation that maps shape A onto shape B

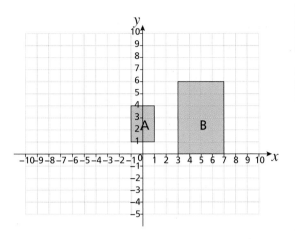

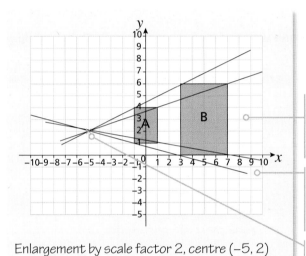

Enlargement by scale factor 2, centre (−5, 2)

Each side of rectangle B is double the length of the corresponding side in rectangle A, so the scale factor for the enlargement is 2

Drawing ray lines through each vertex on the original shape and the corresponding vertex on the enlarged shape can help you to find the centre of enlargement.

Extend the rays so that they meet at a point. This point is the centre of enlargement. In this case it is at (−5, 2)

Practice 12.2B

1 Faith has enlarged shape A by scale factor 4 from point P

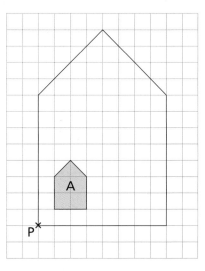

a Explain one thing that Faith has done well and one mistake that Faith has made.

b Beca thinks the enlargement should look like this. Do you agree? Explain your answer.

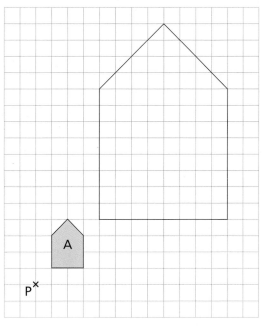

2 Copy each shape onto squared paper and enlarge it by scale factor 2, using point P as the centre of enlargement.

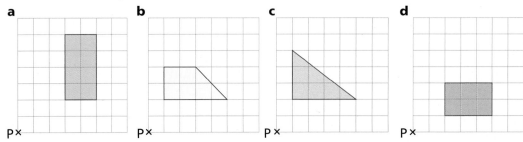

a

b

c

d

3 Copy each diagram and enlarge each shape using the given scale factor and centre of enlargement.

a scale factor 3, centre P

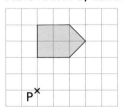

b scale factor 2, centre P

How are these different from the enlargements you drew in question **2**?

4 For each part of this question, you will need to draw a coordinate grid with x- and y-axes numbered from −10 to 10

The tables show the coordinates of the vertices of five quadrilaterals.

Draw each quadrilateral on a separate coordinate grid and enlarge it using the given scale factor and centre of enlargement.

a scale factor 5, centre (0, 0)

Vertex	A	B	C	D
Coordinates	(2, 2)	(1, 2)	(1, 1)	(2, 1)

b scale factor 3, centre (1, 3)

Vertex	A	B	C	D
Coordinates	(3, 3)	(2, 3)	(2, 5)	(3, 5)

c scale factor 2, centre (1, 3)

Vertex	A	B	C	D
Coordinates	(4, 3)	(3, 3)	(3, 5)	(4, 6)

d scale factor 4, centre (0, 0)

Vertex	A	B	C	D
Coordinates	(−1, −1)	(−1, 0)	(−2, −2)	(0, −1)

e scale factor 3, centre (2, 5)

Vertex	A	B	C	D
Coordinates	(1, 3)	(−1, 3)	(−1, 4)	(−2, 2)

5 For each diagram, describe fully the single transformation that maps shape A onto shape B

a
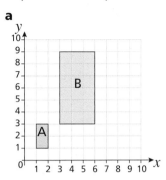

b
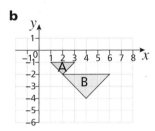

c
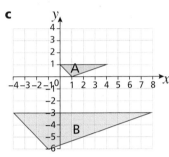

d
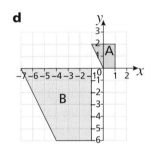

What do you think? 💭

1 What do you predict will happen when you enlarge this shape by scale factor 3, from point P?

On a copy of the grid, complete the enlargement to check whether your prediction was correct.

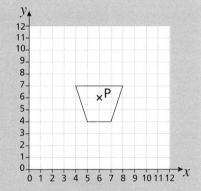

2 The diagram shows the result of an enlargement by scale factor 2, centre (5, 6)

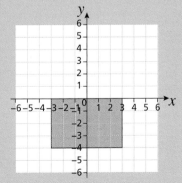

Find the coordinates of the vertices of the original shape.

Consolidate – do you need more?

1 On squared paper, enlarge each shape by each of these scale factors.

    a scale factor 2   **b** scale factor 3   **c** scale factor 4

    i 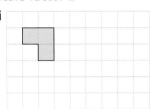   **ii**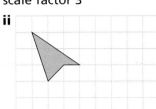

2 For each diagram, find the scale factor of enlargement from A to B

    a 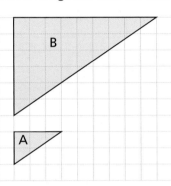   **b**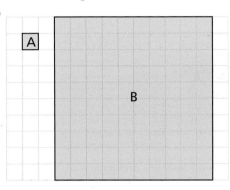

3 On squared paper enlarge each shape by each scale factor using point P as the centre of enlargement.

    a scale factor 2   **b** scale factor 3

    i 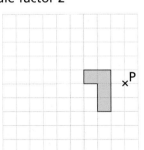   **ii**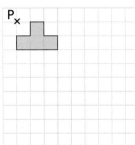

    iii 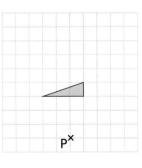   **iv**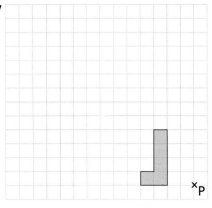

4 Copy each diagram onto squared paper. Enlarge each shape by the given scale factor and centre of enlargement.

a scale factor 2

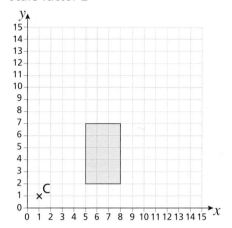

b scale factor 4

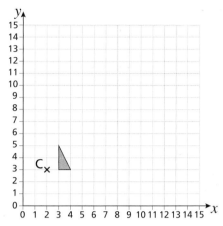

c scale factor 2

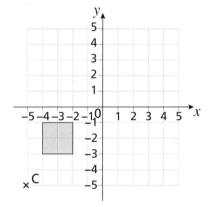

d scale factor 3

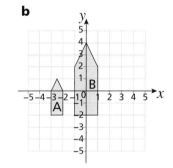

5 For each diagram, describe fully the single transformation that maps shape A onto shape B

a

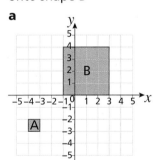

b

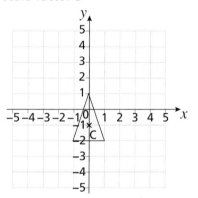

c

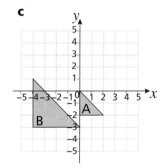

Stretch – can you deepen your learning?

1 ABC is an isosceles right-angled triangle. The vertex at A has coordinates (–4, 3)

DEF is an enlargement of ABC with centre P

The translation from P to D can be described using the vector $\begin{pmatrix} 6 \\ 3 \end{pmatrix}$

The translation from A to D can be described using the vector $\begin{pmatrix} 4 \\ 2 \end{pmatrix}$

Describe fully the single transformation that maps ABC onto DEF

2 Shape A is enlarged by scale factor 8, centre (0, 0) to give shape B

One of the vertices of shape A has coordinates (–2, 5)

Find the coordinates of the corresponding vertex on shape B

3 A parallelogram has vertices at (–2, 2), (1, 2), (2, –1) and (–1, –1)

It is enlarged by scale factor 4, centre (0, 0)

a Show that the area of the enlarged parallelogram is not four times the area of the smaller shape.

b Write the area of the smaller parallelogram as a percentage of the area of the larger parallelogram.

4 The table shows the coordinates of five points.

Point	A	B	C	D	E
Coordinates	(p, q)	$(p + 1, q – 2)$	$(p + 2, q + 1)$	$(p + 2, q – 5)$	$(p + 3, q – 1)$

Points B, C, D and E are joined to form a quadrilateral.

Quadrilateral BCDE is enlarged using point A as the centre of enlargement.

a Find the coordinates of the vertices of the enlarged shape if BCDE is enlarged by

    i scale factor 2   **ii** scale factor 5   **iii** scale factor m

b After an enlargement by scale factor 15, point D has coordinates (9, –26)
Work out the value of p and the value of q

Reflect

What's the same and what's different about enlarging a shape when given a centre of enlargement and enlarging a shape without a centre of enlargement?

Small steps

- Enlarge a shape by a positive fractional scale factor

- Enlarge a shape by a negative scale factor ⓗ

Are you ready?

1 Work out

   a $\frac{1}{2} \times 10$ **b** $\frac{1}{3} \times 15$ **c** $\frac{1}{4} \times 24$ **d** $\frac{3}{4} \times 24$ **e** $\frac{2}{3} \times 27$

2 Complete these calculations.

   a $10 \div 2 = 10 \times \square$ **b** $15 \div 3 = 15 \times \square$ **c** $24 \div \square = \frac{1}{4} \times 24$

3 What is the missing fraction in each calculation?

   a $\square \times 6 = 3$ **b** $20 \times \square = 4$ **c** $\square \times 100 = 10$

Models and representations

Ray lines

Straight lines from the centre of enlargement to each vertex of the shape can help you to draw and describe enlargements.

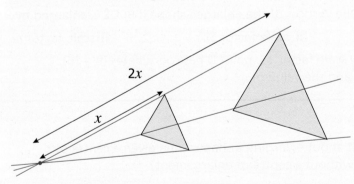

Up to now, all the enlargements that you have seen have made a shape larger. In this section, you will explore enlargements that result in a shape becoming smaller. This happens when the scale factor is a fraction less than one.

For example, when you enlarge a rectangle by scale factor $\frac{1}{2}$, the new dimensions will be $\frac{1}{2}$ the original size.

Example 1

a What is the scale factor of enlargement from shape A to shape B?

b Enlarge the triangle by scale factor $\frac{1}{3}$, centre O

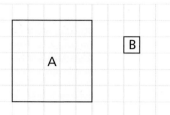

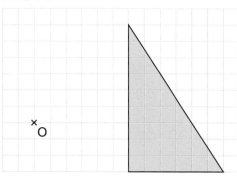

a

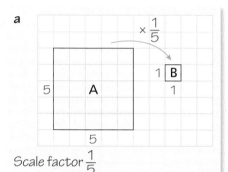

Scale factor $\frac{1}{5}$

The side lengths of the square have been divided by 5

$5 \div 5$ is the same as $5 \times \frac{1}{5}$ so shape A has been enlarged by scale factor $\frac{1}{5}$

b

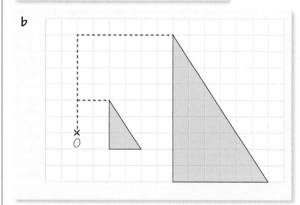

Count how far up and along it is from the centre of the enlargement to one vertex of the shape. In this case it is 6 squares up and 6 squares right.

As you are enlarging by scale factor $\frac{1}{3}$ you need to divide each of these by 3 to find the position of the corresponding vertex on the enlarged shape. You could think of this as $\begin{pmatrix} 6 \\ 6 \end{pmatrix} \times \frac{1}{3} = \begin{pmatrix} 2 \\ 2 \end{pmatrix}$

You can then redraw the rest of the shape.

You could repeat this for every vertex to check.

Practice 12.3A

1. On squared paper, enlarge each shape by the given scale factor.

   a scale factor $\frac{1}{2}$ **b** scale factor $\frac{1}{3}$ **c** scale factor $\frac{1}{4}$

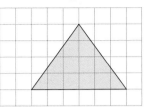 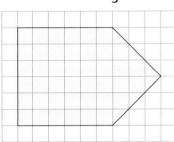 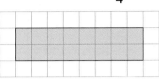

2. On squared paper, enlarge each shape by the given scale factor using point O as the centre of enlargement.

   a scale factor $\frac{1}{2}$ **b** scale factor $\frac{1}{3}$

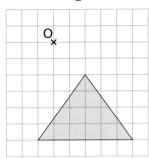 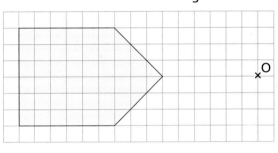

   c scale factor $\frac{1}{4}$

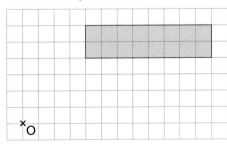

3. Junaid has tried to enlarge rectangle X by scale factor $\frac{2}{3}$, centre P

   a What has Junaid done correctly?

   b What mistake has Junaid made?

   c Enlarge rectangle X correctly by scale factor $\frac{2}{3}$, centre P

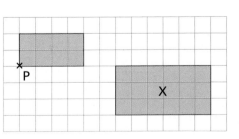

④ Copy the diagrams and enlarge each shape by scale factor $\frac{1}{3}$, centre C

a

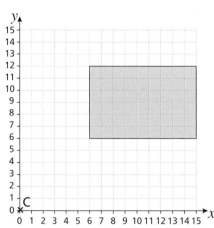

b

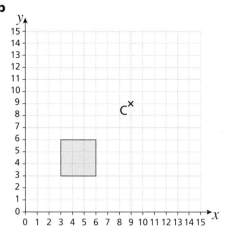

⑤ For each diagram, describe fully the enlargement that maps shape A onto shape B

a

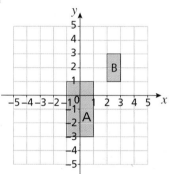

b

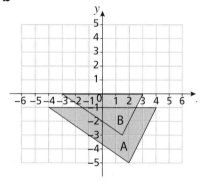

What do you think? 💭

1

If I enlarge a shape by scale factor $\frac{3}{2}$ the resulting shape will be smaller than the original because it is a fractional scale factor.

Do you agree with Zach? Draw some examples to explain your answer.

2 a Draw x- and y-axes numbered from 0 to 10. Plot the points (6, 3), (9, 3), (6, 9) and (9, 9) and join them to form a rectangle. Label your rectangle A

b Enlarge rectangle A by scale factor $\frac{1}{3}$, centre the origin. Label the image B

c Work out these ratios.

i the length of the sides of rectangle A : the length of the sides of rectangle B

ii the perimeter of rectangle A : the perimeter of rectangle B

iii the area of rectangle A : the area of rectangle B

d What do you notice about your answers to part **c**?

Example 2

Enlarge shape A by scale factor −2 centre the origin.

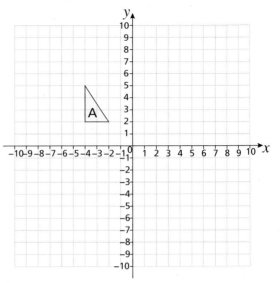

As the scale factor is negative, you need to draw ray lines from each vertex of the shape to the centre of enlargement and then continue them on the other side.

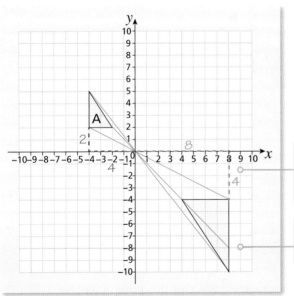

Measured from the origin, each ray line to a vertex of the new shape needs to be twice as long as to the corresponding vertex on the original shape.

As the scale factor is −2, the triangle is double the size but in a different orientation.

Think about how you could do this using **vectors**.

Practice 12.3B

1 Here are three different enlargements.

1

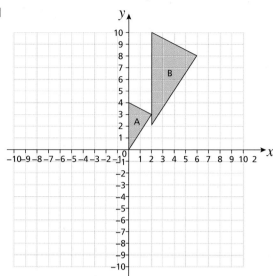

2

3

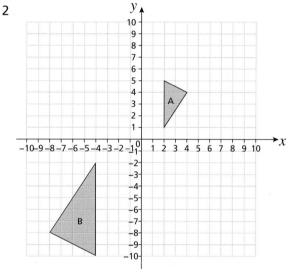

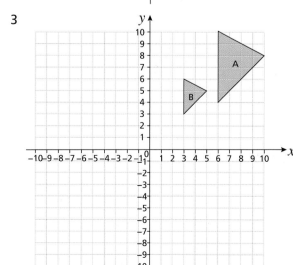

a In which diagram is B an enlargement of A by scale factor 2, centre (−2, −2)?

b Explain how you can identify an enlargement by a positive integer scale factor.

c In which diagram is B an enlargement of A by scale factor $\frac{1}{2}$, centre (0, 2)?

d Explain how you can identify an enlargement by a fractional scale factor using a fraction between 0 and 1

e In which diagram is B an enlargement of A by scale factor −2, centre (0, 0)?

f What are the key features of an enlargement by a negative scale factor?

2 Triangle B is an enlargement of triangle A

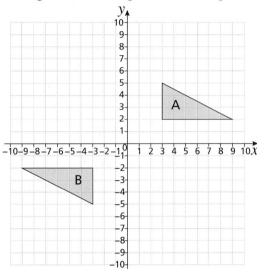

The scale factor of enlargement must be 1 because triangles A and B are the same size.

Abdullah

I don't think the scale factor of enlargement can be 1

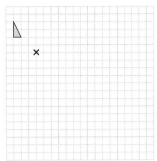

Jackson

Who do you agree with? Explain your answer.

3 Copy each diagram and enlarge the shape by the given scale factor using the cross (×) as the centre of enlargement.

a   scale factor −1

b   scale factor −2

c   scale factor −4

d   scale factor $\frac{1}{3}$

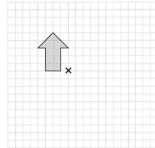

4 Copy each diagram and complete the enlargements.

a scale factor −3, centre (5, 4)

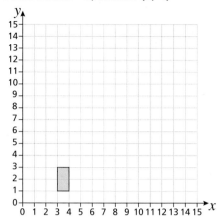

b scale factor −2, centre (11, 5)

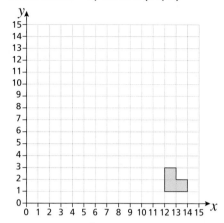

c scale factor −2, centre (8, 9)

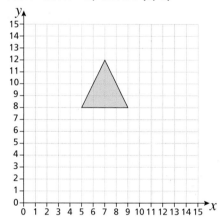

d scale factor $-\frac{1}{2}$, centre (8, 7)

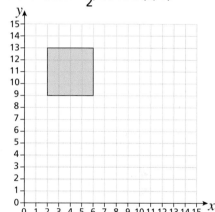

5 Copy each diagram and complete the enlargements.

a scale factor −1, centre (0, 0)

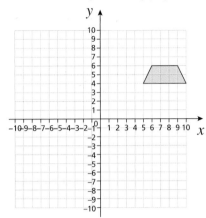

b scale factor −5, centre (−2, −3)

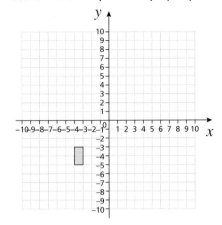

c scale factor $-\frac{2}{3}$, centre (0, 0)

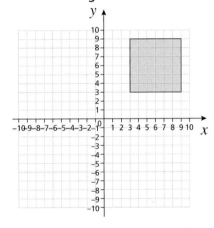

d scale factor $-\frac{3}{2}$, centre (−2, 4)

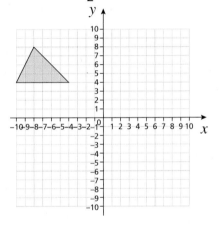

6 Describe fully the single transformation that maps shape A onto shape B

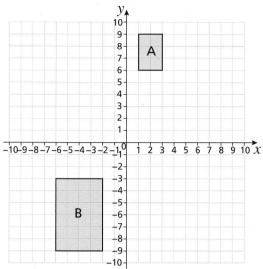

What do you think?

1 Describe fully the single transformation that maps shape A onto shape B

Is there more than one possible answer?

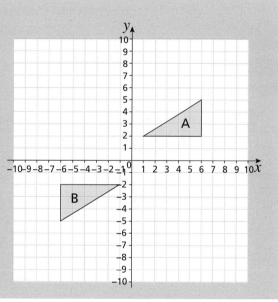

2 **a** Write the vector for the translation that would map point P onto each of these vertices.

    i A   **ii** B   **iii** C

  b Write the vector for the translation that would map point P onto each of these vertices.

    i A′   **ii** B′   **iii** C′

  c Triangle A′B′C′ is an enlargement of triangle ABC by scale factor −2, centre P

    How do the vectors in parts **a** and **b** show this?

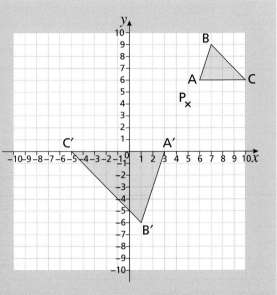

3 A quadrilateral, PQRS, is enlarged by scale factor −5, centre X, to give P′Q′R′S′

The vector for the translation that maps point X onto point P is $\begin{pmatrix} -4 \\ 7 \end{pmatrix}$

Find the vector for the translation that maps point X onto point P′

Consolidate – do you need more?

1 Copy each diagram and enlarge each shape by the given scale factor.

  a scale factor $\frac{1}{2}$

  b scale factor $\frac{1}{3}$

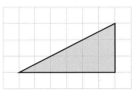

  c scale factor $\frac{1}{4}$

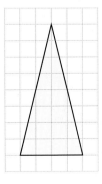

  d scale factor $\frac{1}{5}$

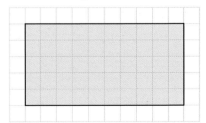

2 Enlarge this shape by scale factor $\frac{1}{2}$, centre O

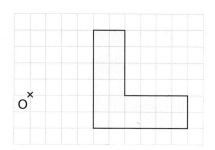

3 For each diagram, describe the enlargement from A to B

a

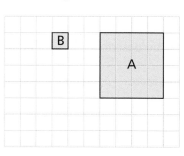

b

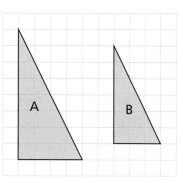

4 Copy each diagram and enlarge each shape by scale factor −2, centre O

a

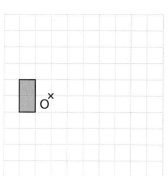

b

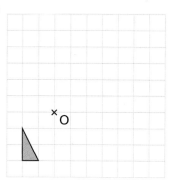

5 Copy each diagram and enlarge each shape by scale factor −3, centre O

a

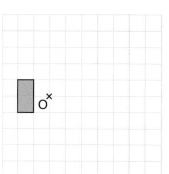

b

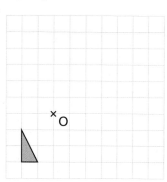

Stretch – can you deepen your learning?

1 B is an enlargement of A by scale factor –2, centre (0, 0)

C is an enlargement of A by scale factor –8, centre (0, 0)

Describe fully the single transformation that maps shape B onto shape C

2 Decide whether each statement is always true, sometimes true or never true.

Use examples to support your reasoning.

a Enlargements make a shape bigger.

b The resulting shape after an enlargement is in the same orientation as the original.

c An enlargement by a negative scale factor can be described as a reflection.

3 A square with side length 3 cm is enlarged by scale factor –4

Ed says, "The area of the enlarged square is –144 cm²"

a Explain why Ed thinks this.

b Give two reasons why Ed is incorrect.

4 The vector that maps point P onto point X is $\begin{pmatrix} 7 \\ -5 \end{pmatrix}$

The vector that maps point P onto point Y is $\begin{pmatrix} 6 \\ -8 \end{pmatrix}$

The vector that maps point P onto point Z is $\begin{pmatrix} 9 \\ -1 \end{pmatrix}$

Triangle XYZ is enlarged with centre P to give triangle X′Y′Z′

For the enlargement with each of the following scale factors, find the vector for the translation that maps point P onto

    i X′   **ii** Y′   **iii** Z′

a scale factor 2   **b** scale factor –4

c scale factor 1.5   **d** scale factor –2.5

5 a Chloe has formed a conjecture.

If I enlarge a shape by scale factor $\frac{a}{b}$ then the side lengths of the object and the image are in the ratio $b:a$

Chloe

Investigate Chloe's conjecture and explain your findings.

b Faith has formed a conjecture.

If I enlarge a shape by scale factor $\frac{a}{b}$ then the area of the image is $\frac{a}{b}$ of the area of the object.

Faith

Give an example that

    i supports Faith's conjecture   **ii** disproves Faith's conjecture.

Summarise your findings.

6 **a** **i** On a copy of the grid, enlarge triangle ABC by scale factor −1 about (2, 2)

   ii Write the coordinates of each vertex of triangle ABC before and after the transformation.

b **i** On a copy of the grid, reflect triangle ABC in the line $x + y = 4$

   ii Write the coordinates of each vertex of ABC before and after the transformation.

c Compare your answers to parts **a** and **b**. What's the same? What's different?

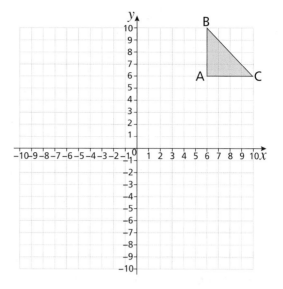

Reflect

Are these statements always true, sometimes true or never true?

a When you enlarge a shape by a fractional scale factor, the resulting shape is smaller than the original.

b When you enlarge a shape by a negative scale factor, the resulting shape is in a different orientation.

Ⓗ 12.4 More similarity

Small steps

- ◼ Solve problems with similar triangles Ⓗ
- ◼ Explore ratios in right-angled triangles Ⓗ

Are you ready?

1 The sides of a triangle are 10 cm, 6 cm and 9 cm long. The triangle is enlarged by scale factor 3.

What are the lengths of the sides of the enlarged triangle?

2 The angles in a triangle are 50°, 85° and 45°. The triangle is enlarged by scale factor 3.

What are the sizes of the angles in the enlarged triangle?

3 Shape B is an enlargement of shape A. Find the values of x, y and z

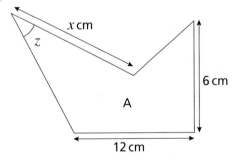
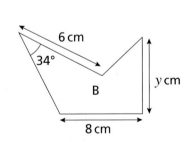

Models and representations

Dynamic geometry

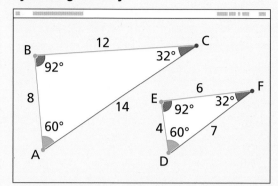

This can be used to explore the relationships between the sides and angles of **similar** triangles.

Enlargements on paper

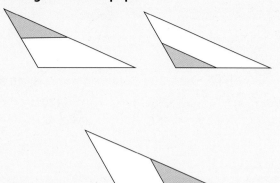

You can cut out a triangle and its enlargement to see where the smaller shape fits inside the larger shape.

Shapes are mathematically **similar** when one is an enlargement of the other.

In this chapter, you will learn more about similar triangles.

If two triangles have equal angles, then they must be similar.

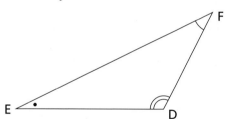

Triangles ABC and DEF are similar because they have the same three angles.

Writing the triangles with corresponding vertices above each other helps you to see which sides are corresponding, for example AC corresponds to FE

ABC

FDE

Example 1

Triangles ABC and PQR are similar.

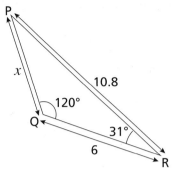

a Find the scale factor of the enlargement from triangle ABC to triangle PQR

b Work out the length of sides labelled x and y

a $\dfrac{QR}{BC} = \dfrac{6}{4} = 1.5$ ○———— QR and BC are corresponding sides

b $x = 4.2 \times 1.5$ ○———— Side PQ corresponds to side AB

$= 6.3\,cm$ ○———— Side AC corresponds to side PR

$y = 10.8 \div 1.5$

$= 7.2\,cm$

Practice 12.4A

1. Triangles ABC and XYZ are similar.

   a Which vertex in triangle ABC corresponds to vertex Z in triangle XYZ?

   b Which vertex in triangle XYZ corresponds to vertex A in triangle ABC?

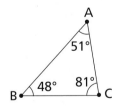
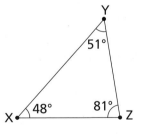

   c Which statement shows the corresponding vertices in the correct order?

A	B	C
Triangles ABC are similar. XYZ	Triangles ABC are similar. YXZ	Triangles ABC are similar. YZX

2 For each pair of similar triangles, identify which vertices are corresponding.

Write your answers in this form:

Triangles ABC, XYZ are similar, where A corresponds to X, B to Y and C to Z.

a

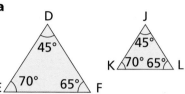

b

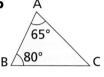

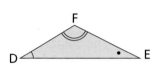

c

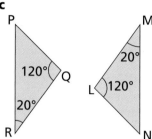

d

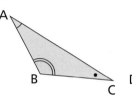

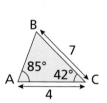

💬 3 **a** Explain why triangles ABC and LMN are similar even though only two of the angles are marked as equal.

b What is the scale factor of the enlargement from triangle ABC to triangle LMN?

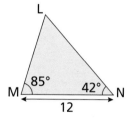

c Which side in triangle ABC corresponds to LN?

d Use your answers to parts **b** and **c** to work out the length of LN

e Copy and complete this statement:

Triangles $\frac{ABC}{***}$ are similar.

4 Explain why triangles PQR and XYZ are similar.

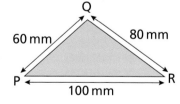

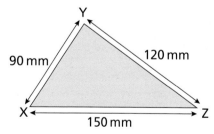

5 Find the value of each letter in these pairs of similar triangles.

a

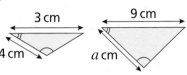

b

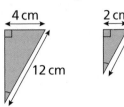

c

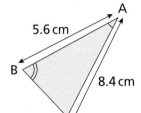

d

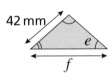

6 The diagram shows a pair of similar triangles.

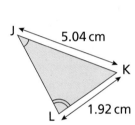

Work out the length of

a JL **b** BC

7 **a** Prove that triangles ABC and EDC are similar.

b Work out the length of BC

c Find the perimeter of triangle EDC

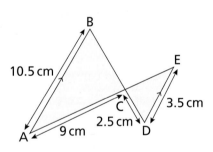

8

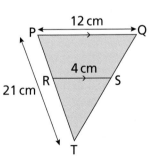

Work out the length of

a RT

b PR

What do you think?

1 If a pair of triangles are similar, they must be congruent too.

Emily

If a pair of triangles are congruent, they must be similar too.

Jackson

Do you agree with either Emily, Jackson, both of them or neither of them?

2 Work out the values of x.

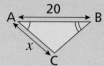 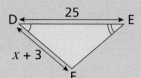

3 Triangles ABC and DEF are similar because $\frac{6}{3} = \frac{8}{4} = \frac{10}{5} = 2$

 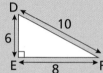

Work out these ratios.

a $\frac{AB}{AC}$ and $\frac{DE}{DF}$ **b** $\frac{AB}{BC}$ and $\frac{DE}{EF}$ **c** $\frac{BC}{AC}$ and $\frac{EF}{DF}$

What do you notice? Investigate the ratios between the sides in other pairs of similar triangles.

4 All right-angled triangles with an angle of 40° are similar.

Is Faith correct? Explain your answer.

In a right-angled triangle, the longest side is called the **hypotenuse**.

The other sides can be labelled as **opposite** or **adjacent** to one of the smaller angles.

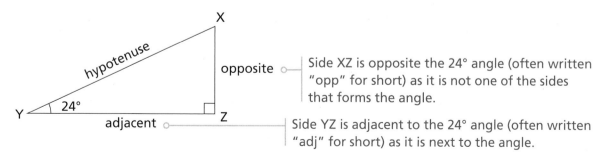

Side XZ is opposite the 24° angle (often written "opp" for short) as it is not one of the sides that forms the angle.

Side YZ is adjacent to the 24° angle (often written "adj" for short) as it is next to the angle.

The study of the side lengths and sizes of angles in triangles is called **trigonometry**.

Example 2

a Use the words hypotenuse, opposite and adjacent to label the sides of these right-angled triangles.

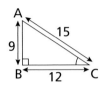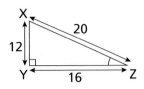

b Show that triangles ABC and XYZ are similar.

c Calculate these ratios.

i $\dfrac{AB}{AC}$ and $\dfrac{XY}{XZ}$

What do you notice?

ii $\dfrac{AB}{BC}$ and $\dfrac{XY}{YZ}$

iii $\dfrac{BC}{AC}$ and $\dfrac{YZ}{XZ}$

a

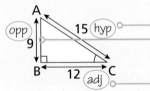

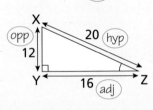

Start by labelling the hypotenuse, which is always opposite the right angle. Then label the side opposite the marked angle.

The remaining side is the side adjacent to the marked angle.

b $\dfrac{XY}{AB} = \dfrac{12}{9} = 1.33\ldots$

$\dfrac{XZ}{AC} = \dfrac{20}{15} = 1.33\ldots$

$\dfrac{YZ}{BC} = \dfrac{16}{12} = 1.33\ldots$

All pairs of corresponding sides are in the same ratio, so the triangles are similar.

You can show that two triangles are similar by proving that they have the same angles or by finding the ratios of the corresponding sides.

c **i** $\dfrac{AB}{AC} = \dfrac{9}{15} = 0.6$

$\dfrac{XY}{XZ} = \dfrac{12}{20} = 0.6$

This is the ratio of $\dfrac{\text{opposite}}{\text{hypotenuse}}$ for each triangle.

ii $\dfrac{AB}{BC} = \dfrac{9}{12} = 0.75$

$\dfrac{XY}{YZ} = \dfrac{12}{16} = 0.75$

This is the ratio of $\dfrac{\text{opposite}}{\text{adjacent}}$ for each triangle.

iii $\dfrac{BC}{AC} = \dfrac{12}{15} = 0.8$

$\dfrac{YZ}{XY} = \dfrac{16}{20} = 0.8$

This is the ratio of $\dfrac{\text{adjacent}}{\text{hypotenuse}}$ for each triangle.

The ratios of each pair of corresponding sides within the triangles are the same.

This is true in all pairs of similar triangles and leads to some very important results in right-angled triangles.

Practice 12.4B

1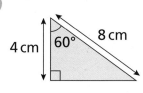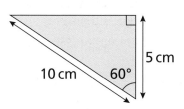

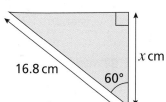

💬 **a** Explain why these three triangles are similar.

b Calculate the ratio of $\dfrac{\text{adjacent}}{\text{hypotenuse}}$ for each of the triangles.

💬 **c** Compare your results with a partner's.

> The ratio of the length of the adjacent side to the length of the hypotenuse is called the **cosine** of the angle. You have shown that the cosine of 60° is $\dfrac{1}{2}$
>
> This is shortened to $\cos 60° = \dfrac{1}{2}$

2 Use the fact that $\cos 60° = \dfrac{1}{2}$ to find the value of x in each of these triangles.

a 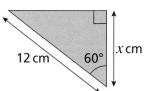 **b** 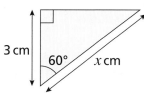 **c**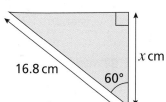

3 In which of these triangles is the labelled angle 60°?

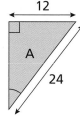 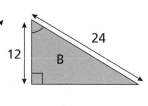 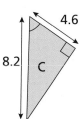 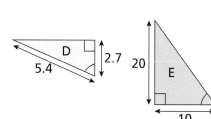

4 **a** Use Pythagoras' theorem to work out the length of the unknown side in each of these right-angled triangles. Give your answers to 1 decimal place.

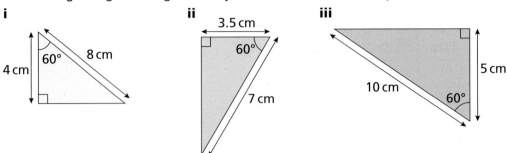

i

ii

iii

b Work out the ratios $\dfrac{\text{opposite}}{\text{hypotenuse}}$ and $\dfrac{\text{opposite}}{\text{adjacent}}$

for each of the triangles in part **a**. Compare your answers with a partner's. Give your answers to 2 decimal places.

> The ratio of the length of the opposite side to the length of the hypotenuse is called the **sine** of the angle. You should have found that the sine of 60° is about 0.866. The exact value is $\dfrac{\sqrt{3}}{2}$. This is shortened to $\sin 60° = \dfrac{\sqrt{3}}{2}$
>
> The ratio of the length of the opposite side to the length of the adjacent side is called the **tangent** of the angle. You should have found that the tangent of 60° is about 1.73. The exact value is $\sqrt{3}$ This is shortened to $\tan 60° = \sqrt{3}$

5 By investigating the ratios $\dfrac{\text{opposite}}{\text{hypotenuse}}$, $\dfrac{\text{adjacent}}{\text{hypotenuse}}$

and $\dfrac{\text{opposite}}{\text{adjacent}}$ for each triangle, find the values of sin 30°, cos 30° and tan 30°

Give your answers as a surd.

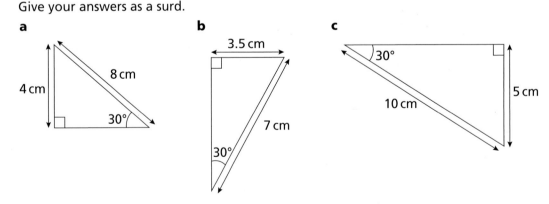

a

b

c

You studied surds earlier in Block 6

What do you think? 💭

1

> The sine of 30° and the cosine of 60° are equal.

Use a diagram to show that Marta is correct.

2

> All isosceles right-angled triangles are similar

Investigate Jakub's claim.

3 **a** Use the diagram to find the value of tan 45°

   b Use Pythagoras' theorem to find the length of AC
   Leave your answer as a surd.

   c Work out the exact value of

   i sin 45°   **ii** cos 45°

   What do you notice about your answers?

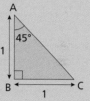

Consolidate – do you need more?

1 For each pair of similar triangles, identify the pairs of corresponding angles and corresponding sides. The first one has been started for you.

a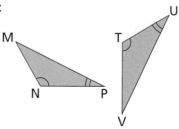

Corresponding angles	Corresponding sides
∠CAB and ∠QRP	BC and PQ
∠ABC and _____	AC and _____
∠BCA and _____	AB and _____

b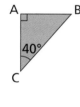

c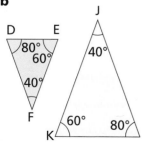

2 Triangles PQR and XYZ are similar.

a What is the scale factor of enlargement from triangle PQR to triangle XYZ?

b Which side in triangle XYZ corresponds to side PR in triangle PQR?

c Work out the length of

i XZ ii QR

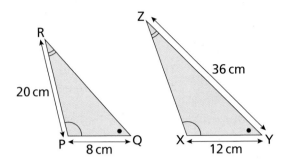

3 Triangles DEF and JKL are similar.

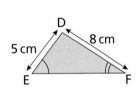
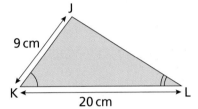

Work out the length of

a JL b EF

4 Are triangles PQR and XYZ similar? Give a reason for your answer.

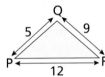
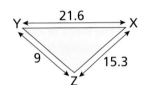

5 $\sin 30° = \dfrac{1}{2}$, $\cos 30° = \dfrac{\sqrt{3}}{2}$ and $\tan 30° = \dfrac{1}{\sqrt{3}}$

Use these facts to find the lengths of the sides labelled with letters.

a b c d

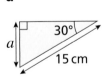
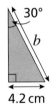

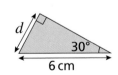

Stretch – can you deepen your learning?

1

Faith

I've drawn a quadrilateral with two angles of 110° and two angles of 70°

I've drawn a quadrilateral with exactly the same angles. My quadrilateral must be similar to yours.

Zach

Show that Zach might be wrong.

2 A man of who is 2 m tall casts a shadow of length 3.75 m. At the same place, at the same time, a tower of height 47 m casts a shadow. How long is the shadow of the tower?

3 **a** Prove that triangles PQR and TSR are similar.

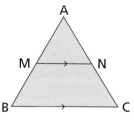

b If PQ = ST, show that triangles PQR and TSR are congruent.

4 M and N are the midpoints of AB and AC, respectively.

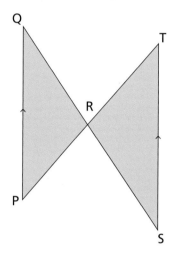

AM = 3, AC = $8y$ and BC = $10z$

Find an expression for the perimeter of trapezium MNCB

5 **a** Sketch an equilateral triangle with side length 2 units. By drawing the line of symmetry, divide the triangle into two right-angled triangles. Use Pythagoras' theorem to find the exact values of the sine, cosine and tangent of 60° and 30°

b Show that $\sin 45° = \cos 45°$. Will this be true for any other angle in a right-angled triangle? Explain why or why not.

Reflect

1 Describe two ways in which you can prove that a pair of triangles are similar.

2 Explain why when two right-angled triangles share another common angle they must be similar.

I have become **fluent** in...	I have developed my **reasoning** skills by...	I have been **problem-solving** through...
▪ recognising enlargement and similarity ▪ enlarging a shape by a positive integer scale factor ▪ enlarging a shape by positive scale factors from a centre of enlargement ▪ enlarging a shape by a negative scale factor. Ⓗ	▪ explaining why shapes are or are not similar ▪ being able to explain why not all enlargements make a shape larger ▪ explaining how to calculate side lengths and angles in similar shapes.	▪ applying enlargement and similarity in different mathematical contexts ▪ proving similarity in different contexts ▪ exploring ratios in right-angled triangles. Ⓗ

Check my understanding

1 A rectangle has length 16 cm and width 8 cm. It is enlarged by scale factor 4
What are the dimensions of the enlarged rectangle?

2 Copy each grid and complete the enlargements.

   a scale factor 2 centre X

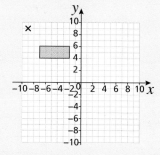

   b scale factor $\frac{1}{4}$ centre (9, –9)

3 **a** Explain why triangles PQR and TUV are similar.

   b Work out the length of

     i PR   **ii** TV

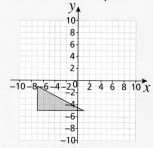

4 Use the fact that $\sin 30° = \frac{1}{2}$ to work out the length of AB Ⓗ

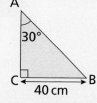

13 Solving ratio and proportion problems

In this block, I will learn...

how to draw and use direct proportion graphs

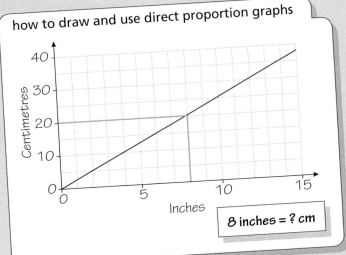

8 inches = ? cm

how to solve problems with direct proportion

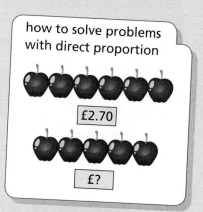

£2.70

£?

how to solve problems with inverse proportion

8 workers take 15 days to build a wall. How long will 4 workers take?

about graphs of inverse proportion H

Lengths and widths of rectangles with area 24 cm²

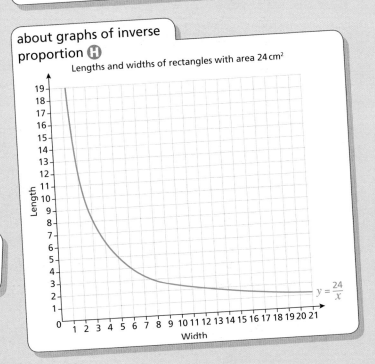

$y = \dfrac{24}{x}$

how to solve ratio problems

£75				
£?	£?	£?	£?	£?

how to solve ratio problems involving algebra H

$2 : x = x : 8$

$\dfrac{2}{x} = \dfrac{x}{8}$

$16 = x^2$

$\pm x = 4$

how to solve best buy problems

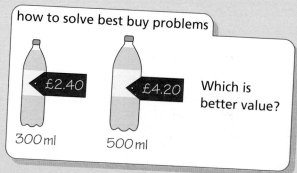

£2.40

£4.20

Which is better value?

300 ml   500 ml

Small steps

- Solve problems with direct proportion ®
- Explore direct proportion and conversion graphs ®

Key words

Direct proportion – two quantities are in direct proportion when, as one increases or decreases, the other increases or decreases at the same rate

Conversion graph – a graph used to change from one unit to another

Unitary method – a technique for solving problems by first finding the value of one unit

Are you ready?

1 Work out the value of x if $4x$ is equal to

    a   48   **b**   24   **c**   140   **d**   8.4

2 One bottle of water costs £0.85. Work out the cost of

    a   2 bottles   **b**   5 bottles   **c**   12 bottles   **d**   17 bottles.

    Assume that each bottle costs the same.

3 Write down the coordinates of the points A, B, C and D

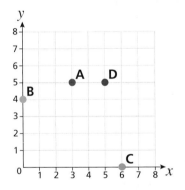

4 Draw x- and y-axes numbered from 0 to 5. Plot the points (4, 3), (0, 2), (4, 0) and (4, 4)

Models and representations

Double number lines

These can be used to show the multiplicative nature of exchange rates.

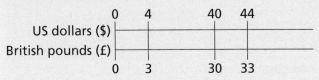

Proportion diagrams

These can be used to show different multiplicative relationships and are helpful when problem-solving.

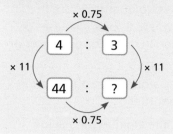

Conversion graphs

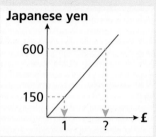

These are useful for converting from one currency to another.

Example 1

8 litres of paint will cover an area of 48 m²

a   What area will be covered by 6 litres of paint?

b   How much paint is needed to cover an area of 72 m²?

a $48 \div 8 = 6$

   1 litre covers 6 m²

   $6 \times 6 = 36\,m^2$

Divide 48 m² by 8 to work out how many square metres 1 litre would cover.

Now you can scale up to 6 litres by multiplying by 6

There are other ways of working this out, for example you could have worked out what area 2 m² covers first or used a proportion diagram.

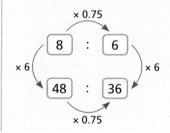

b $72 \div 48 = 1.5$

   The area is 1.5 times greater.

   $8 \times 1.5 = 12$ litres

   You will need 12 litres of paint.

You can work out how many times greater 72 is compared with 48

If the area is 1.5 times greater, you will need 1.5 times as much paint.

Practice 13.1A

1   Rhys pays £3.20 for 5 kg of potatoes.

   a   Work out the cost of 2.5 kg of potatoes.

   b   Work out the cost of 6 kg of potatoes.

   c   Work out the cost of 500 g of potatoes.

2 Six apples cost £1.80. Work out the cost of 8 of these apples.

3 Five identical pencils cost £3.25

    a Work out the cost of 8 of these pencils.

Huda has £10

    b What is the maximum number of pencils she can buy?

    c How much change will she receive?

4 The calorie contents for some types of fruit are given below.

> Melon   34 calories per 100 g
> Mango   60 calories per 100 g
> Pineapple 50 calories per 100 g

    a Work out the number of calories in

        i 150 g of mango   **ii** 75 g of pineapple.

    b Filipo makes a fruit salad with 30 grams of each of the three types of fruit. Work out the total number of calories in the fruit salad.

5 Here are the ingredients for making 12 marshmallow squares.

> 45 g butter
> 180 g puffed rice
> 300 g marshmallows

    a Adapt the recipe so that it makes 18 marshmallow squares.

    b Marshmallows cost £1.40 for a 150 g packet. Work out the cost of the marshmallows needed to make 18 squares.

6 Here are the ingredients needed to make 15 ginger cookies.

> 1 egg
> 150 g sugar
> 3 tbsp ground ginger
> 300 g flour
> 150 g butter

    a Marta makes the recipe using 450 g flour. How many cookies does she make?

    b Bobbie has 700 g of flour and 400 g of sugar. Work out the greatest number of cookies that she can make.

7 Rob's wages are directly proportional to the number of hours that he works.
When Rob works for 7 hours, he earns £84

    a How much does Rob earn when he works for 5 hours?

    b How many hours would Rob need to work to earn more than £100?

    c Show that Rob's total wages, w, can be written $w = 12t$, where t is the
number of hours worked.

What do you think?

1 Faith makes orange paint by mixing red and yellow paint.

She mixes 12 litres of red paint with 15 litres of yellow paint to get the right shade
of orange.

Faith wants to make more of the same shade of orange paint. She uses another 8 litres
of yellow paint. Work out how many more litres of red paint she needs to use.

2 There are 16 children in a playgroup. At least one adult is needed for every 5 children.

    a What is the minimum number of adults needed at the playgroup?

    b Three new children join the playgroup. How many more adults will now be
needed? Explain your answer.

    c Darius says that the number of children and number of adults are directly
proportional because as the number of children increases, so does the number
of adults. Explain Darius' mistake.

In this section, you will review **conversion graphs** and **direct proportion** graphs from Book 2

Example 2

The graph shows how to convert between British pounds and US dollars.

Conversion graph for British pounds and US dollars

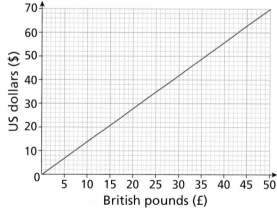

Use the graph to convert

a £20 into dollars   **b** $60 into pounds   **c** £500 into dollars.

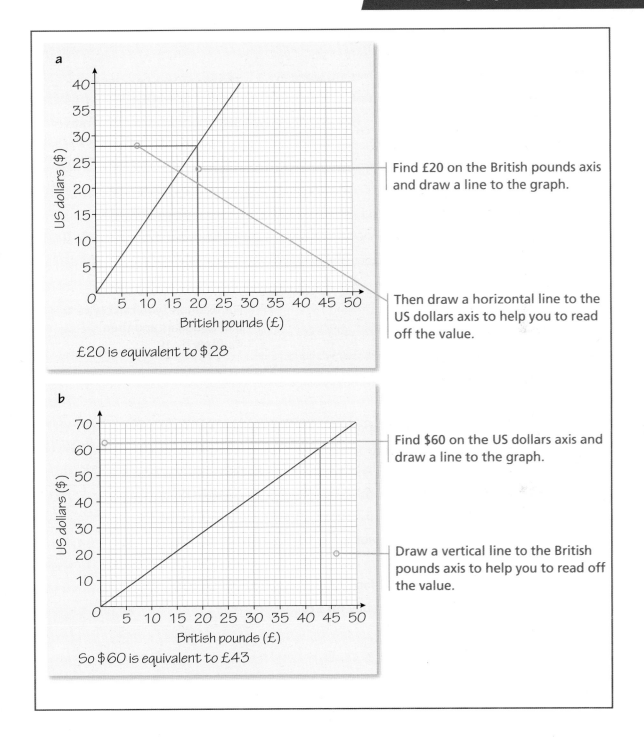

a

US dollars ($) vs British pounds (£)

Find £20 on the British pounds axis and draw a line to the graph.

Then draw a horizontal line to the US dollars axis to help you to read off the value.

£20 is equivalent to $28

b

US dollars ($) vs British pounds (£)

Find $60 on the US dollars axis and draw a line to the graph.

Draw a vertical line to the British pounds axis to help you to read off the value.

So $60 is equivalent to £43

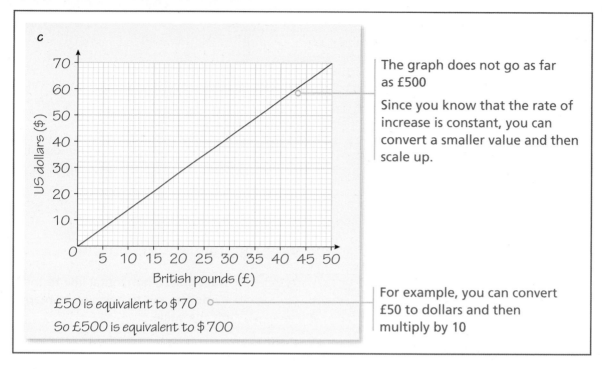

c

US dollars ($)

British pounds (£)

The graph does not go as far as £500

Since you know that the rate of increase is constant, you can convert a smaller value and then scale up.

£50 is equivalent to $70

So £500 is equivalent to $700

For example, you can convert £50 to dollars and then multiply by 10

Practice 13.1B

1 The cost of fabric is directly proportional to its length. Three metres of fabric costs £13.50

a Copy and complete the table to show the costs of different lengths of fabric.

Length of fabric, l (m)	0	1	2	3	4	5	6
Cost of fabric, C (£)	0			£13.50			

b Draw the graph of cost of fabric, C against length of fabric, l

c Use your graph to find

  i the cost of 1.5 metres of fabric

  ii how much fabric you can buy for £20

2 The graph shows the cost of tomatoes.

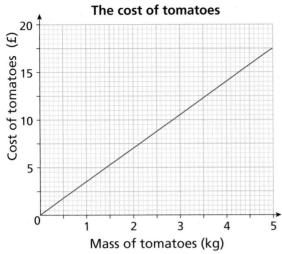

The cost of tomatoes

a Use the graph to work out the cost of 3 kg of tomatoes.

b Kate has £5. How many kilograms of tomatoes can she buy?

c Benji buys 2.5 kg of tomatoes and pays with a £10 note. How much change will he receive?

3 The graph shows how to convert between British pounds and euros (€)

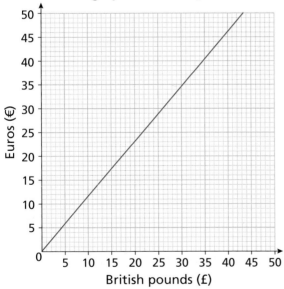

Conversion graph for British pounds to euros

Use the graph to convert

a £30 into euros

b €20 into British pounds.

c Explain how the graph could be used to convert £350 into euros.

d Amina says that this conversion graph is an example of direct proportion. Do you agree? Explain your answer.

4 Graph 1 can be used to convert between masses in pounds and kilograms. Graph 2 shows how to convert between temperatures in Celsius and Fahrenheit.

Graph 1
Conversion graph for pounds and kilograms

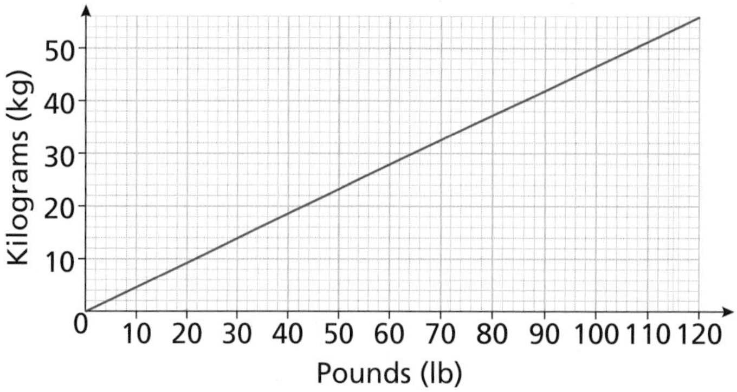
Kilograms (kg)
50
40
30
20
10
0
10 20 30 40 50 60 70 80 90 100 110 120
Pounds (lb)

Graph 2
Conversion graph for temperatures in Celsius and Fahrenheit

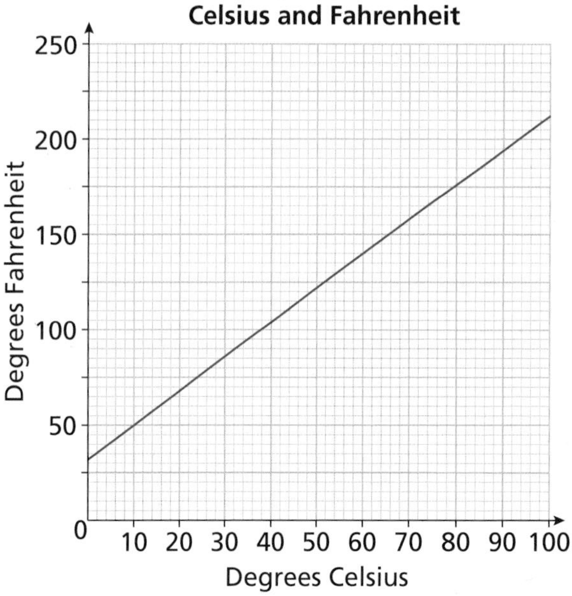
Degrees Fahrenheit
250
200
150
100
50
0
10 20 30 40 50 60 70 80 90 100
Degrees Celsius

Jakub

As the number of pounds increases, so does the number of kilograms, so this is an example of direct proportion.

As the temperature in Celsius increases, so does the temperature in Fahrenheit, so this is an example of direct proportion too!

Chloe

Do you agree with each statement? Explain your answer.

5 Which of these graphs represents two quantities that are directly proportional? Explain your answer.

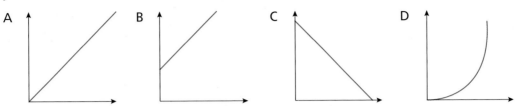

A   B   C   D

What do you think?

1 Graph 1 shows how to convert between British pounds and US dollars. Graph 2 shows how to convert between US dollars and Japanese yen.

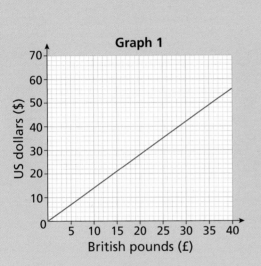

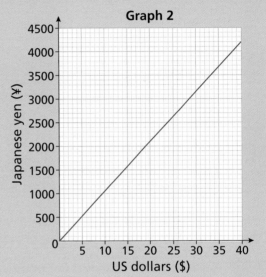

Use the graphs to find the exchange rate between British pounds and Japanese yen.

2 The graph shows how to convert between gallons and litres.

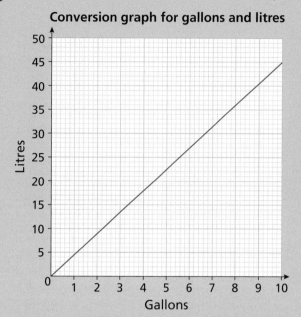

Conversion graph for gallons and litres

A cylindrical milk container has radius 50 cm and a height of 1.5 m

It contains 150 gallons of milk.

Is the container more than or less than half full? Show working to explain how you decide.

Consolidate – do you need more?

1 Lydia pays £30 for 24 litres of petrol.

  a Work out the cost of

    i 12 litres of petrol

    ii 15 litres of petrol.

  b How many litres of petrol can Lydia buy with £50?

2 Five cups of coffee cost £14

  a Work out the cost of three of these cups of coffee.

Three croissants cost £5.40

  b Bobbie buys two cups of coffee and two croissants. Work out how much money she spends.

3 This recipe for mushroom soup serves 4 people.

> 30 g butter
>
> 2 garlic cloves
>
> 450 g mushrooms
>
> 2 small onions
>
> 1 litre vegetable stock

Adapt the recipe to serve 6 people.

4 At a supermarket, 2 kg of oranges cost £4.40

   a Copy and complete the table to show the cost of different quantities of oranges.

Mass of oranges m (kg)	0	1	2	3	4	5
Cost, C (£)			£4.40			

   b Draw the graph of cost, C against mass of oranges, m

   c Use your graph to work out the cost of 3.5 kg of oranges.

   d Can you use your graph to work out the cost of 8 kg of oranges?
      Explain your answer.

5 The graph shows how to convert between euros, €, and US dollars, $

Conversion graph for euros and dollars

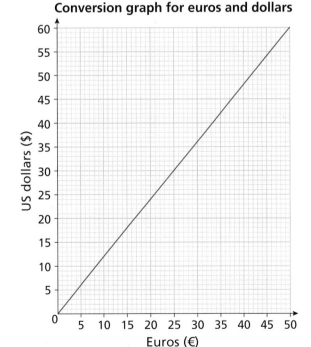

Use your graph to convert

a $40 into euros

b €30 into dollars

c $300 into euros.

Stretch – can you deepen your learning?

1 Squash is made by mixing water and fruit juice. In a 1-litre jug of squash, there is 800 ml of water. How much fruit juice will be needed to make 12 litres of this squash?

2 Quantity a is directly proportional to quantity b and b is directly proportional to quantity c. Is it true that a is directly proportional to c? Explain your answer.

3 Emily is charged 22p for every unit of electricity she uses.

Seb is charged 18p for every unit of electricity he uses, plus a standing charge of £8

  a What would a graph of cost of electricity against units used look like for each person?

  b Do both graphs represent direct proportion? Explain how you decide.

  c Which payment plan, Emily's or Seb's, is better value?

Reflect

1 Explain why a graph that shows direct proportion must go through the origin and be a straight line.

2 "When one value increases so does the other." Explain why this definition of direct proportion is incomplete and misleading.

Small steps

- Solve problems with inverse proportion
- Explore graphs of inverse relationships (H)

Key words

Inverse proportion – if two quantities are in inverse proportion, when one quantity increases, the other decreases at the same rate

Constant – not changing

Are you ready?

1 Write down the inverse of each of these operations.

    a multiply by 4   **b** divide by 3   **c** multiply by 6   **d** divide by 7

2 Write down all the factor pairs for each of these numbers.

    a 16   **b** 24   **c** 36   **d** 50

3 Write down the coordinates of points A, B, C and D

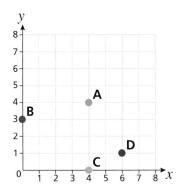

4 Draw x- and y-axes numbered from 0 to 5. Plot the points (5, 2), (0, 3), (1, 0) and (3, 4)

Models and representations

Bar models

These can be used to show relationships involving **inverse proportion**.

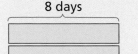

16 days

8 days

Tables of values

x and y are inversely proportional, so the product xy is **constant**.

x	1	2	4	8
y	16	8	4	2

In this chapter, you will explore inverse proportion.

For example, suppose it takes one farmer 24 days to harvest a crop. If there are *more* workers, it will take *less* time.

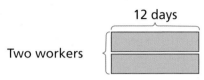

One worker

There are 24 days of work to be done. If there are two workers, the job can be completed in half the time, 12 days.

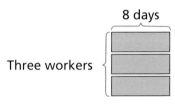

Two workers

If there are three workers, the work bar can be split into three pieces. Each piece represents 8 days.

Three workers

This is an example of inverse proportion. When you increase the number of workers, the time taken decreases at the same rate, that is, twice as many workers means half the time.

Workers	Number of days
1	24
2	12
6	4

×2 (1 → 2) ×3 (2 → 6)

÷2 (24 → 12) ÷3 (12 → 4)

Notice that the number of workers multiplied by the number of days is always 24. This is because 24 units of work are being shared out.

Example 1

Five people can build a wall in 30 days. Assuming that all the people work at the same rate

a   how long would it take 2 people to build the wall

b   how many people would be needed to build the wall in 10 days?

a
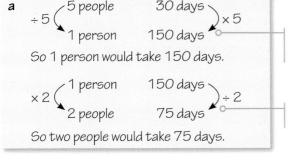

So 1 person would take 150 days.

So two people would take 75 days.

> Start by working out how long one person would take.
>
> Since you have 5 times fewer people, it will take 5 times as long to do the work.
>
> If you now double the number of workers, it will take half the time.

b
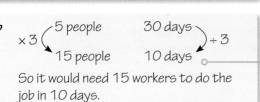

So it would need 15 workers to do the job in 10 days.

> Five people take 30 days.
>
> To finish in 10 days would mean dividing the time taken by 3, so three times as many workers are needed.

Practice 13.2A

1   Five people work together to construct a house in 30 days. Assuming that all the people work at the same rate, how long would it take to construct the house with

    a   one person working

    b   six people working?

2   Six hose pipes are used to fill a swimming pool in 3 hours. Copy and complete these sentences.

    a   12 hose pipes would take _____ hours.

    b   1 hose pipe would take _____ hours.

    c   _____ hose pipes would take 6 hours.

    d   _____ hose pipes would take _____ hours.

3   Twelve bricklayers can build a wall in 30 days. How long would it take to build the same wall if 15 bricklayers worked together?

4   At a speed of 60 miles per hour, it takes 4 hours to drive from Southampton to Manchester.

    a   How long would the same journey take at a speed of 50 miles per hour? Give your answer in hours and minutes.

    b   How fast would you have to travel to complete the journey in 3 hours?

5 Four workers can resurface a road in 6 hours.

    a How long would it take to resurface the same road if 6 people worked together?

    b How many workers are needed if the road is to be resurfaced in 3 hours?

6 Four people take 1 hour and 20 minutes to water a football pitch.

    How long would it take to water the same pitch if five people worked together?
Give your answer in hours and minutes.

7 Chloe carries out an experiment to see how long it takes to load 200 boxes
on to a lorry.

    1 person alone took 4 hours.

    2 people took 3 hours.

    3 people took 2 hours.

> This is an example of inverse proportion because, as
> more people help, it takes less time to load the lorry.

Do you agree with Chloe? Explain your answer.

8 The variables x and y are inversely proportional.

x	8	16		2.4
y	12		10	

    a Copy and complete the table.

    b What is the relationship between each pair of x- and y-values?

    c Express this relationship in the form $xy = k$, where k is a constant to be found.

9 State whether each table shows variables that are directly proportional, inversely
proportional or neither.

    a

x	2	3	4
y	6	4	3

    b

a	4	5	8
b	12	13	16

    c

p	1	5	10
q	1.6	8	16

What do you think? 🌐

1 Three postal workers can deliver 270 letters in 1 hour.

    a How long would it take

        i one postal worker to deliver 540 letters

        ii two postal workers to deliver 180 letters?

    b How many postal workers would be needed to deliver 900 letters in 2 hours?

2 Eight people can construct a house in 20 days.

For the first 4 days, three of the workers are off sick. For the remaining days, all eight people work on the construction of the house. How long will it take to build the house?

In this section, you will explore graphs of inverse relationships.

Example 2

The time it takes to travel 60 km depends on the speed at which you travel.

Speed (km/h)	1	2	5	10	20	30	60
Time (h)							

a Copy and complete the table.

b Plot a graph of speed against time for the data in the table. Plot time on the horizontal axis and speed on the vertical axis. Describe the shape and features of the graph.

c Use your graph to find the speed needed for the journey to be completed in 8 hours.

a

Speed (km/h)	1	2	5	10	20	30	60
Time (h)	60	30	12	6	3	2	1

If you travel 1 km every hour, then it would take 60 hours to cover 60 km

If you travel 2 km every hour, then it would take 30 hours to cover 60 km

You can fill in the table by dividing the distance (60 km) by each speed.

b

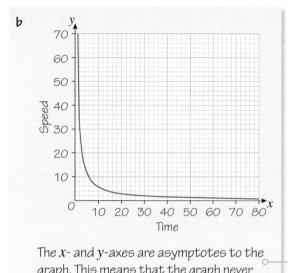

Time is on the x-axis and speed is on the y-axis.

Plot each point and join them up with a smooth curve.

You met graphs like this in Chapter 1.4

The x- and y-axes are asymptotes to the graph. This means that the graph never touches either axis.

This is because speed cannot be 0 (as you would never travel anywhere!) and time cannot be 0 (because no matter how fast you travel it will take more than 0 hours!)

c

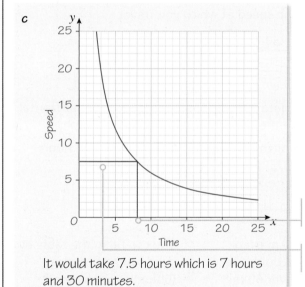

It would take 7.5 hours which is 7 hours and 30 minutes.

Find 8 on the x-axis. Draw a vertical line up to the graph.

Draw a line across to the speed axis and read off the value.

Practice 13.2B

1 A rectangle has area 48 cm². The table shows some possible lengths and widths of the rectangle.

Length, x (cm)	1	2	4	6	8	10	12	16	48
Width, y (cm)	48	24	12	8	6	4.8	4	3	1

a Explain why this is an example of inverse proportion.

b Plot a graph of length against width for the data in the table. Plot width on the horizontal axis and length on the vertical axis.

c Use your graph to find the length of a rectangle with area 48 cm² and width 5 cm

2 The table shows the time it takes to travel 120 km at different speeds.

Speed (km/h)	2	5	10	15	20	30	40	60	80	120
Time (h)	60	24	12				3		1.5	

a Copy and complete the table.

b Explain why this is an example of inverse proportion.

c Draw a graph of a time against speed for the data in the table. Plot speed on the horizontal axis and time on the vertical axis.

d Use your graph to find the time taken when travelling at a speed of 55 km/h

3 The graph shows the relationship between the base length and perpendicular height of a parallelogram with area 16 cm²

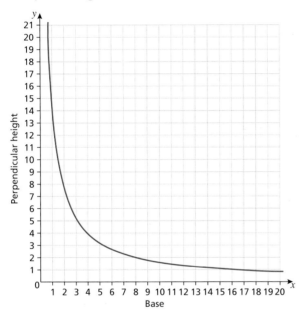

a Use the graph to find

i the height of a parallelogram with base length 8 cm

ii the height of a parallelogram with base length 10 cm

iii the base length of a parallelogram with height 6 cm

b Explain why the equation of the graph is given by $xy = 16$

④ The graph below shows Boyle's law. This is an inversely proportional relationship between the pressure and volume of a gas.

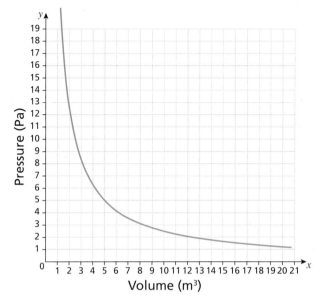

You may meet Boyle's law in your Science lessons.

Pa stands for "pascal", which is one of the units used for measuring pressure.

a Use the graph to find

   i the volume when the pressure is 15 Pa

   ii the pressure when the volume is 4 m³

b Write the equation of the graph in the from $xy = k$, where k is a constant to be found.

⑤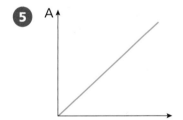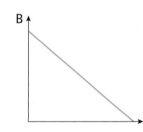

If graph A represents direct proportion, then graph B represents inverse proportion.

Do you agree with Benji? Explain your answer.

What do you think? 🌐

1 Sven is investigating the perimeter of rectangles. He is looking at rectangles that have a perimeter of 12 cm. He draws this table of values to record his findings.

Length (cm)	1	2	3	4	5
Width (cm)	5	4		2	

a Fill in the missing values in the table.

b Sven says that this is an example of inverse proportion because as the length increases, the width decreases. Explain Sven's mistake.

c Draw a graph of length against width for Sven's rectangles. Plot width on the horizontal axis and length on the vertical axis.

Explain how this graph is different from those that show inverse proportion.

Consolidate – do you need more?

1 Eight people can build a wall in 12 days.

   a How long would it take

     i 4 people   **ii** one person?

   b How many people would be needed to build the wall in 16 days?

2 Three workers can resurface a road in 10 hours 30 minutes.

How long would it take to resurface the same road if 5 people worked together? Give your answer in hours and minutes.

3 p and q are inversely proportional.

p	8		5	2
q	5	10		

Copy and complete the table.

4 Three people can deliver 500 leaflets in 30 minutes. How long would it take five people to deliver the same number of leaflets?

5 Bobbie is investigating the dimensions of triangles with area 32 cm²

The table shows some possible values for the base and perpendicular height.

Base (cm)	2	4		10	16	64
Height (cm)	32	16	8	6.4		1

   a Copy and complete the table.

   b Plot a graph of base against height for these triangles.

   c Use your graph to find

     i the base of a triangle with height 12 cm

     ii the height of a triangle with base 20 cm

Stretch – can you deepen your learning?

1 Two factory workers can pack 200 boxes in 5 hours.

If 6 workers work for 2 hours, how many of these boxes can they pack?

2 It takes one day for 2 workers to harvest a crop from a field with an area of $40\,m^2$

Mario's farm has an area of $320\,m^2$. He hires 8 workers and pays each of them £90 per day. How much will it cost him to harvest his crop?

3 Abdullah takes 4 hours to paint a room. Jakub takes 6 hours to paint a room of the same size.

If they work together, how long will it take them to paint one room of this size? Give your answer in hours and minutes.

Reflect

1 What is the difference between inverse and direct proportion?

2 Sketch graphs showing direct and inverse proportion. Identify the differences between the two graphs.

Small steps

- Solve ratio problems given the whole or part **R**
- Solve problems involving ratio and algebra **H**

Key words

Ratio – a ratio compares the sizes of two or more values

Divide in a ratio – share a quantity into two or more parts so that the shares are in a given ratio

Multiplier – a number you multiply by

Are you ready?

1 A fruit bowl contains only apples and pears. $\frac{1}{3}$ of the fruit are apples.

    a What fraction of the fruit are pears?

    b Write the ratio of

        i apples to pears   **ii** pears to apples.

2 The ratio of red sweets to green sweets in a jar is $3:5$

    What fraction of the sweets are

    a red   **b** green?

3 Simplify each of these ratios.

    a $5:15$ **b** $24:16$ **c** $35:49$ **d** $32:80$

4 The ratio of red beads to white beads on a necklace is $3:4$. How many red beads are there if there are

    a 8 white beads   **b** 24 white beads   **c** 44 white beads?

Models and representations

Bar models

| 14 | 14 | 14 | 14 |

| 14 | 14 | 14 |

Double number lines

These can show the multiplicative relationship between **ratios**.

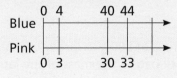

Ratio diagrams

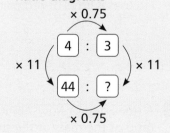

Example 1

Marta and Beth share some sweets in the ratio $3:4$

a If there are 28 sweets altogether, how many do they each get?

b If Beth gets 28 sweets, how many does Marta get?

a Marta

Beth

28 sweets

$28 \div 7 = 4$

Each part is worth 4 sweets

Marta gets $3 \times 4 = 12$ sweets

Beth gets $4 \times 4 = 16$ sweets

You can draw a bar model to help.

Marta gets 3 parts and Beth gets 4 parts.

There are 7 parts altogether.

So the 28 sweets need to be shared between 7 parts.

Once you have the value of one part, you can work out the values of 3 parts and 4 parts.

b Marta

Beth

28 sweets

$28 \div 4 = 7$

Each part is worth 7 sweets.

Marta gets 3 parts, so $3 \times 7 = 21$ sweets

Draw a bar model to help.

This time, Beth's share is worth 28 sweets.

This means that 4 parts are 28 sweets.

So you can work out what one part is worth.

Now you can work out Marta's share, which is three parts.

You can find more practice and examples in Book 2 Block 1

Practice 13.3A

1 A shade of pink paint is made by mixing red and white paint in the ratio $2:5$

Flo uses 8 litres of red paint. How much white paint will she need?

R   W

$\times ? \begin{pmatrix} 2 & : & 5 \\ 8 & : & \boxed{} \end{pmatrix} \times ?$

2 A small bag of flour has a mass of 500 g. A larger bag has a mass of 2.5 kg. Write the ratio of the mass of the small bag to the mass of the large bag in its simplest form.

3 Abdullah and Samira share £45 in the ratio $2:7$. Work out how much money they each get.

Abdullah

Samira

£45

4 Marta and Faith share some sweets in the ratio 3:2
   a If there are 30 sweets altogether, how many do they each get?
   b If Marta gets 30 sweets, how many sweets were there altogether?
   c If Faith gets 30 sweets, how many sweets does Marta get?
   d If Marta gets 30 more sweets than Faith, how many sweets were there altogether?

5 At the school cake sale, the ratio of cupcakes sold to cookies sold was 5:4
   a If 15 cupcakes were sold, work out the number of cookies sold.
   b If 15 more cupcakes than cookies were sold, work out the number of cookies sold.
   c 32 cookies were sold. Each cookie sold for 50p and each cupcake sold for 80p.
      Work out the total amount of money raised.

6 In a test, the ratio of Flo's mark to Seb's mark was 1:4. Which of the following
   statements are true?
   A For every mark Flo scored, Seb scored 4 marks.
   B For every mark Seb scored, Flo scored 4 marks.
   C Seb scored 4 marks.
   D Flo's mark is $\frac{1}{4}$ of Seb's mark.

7 The ratio of men to women to children in a theatre audience is 4:5:3
   There are 216 adults in the theatre. Work out the number of children.

8 Some friends are sharing bars of chocolate.
   In group A, 6 bars are shared equally between 14 people.
   In group B, 4 bars are shared equally between 10 people.
   a Write, giving your answers in the form 1:n, the ratio of people to chocolate bars in
      i group A   ii group B
   b What conclusion can you draw about which group gets the larger portions?

9 The perimeter of a rectangle is 50 cm. The dimensions of the rectangle are in
   the ratio 4:1. Work out the area of the rectangle.

10 Seb, Benji and Chloe shared some money.
    Seb received $\frac{1}{4}$ of the money. Benji and Chloe shared the remaining money in the
    ratio 3:5. Benji received £9.
    Work out the total amount of money that was shared between Seb, Benji and Chloe.

What do you think?

1 x and y are both prime numbers. Explain why the ratio $x:y$ cannot be simplified.

2 A bag contains yellow and purple counters in the ratio $3:5$. There are no other counters in the bag.

> There could be 25 counters in the bag.

Is Zach right? Explain how you know.

3 In a bunch of flowers, the ratio of roses to daisies is $3:4$ and the ratio of daisies to violets is $3:5$.

Work out the ratio of roses to daisies to violets.

In this section, you will explore ratio problems that need the use of algebraic notation and manipulation.

Example 2

$x:y = 5:6$

a Find a formula connecting x and y

b Show that $2x + y : 2y = 4:3$

a x [| | | | |]   Draw a bar model to help.
                               x is worth 5 parts.

   y [| | | | | |] y is worth 6 parts.

   $x = \frac{5}{6}y$ $\frac{x}{y} = \frac{5}{6}$ so x is $\frac{5}{6}$ of y You can write this using algebraic notation.

                               Other formulae could be:

                               $y = \frac{6}{5}x$ or $6x = 5y$

b $2x + y : 2y$ x is 5 parts and y is 6 parts, so you can substitute these values into the expression.

   $2(5) + 6 : 2(6)$

   $16 : 12$
   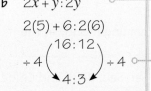
   $\div 4$ ( ) $\div 4$ You need to simplify the ratio by dividing both numbers by 4,
   $4 : 3$ since 4 is the highest common factor of 16 and 12

Example 3

For three quantities p, q and r,

p is 20% of q and the ratio of $q:r = 10:3$

Work out the ratio of $p:q:r$

$p: q :r$

$p:10:3$ ——————— You know the ratio of $q:r$, so you can substitute this.

You can do this algebraically or using bar models.

p is 20% of q, so $p = 0.2q$ ——— 20% as a **multiplier** is 0.2

$p = 0.2 \times 10 = 2$ ——— q is 10 parts so you can work out how many parts p is worth.

$2:10:3$ ——— Now substitute $p = 2$ into the ratio $p:q:r$

Example 4

The ratio of Huda's age to Samira's age is $1:7$

In four years, the ratio of their ages will be $1:4$

Work out how old Huda and Samira are now.

Huda : Samira

$x:7x$ ——— Huda's age is one part, and Samira's age is 7 parts.
Since you do not know the value of each part, you can call these x and $7x$

In four years

$x + 4:7x + 4$ ——— To find the ages in 4 years, you add 4 to their current ages.

$1:4$

Samira is 4 times as old as Huda.

$(x + 4) \times 4 = 7x + 4$ ——— Huda's age × 4 = Samira's age. You can set up an equation using this fact.

$4(x + 4) = 7x + 4$

$4x + 16 = 7x + 4$

$16 = 3x + 4$

$12 = 3x$

$4 = x$ ——— You can solve the equation to find x

If $x = 4$, then Huda is 4 years old, and Samira is $7 \times 4 = 28$ years old.

Substitute $x = 4$ to find Huda's and Samira's ages. Check: in 4 years Huda will be 8 and Samira will be 32, and $32 = 4 \times 8$

Practice 13.3B

1 $x:y = 5:1$

x ⬚⬚⬚⬚⬚

y ⬚

Which of the following statements are true?

A $y = \dfrac{x}{6}$ B $y = \dfrac{x}{5}$ C $y = 5x$ D $x = 5y$

2 $x:y = 1:4$

Write a formula connecting x and y

3 $x:y = 2:3$

a Write a formula connecting x and y

b Show that $x + 2y:4y - x = 4:5$

4 Work out the value of x in each of these ratios.

a $5x:8 = 9:2$ **b** $4x:9 = 5:3$ **c** $3x:9 = 4:5$

5 $x:9 = 4:x$

Work out the value of x

6 w is 20% of x. y is $\dfrac{2}{5}$ of x

Find the ratio $w:x:y$

7 $m:n = 3:8$. r is 25% of n

Find the ratio $m:r$

8 A regular polygon has exterior angles and interior angles in the ratio $2:7$

How many sides does it have?

9 A cuboid has dimensions in the ratio $1:3:5$. It has a volume of $120\,\text{cm}^3$

Find

a the dimensions of the cuboid

b the surface area of the cuboid.

10 The ratio of blue counters to orange counters in a bag is $1:4$. Five more blue counters are added to the bag. The ratio of blue counters to orange counters is now $1:3$

Work out the total number of counters in the bag at the start.

11 The ratio of Rob's age to Ed's age is $4:5$. In 4 years, the ratio of their ages will be $5:6$

Work out how old Rob and Ed are now.

What do you think? 💭

1 The ratio of the number of marbles in box A to the number of marbles in box B is $7:3$. Jackson removes 3 marbles from box A and places them into box B. The ratio is now $5:3$

How many marbles were in each box originally?

2 P and Q are in the ratio $4:1$. R and S are in the ratio $2:3$

Given that $P + Q = 2(R + S)$, find $P:Q:R:S$

Consolidate – do you need more?

1 In a class, the ratio of boys to girls is $2:3$. There are 12 boys in the class. Work out the number of girls in the class.

2 A small bottle of cola contains 250 ml and a large bottle contain 2 litres. Write the ratio of the volume of cola in the small bottle to the volume of cola in the large bottle. Give your answer in its simplest form.

3 Huda, Mario and Rob share £60 in the ratio $5:3:2$. Work out how much they each get.

4 Faith and Kate share some sweets in the ratio $5:3$. Faith gets 8 more sweets than Kate. Work out how many sweets they each get.

5 A bag contains red, yellow and blue counters. 20% of the counters are red. The ratio of yellow counters to blue counters is $5:4$. There are 36 red counters in the bag.

Work out the number of

a yellow counters   **b** blue counters.

6 The angles in a triangle are in the ratio $2:5:3$

a Work out the size of each angle.

b Give the mathematical name for this type of triangle.

Stretch – can you deepen your learning?

1 A box of chocolates contains dark chocolates, milk chocolates and white chocolates in the ratio $1:3:2$

There are fewer than 40 chocolates in the box. What is the maximum possible number of chocolates in the box?

2 Bobbie and Emily share some money in the ratio $2:5$. Bobbie gives 20% of her share to Sven. Sven receives £24.

Work out the original amount of money that was shared.

3 A bowl contains apples and oranges in the ratio $1:3$. Ed adds 4 more apples to the bowl.

The ratio of apples to oranges is now $1:2$

Work out the number of oranges in the bowl.

4 a and b are two integers. The ratio of the sum of a and b to the difference between a and b is $7:1$

Find some possible values of a and b. How many solutions can you find?

5 The ratio of Chloe's age to Beca's age is $3:4$

a In 7 years, the ratio of their ages will be $4:5$. Work out how old Chloe and Beca are now.

b When will the ratio of their ages be $5:6$?

Reflect

1 Using 120 sweets and the ratio $2:3$, make up four different ratio sharing problems. Draw bar models to show the differences between them.

2 If the ratio A:B is $1:6$, write down some formulae to describe the relationship between A and B. How many can you think of?

Small steps

■ Solve best buy problems

Key words

Unit cost/price – the cost or price of 1 item or 1 unit of an item

Best buy (or **best value**) – the item which is cheapest when equal-sized amounts of different items are compared

Are you ready?

1 Calculate

   a £3.44 ÷ 6   **b** £24.99 ÷ 8   **c** £2.49 ÷ 5   **d** £32.86 ÷ 12

2 Work out the cost of 1 cookie if

   a 3 cookies cost 75p   **b** 5 cookies cost £1.55

   c 8 cookies cost £4.32   **d** 10 cookies cost £6

3 Round each number to 2 decimal places.

   a 3.2242   **b** 0.41863   **c** 1.3485   **d** 0.15242

Models and representations

Bar models

2 items for 70p

3 items for £1.20

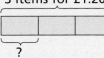

In Chapter 8.4, you used the unitary method and found the cost of one item or **unit cost**.

In this chapter, you will explore other methods of finding **best buys** such as scaling up or scaling down.

Example 1

A supermarket sells milk in 2-litre or 3-litre cartons.

The price for 2 litres is 88p

The price for 3 litres is £1.25

Which offer is better value for money?

Method A

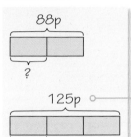

You can work out the cost per litre for each carton.

You can draw bar models to help.

Change pounds to pence so that the units are the same.

88p ÷ 2 = 44p per litre

125p ÷ 3 = 42p per litre   ○————| You round to 2 d.p. for money.

3 litres for £1.25 is better value. ○——| The better value is the one that costs less per litre.

Method B

You can also work out the amount of milk that you can buy with 1p

Option 1

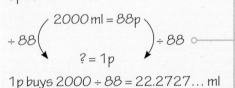

To get from 88 to 1, you divide by 88

So you also need to divide 2000 by 88

1p buys 2000 ÷ 88 = 22.2727... ml

Option 2

$$3000\,ml = 125p$$
$$\div 125 \qquad \qquad \div 125$$
$$? = 1p$$

To get from 125 to 1, you divide by 125

So you also need to divide 3000 by 125

1p buys 3000 ÷ 125 = 24 ml

3 litres for £1.25 is better value. ○——| The better value is the option that gives you more millilitres of milk for each 1p spent.

Method C

Option 1

2 litres = 88p

6 litres = 3 × 88p = 264p, or £2.64

You can also scale the quantities up to compare the cost of the same amount of milk with each option.

In this case, 6 is a multiple of 2 and 3, so you can work out the cost of 6 litres for each option.

Multiply the cost of 2 litres by 3 to find the cost of 6 litres.

Option 2

3 litres = £1.25

6 litres = 2 × £1.25 = £2.50

Multiply the cost of 3 litres by 2 to find the cost of 6 litres.

£2.50 < £2.64

3 litres for £1.25 is better value.

The better value is the cheaper option.

Practice 13.4A

1. In a supermarket, eggs are sold in boxes of 6 and in boxes of 8

   6 eggs cost £2.55 and 8 eggs cost £3.50

   a For each option work out

      i the cost of one egg

      ii how many eggs can be bought for £1

      iii the cost of 24 eggs.

   b Determine which option is better value for money.

2. In a shop, potatoes are sold in 2 kg bags for £3.50

   In a wholesale market, potatoes are sold in 7 kg bags for £12.50

   a For each option work out

      i the cost of 1 kg of potatoes

      ii the number of kilograms of potatoes that can be bought for £1

      iii the cost of 14 kg of potatoes.

   b Determine which option is better value for money.

3. Coffee is sold in two jars. A 200 g jar costs £1.55 and a 500 g jar costs £3.70

   Determine which jar offers better value for money.

4. A 200 g packet of biscuits costs £0.95 and a 300 g packet costs £1.25. Determine which is the better value.

5 Sven is trying to work out which offer is better value:

2 litres of cola for £1.55 or 500 ml for 45p

He decides to work out the amount of cola he can buy for 1p with each option.

> 2000 ÷ 1.55 = 1290 ml per p
>
> 500 ÷ 45 = 11 ml per p

Explain Sven's mistake and correct his working.

6 Chocolate bars cost 70p each. A shop has these offers:

Darius wants to buy 35 chocolate bars. Work out the cheapest way that he can do this.

7 Orange juice is sold in two sizes of carton: 300 ml for £3.55 or 500 ml for £5.75

   a Which is better value for money? How do you know?

   b Ali needs to buy 1.2 litres of orange juice. Which combination of cartons should he buy to minimise his costs?

8 A shop has these special offers on soda drinks.

Which is better value for money? How do you know?

9 At a cake sale, there are three options for buying cupcakes.

 A 4 cupcakes for £3.60

 B 6 cupcakes for £5.50

 C 10 cupcakes for £8.75

> To work out the best buy, I did: £3.60 × 15, £5.50 × 10 and £8.75 × 6

Explain what Ed is calculating and how it will help him to work out which option is best value.

What do you think? 💬

1 60 teabags cost £3. Explain in words what each of these calculations works out.

 a $300 \div 60$ **b** $3 \div 6$ **c** $60 \div 3$

2 In the UK, a box of 6 macarons costs £7.40. In France, a box of 8 macarons costs €11.20

 The exchange rate is £1 = €1.15. Determine in which country the macarons are better value.

3 At a local shop, DVDs cost £20. The shop has three special offers:

 Offer 1  Offer 2 Offer 3

 Offers cannot be combined.

 a Amina wants to buy 3 DVDs. Which offer would be cheapest?

 b Rhys wants to buy 6 DVDs. Which offer would be cheapest?

 c

> I want to buy n DVDs, and all the offers come to exactly the same price!

 Given that $n \leqslant 10$, how many DVDs does Flo want to buy?

567

Consolidate – do you need more?

1 Which option offers better value for money? How do you know?

A   5 apples for £1.25

B   8 apples for £1.95

2 A bakery sells doughnuts in different-sized packages.

| 4 for £1.55 | 6 for £2.30 | 10 for £3.75 |

a   Which offers best value for money? How do you know?

b   Beth wants to buy as many doughnuts as she can for £10. Which combination of offers should she buy?

3 Pencils come in three different-sized boxes:

Box A 12 pencils for £2.45   Box B 20 pencils for £4.20   Box C 35 pencils for £7.00

a   Work out the price per pencil for each box.

b   Work out the number of pencils that can be bought for £1 from each box.

c   Determine the best value box. Which method did you prefer?

4 Which is better value for money, a 400 g loaf of bread for £1.20 or a 600 g loaf for £1.95?

5 Dinner plates are sold in two different-sized packages.

A box of 6 plates costs £32 and a box of 10 plates costs £52

a   Which option is better value for money?

The price of the 6 plate box is reduced by 15% and the price of the 10 plate box is reduced by 12%

b   Does this change which box is better value for money?

6   Bobbie is making chocolate brownies. She needs 1.2 kg of sugar for her recipe.

In the supermarket, she sees these bags of sugar.

a   Which bag offers better value for money?

b   Which combination of bags will give the minimum cost for Bobbie?

c   Why is the better value option sometimes not the one that you should purchase?

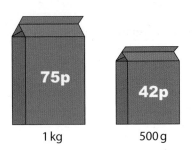

75p

42p

1 kg   500 g

Stretch – can you deepen your learning?

1   Laundry detergent is sold in 750 ml bottles costing £5.50.

The supermarket has two different offers.

Offer 1   Offer 2

Special bottles
with 50% extra free

Buy one,
get one
HALF
PRICE

Which offer is better value for money?

2   A shop sells two different-sized jars of hot chocolate. Ali and Samira are trying to calculate which is better value for money. Here are some of their calculations:

| 150 ÷ 2.70 | 2.70 × 5 | 3.50 × 3 | 250 ÷ 3.50 |

Work out the possible size and price of each jar.

3   In the UK, a 25-mile taxi trip costs £40

In Spain, a 45 km taxi trip costs €50

Given that £1 = €1.12 and 8 km = 5 miles, work out which taxi journey is better value for money.

Reflect

1   What is the difference between the best value and the cheapest option when shopping? When might you choose the cheapest option rather than the better value?

2   Make a poster or revision card showing the different ways in which you can determine the best buy.

13 Solving ratio and proportion problems
Chapters 13.1–13.4

I have become fluent in...

- solving ratio problems when given a part or whole
- solving direct proportion problems
- finding the best value using the unit cost
- drawing and using direct proportion graphs
- solving simple inverse proportion problems.

I have developed my reasoning skills by...

- representing ratios in a variety of forms
- solving best buy problems using different strategies
- recognising whether two quantities are directly or inversely proportional to each other
- using conversion graphs to find values beyond the range of the graph.

I have been problem-solving through...

- solving more complex direct and inverse proportion problems
- exploring inverse proportion graphs **H**
- identifying which methods to apply in a given problem
- solving problems using algebra. **H**

Check my understanding

1 Carrots cost £2.55 for 3 kg

  a What is the cost of

    i 1 kg of carrots   **ii** 2.5 kg of carrots?

  b How many kilograms of carrots can be bought for £3.60?

2 Use the graph to convert

  a 10 miles into kilometres

  b 12 kilometres into miles

  c 150 kilometres into miles.

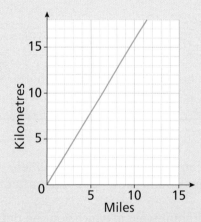

3 Three workers take 8 days to build a wall.

  a How long would it take to build the same wall using

    i 1 worker

    ii 12 workers?

  b If the wall needs to be completed in 4 days, how many workers would be needed?

4 Huda and Sven share £35 between them in the ratio 5:2. How much do they each get?

5 Rhys and Darius share some money between them in the ratio 7:4. Rhys receives £18 more than Darius. How much do they each receive

14 Rates

In this block, I will learn...

how to solve problems involving speed, distance and time

3 hours at 40 mph = 120 miles

80 metres in 5 seconds = 16 metres per second

600 km at 150 km/h = 4 hours

how to interpret and draw distance–time graphs

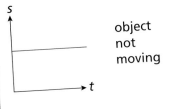

object not moving

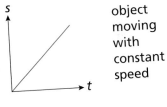

object moving with constant speed

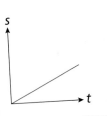

object moving with acceleration

how to solve problems involving density, mass and volume

Density

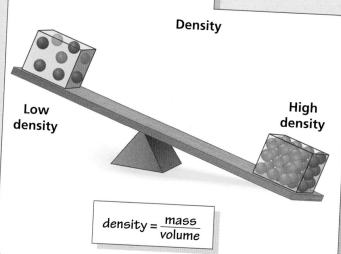

Low density

High density

$$density = \frac{mass}{volume}$$

how to convert compound units

12 metres per second = 720 metres per minute = 43 200 metres per hour = 43.2 km/h

to understand rates and their units

| miles per gallon | litres per 100 km |

| words per minute | grams per m^2 |

how to solve flow problems

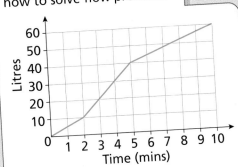

Small steps

- Solve speed, distance and time problems without a calculator
- Solve speed, distance and time problems with a calculator
- Use distance–time graphs

Key words

Rate – a comparison between two quantities

Speed – the rate at which an object is moving

Gradient – the steepness of a line

Constant – not changing

Are you ready?

1 Solve these equations.

   a $3x = 150$ **b** $200 = \frac{m}{12}$ **c** $60 = \frac{300}{t}$

2 Rearrange these formulae to make d the subject.

   a $a = 6d$ **b** $b = \frac{d}{4}$ **c** $c = \frac{10}{d}$

3 How many minutes are there in

   a   1 hour   **b**   4 hours   **c**   one-third of an hour?

4 Round each of these numbers to 3 significant figures.

   a   12.8769   **b**   1.28373   **c**   2.35678   **d**   35.411178

Models and representations

Distance–time graphs

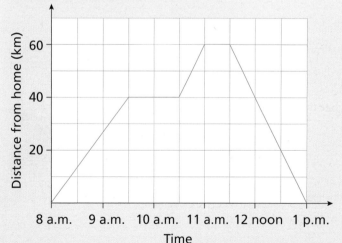

A distance–time graph shows the distance of an object from a given point at different times.

In this block you will explore **rates**. Rates are used to compare quantities and are often expressed in the form "something per something", such as "cost per litre" or "dollars per pound".

Speed is the rate at which an object is moving.

You can work out speed using the formula

speed = $\dfrac{\text{distance}}{\text{time}}$ or $s = \dfrac{d}{t}$

Common units of speed are miles per hour (mph) and kilometres per hour (km/h).

> Speed compares a distance with a time, so that "30 miles per hour" means that a distance of 30 miles is covered every hour.

Example 1

A coach travels 200 miles in 4 hours.

a Work out the average speed of the coach in miles per hour (mph).

b If the coach continues at this speed, how far will it travel in 10 hours?

c If the coach continues at this speed, how long will it take to travel 120 miles? Give your answer in hours and minutes.

a $s = \dfrac{d}{t} = \dfrac{200}{4} = 50 \, \text{mph}$

Substitute the given numbers into the formula for speed.

b $10 \times 50 = 500 \, \text{miles}$

50 mph means "50 miles in every hour", so in 10 hours the coach travels $10 \times 50 = 500$ miles

You could also work this out by forming an equation from the formula $s = \dfrac{d}{t}$

$$s = \dfrac{d}{t}$$

$$50 = \dfrac{d}{10}$$

$$50 \times 10 = d$$

$$500 = d$$

This method is useful with difficult numbers, but you don't need to use the formula if the numbers are simple and you understand "... per...".

c $120 \div 50 = 2.4 \, \text{hours}$

$= 2 \text{ hours and } 24 \text{ mins}$

The coach is travelling at 50 mph (that is, 50 miles in each hour), so you need to work out how many times 50 goes into 120

0.4 hours = 0.4 × 60 mins = 24 mins

You could also work this out by forming an equation from the formula $s = \dfrac{d}{t}$

$$s = \dfrac{d}{t}$$

$$50 = \dfrac{120}{t}$$

$$50t = 120$$

$$t = \dfrac{120}{50}$$

$$= 2.4 \, \text{hours}$$

Practice 14.1A

1 A car is travelling at a speed of 30 metres per second. How far will it travel in

a 2 seconds **b** 10 seconds

c 1 minute **d** half a second?

2 A train is travelling at an average speed of 120 miles per hour. How far will it travel in

a 3 hours **b** 2 hours **c** half an hour

d a quarter of an hour **e** 1.5 hours?

3 A boat is travelling at constant speed. The double number line shows the distance travelled by a boat at different times.

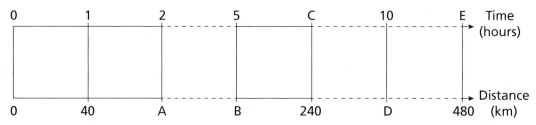

a Find the value of each letter (A to E).

b What is the speed of the boat?

4 A car travels 200 miles in 4 hours.

a Work out the speed of the car.

b Is it likely that the car will have travelled at a constant speed? Why or why not?

5 A cyclist rides at an average speed of 12 km/h.

a What does "km/h" mean?

b How long does it take for the cyclist to travel 60 km?

6 A runner runs 5000 m in 20 minutes.

a How far does the runner run in 1 minute?

b What is the runner's speed in metres per minute?

7 Work out the speed in metres per second of each of these athletes. Give your answers to 3 significant figures.

Name	Year	Distance	Time	Speed
Stanisława Walasiewicz	1937	200 m	23.6 s	**a**
Wyomia Tyus	1968	100 m	11.08 s	**b**
Marie-José Pérec	1996	400 m	48.25 s	**c**
Elaine Thompson	2016	60 m	7.06 s	**d**

8 Work out how long it would take to run 1000 m at a speed of

a 10 metres per second **b** 8 metres per second **c** 7.5 metres per second

9 A boat travels 65 miles in 1 hour 30 mins. Zach works out the average speed of the boat:

$$65 \div 1.3 = 50 \text{ mph}$$

  a Zach has made a mistake. Explain what he has done wrong.

  b Work out the correct speed of the boat.

  c Work out the average speed of the boat if the same journey had taken

   i 1 hour and 15 minutes   **ii** 1 hour and 3 minutes

What do you think?

1 Ed and Chloe are working out the speed, in miles per hour, of a train that travels 45 miles in 20 minutes.

I'm going to divide 45 by 20 and then multiply by 60

Ed

I'm going to multiply 45 by 3

Chloe

  a Discuss their methods. Are they both correct?

  b Work out the speeds of these journeys, giving your answers in miles per hour.

	Distance	Time	Speed
i	30 miles	15 minutes	
ii	25 miles	10 minutes	
iii	20 miles	12 minutes	
iv	8 miles	5 minutes	

2 Which is faster: 100 mph or 100 km/h? How do you know?

3 A train leaves a station at 11:47 and arrives at its destination, 212 km away, at 13:24. Find the average speed of the train, correct to 3 significant figures.

4 Investigate the speeds of trains between stations in your local area.

Many journeys are broken up into parts, and these can be represented on distance–time graphs.

The magnitude of the **gradient** of each section of the distance–time graph gives the speed for that section of the journey.

The average speed for a whole journey can be found using the formula

$$\text{average speed} = \frac{\text{total distance travelled}}{\text{total time taken}}$$

Example 2

Marta drove from Leeds to Liverpool to deliver a parcel, and then she returned home. The graph shows her journey.

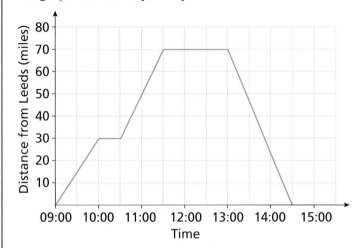

a Marta stopped at a service station on the journey from Leeds to Liverpool. For how long did she stop?

b How far is the service station from

  i Leeds   **ii** Liverpool?

c What was Marta's speed on the return journey from Liverpool to Leeds?

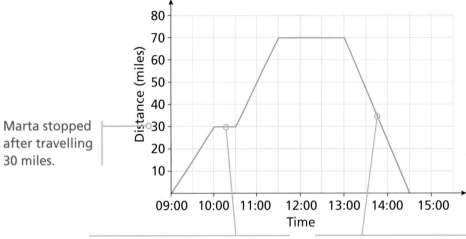

Marta stopped after travelling 30 miles.

The graph is horizontal between 10:00 and 10:30, which means that the car was not moving.

The distance travelled was 70 miles and took 1.5 hours (from 13:00 to 14:30): use the formula speed = $\dfrac{\text{distance}}{\text{time}}$

a 30 minutes

b i 30 miles

  ii 40 miles ○──┤ The furthest point from Leeds is 70 miles, so the distance from the service station is 70 − 30 = 40 miles

c 70 ÷ 1.5 = 47 mph

Practice 14.1B

1 Faith travelled 240 miles to visit her father. The graph shows her journey.

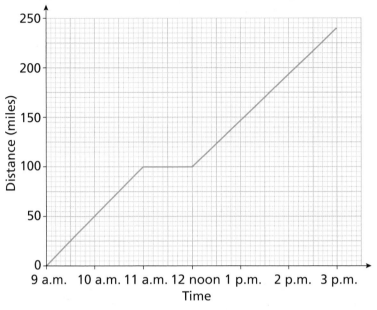

a What was Faith's speed between 9 a.m. and 11 a.m.?

b Faith made one stop on the journey. For how long did she stop?

c What was Faith's speed between 12 noon and 3 p.m.?

d How long did Faith spend driving altogether?

e Find Faith's average speed between 9 a.m. and 3 p.m.

f Find Faith's average speed for the time she was actually driving.

2 Flo and Chloe cycled 25 km. The graph shows their journeys.

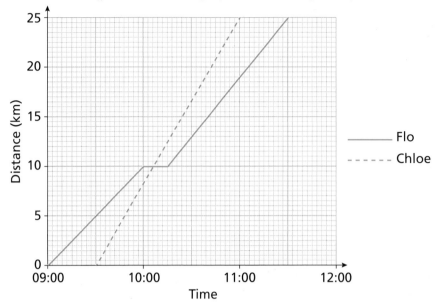

a Explain how the graph shows that Chloe cycled at a constant speed for the whole journey.

b Work out Chloe's speed, giving your answer to 3 significant figures.

c Flo stopped to fix a puncture. For how long did she stop?

d Work out Flo's speed for the remainder of the journey after she fixed the puncture.

e Work out Flo's average speed for the whole journey.

3 Seb went on a day trip to the beach. Here is a description of his journey:

- He left home at 9 a.m.
- He arrived at the beach, 60 km away, at 10:30 a.m.
- He stayed at the beach for 3 hours.
- He drove straight home, arriving back at 3:30 p.m.

a Represent this information on a distance–time graph.

b Work out

   i the total distance Seb drove

   ii the total time he spent driving

   iii his average speed when driving.

4 A cyclist travels on horizontal ground for 3 hours at a speed of 22 km/h and then travels uphill for 1 hour at 12 km/h.

a How far has she cycled altogether?

b Find her average speed over the four hours.

5 When Usain Bolt broke the world record for running 100 m in 2009, he ran the first 40 m in 4.64 seconds and completed the whole race in a total time of 9.58 seconds.

Find his average speed

a for the first 40 m of the race

b for the remainder of the race

c for the whole race.

6 A jogger runs for 1 hour at a speed of 6 mph and then slows down to 5 mph for another 2 hours. Find his average speed for the entire run.

7 A ship makes a 240-mile voyage. It travels at 40 mph for the first 3 hours. How fast must the ship travel for the remainder of the journey to achieve an average speed of 48 mph?

8 Abdullah, Emily and Jackson set off for a 24-mile walk at the same time. Abdullah walked at an average speed of 3 mph.

a Emily completed the route 2 hours quicker than Abdullah. How fast did Emily travel?

b Jackson completed the walk 2 hours slower than Abdullah. How fast did Jackson travel?

c Sketch a graph showing the journeys of Abdullah, Emily and Jackson on the 24-mile challenge.

What do you think? 💭

1 This distance–time graph of a car journey has no numbers.

a Between which two points was the car travelling fastest?

b Describe the journey in as much detail as you can.

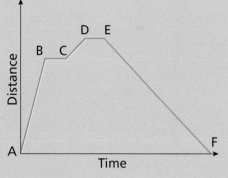

2 This distance–time graph shows the first part of a journey.

a Describe how the speed changes over the first part of the journey.

b What might the scales on the graph be? What type of transport could be being used? Discuss possibilities with a partner.

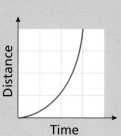

Consolidate – do you need more?

1 A plane is travelling at an average speed of 480 miles per hour. How far will the plane travel in

a 2 hours   **b** 2.5 hours   **c** 3.25 hours   **d** $\frac{1}{3}$ hour?

2 The West Coast Mainline from London to Glasgow is 400 miles long. A train travels at an average speed of 80 mph. How long will it take this train to travel from London to Glasgow?

3 A van travels at an average speed of 36 km/h

  a Find the distance travelled by the van in

    i $\frac{1}{2}$ hour   **ii** $\frac{1}{4}$ hour   **iii** $\frac{1}{3}$ hour   **iv** $2\frac{1}{2}$ hours.

  b Find the time taken for the van to travel

    i 108 km   **ii** 27 km

4 Beca walks at a speed of 0.7 metres per second.

  a How far can she walk in a minute?

  b How far can she walk in one hour?

  c How long will it take her to walk 2.8 km?

5 The graph shows Emily's journey as she walked to the gym, 2.5 km from her home.

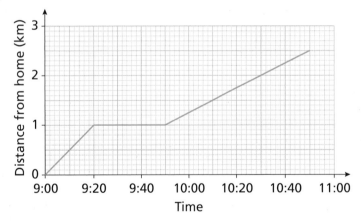

  a After walking 1 km from her house, Emily stopped to pick up her friend Ed, but he wasn't ready. How long did she have to wait for Ed?

  b How long did it take Emily and Ed to walk the rest of the way to the gym?

  c What was their average speed for this part of the journey?

  d Did Emily walk faster when she was on her own or when she was with Ed? Explain your answer.

6 An athlete runs 200 m in 20 seconds and then another 200 m in a further 25 seconds. Find, in metres per second, his average speed over

   a the first 200 m   **b** the last 200 m   **c** the full 400 m

Stretch – can you deepen your learning?

1 The speed of light is about 186 000 miles per second. A light year is the distance that light travels in a year.

   a Find the length, in miles, of a light year. Give your answer in standard form.

   b The speed of light can also be given as approximately 3×10^8 metres per second. Use the information in this question to work out many metres are equivalent to 1 mile.

2 These distance–time graphs represent parts of four car journeys.

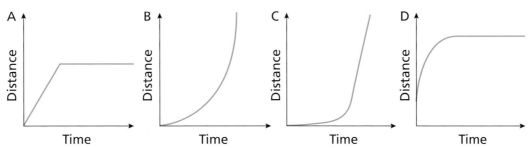

   a Which graph does the following statement represent?

   The car accelerated for a while and then travelled at a constant speed.

   b Write statements to describe the motion represented by each of the other three graphs.

3 Benji travels for a hours at x mph, followed by b hours at y mph.

   a Explain why Benji's average speed is not necessarily $\frac{x+y}{2}$

   b Write a correct expression for Benji's average speed.

   c If Benji's average speed is $\frac{x+y}{2}$, what can you conclude about a and b?

4 Investigate which pairs of the variables d (distance), s (speed) and t (time) can be directly proportional and which can be inversely proportional.

Reflect

1 In how many ways can you express the connection between speed, distance and time?

2 Explain why the gradient of each part of a distance–time graph gives the speed of that part of the journey.

Small steps

- Solve problems with density, mass and volume
- Convert compound units **H**

Are you ready?

1 How many grams are there in a kilogram?

2 Find the volume of a cuboid with dimensions 10 cm by 8 cm by 5 cm

3 Solve these equations.

  a $8p = 1000$ **b** $60 = \dfrac{h}{7}$ **c** $12 = \dfrac{60}{k}$

4 Which of these are measures of speed?

  A miles per minute   B hours per kilometre

  C metres per second   D cm per km

Models and representations

Objects

You can compare the densities of real objects by seeing whether they float or sink in water.

Cork   Wood   Aluminium

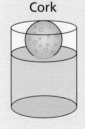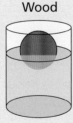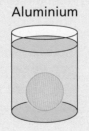

Double number line

The relationship between **mass** and **volume** can be shown on a double number line.

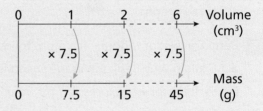

1 cm³ of gold has mass 19.3 g, but 1 cm³ of silver has mass 10.49 g. Gold has a higher **density** than silver because it has more mass per cm³.

You can work out the density of an object by dividing its mass by its volume. This can be written as a formula:

$$\text{density} = \frac{\text{mass}}{\text{volume}} \quad \text{or} \quad D = \frac{M}{V}$$

Example 1

450 cm³ of brass has mass 3834 g. 350 cm³ of wrought iron has mass 2765 g. Which metal has the greater density? Justify your answer.

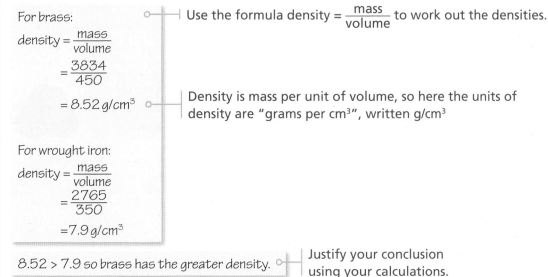

For brass:

$$\text{density} = \frac{\text{mass}}{\text{volume}}$$

$$= \frac{3834}{450}$$

$$= 8.52 \, g/cm^3$$

Use the formula density $= \frac{\text{mass}}{\text{volume}}$ to work out the densities.

Density is mass per unit of volume, so here the units of density are "grams per cm³", written g/cm³

For wrought iron:

$$\text{density} = \frac{\text{mass}}{\text{volume}}$$

$$= \frac{2765}{350}$$

$$= 7.9 \, g/cm^3$$

8.52 > 7.9 so brass has the greater density.

Justify your conclusion using your calculations.

Example 2

The density of a liquid is 1.5 g per cm³

a Work out the mass of $\frac{1}{4}$ litre of the liquid

b What volume of the liquid has mass 600 g?

a $1000 \, cm^3 \div 4 = 250 \, cm^3$

$$\text{density} = \frac{\text{mass}}{\text{volume}}$$

$$1.5 = \frac{\text{mass}}{250}$$

$$1.5 \times 250 = \text{mass}$$

$$\text{mass} = 375 \, g$$

Remember: 1 litre = 1000 cm³, so for $\frac{1}{4}$ litre you divide by 4

Substitute the given information into the formula for density.

Rearrange to find the mass.

b $\text{density} = \frac{\text{mass}}{\text{volume}}$

$$1.5 = \frac{600}{\text{volume}}$$

$$1.5 \times \text{volume} = 600$$

$$\text{volume} = \frac{600}{1.5}$$

$$\text{volume} = 400 \, cm^3$$

Substitute the given information into the formula for density.

Rearrange to find the volume. Notice that this takes two steps.

Practice 14.2A

1. A piece of lead with volume 150 cm³ has a mass of 1694 g. Work out the density of lead.

2. A can of drink holds 0.33 litres. The drink has a mass of 390 g. Find the density of the drink, in grams per millilitre.

3. The table shows some information about two types of sand.

Type	Volume	Mass
Quartz sand	230 cm³	1610 g
Silica sand	800 cm³	2080 g

The density of quartz sand is more than double the density of silica sand.

Is Faith correct? Show working to justify your answer.

4. The density of steel is 7.8 g/cm³. A steel bar has volume 450 cm³

Work out the mass of the steel bar.

5. A gold ring has mass 15 g. The density of gold is 19.3 g/cm³

   a What is the volume of the ring?

   b The density of silver is 10.4 g/cm³. How much less mass would the ring have if it was made of silver instead of gold?

6. The diagram shows a solid triangular prism. The prism is made from metal with a density of 6.2 g/cm³

   Work out the mass of the prism.

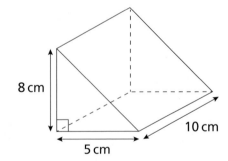

7. Lydia's bicycle frame is made from carbon fibre. It has a mass of 5.76 kg

   a The density of carbon fibre is 1.8 g/cm³. Work out the volume of Lydia's bicycle frame.

   b Emily's bicycle frame is identical to Lydia's, but is made of aluminium. Given that the density of aluminium is 2.7 g/cm³, work out the difference in mass between Emily's bicycle frame and Lydia's bicycle frame.

8 An object will float on a liquid if its density is less than the density of the liquid. The density of water is 0.997 g/cm³. Which of these objects will float on water?

Object	Mass	Volume
A	30 g	45 cm³
B	80 g	82 cm³
C	1.2 kg	250 cm³
D	0.06 kg	50 cm³

What do you think? 💭

1 Compare the three formulae.

$$\text{density} = \frac{\text{mass}}{\text{volume}}$$

$$\text{speed} = \frac{\text{distance}}{\text{time}}$$

$$\text{pressure} = \frac{\text{force}}{\text{area}}$$

What's the same and what's different about them? Which variables are easy to calculate from the formulae and which are more difficult? Why is this so?

2 Explain the difference between mass and weight. How does this affect calculations involving density?

3 200 ml of liquid A are mixed with 200 ml of liquid B. The resulting liquid is called liquid C. Assuming there is no chemical reaction between liquids A and B, are these statements always true, sometimes true or never true?

   a Liquid C will have a greater density than liquid A

   b Liquid B will have a greater density than liquid C

   c The density of liquid C will be between the density of liquid A and the density of liquid B

   d The density of liquid C will be the mean of the density of liquid A and the density of liquid B

Although density is usually measured in "g per cm³", it can also be measured in other units such as "kg per m³" or "pounds per cubic foot". In this next section, you will find out how to change from one compound unit to another.

Example 3

a Explain why kg per m³ is a valid unit for measuring density.

b The density of gold is 19.3 g/cm³, so it is also 19.3 kg/m³

Show that Zach is wrong. Calculate the correct density of gold in kg/m³

a kg is a unit of mass and m³ is a unit of volume,
so kg per m³ compares mass with volume.
It represents the mass of each unit of volume.

> Explain why the units make sense for the measurement of density.

Density also compares mass with volume,
so kg per m³ is a valid unit for measuring density.

b $1 m^3 = 1 m \times 1 m \times 1 m$

$= 100 cm \times 100 cm \times 100 cm$

$= 1\,000\,000\, cm^3$

> To convert g per cm³ to kg per m³, convert each unit, one at a time.
> 1 m = 100 cm
> so 1 m³ = 100 × 100 × 100 cm³

$19.3 g/cm^3 = 19.3 \times 1\,000\,000\, g/m^3$

$= 19\,300\,000\, g/m^3$

> The mass of 1 m³ will be 1 000 000 times greater than the mass of 1 cm³

Every 1000 g is 1 kg

So $19\,300\,000\, g/m^3 = \dfrac{19\,300\,000}{1000} kg/m^3$

$= 19\,300\, kg/m^3$

> Divide by 1000 to convert grams to kilograms.

So Zach is wrong. The correct answer is 19 300 kg/m³

> Remember to answer the question and state your conclusion clearly.

Practice 14.2B

1 A sprinter runs at a speed of 10 metres per second. Assume that they always run at the same rate.

a How many metres will the sprinter run in

i 1 minute   **ii** 1 hour?

b How many kilometres will the sprinter run in

i 1 minute   **ii** 1 hour?

c Find the sprinter's speed in km/h

2 A bus travels 8 km in 10 minutes. Work out the average speed of the bus in

    a kilometres per minute   **b** kilometres per hour

    c metres per hour   **d** metres per minute

    e metres per second.

3 Convert 12 m/s to km/h

4 Convert 120 km/h to m/s

5 Which is faster, 20 m/s or 70 km/h? Show working to justify your answer.

6 Density can be measured in troy per cubic inch. Find the missing values in the table.

Metal	Density (g/cm³)	Density (troy/in³)
Silver	10.5	5.5
Bronze	**a**	3.56
Gold	19.3	**b**

7 Convert 900 kg/m³ to g/cm³

8 1 mile = 1760 yards and 1 yard = 3 feet. Convert 24 mph to feet per second.

What do you think? 💭

1 5 miles ≈ 8 km

Abdullah

To convert a speed in mph to km/h you multiply by $\frac{5}{8}$

To convert a speed in mph to km/h you multiply by $\frac{8}{5}$

Beca

    a Who is correct? How do you know?

    b Which speed is faster, 30 mph or 50 km/h? Justify your answer.

    c The maximum speed limit in Australia is 130 km/h. How much faster or slower is this than the UK speed limit of 70 mph? Give your answers in km/h and in mph.

2 You can convert a speed in m/s to a speed in km/h by multiplying by a single number. Find this number.

3 1 mile ≈ 1.6 km. Convert

    a 40 mph to m/s   **b** 10 m/s to mph

Consolidate – do you need more?

1 $85\,cm^3$ of copper has mass $748\,g$

   a Find the density of copper.

   b Find the mass of $150\,cm^3$ of copper.

2 A liquid has density $1.15\,g/ml$. Calculate the mass of

   a $330\,ml$ of the liquid

   b $\frac{1}{2}$ litre of the liquid.

3 Find the missing values in this table. Give units with your answers.

Object	Density	Mass	Volume
A		60 g	48 cm³
B		150 g	80 ml
C	1.4 g/cm³		250 cm³
D	3.5 g/ml		70 ml
E	1.25 g/cm³	80 g	
F	2.7 g/ml	351 g	

4 The edges of a cube are $2\,cm$ long. The mass of the cube is $60\,g$. Find the density of the cube.

5 The density of petrol is $0.8\,g/cm^3$. The petrol in a can has mass $1600\,g$. Calculate the volume of petrol in the can. Give your answer in both cm^3 and litres.

6 A plank of wood measures $2\,cm$ by $4\,cm$ by $1\,m$ and has a mass of $680\,g$. Work out

   a the volume of the plank

   b the density of the wood.

7 Convert a speed of $15\,m/s$ to km/h.

8 Pressure is usually measured in newtons per m^2 (N/m^2). Convert

   a $12\,N/m^2$ to N/cm^2

   b $80\,000\,N/m^2$ to N/mm^2

   c $0.04\,N/cm^2$ to N/m^2

Stretch – can you deepen your learning?

1   Liquid A has density 1.2 g/cm³ and liquid B has density 1.1 g/cm³

    240 cm³ of liquid A and 220 cm³ of liquid B are mixed to make liquid C

    a   Find the density of liquid C

    240 g of liquid A and 220 g of liquid B are mixed to make liquid D

    b   Find the density of liquid D

2   Metal X has a density of 4.8 g/cm³. 300 cm³ of metal X and some of metal Y are melted together to produce an alloy Z. The alloy Z has a mass of 3.24 kg and a density of 6 g/cm³

    Find the density of metal Y

3   The diagram shows the uniform cross-section of a metal beam. The length of the beam is 4 m and its mass is 765 kg. Calculate the density of the metal, giving your answer in g/cm³

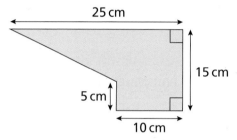

25 cm

15 cm

5 cm

10 cm

Reflect

Describe how to work out each of density, mass and volume given the other two measurements.

14.3 Rates and graphs

Small steps

- Solve flow problems and their graphs
- Calculate with rates of change and their units

Key words

Rate – a comparison between two quantities

Gradient – the steepness of a line

Are you ready?

1 £1 can buy 1.40 US dollars.

  a How many US dollars can be bought with £70?

  b How many pounds can be bought with US$70?

2 A drone flies at 8 metres per second.

  a How far does the drone fly in one minute?

  b How long does it take for the drone to fly 1 km?

3 A straight line goes through the points (0, 0) and (12, 60). Work out the gradient of the line.

Models and representations

Double number lines

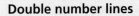
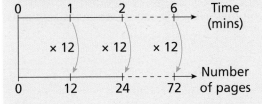

Graphs

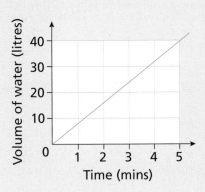

Rates compare related quantities.

Rates are often expressed in the form "one unit per another unit" such as "miles per hour" or "cost per kilogram".

You have already met some examples of rates:

- speed is the rate of change of distance over time (Chapter 14.1 of this book)
- exchange rates for currencies (Book 2 and Chapter 8.4 of this book).

Example 1

Water from a kitchen tap can fill a 6-litre bowl in $1\frac{1}{2}$ minutes.

Water from a bathroom tap can fill an 80-litre bath in 10 minutes.

From which tap does water flow at the faster rate?

Kitchen tap:

$\dfrac{6 \text{ litres}}{1.5 \text{ minutes}} = 4$ litres/min

Bathroom tap:

$\dfrac{80 \text{ litres}}{10 \text{ minutes}} = 8$ litres/min

Work out the rate of flow for each tap.

Divide the capacity of the container by the time taken to fill it.

$8 > 4$, so the bathroom tap is faster.

Compare the rates and state your conclusion.

Example 2

A photocopier makes 25 copies per minute.

a How many copies can be produced in 1 hour?

b How long will the photocopier take to produce 10 000 copies?

a $25 \times 60 = 1500$

Multiply the rate by 60, as 60 is the number of minutes in an hour.

b $10\,000 \div 25 = 400$ minutes

If the photocopier produces 25 copies each minute, you need to work out how many 25s there are in 10 000. This gives the time in minutes.

$= 6$ hours and 40 minutes

It makes sense to convert your answer to hours and minutes.

Practice 14.3A

1 Benji types at an average rate of 50 words per minute.

   a How many words can Benji type in an hour?

   b How long will it take Benji to type a 10 000-word article? Give your answer in hours and minutes.

   c Why does the question say "an average rate of 50 words per minute"?
   Do you think Benji will type exactly 50 words every minute? Why or why not?

2 The volume of water in a reservoir falls from 430 000 litres to 180 000 litres over 5 days.

   a Calculate the rate at which the reservoir is emptying, giving your answer in litres per day.

   b Assume that the reservoir continues to empty at the same rate. How many more days will it take for the reservoir to empty completely?

3 A call centre receives 2400 calls in an hour.

    a What is the rate of calls per minute?

    b At the same rate, how many calls can the call centre expect to receive in a day?

    c Is it likely that the call centre will receive calls at a constant rate? Why or why not?

4 A car travels 805 miles on a full tank of petrol. The petrol tank holds 63 litres of petrol. Work out the average rate at which the car uses petrol, giving your answer in miles per litre.

5 Emily pays £59.45 for 50 litres of petrol. At a different petrol station, Ed pays £47.84 for 40 litres of petrol.

    a Compare the costs of petrol, in pence per litre, at the two petrol stations.

    b Find the difference between the cost of 60 litres of petrol at the two petrol stations.

6 A person's heart beats at an average rate of 70 beats per minute.

    a How many times will a heart beat in a lifetime of 80 years? Give your answer in standard form.

    b How long will it take for a heart to beat 1 million times? Give your answer in days, to 1 decimal place.

7 A water tank has a capacity of 500 litres.

    a How long will it take to fill the tank at the rate of 40 litres per minute?

If you double the volume of the tank, it will take twice as long to fill.

Seb

If you double the rate, it will take half as long to fill the tank.

Flo

    b Are Seb and Flo correct? Show working to justify your answers.

8 Each cleaner at a hotel cleans 12 rooms in a 8-hour shift.

    a Find the rate at which the cleaners work, giving your answer in minutes per room.

    b The hotel has 90 rooms. How many cleaners are needed to clean all the rooms in 8 hours?

What do you think?

1 Beca takes 3 hours to paint a room. Jackson takes 5 hours to paint a room of the same size.

> The average of 3 and 5 is 4, so it would take Beca and Jackson 4 hours if they work together to paint the room.

a Explain why Faith is wrong.

b Work out how long it would take to paint the room if Beca and Jackson work together.

2 Emily measures the flow of water out of a pipe. She measures 450 litres in 2 minutes. She records the number of litres to the nearest 10 litres and the time to the nearest second.

Calculate the minimum and maximum rates at which the water could be flowing out of the pipe. Give your answers in litres per second, correct to 3 significant figures.

3 Abdullah can type x words in a minute.

a How many words can Abdullah type in

i an hour   **ii** y hours?

b How long will it take Abdullah to type a 3000-word essay? Give your answer

i in minutes   **ii** in hours.

You can explore rates using graphs.

This graph shows the number of pages printed in 6 minutes. The printer works at a constant rate.

The number of pages printed is directly proportional to the time.

> The graph is a straight line through the origin. You looked at graphs that show direct proportion in Chapter 13.2

The **gradient** of the graph is the rate at which the printer is printing.

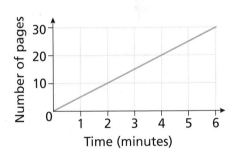

$$\text{gradient} = \frac{30 \text{ pages}}{6 \text{ minutes}}$$

$$= 5 \text{ pages/min}$$

Example 3

Water flows into a tank at a constant rate. The graph shows how the volume of water in the tank changes with time.

a Find the rate of flow in litres/min.

b The capacity of the tank is 2000 litres. How long will it take to completely fill the tank at the rate you found in part **a**?

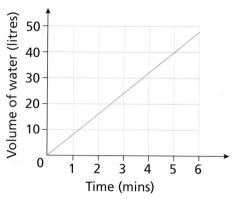

a

The rate of flow is the gradient of the line.

Use (0, 0) and a point whose coordinates you can easily read to find the rate.

The point (5, 40) is chosen because both of the coordinates are on grid lines.

$$\text{rate of flow} = \frac{40 \text{ litres}}{5 \text{ minutes}}$$

$$= 8 \text{ litres/min}$$

Divide the number of litres by the time taken.

b 2000 ÷ 8 = 250 minutes

= 4 hours 10 mins

A rate of 8 litres per minute means that 8 litres flow into the tank each minute.

Work out how many lots of 8 litres there are in 2000 litres.

Practice 14.3B

1 Water flows into a tank at a constant rate. The graph shows how the volume of water in the tank changes with time.

a Find the rate of flow in litres/min.

b How much water will flow into the tank in 1 hour?

c The capacity of the tank is 1800 litres. How long will it take to completely fill the tank at the rate you found in part **a**?

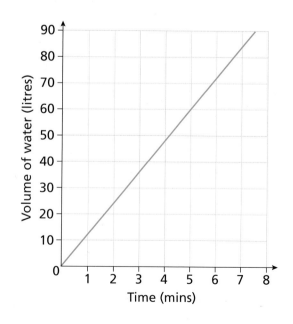

2 A fire hydrant delivers water at a rate of 8 litres per second.

a Copy and complete the table.

Time (s)	0	1	2	3	4	5
Volume of water (litres)	0	8				

b Draw a graph of volume of water delivered against time.

c Find the gradient of your graph. How does this connect with the rate of water delivery?

d Work out the volume of water delivered by the fire hydrant in 5 minutes.

e A second fire hydrant delivers water at a rate of 10 litres per second. Describe how the graph of volume of water delivered against time for this fire hydrant will be different from that of the first fire hydrant.

3 Here are four graphs showing water flow.

a

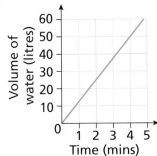

b

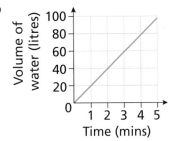

c

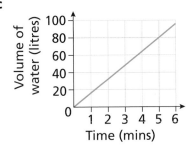

d

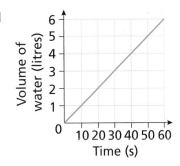

For each graph, work out

i the rate of flow in litres per minute

ii the volume of water delivered in 30 minutes

iii the time it takes to deliver 1000 litres.

4 The graph shows how the mass of a piece of titanium is related to its volume.

a Calculate the gradient of the graph. State the units with your answer.

b What does your answer to part **a** tell you?

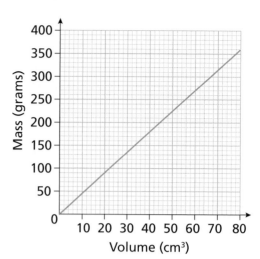

5 Here are some straight-line graphs. Explain what the gradient of each line represents.

a

Cost (£) vs Area of land (m²)

b

Distance travelled (km) vs Petrol used (litres)

c

Distance travelled (m) vs Time (seconds)

6 This graph shows a tank being filled at different rates over a 10-minute period.

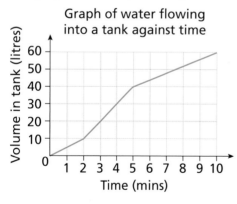

Graph of water flowing into a tank against time

Rates are not always constant and can change over time. One way to explore how rates change is to look at graphs.

a How can you tell when the rate of flow was greatest without doing any calculations?

b Work out the rate at which the tank was being filled

   i in the first 2 minutes

   ii between 2 minutes and 5 minutes

   iii between 5 minutes and 10 minutes.

c Work out the average rate at which the tank was filled over the 10-minute period.

7 The graph shows how the temperature of a cup of tea falls over time.

How do you know that tea does not cool at a constant rate?

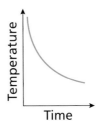

What do you think? 💡

1 Jakub is folding pizza boxes. For the first 10 minutes, he folds the boxes at a rate of 5 per minute. For the next 20 minutes he folds the boxes at a rate of 4 boxes a minute.

   a Draw a graph to represent this information.

   b

 My average rate was $\frac{4+5}{2}$ = 4.5 boxes/min

   Explain why Jakub is wrong. Find his correct average rate.

2 Water is poured into tanks A, B, C and D at a constant rate.

   A   B   C   D

   a This graph shows how the depth of water in a tank changes with time. Which of the tanks could the graph represent? Which could it not represent? Explain your answer.

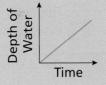

   b Sketch a graph to show how the depth of water changes with time for the other tanks.

   c Investigate the graphs of depth of water against time for other shapes of container.

Consolidate – do you need more?

1 A mechanic can change a tyre in 12 minutes.

   a How many tyres can the mechanic change in 2 hours?

   b How long will it take the mechanic to change 15 tyres?

2 Paper is sometimes classified by its mass per area. For example, tracing paper has a mass of 40 grams per square metre (g/m²).

   a Find the mass of 80 m² of tracing paper.

   b What area of tracing paper has mass 1 kg?

3 A machine fills 90 bottles in 20 minutes.

   a Work out the rate, in bottles per minutes, at which the machine works.

   b How long will the machine take to fill 144 bottles?

   c How many bottles will the machine fill in 24 hours?

4 A 5-litre tin of paint covers a wall 8.25 m long and 4 m high.

   a How many litres of paint are needed to cover an area of 330 m²?

   b How many tins of paint are needed to cover an area of 300 m²?

5 An organisation charges people 6p per minute to call its helpline.

   a Copy and complete the table.

Length of call (mins)	0	1	2	3	4	5
Cost (pence)	0	6				

   b Draw a graph of the cost of the call against time.

   c Find the gradient of your graph. What does it tell you?

   d Chloe has £3 credit on her phone. What is the longest call she can make to the helpline?

Stretch – can you deepen your learning?

1 £1 is equivalent to US$1.40. £1 is also equivalent to €1.15

   a How many euros are equivalent to US$10?

   b How many dollars are equivalent to €x?

2 A cylindrical tank has height 6 m and diameter 5 m. The tank is empty. Water is pumped into the tank at a rate of 500 litres per minute. How long will it take to completely fill the tank? Give your answer to the nearest minute.

3 A water tank has capacity 800 litres. It takes 40 minutes to fill the tank using tap A. It takes 10% less time using taps A and B together. How long would it take to fill the tank using tap B only?

4 The diagram shows a swimming pool. The pool is empty.

   a Water flows into the pool at a rate of 600 litres per minute. How long, in hours, will it take to fill the pool?

   b Sketch a graph that shows how the depth of water at the deep end changes over time when it is filled from being empty.

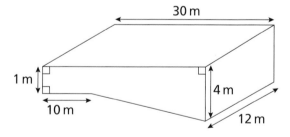

Reflect

1 How is calculating with rates similar to calculating with speed, distance and time?

2 What does the shape of a graph tell you about a rate? Draw some example graphs and annotate them to explain your answer.

I have become **fluent** in...	I have developed my **reasoning** skills by...	I have been **problem-solving** through...
▪ performing calculations involving speed, distance and time ▪ performing calculations involving density, mass and volume ▪ working out rates in different contexts ▪ converting compound units. Ⓗ	▪ interpreting distance–time graphs ▪ applying ratio in new contexts ▪ identifying variables and expressing relationships between variables algebraically.	▪ using graphs to solve problems ▪ representing information in different forms ▪ solving flow problems ▪ solving multi-step problems ▪ modelling situations mathematically.

Check my understanding

1 **a** A plane flies 480 km at an average speed of 320 km/h. How long does the journey take?

    b A train travels 60 miles in 20 minutes. Find the average speed of the train in miles per hour.

2 Which has a greater mass, 25 cm³ of brass (density 8.2 g/cm³) or 22 cm³ of nickel (density 8.9 g/cm³)? Justify your answer.

3 The density of aluminium is 2710 kg/m³. Express this density in g/cm³ Ⓗ

4 A machine packs 840 boxes in an hour.

    a Find the rate at which the machine works, giving your answer in boxes packed per minute.

    b How long will the machine take to pack 7000 boxes?

5 Water is poured at a constant rate into these two containers.

Sketch graphs to show how the depth of water in each container changes with time.

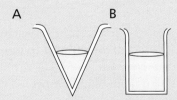

6 A cylinder with base diameter 80 cm and height 2 m is filled from empty at a rate of 12 litres per minute. How long will it take to fill the cylinder completely? Give your answer to the nearest minute.

15 Probability

In this block, I will learn...

how to find the probability of something happening

$P(\text{orange}) = \frac{1}{2}$

$P(\text{blue}) = \frac{1}{6}$

$P(\text{green}) = \frac{1}{3}$

$P(\text{orange or blue}) = \frac{2}{3}$

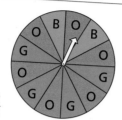

that relative frequencies add up to 1

Outcome	G	R	A	P	H
Frequency	8	12	7	14	9
Relative frequency	0.16	0.24	0.14	0.28	0.18

how to use sample space diagrams

		Notebook			
		A	B	C	D
Pen	1	A1	B1	C1	D1
	2	A2	B2	C2	D2
	3	A3	B3	C3	D3

how to find probabilities from tables and other diagrams

	Year 7	Year 8	Year 9	Total
Dogs	78	65	67	210
Cats	72	75	43	190
Total	150	140	110	400

$P(\text{dog}) = \frac{210}{400}$

how to use tree diagrams to work out probabilities **H**

1st counter	2nd counter	Outcome

$\frac{12}{20}$ green

$\frac{12}{20}$ green — GG

$\frac{8}{20}$ red — GR

$\frac{8}{20}$ red

$\frac{12}{20}$ green — RG

$\frac{8}{20}$ red — RR

Small steps

- Calculate single event probabilities Ⓡ
- Understand relative frequency
- Calculate expected outcomes

Key words

Probability – how likely an event is to occur

Relative frequency – the number of times an event occurs divided by the number of trials

Expected outcomes – an estimate of how many times a possible outcome will occur

Fair – equally likely to land on any possible outcome

Are you ready?

1 A fair 6-sided dice is rolled. What are the possible outcomes?

2 A fair coin is flipped. What are the possible outcomes?

3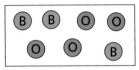

What fraction of the counters are

  a blue   **b** orange   **c** pink?

4 A box contains 10 red pens, 15 blue pens and 7 black pens. What fraction of the pens are

  a blue   **b** blue or black   **c** red, blue or black?

Models and representations

Probability scale

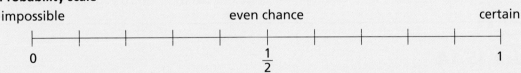

impossible   even chance   certain

0                         $\frac{1}{2}$ 1

Dice, spinners, counters and coins

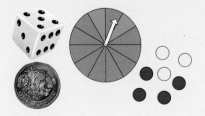

Probability notation

P(A) means "the **probability** of A"

In this section, you will calculate the probability of single events happening.

Probabilities can be given as fractions, decimals or percentages.

Probabilities are always between 0 and 1: 60% is a possible probability, but 60 is not.

Example 1

Here are two fair spinners.

Spinner A

Spinner B

a Explain why it is important that the spinners are fair.

b Spinner A is spun once. What is the probability that it lands on orange?

c Spinner B is spun once. Find P(orange) on this spinner.

a It is important that the spinners are fair so that each outcome is equally likely.

Each spinner has two possible outcomes: orange and blue. If the spinner was not **fair**, you couldn't calculate probabilities because you couldn't assume that the spinner is equally likely to land on each sector.

b $\frac{1}{2}$ or 0.5 or 50%

Spinner A is split into two equal sections, one of which is orange. The probability of it landing on orange is 1 out of 2, which you can write as

$\frac{1}{2}$, 0.5 or 50%

c $\frac{5}{7}$

P(orange) means "the probability that the spinner lands on orange". Spinner B is split into seven equal sections, five of which are orange.

The probability of it landing on orange is 5 out of 7, which you can write as $\frac{5}{7}$

Practice 15.1A

1 A fair coin is flipped.
  a Explain why the probability of the coin landing on heads is $\frac{1}{2}$
  b Write the probability of the coin landing on tails.
  c What do you notice about the sum of the probabilities? Why does this happen?

2 A fair 6-sided dice is rolled.
  Write the probability that the dice lands on

  a 1   **b** 5   **c** an even number   **d** 7

  e a number less than 7

  f Talk about your answers to parts **d** and **e** with a partner. What do you notice?

3 This fair spinner is spun. Find these probabilities.

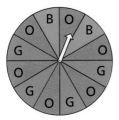

    a P(orange)   **b** P(blue)   **c** P(green)   **d** P(orange or blue)

4 Here are some counters.

A counter is selected at random. Calculate these probabilities.

    a P(green)   **b** P(purple)   **c** P(blue)

Why is it important that the counter is selected at random?

5 Here are some counters.

A counter is selected at random. Calculate these probabilities.

    a P(orange)   **b** P(pink)   **c** P(blue)

    d P(orange or pink)   **e** P(not blue)

What do you notice about your answers to parts **d** and **e**? Why does this happen?

6 A box contains 30 chocolate bars. Each bar is either white chocolate, milk chocolate or dark chocolate.

There are 17 white chocolate bars and 9 milk chocolate bars.
A chocolate bar is selected at random.

Work out the probability that this chocolate bar is dark chocolate.

7 There are 29 students in a class. A student is selected at random. The probability that this student is a girl is $\frac{15}{29}$. Write the probability that the student selected is a boy.

8 The probability of Chloe winning a game is $\frac{3}{5}$. What is the probability that Chloe does not win the game?

9 A bag contains some orange sweets and some green sweets. A sweet is selected at random. P(orange) = 0.3. Find P(green).

10 The probability that Beca scores a penalty is 68%. Write the probability that she does not score a penalty.

What do you think?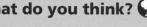

1 Here are two fair spinners.

Spinner A

Spinner B

The probability of each spinner landing on orange is 0.5 because they can either land on orange or on blue.

Do you agree with Abdullah? Explain your answer.

2 A counter is selected at random from each bag.

From which bag are you more likely to select a blue counter? Justify your answer.

Bag A   Bag B

3 Copy the spinner and add numbers to the spinner such that

▥ P(square number) = 0.5

▥ P(prime number) = 10%

▥ P(multiple of 3) = $\frac{2}{5}$

4 The ratio of apples to oranges in a fruit bowl is 4:1. An apple or an orange is selected at random. Find

a P(orange)   **b** P(apple)

Discuss your method with a partner.

In this section, you will learn about **relative frequency** and **expected outcomes**.

For example, if a coin is flipped ten times and it lands on tails three times, then the relative frequency of tails is 3 ÷ 10 which is $\frac{3}{10}$ or 0.3

For example, if a fair coin is flipped ten times you would expect it to land on tails $\frac{1}{2}$ × 10 = 5 times.

Example 2

A fair 6-sided dice is rolled 60 times.

a If the dice is fair, how many times would you expect it to land on 4?

Here are the results.

Outcome	1	2	3	4	5	6
Frequency	18	15	6	12	3	6

b What is the relative frequency of 4?

The same dice is rolled 300 times.

c Use your answer to part **b** to estimate the number of times the dice will land on 4

a $\frac{1}{6} \times 60 = 10$ — If the dice is fair, then the probability of it landing on each outcome is $\frac{1}{6}$

The dice is rolled 60 times; this means that there were 60 trials.

b $12 \div 60 = \frac{1}{5}$ or 0.2 — The relative frequency of 4 is the number of times the dice landed on 4 divided by the number of trials.

This could suggest that the dice is not fair because $0.2 \neq \frac{1}{6}$

c $0.2 \times 300 = 60$ — Multiply the relative frequency by the number of trials.

Practice 15.1B

1 A fair coin is flipped 500 times. How many times would you expect it to land on

a heads **b** tails?

2 A fair 6-sided dice is rolled 600 times. How many times would you expect it to land on

a 2 **b** 5 **c** an even number?

3 Each spinner is spun 1400 times.

Spinner A   Spinner B

Estimate how many times

a spinner A will land on blue

b spinner B will land on orange.

4. A biased spinner is divided into four equal-sized sections, with colours orange, blue, green and yellow. The table shows the probability of the spinner landing on each possible outcome.

Outcome	Probability	Expected frequency after 800 rolls
orange	$\frac{1}{5}$	
blue	$\frac{9}{25}$	
green	0.37	
yellow	7%	

Copy and complete the table.

5. A box contains black pens, blue pens and green pens. Abdullah selects a pen at random, notes its colour then puts it back in the box. He does this 20 times. The results are shown in the table.

Outcome	black	blue	green
Frequency	7	5	8

   a Explain why the relative frequency of selecting a blue pen is 0.25

   b What is the relative frequency of selecting a pen that is

   i black

   ii green?

   c What do you notice about the sum of your relative frequencies? Why does this happen?

6. These letter cards are placed in a bag.

| G | R | A | P | H |

   A card is selected at random. Its letter is noted and then it is returned to the bag.

   This is carried out 50 times. The results are shown in the table.

Outcome	G	R	A	P	H
Frequency	8	12	7	14	9
Relative frequency					

   a Copy and complete the table.

   b Compare the relative frequency of each outcome with the theoretical probability. Discuss your findings with a partner.

   c Jakub carries out this process 200 times. Estimate how many times he will select the letter H

What do you think? 🗨

1 A biased 8-sided dice is rolled 900 times. The relative frequency of it landing on 7 is 0.4. State how many times the dice does not land on 7

2 Benji flips a coin twice. It lands on heads both times. Benji says, "The coin must be biased. I predict that it will never land on tails."

Do you agree with Benji? Explain your answer.

3 Zach and Beca spin the same spinner to decide whether or not it is biased. Their results are shown in the tables.

Zach

Outcome	A	B	C	D
Frequency	18	13	19	10
Relative frequency				

Beca

Outcome	A	B	C	D
Frequency	49	53	47	51
Relative frequency				

a Copy and complete both tables.

b Whose results are more accurate? Explain why.

c Do you think the spinner is biased? Explain your answer.

d Jakub then spins the same spinner 1000 times and records his results. Will his data be more or less accurate than Zach's or Beca's? Explain your reasoning.

Consolidate – do you need more?

1 A counter is selected at random from this box.

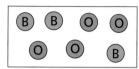

a What is the probability that the counter is

i blue

ii orange?

b Explain why the probability that the counter is green is 0

c Explain why P(blue) + P(orange) = 1

2 A fruit bowl contains 8 bananas, 2 apples, 3 pears and 7 oranges. A piece of fruit is selected at random. Calculate these probabilities, giving your answers as fractions, decimals and percentages.

a P(banana)

b P(apple)

c P(pear)

d P(orange)

e P(banana or apple)

f P(not orange)

g P(banana or apple or pear or orange)

3 A fair spinner is split into five equal sections labelled A, B, C, D and E. The spinner is spun 500 times.

a Estimate how many times the spinner will land on

i B

ii A or E

b The spinner lands on D 85 times. Find the relative frequency of D

c The relative frequency of C is 0.21. How many times did the spinner land on C?

Stretch – can you deepen your learning?

1 The probability of a spinner landing on purple is x. Write an expression for the same spinner not landing on purple.

2 There are p crayons in a tub. a of the crayons are red. Find P(red).

3 There are w red, x green and y orange sweets in a packet. A sweet is selected at random.

    a Write an expression for how many sweets there are in total.

    b Write an expression for each of these probabilities.

       i P(red)

       ii P(green)

       iii P(orange)

       iv P(red or green)

       v P(not orange)

       Could you record your answers to parts **iv** and **v** in a different way?

    c One green sweet is removed and eaten. Write a new expression for P(green).

4 Here is some information about a biased 6-sided dice.

    ■ P(1) = 2 × P(2)

    ■ The probability of the dice landing on 3 is half of the probability of it landing on 4

    ■ P(4) = P(1 or 2)

    ■ The probability of the dice landing on 5 is equal to the probability of it landing on 6

    ■ P(5 or 6) = 25%

    Calculate these probabilities.

    a P(1)   **b** P(2)   **c** P(3)   **d** P(4)   **e** P(5)   **f** P(6)

5 A biased coin is flipped k times and lands on tails t times. Write an expression for the relative frequency of it landing on tails.

6 The total number of trials in an experiment is given by the expression $10(140x + 3)$. The number of times the outcome is orange is $500x$. The relative frequency of orange is $7x - 1$

    Show that $980x^2 - 169x - 3 = 0$

Reflect

1 Why is it important to know whether or not a trial is fair?

2 What is the difference between probability and relative frequency?

Small steps

- Calculate with independent events
- Use diagrams to work out probabilities

Key words

Independent events – not affected by other events

Are you ready?

1 A fair coin is flipped. What is the probability that it lands on heads?

2 A fair 6-sided dice is rolled. What is the probability that it lands on 5?

3 Here are some number cards.

A card is selected at random. Find the probability that

a the card is blue

b the card has the number 6 on it

c the card has a number greater than 2 on it.

4 Complete the calculations.

a $\frac{1}{2} \times \frac{1}{6}$
   b $\frac{1}{2} \times \frac{1}{4}$
   c $\frac{1}{2} \times \frac{1}{2}$
   d $\frac{1}{4} \times \frac{5}{6}$

Models and representations

Two-way tables

A two-way table has two categories.

Here, the categories are whether you own a pet and the year group you are in.

	Owns a pet	Does not own a pet	Total
Year 7			
Year 8			
Total			

Sample space diagrams

A sample space diagram helps you to record all the possible outcomes from two events.

This diagram represents the possible outcomes when these two spinners are spun.

Spinner 1

Spinner 2

		Spinner 2				
		2	2	2	5	5
Spinner 1	1					
	3					
	3					
	4					

In this section, you will calculate probabilities of **independent events**.

Independent events are events that are not affected by other events. For example, if you roll a dice and flip a coin, the outcome of the dice roll has no effect on the outcome of the coin flip.

Example 1

Four notebooks are labelled A, B, C and D. Three pens are numbered 1, 2 and 3
A notebook and a pen are selected at random.

Find the probability that it is notebook B and pen 3 that are selected.

Method A

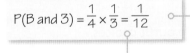

		Notebook			
		A	B	C	D
Pen	1	A1	B1	C1	D1
	2	A2	B2	C2	D2
	3	A3	B3	C3	D3

$\frac{1}{12}$

These two events are independent because the choice of notebook doesn't affect the choice of pen.

A sample space diagram is a way of listing these outcomes in a systematic and organised way.

There are 12 different possible outcomes, and one of these is B3, so the probability is $\frac{1}{12}$

Method B

$$P(B \text{ and } 3) = \frac{1}{4} \times \frac{1}{3} = \frac{1}{12}$$

These two events are independent because the choice of notebook doesn't affect the choice of pen.

To find the probability of selecting notebook B and pen 3 you can multiply the probabilities of the two separate outcomes.

Practice 15.2A

1 A fair coin is flipped and a fair dice is rolled.

  a What is the probability that the coin lands on heads?

  b What is the probability that the dice lands on 6?

  c Copy and complete this sample space diagram.

	1	2	3	4	5	6
H	(1, H)	(2, H)				
T						

  d Use your sample space diagram to find the probability that the coin lands on heads and the dice lands on 6

  e Show that P(H) × P(6) = P(H and 6)

  f Using your sample space diagram or otherwise, find these probabilities.

    i P(H and 3)

    ii P(T and 3)

    iii P(T and even)

    iv P(H and prime)

  Compare your methods with a partner's.

2 Beca has five different tops: plain, striped, spotty, sleeveless and long-sleeved. She has four different pairs of bottoms: jeans, shorts, leggings and joggers.

  a Draw a sample space diagram to show Beca's possible outfit choices.

  b Beca selects a top and pair of bottoms at random. Find the probability that she selects the plain top and shorts.

  c Show that P(sleeveless top) × P(leggings) = P(sleeveless top and leggings)

  Does this always happen?

3 Each spinner is spun once. Find P(3 and 5)

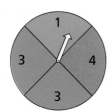 

4 **a** Events A and B are independent.

    P(A) = $\frac{1}{10}$, P(B) = $\frac{1}{5}$. Show that P(A and B) = $\frac{1}{50}$

  b Events C and D are also independent.

    P(C) = $\frac{7}{10}$, P(D) = $\frac{2}{5}$. Show that P(C and D) = $\frac{7}{25}$

5 X, Y and Z are outcomes from separate independent events such that P(X) = 0.3, P(Y) = $\frac{1}{2}$ and P(Z) = 25%

   a Why do the probabilities add up to more than 1?

   b Calculate these probabilities.

      i P(X and Y)   **ii** P(X and Z)   **iii** P(Y and Z)

What do you think?

1 Abdullah is flipping two separate fair coins. He says, "The probability of getting one head and one tail is $\frac{1}{4}$ because P(H) = $\frac{1}{2}$ and P(T) = $\frac{1}{2}$ and $\frac{1}{2} \times \frac{1}{2} = \frac{1}{4}$." Do you agree with Abdullah? Explain your answer.

2 Beca rolls two fair dice. Find the probability that she scores a 3 and an even number.

3 Marta has used a sample space diagram to find the probability that both spinners A and B land on orange. Here is her working.

Spinner A   Spinner B

	Orange	Blue
Orange	OO	BO
Blue	OB	BB

P(OO) = $\frac{1}{4}$

   a Explain the mistake Marta has made.

   b Calculate P(OO), show your working to justify your answer.

In this section, you will calculate probabilities from some diagrams that you should already be familiar with.

Example 2

Students in Years 7, 8 and 9 were asked whether they prefer lions or tigers. The results are shown in this two-way table.

	Year 7	Year 8	Year 9	Total
Tigers	78	65	67	210
Lions	72	75	43	190
Total	150	140	110	400

a A student is selected at random. Find the probability that this student is a Year 8 student who prefers tigers.

b A Year 8 student is selected at random. Find the probability that this student prefers tigers.

c A student who prefers tigers is selected at random. Find the probability that this student is in Year 8

a

	Year 7	Year 8	Year 9	Total
Tigers	78	(65)	67	210
Lions	72	75	43	190
Total	150	140	110	(400)

$\frac{65}{400}$

A **student** is selected at random. This means that the person is being selected from the total number of students, so the probability is out of 400

There are 65 Year 8 students who prefer tigers.

The probability is therefore $\frac{65}{400}$

This can be simplified to $\frac{13}{80}$, but be careful – the question hasn't asked for this so don't risk making numerical errors if simplification isn't needed.

b

	Year 7	Year 8	Year 9	Total
Tigers	78	(65)	67	210
Lions	72	75	43	190
Total	150	(140)	110	400

$\frac{65}{140}$

A **Year 8 student** is selected at random. This means that the person is being selected from the total number of Year 8 students, so the probability is out of 140

65 of these 140 students prefer tigers, so the probability is $\frac{65}{140}$

c

	Year 7	Year 8	Year 9	Total
Tigers	78	(65)	67	(210)
Lions	72	75	43	190
Total	150	140	110	400

$\frac{65}{210}$

A **student who prefers tigers** is selected at random. This means that the person is being selected from the total number of students who prefer tigers, so the probability is out of 210

65 of these students are in Year 8, so the probability is $\frac{65}{210}$

Make sure that you always pay close attention to the wording of the question. Note how a slight difference in wording changes the answers here.

If you are asked to, you should simplify your fractions.
If not, you can leave the answers as they are.

Practice 15.2B

1 320 people were asked to name their favourite type of food. The results are shown in the table.

	Boys	Girls	Total
Italian	27	18	45
Indian	31	40	71
Chinese	27	23	50
Other	18	26	44
No preference	48	62	110
Total	151	169	320

a A person is selected at random. Calculate these probabilities.

   i P(Italian)   **ii** P(other)   **iii** P(girl)

   iv P(boy and no preference)

b A boy is selected at random. Calculate these probabilities.

   i P(Indian)   **ii** P(Chinese)   **iii** P(no preference)

   iv P(not Italian)

c A person who named Chinese as their favourite food is selected at random. Calculate these probabilities.

   i P(boy)   **ii** P(girl)

2 The frequency tree shows information about members of a running club.

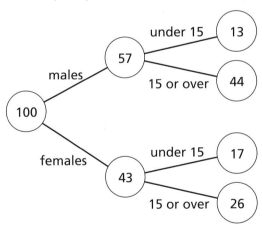

a A member of the club is selected at random. Calculate these probabilities.

   i P(male)   **ii** P(under 15)   **iii** P(female and 15 or over)

b A female is selected at random. Find the probability that she is under 15

3 The Venn diagram shows information about the sports played by 120 students.

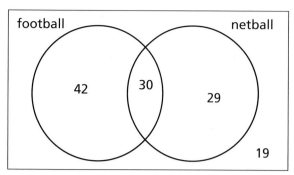

football   netball

42   30   29

19

a A student is selected at random. What is the probability that they play

i football and netball

ii only football

iii exactly one of the two sports?

b A student who plays netball is selected at random. What is the probability this student also plays football?

4 A group of people were asked whether or not they wear glasses. Some of the results are shown in the two-way table.

	Under 21	21–40	Over 40	Total
Glasses		92	78	
No glasses	125			
Total	230		300	800

a Copy and complete the two-way table.

b A person is selected at random. Calculate these probabilities.

i P(under 21)   **ii** P(glasses)   **iii** P(under 21 and wears glasses)

c A person who wears glasses is selected at random. What is the probability that they are over 40?

d A person in the 21–40 category is selected at random. What is the probability that they do not wear glasses?

5 Flo has collected some information about 300 bus and train journeys. She has started to record her findings in a frequency tree.

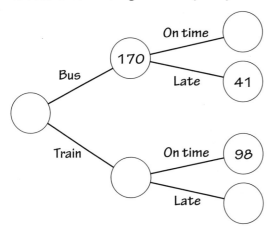

a Copy and complete the frequency tree.

b One of the journeys is selected at random. What is the probability that it was a bus journey that was on time?

c A bus journey is selected at random. What is the probability that it was on time?

 d A late journey is selected at random. What is the probability that it is a bus journey?

6 The Venn diagram shows some information about the languages studied by 100 students.

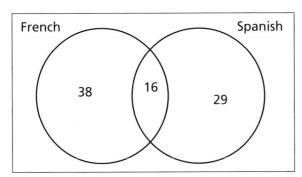

 a Explain why the numbers in the Venn diagram don't sum to 100

b A student is selected at random. What is the probability that they study

   i French and Spanish   **ii** neither French nor Spanish

   iii Spanish?

c A student who studies French is selected at random. What is the probability that they also study Spanish?

What do you think? 🌑

1 50 students were asked whether they like school dinners or not. 28 of the students were girls. 17 girls said they like school dinners and 15 boys said they dislike school dinners.

    a Use this information to draw a frequency tree.

    b A student is selected at random. Find the probability that this student is a girl who dislikes school dinners.

    c A boy is selected at random. Find the probability that this boy likes school dinners.

2 A group of 136 people were asked whether they like ketchup and mayonnaise. 28 people said they like both mayonnaise and ketchup. 52 people like ketchup. 9 people like neither ketchup nor mayonnaise.

    a Draw a Venn diagram to represent this information.

    b A person is selected at random. Find the probability that this person likes exactly one of the sauces.

3 A bag contains red shapes and blue shapes. There are 80 shapes in total. Each shape is either a circle, a rectangle or a triangle. 18 of the 30 circles are blue. 11 of the 17 triangles are red. There are 4 blue rectangles. A shape is selected at random.

    a Find the probability that this shape is a blue circle.

    b Compare your method with a partner's.

Consolidate – do you need more?

1 A counter is selected at random from a large pile of counters that are red, blue, yellow or green; there are equal numbers of each colour. A fair spinner that is divided into three equal sections numbered 1–3 is then spun.

    a Explain why the events are independent.

    b Copy and complete the sample space diagram for possible outcomes.

	Red	Blue	Yellow	Green
1	(1, red)	(1, blue)		
2				
3				

    c Find the probability that the counter is yellow and the spinner lands on 3

    d Show that P(2 and blue) = P(2) × P(blue)

2 Given that $P(G) = \frac{1}{5}$ and $P(H) = \frac{2}{7}$, and given that G and H are independent events, find P(G and H), showing your working clearly.

3 The table shows some information about number cards in a bag.

	1	2	3	Total
Black	21	15	7	43
White	14	14	29	57
Total	35	29	36	100

Each card is either black or white, and has 1, 2 or 3 written on it.

a A number card is selected at random from the bag. Calculate these probabilities.

i P(1)　　　　**ii** P(white)　　　　**iii** P(1 and white)

iv P(odd)　　　　**v** P(odd and black)

b A white number card is selected at random from the bag. Calculate these probabilities.

i P(1)　　　　**ii** P(3)　　　　**iii** P(greater than 1)

iv P(prime)　　　　**v** P(less than 4)

Stretch – can you deepen your learning?

1 The probability that Ed passes his driving test first time is 0.47. The probability that Ed scores a goal in his football match on Saturday is $\frac{3}{11}$. Find the probability that Ed passes his driving test first time and scores a goal in his football match on Saturday.

2 The ratio of men to women in a chess club is $5:7$. There are 40 more women than men. $\frac{2}{5}$ of the women play competitively; the rest do not. 10 more men than women play competitively. A person from the chess club is selected at random. Find the probability that this person is a male who does not play competitively.

3 You are told that A and B are independent. You are also told that P(A) = $\frac{3}{8}$, P(A and B) = $\frac{11}{15}$
Explain why this scenario is impossible.

4 J and K are possible outcomes of independent events. P(J) = $9a^6b^2$, P(K) = $\frac{1}{6a^3b}$
Find P(J and K)

5 A bag contains 10 black pens, 15 blue pens and 5 red pens. Seb takes a pen from the bag and keeps hold of it. Zach then takes a pen from the bag. Find the probability that

a Seb takes a black pen and Zach takes a red pen

b both Zach and Seb take a black pen.

Reflect

1 In your own words, explain what it means for two events to be independent.

2 How do you know from the wording in a question what number a probability will be out of?

15.3 Probabilities and tree diagrams

Small steps

- Use tree diagrams (H)
- Use tree diagrams to solve "without replacement" problems (H)

Key words

Tree diagram – a way of recording possible outcomes that can be used to find probabilities

Replacement – putting back – when an item is replaced, the probabilities do not change; when an item is not replaced, the probabilities do change

Are you ready?

1 A box contains 8 white chocolates and 5 dark chocolates. A chocolate is selected at random. What is the probability that the chocolate is a white chocolate?

2 A bag contains 12 green counters and 8 yellow counters. Beca removes a counter, writes down its colour and then puts it back. How many of each colour counter are there now?

3 There are 4 cheese sandwiches and 3 chicken sandwiches on a shelf. Marta buys one of the sandwiches.

    a How many sandwiches are left on the shelf?

    b How many of each type of sandwich are left if Marta bought

        i a cheese sandwich

        ii a chicken sandwich?

Models and representations

Probability tree diagram

A probability **tree diagram** helps to record all the possible outcomes from two events.

This diagram represents the possible outcomes when these two spinners are spun.

Spinner 1   Spinner 2

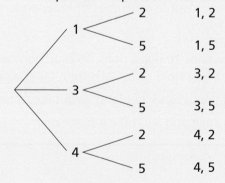

In this section, you will calculate probabilities from tree diagrams.

A tree diagram is a way of representing all the possible outcomes of two or more events. Each "branch" is labelled at its end point with an outcome, and the probabilities are written alongside the branches.

This section will focus on independent events where the outcome of the first event does not affect the outcomes of later events.

Example 1

A bag contains 12 green counters and 8 red counters. Faith removes a counter, writes down its colour then puts it back in the bag. She then selects another counter and writes down its colour.

a Draw a probability tree diagram to represent all the possible outcomes.

b Find the probability that Faith selects two green counters.

c Find the probability that Faith selects at least one green counter.

A probability tree diagram is used to represent the possible outcomes of two or more events. In this case, the events are selecting counters. This is shown in the headings of the tree diagram "1st counter" and "2nd counter".

a

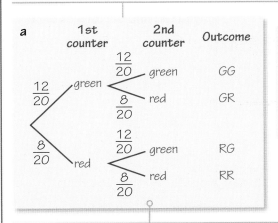

If you trace your finger along the branches, you can see that the outcomes are green then green (GG), green then red (GR), red then green (RG) and red then red (RR). It is a good idea to list the outcomes at the side, as shown here, to make sure that none are missed.

The probabilities do not change for the 2nd counter because Faith puts the 1st counter back in the bag.

There are still 20 counters, 12 of which are green and 8 are red.

Notice that the probabilities on any pair of branches sum to 1; this shows that you've covered all possible outcomes.

b

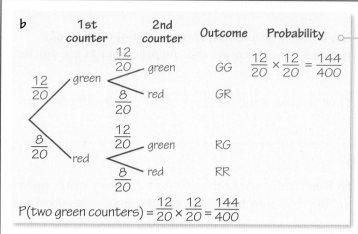

$$P(\text{two green counters}) = \frac{12}{20} \times \frac{12}{20} = \frac{144}{400}$$

The probability of selecting two green counters is represented in the top "row" of outcomes.

$$P(\text{first counter is green}) = \frac{12}{20}$$

$$P(\text{second counter is green}) = \frac{12}{20}$$

So the probability of both counters being green is equal to $\frac{12}{20} \times \frac{12}{20} = \frac{144}{400}$

$$P(GG) = P(G) \times P(G)$$

There are three outcomes that contain at least one green counter; these are GG, GR and RG

c

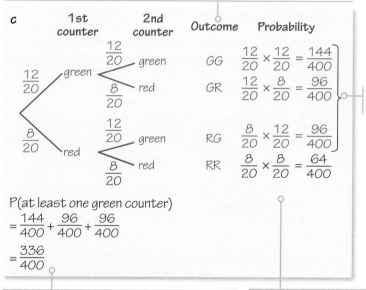

To find the required probability, you can find the total of these.

$P(\text{at least one green counter})$

$$= \frac{144}{400} + \frac{96}{400} + \frac{96}{400}$$

$$= \frac{336}{400}$$

You could also answer this question by finding P(not RR) = 1 − P(RR)

Check that this gives the same answer.

Notice how the probabilities on the right sum to 1. This shows you that you have covered all possible outcomes. If they don't sum to 1 then you've made a mistake.

Practice 15.3A

1 Jackson rolls a fair 6-sided dice and notes whether it lands on an odd or an even number. He then rolls a second fair 6-sided dice and notes whether it lands on an odd or an even number.

    a Draw a probability tree diagram to represent all the possible outcomes of Jackson's experiment.

    b Find the probability that Jackson

        i rolls two even numbers

        ii rolls at least one even number

        iii does not roll an even number.

2 Chloe is playing a game. She rolls a fair 6-sided dice. She then flips a fair coin.

    a Draw a probability tree diagram to model all the possible outcomes of Chloe's game.

    b Find these probabilities.

        i P(5 and heads)

        ii P(2 and tails)

        iii P(5 and heads or 2 and tails)

    c Chloe wins the game if she rolls an even number and flips a head. What is the probability that Chloe

        i wins the game

        ii does not win the game?

    d What do you notice about your answers to part **c**?

3 A bag contains 10 counters. 3 of the counters are red and the rest are blue. Abdullah selects a counter at random, notes its colour and puts it back in the bag. He then selects another counter at random and notes its colour.

    a Draw a probability tree diagram to represent all the possible outcomes.

    b Find the probability that

        i both counters are the same colour

        ii Abdullah selects exactly one red counter.

4 The probability that Beca passes her piano exam is $\frac{3}{5}$. The probability that Beca passes her spelling test is $\frac{6}{7}$

    a Draw a probability tree diagram to represent all the possible outcomes of Beca's tests.

    b Find the probability that Beca passes at least one of her tests.

5 A bag contains 100 counters. 60 of the counters are red and the rest are blue. Seb removes a counter, notes its colour and puts it back in the bag. He does this three times.

   a Draw a probability tree diagram to represent all the possible outcomes.

   b Find the probability that

      i all three counters are the same colour

      ii Seb removes exactly one red counter.

6 The probability that Emily wins a game of tennis is 0.7. The probability that Emily loses a game of chess is 0.1

   a Draw a probability tree diagram to model all the possible outcomes of Emily's games.

   b Find the probability that Emily wins at least one game.

What do you think? 💭

1 Beca has drawn this probability tree diagram and written the following probabilities.

The probability that Abdullah is late to school on Monday is $\frac{3}{8}$

The probability that Abdullah is late to school on Tuesday is $\frac{2}{9}$

Monday	Tuesday	Outcome	Probability
	$\frac{2}{9}$ late	LL	$\frac{3}{8} \times \frac{2}{9} = \frac{1}{12}$
$\frac{3}{8}$ late	$\frac{7}{9}$ not	LN	$\frac{3}{8} \times \frac{7}{9} = \frac{7}{24}$
	$\frac{2}{9}$ late	NL	$\frac{5}{8} \times \frac{2}{9} = \frac{5}{36}$
$\frac{5}{8}$ not	$\frac{7}{9}$ not	NN	$\frac{5}{8} \times \frac{7}{9} = \frac{35}{72}$

   a Is Beca correct? Discuss this with a partner.

   b Beca now wants to find the probability that Abdullah is late on at least one day. Explain why simplifying the fractions has made this question harder for Beca.

2 Marta has drawn a probability tree diagram to represent the outcomes of flipping two coins.

I must have done something wrong because the probabilities on my second flip sum to 2

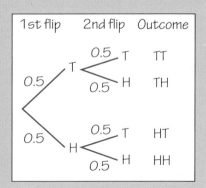

Do you agree with Marta's comment? Explain your reasoning.

In this section, you will continue to calculate probabilities from tree diagrams. However, you will now focus on dependent events.

Example 2

A bag contains 12 green counters and 8 red counters. Faith selects a counter at random from the bag, writes down its colour and puts the counter in her pocket. She then randomly selects another counter from the bag and writes down its colour.

a Draw a probability tree diagram to represent all the possible outcomes.

b Find the probability that Faith selects two green counters.

Although this example is very similar to Example 1 in Practice 15.3A, there is a key difference. Here the counter is not replaced. These types of question are often referred to as "without **replacement**".

a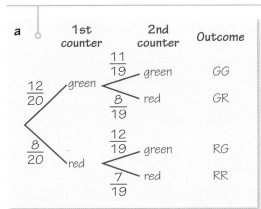

The events are dependent; the probability of the second event depends on the outcome of the first event.

This means that the probabilities on the second set of branches are not the same as on the first set.

b

	1st counter	2nd counter	Outcome	Probability

$$\frac{12}{20} \text{ green} \begin{cases} \frac{11}{19} \text{ green} & GG & \frac{12}{20} \times \frac{11}{19} = \frac{132}{380} \\ \frac{8}{19} \text{ red} & GR \end{cases}$$

$$\frac{8}{20} \text{ red} \begin{cases} \frac{12}{19} \text{ green} & RG \\ \frac{7}{19} \text{ red} & RR \end{cases}$$

$$P(\text{two green counters}) = \frac{12}{20} \times \frac{11}{19} = \frac{132}{380}$$

Practice 15.3B

1 A bag contains 10 counters. Three of the counters are red and the rest are blue. Abdullah selects a counter at random, notes its colour and puts it to the side. He then randomly selects another counter and notes its colour.

    a Draw a probability tree diagram to represent all the possible outcomes.

    b Find the probability that

        i both counters are the same colour

        ii Abdullah removes exactly one red counter.

2 There are 4 cheese sandwiches and 3 chicken sandwiches for sale in the canteen. Flo buys one of the sandwiches. Then Seb buys one of the sandwiches.

    a Draw a probability tree diagram to represent the possible outcomes.

    b Find the probability that Seb and Flo have sandwiches with the same filling.

3 A pencil case contains 8 black pens, 5 blue pens and 2 red pens. Ed randomly takes a pen from the pencil case and does not replace it. Then Benji randomly takes a pen from the pencil case.

    a Draw a probability tree diagram to model all the possible outcomes.

    b Find the probability that

        i there are no red pens left in the pencil case

        ii Ed and Benji select the same colour pen

        iii at least one blue pen is selected.

4 A box contains 16 identical stripy socks and 12 identical spotty socks. Emily selects two socks at random.

    a Draw a probability tree diagram to model all the possible outcomes.

    b Find the probability that Emily selects a matching pair of socks.

5 Letter tiles that spell the word MATHEMATICAL are placed in a bag. Zach randomly selects a letter tile and does not replace it. Faith then randomly selects a letter tile from the bag. Find the probability that exactly one person selects a vowel.

6 Lydia plays two tennis matches. The probability that she wins the first game is 0.6. If she wins the first game, then the probability that she wins the second game is 0.8. If she loses the first game, then the probability that she loses the second game is 0.3. Find the probability that Lydia wins at least one game.

What do you think? 🌐

1 Marta has answered a probability question.

A bowl contains 5 orange sweets and 4 lemon sweets. Two sweets are taken from the bowl. Find the probability that both sweets are orange.

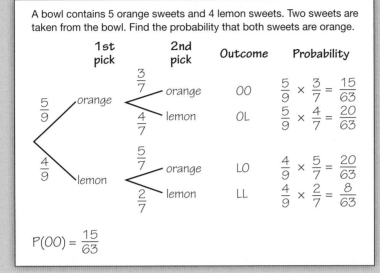

$$P(OO) = \frac{15}{63}$$

Find and explain Marta's mistake.

2 The probability of passing a writing test is 0.8. If the writing test is passed, then the probability of passing a spelling test is 0.7. The probability of passing at least one of the tests is 0.88. Find the probability that the spelling test is passed but the writing test is not.

Consolidate – do you need more?

1 A freezer contains 3 lemon ice lollies and 5 strawberry ice lollies. An ice lolly is selected at random. Its flavour is noted and it is then replaced in the freezer.

a Copy and complete the probability tree diagram to represent all the possible outcomes and their probabilities.

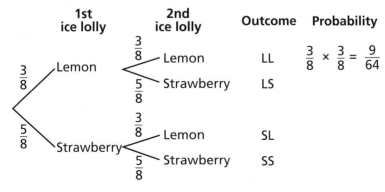

b Find the probability that

 i both ice lollies are strawberry

 ii at least one ice lolly is strawberry.

2 The probability of Jakub winning a hockey match is $\frac{1}{3}$. The probability of Jakub winning a rugby match is $\frac{2}{5}$

   a Copy and complete the probability tree diagram.

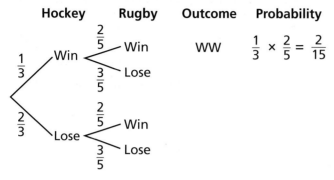

   b Find the probability that Jakub

     i wins exactly one game

     ii does not win either game.

3 A freezer contains 3 lemon ice lollies and 5 strawberry ice lollies. Two ice lollies are selected at random without replacement.

   a Copy and complete the probability tree diagram to represent all the possible outcomes and their probabilities.

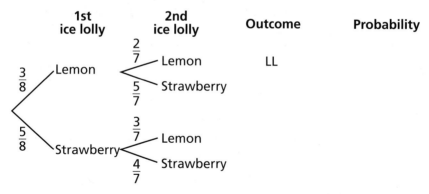

   b Find the probability that

     i both ice lollies are strawberry

     ii at least one ice lolly is strawberry.

Stretch – can you deepen your learning?

1 Seb is organising a tombola. He has 3000 red tickets numbered 1–3000 and 2000 green tickets numbered 1–2000. A player selects one red ticket and one green ticket at random. They win a prize if both tickets end in either 0 or 5

   a Find the probability that a player wins a prize.

   b 1000 people buy tickets for the tombola. Estimate the number of prize winners.

   c Each player pays 10p. Each winner gets £5. Predict Seb's amount of profit/loss when 1000 people buy tickets.

2 Here is a spinner.

The probability of the spinner landing on 1 is $5y$. The probability of the spinner landing on 2 is $3y$. The probability of the spinner landing on 3 is $6y$. The spinner is spun twice and the outcomes are added together. Find the probability that the total score is odd.

3 Two spinners each have two sections, which are blue and orange. The probability that spinner A lands on blue is $\frac{3}{4}$. The probability that spinner B lands on blue is x. The probability that exactly one spinner lands on blue is 0.56. Work out the value of x

4 There are p pencils in a tub. Five of the pencils are blunt and the rest are not.

   a If two pencils are selected at random from the tub with replacement, then the probability that both pencils are not blunt is $\frac{169}{324}$. Show that
$$31p^2 - 648p + 1620 = 0$$

   b If two pencils are selected at random without replacement, then the probability that
exactly one pencil is blunt is $\frac{65}{153}$. Show that $13p^2 - 319p + 1530 = 0$

Reflect

1 What is the difference between a selection made with replacement and a selection made without replacement? How does it affect your working and your answers?

2 Why do the probabilities on any set of branches sum to 1?

I have become fluent in...

- calculating the probability of a single event
- using diagrams to calculate probabilities
- calculating relative frequency
- drawing probability tree diagrams to model possible outcomes. (H)

I have developed my reasoning skills by...

- being able to explain what is meant by dependent and independent events
- explaining why expected outcomes are only estimates
- comparing and contrasting different diagrams for calculating probabilities.

I have been problem-solving through...

- exploring probability in different contexts
- working backwards when solving probability problems
- calculating complex probabilities or missing values using diagrams.

Check my understanding

1 One of the number cards is selected at random.

| 1 | 2 | 3 | 4 | 5 | 6 | 7 | 8 | 9 | 10 |

Find

a P(odd) **b** P(greater than 4) **c** P(multiple of 3) **d** P(less than 1)

2 A coin is flipped 300 times.

It lands on heads 240 times.

a Find the relative frequency of heads.

b The same coin is flipped 700 times. How many times would you expect it to land on tails?

3 The probability of a spinner landing on red is 0.3

The probability of a biased dice landing on 3 is 0.25

The spinner is spun and the dice is rolled.

Find P(red and 3)

4 A box contains 8 pens. 5 are blue and the rest are red.

A pen is selected from the box at random. It is replaced and then another pen is selected.

Calculate the probability that the pens are different colours.

5 A box contains 8 pens. 5 are blue and the rest are red.

2 pens are selected at random.

Calculate the probability that both pens are the same colour. (H)

In this block, I will learn...

how to draw and interpret quadratic graphs

x	−4	−3	−2	−1	0	1	2
$y = x^2 + 2x - 3$	5	0	−3	−4	−3	0	5

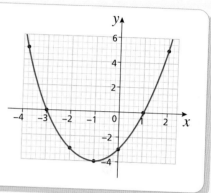

how to interpret piece-wise and other graphs

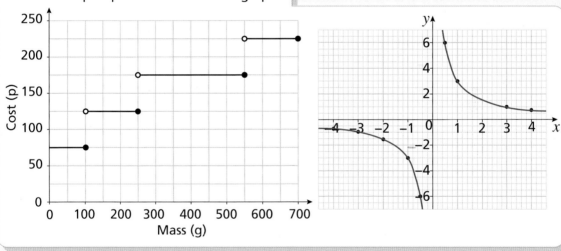

how to use graphs to solve simultaneous equations Ⓗ

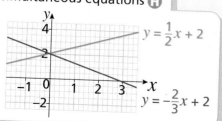

how to represent inequalities

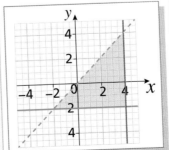

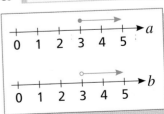

16.1 Further graphs

Small steps

- Draw and interpret quadratic graphs
- Interpret graphs, including reciprocal and piece-wise

Key words

Quadratic – an expression is quadratic when the highest power of the variable is 2

Parabola – a type of curve that is approximately U-shaped and has a line of symmetry

Piece-wise graph – a graph that consists of more than one straight line

Are you ready?

1 **a** Copy and complete the table of values for $y = 2x + 3$

x	−2	−1	0	1	2
$y = 2x + 3$		1			

b Draw the graph of $y = 2x + 3$ from $x = -2$ to $x = 2$

2 Copy and complete the table of values for $y = 3 - 2x$

x	−2	−1	0	1	2
$y = 3 - 2x$		5			−1

3 Work out the values of the expressions when **i** $x = 3$ **ii** $x = -3$

a x^2 **b** $x^2 + x$ **c** $x^2 - x$ **d** $x^2 + 2x - 1$

Models and representations

Graphing software

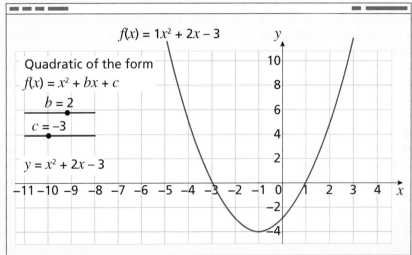

This can help you to see how the position and shape of a graph change when you change the coefficients in its equation.

Coordinate grids

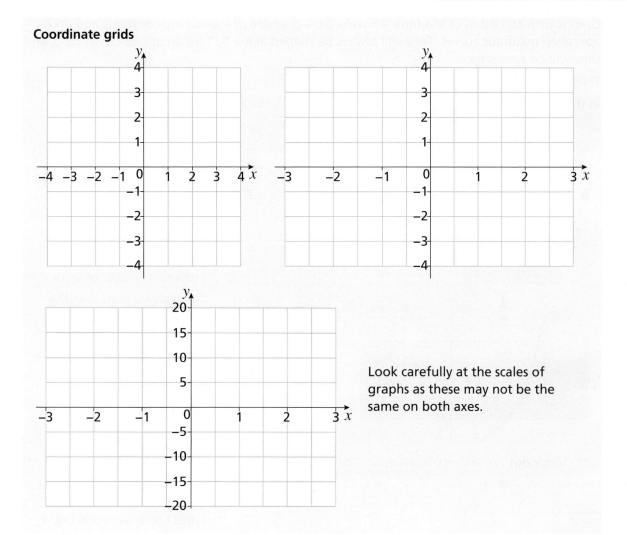

Look carefully at the scales of graphs as these may not be the same on both axes.

Graphs with equations of the form $y = ax^2 + bx + c$, where a, b and c are constants and $a \neq 0$, are called **quadratic** curves. They will always be shaped like a "U" (or an upside-down "U"), and they will be symmetrical.

These curves are called **parabolas**.

In this chapter, the quadratic graphs you will explore will all have $a = 1$

Example 1

a Draw the graph of $y = x^2 + 2x - 3$ for $x = -4$ to $x = 2$

b Use the graph to estimate the values of x when $y = 3$

a

x	−4	−3	−2	−1	0	1	2
$y = x^2 + 2x - 3$	5	0	−3	−4	−3	0	5

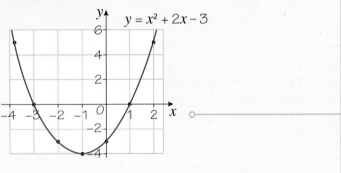

Work out the value of y for each value of x in the given range and write them in a table of values.

Be careful with negative values. For example, when $x = -2$, $y =$ $(-2)^2 + 2 \times -2 - 3 =$ $4 - 4 - 3 = -3$

Plot the points shown in your table of values $(-4, 5)$, $(-3, 0)$ and so on, and join them with a smooth curve.

Do not worry if it does not look perfect first time – this takes some practice.

b

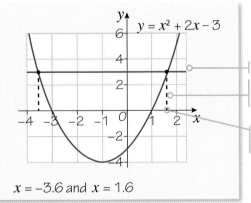

Add the line $y = 3$ to your graph
Draw vertical lines from where the line and curve meet down to the x-axis

Read the values where the vertical lines meet the x-axis.

$x = -3.6$ and $x = 1.6$

Practice 16.1A

1 Here is the graph of $y = x^2 - 3x - 2$

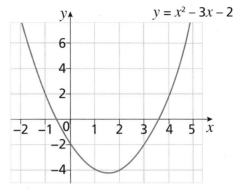

Use the graph to

a find the value of y when $x = 2$

b find the value of x when $y = 2$

c estimate the value of y when $x = 0.5$

d estimate the values of x when $y = 3$

2 **a** Draw axes numbered from –3 to 3 in the x-direction and from –2 to 10 in the y-direction.

b Copy and complete this table of values for $y = x^2$

x	–3	–2	–1	0	1	2	3
$y = x^2$		4		0			9

c Use the values from your table to plot the graph of $y = x^2$ on your pair of axes. Join the points with a smooth curve.

3 **a** Copy and complete this table of values for $y = x^2 + 1$

x	–3	–2	–1	0	1	2	3
$y = x^2 + 1$	10			1			

b Use the values from your table to plot the graph of $y = x^2 + 1$ on the same set of axes as you used in question **2**. Join the points with a smooth curve.

c Compare your graphs. What's the same and what's different?

d Discuss what you think these graphs will look like.

$y = x^2 + 2$ $y = x^2 - 1$ $y = x^2 - 2$

e Check your answers by drawing the graph of $y = x^2 + a$ for different values of a

4 **a** Copy and complete these tables of values.

x	–3	–2	–1	0	1	2	3
$y = x^2 + 2x + 1$	4			1			

x	–3	–2	–1	0	1	2	3
$y = x^2 - 2x - 1$	14			–1			

b Draw the graphs of $y = x^2 + 2x + 1$ and $y = x^2 - 2x - 1$ on the same set of axes.

c Compare the graphs with those you drew in questions **2** and **3**

5 Draw the graph of $y = x^2 - 4x + 2$ for values of x from –1 to 5

6 Rhys is drawing the graph of $y = x^2 - 5x + 3$

x	−1	0	1	2	3	4	5	6
$y = x^2 - 5x + 3$	9	3	−1	−3	−3	−1	3	9

He notices that the x-values 2 and 3 have the same y-value and thinks that the graph will look like this:

Rhys is wrong.

a At what value of x will the lowest point of the graph be? What will the y-value be at this point?

b Draw the graph of $y = x^2 - 5x + 3$ for values of x from −1 to 6

What do you think?

 1 When you square a negative number you get a positive number, so the graph of $y = -x^2$ will be the same as the graph of $y = x^2$

Show that Beca is wrong.

2 Which of the cards show the equation of a straight line and which show the equation of a parabola? How do you know?

A $y = 3x + 5$ B $y = 3x + 5x^2$ C $y = 3 + x$ D $y = 3 + x^2$

E $y = 3 + x + x^2$ F $y = 6 - 2x$ G $y = 6 - 2x^2$ H $y = (6 - 2x)^2$

 3 There's no squared term in the equation $y = \frac{1}{x}$ so the graph will be a straight line.

You will plot graphs with equations like this in the next section, but think about some numbers and decide for yourself first.

Do you agree with Benji?

Now you will explore some graphs with different shapes, including those that have more than one part.

This is an example of a **piece-wise graph**.

You will learn how to read this graph in Practice 16.1B, question **4**

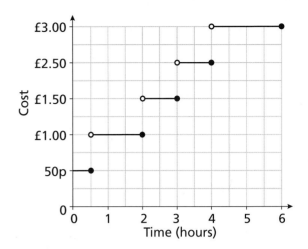

Example 2

a Copy and complete the table of values and then draw the graph of $y = \frac{3}{x}$ for values of x from -4 to 4

x	-4	-3	-2	-1	-0.5	0.5	1	2	3	4
$y = \dfrac{3}{x}$										

b Describe the shape of the graph.

c Why is $x = 0$ not included in the table of values?

Remember:
$\frac{3}{x}$ means $3 \div x$

a

x	-4	-3	-2	-1	-0.5	0.5	1	2	3	4
$y = \dfrac{3}{x}$	-0.75	-1	-1.5	-3	-6	6	3	1.5	1	0.75

Work out the value of y for each value of x in the given range and write them in the table of values.

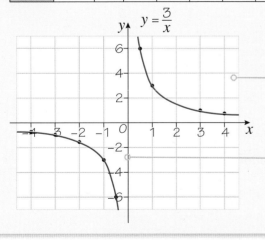

Join through the points in the first quadrant with a smooth curve

Join through the points in the third quadrant with a smooth curve.

b The graph is in two sections. Both sections get closer and closer to the axes for very small and very large values of x, but they never touch them.

You can check these observations by using graphing software to draw the graph.

c It is impossible to work out $3 \div 0$

You cannot divide a number by 0

637

Practice 16.1B

1 Here is the graph of $y = \frac{2}{x}$

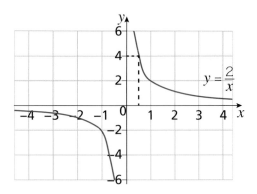

a Explain how the dotted lines show that when $x = 0.5$, $y = 4$

b Find the value of

    i y when $x = 4$ **ii** y when $x = -4$ **iii** x when $y = -2$

c Is it easier to use the graph or the equation to find a value of y given a value of x? Is it easier to use the graph or the equation to find a value of x given a value of y?

2 Here are the graphs of $y = \frac{1}{x}$, $y = \frac{2}{x}$, $y = \frac{3}{x}$, $y = \frac{4}{x}$ and $y = \frac{5}{x}$

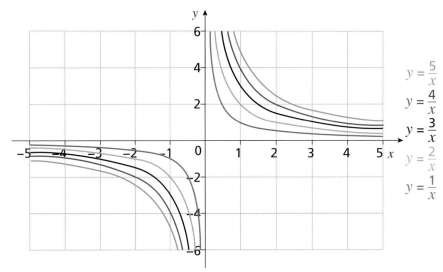

What's the same and what's different about the graphs?

3 Here is the graph of $y = 2^x$

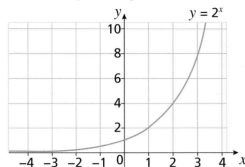

Graphs of equations of the form $y = a^x$ are called "exponential graphs".

a Create a table of values and make your own copy of the graph of $y = 2^x$ for values of x from −4 to 3

b Which parts of the graph are most difficult to draw? Why?

c Use your graph to estimate the value of 2^x when

   i $x = 0.5$ **ii** $x = 2.5$

d Create another table of values and draw the graph of $y = 3^x$ for values of x from −2 to 2. Compare your two graphs.

4 The graph shows the cost of parking in a car park for different lengths of time.

Parking for up to half an hour costs 50p. After this time, the cost increases to £1

a How much does it cost to park for

   i $2\frac{1}{2}$ hours

   ii 4 hours

   iii 5 hours?

b Zach has £1.75. For how long can he park?

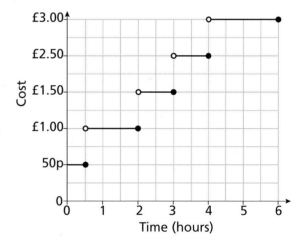

5 The graph shows the amount charged by a company to deliver letters of different masses.

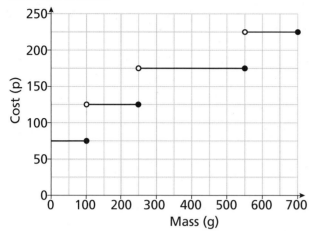

a How much does it cost to post a letter of mass

   i 150 g   **ii** 250 g   **iii** 255 g?

b Sven wants to post four magazines to his friend. Each of the magazines has mass 250 g. Investigate the different ways in which Sven can send the magazines, and how much it will cost him.

6 Describe the journey shown by this distance–time graph.

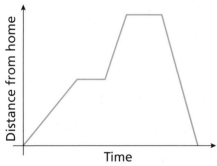

What do you think? 🔮

1 The table shows the amount charged by an electrician for different lengths of jobs.

Duration	Up to 1 hour	Up to 2 hours	Up to 3 hours	Up to 5 hours	Up to 8 hours
Cost	£50	£80	£110	£140	£250

Represent this information as a piece-wise graph.

2 Draw a graph of each of these functions.

a
$$y = \begin{cases} 2 \text{ for } 0 \leqslant x < 3 \\ 4 \text{ for } 3 \leqslant x < 5 \\ 5 \text{ for } 5 \leqslant x < 8 \end{cases}$$

b
$$y = \begin{cases} x \text{ for } 0 \leqslant x < 2 \\ 2 \text{ for } 2 \leqslant x < 4 \\ x - 2 \text{ for } 4 \leqslant x < 8 \end{cases}$$

3 Draw some piece-wise linear functions of your own like those in question **2**. Challenge a friend to identify the rules for each region.

Consolidate – do you need more?

1 **a** Draw axes numbered from −3 to 3 in the x-direction and from −2 to 10 in the y-direction.

b Copy and complete this table of values for $y = 2x^2 + 1$

x	−2	−1	0	1	2
$y = 2x^2 + 1$	9		1		

c Use the values from your table to draw the graph of $y = 2x^2 + 1$ for values of x from −2 to 2 on your set of axes. Join the points with a smooth curve.

2 **a** Draw axes numbered from −5 to 1 in the x-direction and from −6 to 4 in the y-direction.

b Copy and complete this table of values for $y = x^2 + 4x - 2$

x	−5	−4	−3	−2	−1	0	1
$y = x^2 + 4x - 2$	3			−6		−2	

c Use the values from your table to draw the graph of $y = x^2 + 4x - 2$ for values of x from −5 to 1 on your set of axes. Join the points with a smooth curve.

3 **a** Copy and complete the table of values.

x	−4	−2	−1	−0.5	0.5	1	2	4
$y = \dfrac{4}{x}$								

b Draw the graph of $y = \dfrac{4}{x}$

4 The graph shows the cost of making an international phone call. The cost of a call that lasts less than a minute is £5

   a How much does it cost to make a call that lasts

      i 2 minutes

      ii $2\frac{1}{2}$ minutes

      iii 3 minutes?

   b Which is cheaper, an 8-minute call or two 4-minute calls? How much cheaper is it?

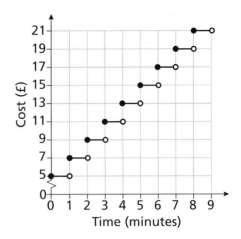

Stretch – can you deepen your learning?

1 The equation of the line of symmetry of the graph of $y = x^2 + 2x - 3$ is $x = -1$

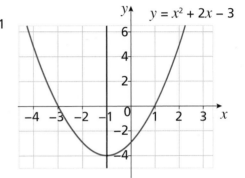

   a Look back at all the quadratic graphs that you have drawn or have seen in this chapter. Investigate the connection between the equation of each graph and the equation of its line of symmetry.

   b Use your findings to predict the equation of the line of symmetry of each of these graphs.

      i $y = x^2 + 6x + 1$ **ii** $y = x^2 - 6x - 1$ **iii** $y = x^2 + 3x + 1$ **iv** $y = 10x + x^2$

      Check your answers using graphing software.

   c State the lines of symmetry of the graph of $y = \frac{4}{x}$

2 **a** Use graphing software to explore how the shape of the graph of $y = x^2 + bx + c$ changes as the values of b and c are varied.

   b Extend your investigation to look at the graphs of $y = ax^2 + bx + c$

3 Use graphing software to investigate how each of these graphs change for different values of a

   a equations of the form $y = \frac{1}{x} + a$ **b** equations of the form $y = \frac{1}{x + a}$

Reflect

1 How can you tell from the equation of a graph whether it will be quadratic? Describe the features of a quadratic curve.

2 How is a piece-wise graph different from other graphs?

Small steps

■ Investigate graphs of simultaneous equations ⓗ

Key words

Simultaneous – at the same time

Graphical – using a graph

Point of intersection – the point where two graphs cross each other

Are you ready?

1 Draw axes numbered from –5 to 5 in both the x- and the y- directions. Draw straight lines with these equations on the same set of axes.

   a $x = 3$ **b** $y = -2$ **c** $x = -1$ **d** $y = 4$

2 **a** Copy and complete this table of values for $y = 3x - 1$

x	–2	–1	0	1	2
$y = 3x - 1$		–4			

   b Draw the graph of $y = 3x - 1$ for values of x from –2 to 2

3 Write down the equation of a line that is parallel to each of these.

   a $x = 7$ **b** $y = 2$ **c** $y = 2x + 4$

Models and representations

Coordinate grids

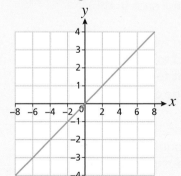

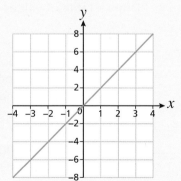

Look carefully at the scales.

Straws

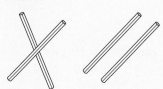

You can use straws to model straight lines and investigate when/if they cross.

You can also use geoboards or graphing software to represent straight lines.

Equations in two variables, such as $x + y = 10$, have an infinite number of solutions, for example $x = 4$ and $y = 6$ or $x = -4$ and $y = 14$. All of these solutions are points on the straight line with equation $x + y = 10$

A pair of equations such as $x + y = 10$ and $x - y = 4$ are called **simultaneous** equations.

A solution of the simultaneous equations is a pair of values of x and y that satisfies both equations.

The solution to the simultaneous equations lies on both of the straight lines $x + y = 10$ and $x - y = 4$

The solution is the values of x and y where the straight lines meet, which is called their **point of intersection**. Here the point of intersection is (7, 3).

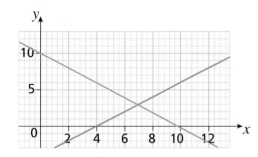

This is a **graphical** method of solving the pair of simultaneous equations. You will study other methods in Key Stage 4

The solution is $x = 7$, $y = 3$

Example 1

Use a graphical method to solve the simultaneous equations

$y = x + 4$

$y = 10 - 2x$

x	−1	0	1
$y = x + 4$	3	4	5

x	−1	0	1
$y = 10 - 2x$	12	10	8

Draw a table of values for each graph.

Although you only need two points to determine a straight line, it is good to do another point as a check. If all three do not form a straight line, you know you have made a mistake.

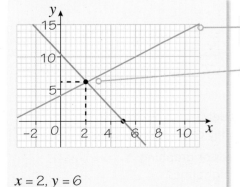

$x = 2$, $y = 6$

Draw the graphs.
Remember to draw through the points and not to just join up the points you plotted.

The solution is given by the point of intersection of the graphs. This is the point (2, 6)
You can check that the solution satisfies both equations.

If $x = 2$
for $y = x + 4$ and for $y = 10 - 2x$
$\quad y = 2 + 4 = 6$ $y = 10 - 2 \times 2$
so $y = 6$ ✓   so $y = 6$ ✓

Practice 16.2A

1 **a** Write down the equations of the lines labelled A and B

b What are the coordinates of the point of intersection of the lines A and B?

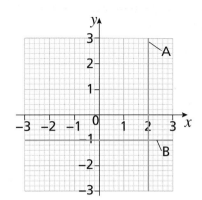

2 **a** On a pair of axes numbered from – 5 to 5 in both directions, draw the lines $x = 4$ and $y = -3$

Why do you not need a table of values for these lines?

b What are the coordinates of the point of intersection of the lines $x = 4$ and $y = -3$?

3 Write down the coordinates of the points of intersection of these pairs of lines.

a $x = 1, y = 4$ **b** $x = -2, y = 3$ **c** $y = 4, x = -1$ **d** $x = a, y = b$

4 Here are the graphs of $y = 2x + 1$, $x = 3$ and $y = 3$

Write down the coordinates of the points of intersection of each pair of lines.

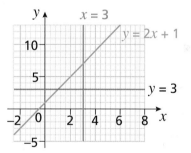

a $y = 2x + 1$ and $x = 3$

b $y = 2x + 1$ and $y = 3$

c What do you notice when you substitute $x = 3$ and $y = 3$ into the equation $y = 2x + 1$?

5 Here are the graphs of $y = 2x - 1$ and $y = x - 3$

Use the graphs to solve the simultaneous equations

$y = 2x - 1$

$y = x - 3$

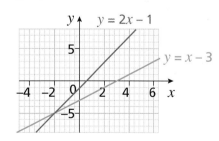

6 Use a graphical method to solve these pairs of simultaneous equations.

a $y = 2x + 3$
$y = 6 - x$

b $y = \frac{1}{2}x - 4$
$y = 2 - x$

Hint: use even values for x in your tables of values.

What do you think? 💭

1

> The simultaneous equations $y = 2x - 1$ and $y = 3 + 2x$ have no solutions.

Draw the graphs of $y = 2x - 1$ and $y = 3 + 2x$ to help you to explain why Marta is right.

2 Darius is solving the simultaneous equations $y = 5 - x$ and $y = 2x - 3$. He draws these graphs.

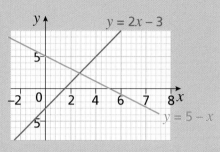

Darius reads off the solution as $x = 2.7$, $y = 2.3$

a Substitute $x = 2.7$ and $y = 2.3$ into the equations. What do you notice?

b Verify that the actual solutions are $x = \dfrac{8}{3}$ and $y = \dfrac{7}{3}$

c When is a graphical method reliable for solving pairs of simultaneous equations and when is it not?

In the next example, you will see a quick way of drawing graphs of equations given in the form $ax + by = 0$

Example 2

Use a graphical method to solve the simultaneous equations.

a $3x + 2y = 12$ **b** $y - x = 1$

a $3x + 2y = 12$ ○───┤ Substitute $x = 0$ into the equation to find where the graph cuts the y-axis.

When $x = 0$, $0 + 2y = 12$

$2y = 12$

$y = 6$

When $y = 0$, $3x + 0 = 12$ ○───┤ Substitute $y = 0$ into the equation to find where the graph cuts the x-axis.

$3x = 12$

$x = 4$

The graph goes through $(0, 6)$ and $(4, 0)$ ○───┤ Use these two points to draw the graph of $3x + 2y = 12$

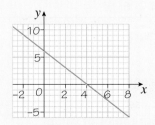

b $y - x = 1$ ○───── Use the same method to find the coordinates of the points where $y - x = 1$ cuts the axes.

When $x = 0, y - 0 = 1$

$y = 1$

When $y = 0, 0 - x = 1$ ○───── Be careful not to "lose" the − sign.

$-x = 1$

$x = -1$

The graph goes through $(0, 1)$ and $(-1, 0)$

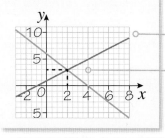

○───── Add the second graph to your set of axes.

○───── The point of intersection has coordinates $(2, 3)$

$x = 2, y = 3$ ○───── Read off the solution from the point of intersection.

Practice 16.2B

1 Find the coordinates of the points where these graphs meet the coordinate axes.

a $x + 2y = 6$ **b** $2x + y = 8$ **c** $3y - 2x = 6$ **d** $2x - y = 4$

2 Use graphs to solve these pairs of simultaneous equations.

a $x + y = 5$ and $2x + y = 6$

b $3y + x = 9$ and $2x - y = 4$

c $x + y = -4$ and $x - y = 2$

How can you check your answers?

3 The sum of two numbers is 6 and their difference is 2

a Explain why this can be modelled using the equations $x + y = 6$ and $x - y = 2$

b Solve the simultaneous equations $x + y = 6$ and $x - y = 2$ to find the two numbers.

c Would the simultaneous equations $y + x = 6$ and $y - x = 2$ have the same solution or a different solution? What does this mean about the two numbers in the original problem?

4 Two numbers have a sum of 8 and a difference of 3. Find the two numbers.

5 Emily is finding where graphs meet the axes to solve the simultaneous equations
$x - y = 2$ and $y - 2x = 0$. Here is her working.

$\underline{x - y = 2}$

When $x = 0$, $0 - y = 2$

                      $y = -2$

When $y = 0$ $x - 0 = 2$

                      $x = 2$

The graph goes through $(0, -2)$ and $(2, 0)$

$\underline{y - 2x = 0}$

When $x = 0$, $y - 0 = 0$

                      $y = 0$

When $y = 0$, $0 - 2x = 0$

                    $2x = 0$

                      $x = 0$

The graph goes through $(0, 0)$ and $(0, 0)$

> I can only find one point for the second equation so I can't draw the graph.

a Explain how Emily can find another point on the graph.

b Use a graphical method to solve the simultaneous equations $x - y = 2$ and $y - 2x = 0$

6 Use graphs to solve these pairs of simultaneous equations. Think carefully about whether to use a table of values or another method.

a $x = 2y$
$2x + 4y = 8$

b $y = 2$
$x + 3y = 9$

c $y + x = 0$
$y - 2x = 6$

What do you think? 💭

1 Two straight lines meet at the point $(3, 2)$

What might the equations of the lines be? How many possibilities can you find?

2 Repeat question **1** for the points $(3, -2)$, $(2, -3)$ and $(-3, -2)$

Consolidate – do you need more?

1 Use the graphs to solve the simultaneous equations $y + 2x = -5$ and $y = x + 4$

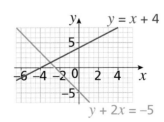

2 a Draw the graphs of $y = 2x - 1$ and $y = x + 3$ on the same axes.

b Use your graphs to solve the simultaneous equations $y = 2x - 1$ and $y = x + 3$

3 Find the coordinates of the points where these graphs meet the coordinate axes.

   a $x + 2y = 8$ **b** $2x + 4y = 8$ **c** $2y - 3x = 6$ **d** $3x - y = 9$

4 Use a graphical method to solve these pairs of simultaneous equations.

   a $y + x = 2$ and $y + 2x = 6$

   b $y + 2x = 8$ and $y - x = -1$

   c $y + x = 1$ and $y - x = 5$

Stretch – can you deepen your learning?

1 You can solve simultaneous equations by substitution. Here are two ways of solving the simultaneous equations $y = 3x + 1$ and $y = 2x$ by substitution.

Method A	Method B
▧ Replace y in the first equation with $2x$	▧ Replace x in the first equation with $\frac{y}{2}$
▧ Solve the resulting equation to find x	▧ Solve the resulting equation to find y
▧ Substitute this value of x in the second equation to find y	▧ Substitute this value of y in the second equation to find x

Why do the methods work?

Compare these methods with the graphical method.

2 Solve the simultaneous equations $y = 2x + 1$ and $4x - 3y = 1$ using the substitution method.

3 a The bar models show the simultaneous equations $a + b = 12$ and $a + 2b = 19$ Deduce the value of b and hence work out the value of a

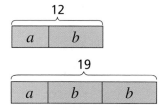

   b $p + 2q = 8$ and $p + 4q = 13$. Find the value of p and the value of q

   c $2c + d = 14$ and $c + 2d = 13$. Find the value of c and the value of d

Reflect

1 Describe how to solve a pair of simultaneous equations using a graphical method.

2 When is the graphical method reliable and when might it be unreliable?

Small steps

■ Represent inequalities

Are you ready?

1 Solve these inequalities.

    a $t + 5 > 12$ **b** $2p - 3 \leqslant 7$ **c** $10 < \frac{x}{4} - 2$

2 Which of these inequalities are satisfied by the value $x = 4$?

    $\boxed{x + 3 > 6}$ $\boxed{x + 3 \geqslant 7}$ $\boxed{2x < 9}$ $\boxed{5 + x < 3}$ $\boxed{11 < 3x - 1}$

3 On separate coordinate grids, draw the straight lines with these equations.

    a $x = 2$ **b** $y = -3$ **c** $y = x$

Models and representations

Graphing software

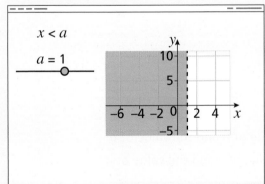

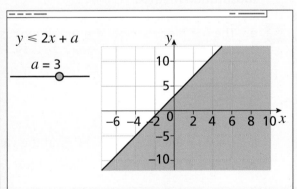

You can use graphing software to investigate regions formed by inequalities.

Number lines

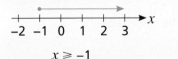

You can use number lines to represent the solutions to inequalities.

You have already learned to solve linear inequalities in one variable. In this chapter, you will explore regions of graphs that represent inequalities in one and two variables.

Example 1

Draw graphs to show the regions that satisfy

a $x \geqslant 2$ **b** $y < 3$ **c** $y \leqslant x$

a

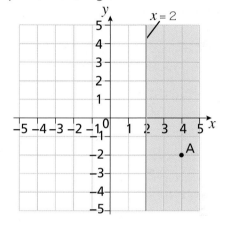

Start by drawing and labelling the line $x = 2$ and then shade the region to the right, as this is where the values of x are greater than 2

The line is solid to show that all points on the line $x = 2$ are included in the required region (the **inequality** is $x \geqslant 2$, not $x > 2$)

You can check that you have shaded the correct region by testing the coordinates of a point in the region.

The coordinates of A are (4, −2). The x-value of point A is 4, which is greater than 2 so the correct side of the line has been shaded.

b

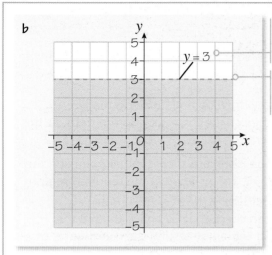

Start by drawing and labelling the line $y = 3$ and then shade the region below the line, as this is where the values of y are less than 3

The line is dotted to show that points on the line $y = 3$ are not included (the inequality is $y < 3$, not $y \leqslant 3$)

You can check that you have shaded the correct region by testing the coordinates of a point in the region.

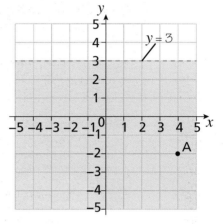

The coordinates of A are (4, −2). The y-value of point A is −2, which is less than 3, so the correct side of the line has been shaded.

c

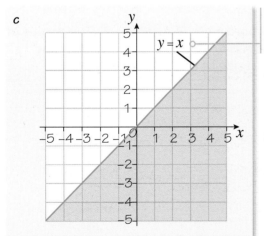

Start by drawing and labelling the line $y = x$ and then shade the region below the line, as this is where the values of y are less than the values of x

You can check that you have shaded the correct region by testing the coordinates of a point in the region.

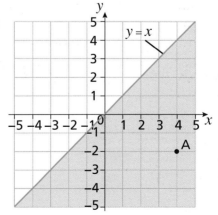

The coordinates of A are (4, −2). The y-value of point A is less than the x-value so the correct side of the line has been shaded.

Practice 16.3A

1. Write the inequality shown by each of these graphs.

a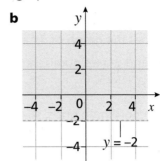

b

c

d

2. Draw x- and y-axes numbered from -5 to 5 for each question part. On your grids, shade the regions that satisfy these inequalities.

a $x < 1$ **b** $y > 2$ **c** $x > -1$ **d** $y \leqslant -1$

3. The number lines show the inequalities $x \geqslant 3$ and $x > 3$

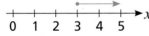

a What's the same and what's different?

b Draw a number line to represent each of these inequalities.

   i $x < 4$ **ii** $p > -2$ **iii** $t \leqslant -1$ **iv** $g \geqslant 2$

4. Here is the graph of $y = x + 2$

   a Which of these points satisfy the inequality $y \leqslant x + 2$?

     A $(2, 3)$ B $(-2, 1)$

     C $(-4, -1)$ D $(-1, -3)$

   b Copy the graph and shade the region that satisfies $y \leqslant x + 2$

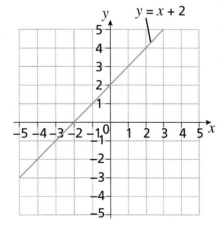

5 Draw x- and y-axes numbered from –5 to 5 for each question part. On your grids, shade the regions that satisfy these inequalities.

a $y \geqslant x$ **b** $y < x - 1$ **c** $y > 2x + 1$ **d** $y \leqslant 3 - x$

6 a Where does the graph of $2x + 3y = 6$ meet the coordinate axes?

b Draw x- and y-axes numbered from –5 to 5. On your grid shade the region that satisfies $2x + 3y \leqslant 6$

7 On separate coordinate grids, shade the regions that satisfy

a $x + 3y > 6$ **b** $2x - y \geqslant 4$

What do you think? 💭

1 Draw x- and y-axes numbered from –5 to 5 for each question part.

On your grids, shade the regions that satisfy these inequalities.

a $x \geqslant 2$ and $y \geqslant 2$ **b** $x < 1$ and $y \leqslant -2$ **c** $y \geqslant -1$ and $x < 3$

2 The graph shows the region $-1 < x < 2$

a Explain what is meant by $-1 < x < 2$

b On separate coordinate grids, show the regions that satisfy

   i $-4 < y \leqslant 3$ **ii** $-2 < x < 0$

   iii $-1 < x < 2$ and $-1 < y < 3$

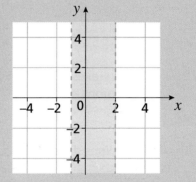

3 Describe the shaded region using three inequalities.

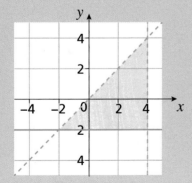

Consolidate – do you need more?

1 Inequalities can be represented in four different forms:

- in words, for example "x is greater than or equal to 2"

- in symbols, for example "$x \geqslant 2$"

- graphically, for example the graph shows the region that satisfies $x \geqslant 2$

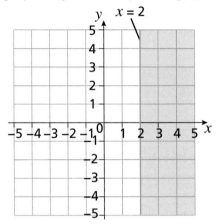

- on a number line, for example $x \geqslant 2$

For each part of this question, you are given an inequality in one of these forms. Represent the inequality in the other three ways.

a y is more than 3

b $x \leqslant 1$

c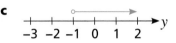

d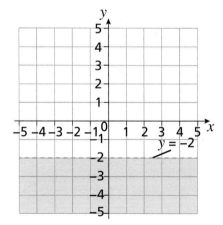

2 Solve these inequalities and represent each solution graphically.

a $3x - 1 \geqslant 11$ **b** $2y - 4 < -6$

3 Write the inequality which describes each shaded region.

a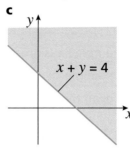

b

c

d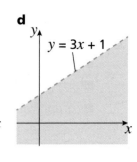

4 **a** Where does the graph of $4x + 3y = 12$ meet the coordinate axes?

b Draw x- and y-axes numbered from −5 to 5. On your grid, shade the region that satisfies $4x + 3y < 12$

Stretch – can you deepen your learning?

1 Using set notation, the region shown can be described as $\{x > 2\} \cup \{y < 3]$

a Sketch the region given by $\{x > 2\} \cap \{y < 3]$

b Challenge a partner to create and describe regions using set notation.

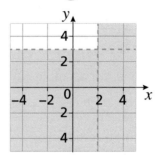

2 At the school shop, revision guides cost £6 and workbooks cost £8

Seb can spend up to £72. He wants to buy more than two revision guides and more than three workbooks.

a Explain how this graph models Seb's situation.

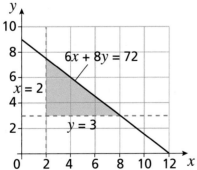

b List the possible combinations of revision guides and workbooks that Seb can buy.

Reflect

How do you use a graph to represent an inequality? What is different about regions bounded by dotted lines and those bounded by solid lines?

16 Algebraic representations
Chapters 16.1–16.3

White Rose Maths

I have become **fluent** in...	I have developed my **reasoning** skills by...	I have been **problem-solving** through...
■ reading quadratic graphs ■ drawing parabolas ■ interpreting piece-wise graphs ■ solving simultaneous equations using graphs Ⓗ ■ showing inequalities using number lines and graphs.	■ recognising shapes of graphs ■ comparing sets of graphs ■ deciding when to draw straight lines and when to draw curves ■ interpreting results read from graphs ■ making connections between algebraic and graphical representations.	■ showing information in a variety of forms ■ selecting appropriate concepts, methods and techniques ■ choosing the best approach to graph simultaneous equations. Ⓗ

Check my understanding

1 Here is the graph of $y = x^2 - 3x + 1$

Estimate

   a the value of y when $x = 1.5$

   b the values of x when $y = 1.5$

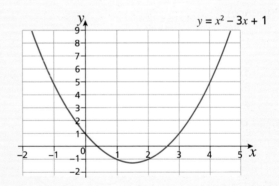

2 The graph shows the cost of taxi fares for journeys of different lengths.

   a How much does a 2.5-mile journey cost?

   b Seb has £4.75. What is the maximum distance he can travel by taxi?

3 Draw the graph of $y = x^2 + 2x + 3$ for values of x from -3 to 2

4 Represent these inequalities

     i on a number line   **ii** on a graph.

   a $x \geqslant -2$ **b** $-3 < y < 2$

5 Use a graphical method to solve the simultaneous equations $4x + 2y = -8$ and $x - y = 1$ Ⓗ

In this block, I will review and extend my learning about...

Handling data

Number of visits	Frequency	Subtotals
0	5	0
1	8	8
2	17	34
3	6	18
4	3	12
5	1	5
	40	77

$$\text{Mean} = \frac{\text{total}}{\text{number of items}} = \frac{77}{40} = 1.925$$

"The words used in the local newspaper are shorter on average than the words used in a national newspaper."

"Salt and vinegar crisps are the easiest to identify in a blind tasting."

"Goals are more likely in the last five minutes of a football match than in any other five-minute period."

Sequences

How many sticks are needed for the 4th pattern? The 10th pattern? The nth pattern? The 100th pattern?

First differences

Second differences

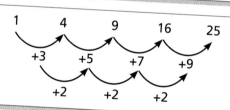

Trigonometry

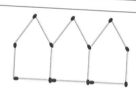

$$\sin 25° = \frac{DE}{AE} = \frac{CF}{AF} = \frac{BG}{AG}$$

$$\cos 25° = \frac{AD}{AE} = \frac{AC}{AF} = \frac{AB}{AG}$$

$$\tan 25° = \frac{DE}{AD} = \frac{CF}{AC} = \frac{BG}{AB}$$

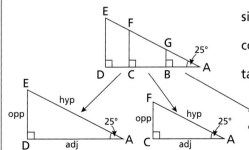

17.1 Handling data review

Reminders and links

Here are some of the statistical diagrams you drew in Books 1 and 2

Bar charts and pie charts show information about one variable, while scatter diagrams look at two variables and the relationship between them.

Bar chart

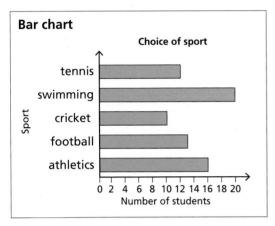

Pie chart

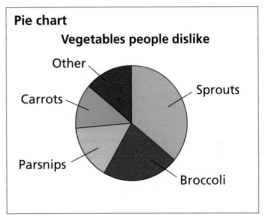

Scatter graph

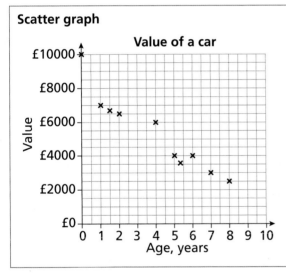

This graph shows that there is a negative correlation between the age of a car and its value.

You learned how to calculate these **measures of location**.

Mean – the total of a set of items divided by the number of items

Median – the middle value of a set of numbers when they are arranged in order

If there is an even number of items, then the median is the mean of the middle pair.

If there are n items, the median is in the $(\frac{n+1}{2})^{\text{th}}$ position

Mode – the most common item in a set of data

You also calculated this **measure of spread**.

Range – the difference between the lowest and the highest values in a set of data
The greater the range, the more spread out the data set is.

Moving into Key Stage 4

You will continue to use the diagrams and measures of location and spread that you used in Key Stage 3. You will also learn how to draw and interpret the diagrams below:

Frequency diagrams

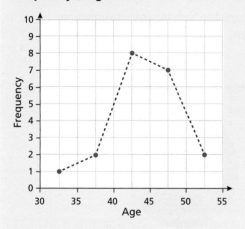

Histograms

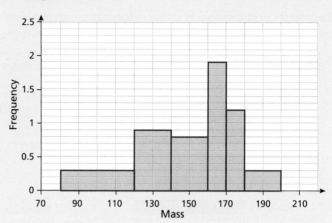

Cumulative frequency diagrams

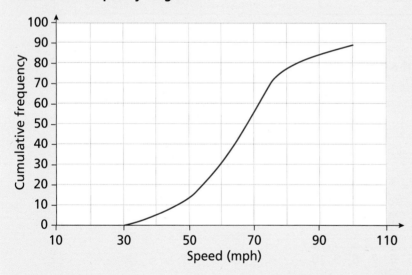

In addition, you will find out about **quartiles**, which split a distribution into four equal parts; this is similar to the way in which the median splits a distribution into two equal parts. These can be used to construct **box plots**, which are very useful for comparing sets of data.

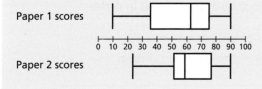

In this chapter, you will look at data in frequency tables. Some of this content may be familiar to you if you completed all the Higher Ⓗ steps in Book 2

Example 1

The table shows how many times a group of 40 people visited a dentist in the past year.

Number of visits	Frequency
0	5
1	8
2	17
3	6
4	3
5	1

Work out

a the mode   **b** the median   **c** the mean

number of visits to the dentist made by the group.

a   2

The mode is the data value with the highest frequency, so the modal number of visits was 2

Number of visits	Frequency
0	5
1	8
2	17
3	6
4	3
5	1

b

Number of visits	Frequency	Cumulative frequency
0	5	5
1	8	$5 + 8 = 13$
2	17	$13 + 17 = 30$
3	6	$30 + 6 = 36$
4	3	$36 + 3 = 39$
5	1	$39 + 1 = 40$

The median is the $(\frac{40 + 1}{2})^{th} = 20.5^{th}$ item.

Median = 2

You can find the median from a table using the cumulative frequency. This is the total of all frequencies so far in a frequency distribution.
You find the cumulative frequencies by adding each frequency to the total so far.

The median is in the $(\frac{n + 1}{2})^{th}$ position
This is halfway between the 20th and 21st items.
Both the 20th and 21st items are 2 so the median is 2

c

Number of visits	Frequency	Subtotals
0	5	0
1	8	8
2	17	34
3	6	18
4	3	12
5	1	5
	40	77

Instead of working out the mean by adding $0 + 0 + 0 + 0 + 0 + 1 + 1 + 1 +...$ and so on, you can add an extra column to the table and find subtotals

You find the subtotals by multiplying. For example, 2 occurs 17 times so these data items add up to $2 \times 17 = 34$. In the same way, 3 occurs 6 times so these data items add up to $3 \times 6 = 18$

You add the subtotals together to find the overall total of all 40 items.

$$\text{Mean} = \frac{\text{total}}{\text{number of items}} = \frac{77}{40} = 1.925$$

Use the formula for the mean. The number of items is the total frequency.

Practice 17.1A

1 Find the mean, median, mode and range of each of these sets of data.

   a   5   7   10   12   12

   b   21   23   8   12   50   12   46   30

2 A group of people were asked how many holidays they had taken over the last three years. The table shows the results.

Number of visits	Frequency
0	37
1	23
2	28
3	18
4	8
5	6

   a   How many people were asked altogether?

   b   Explain why it would be easy to represent this data on a pie chart.

   c   Write down the modal number of holidays taken.

   d   Show that the median number of holidays taken was 1.5

   e   Work out the mean number of holidays taken.

3 For the data in each of the tables, find the

    i   mode

    ii   median

    iii   mean

    iv   range.

a

Number of cars	0	1	2	3	4
Frequency	4	17	8	3	1

b

Score in test	6	7	8	9	10
Number of students	2	6	9	6	5

4 The grouped frequency table shows the masses of some plants.

Mass (g)	Frequency
$100 < w \leqslant 150$	36
$150 < w \leqslant 200$	42
$200 < w \leqslant 250$	53
$250 < w \leqslant 300$	28
Total	159

    a   Identify the modal class.

    b   In which class does the plant with median mass lie?

5 The grouped frequency table shows the lengths of time that a group of 150 people spent exercising last week.

Time (hours)	Frequency	Midpoint	Subtotals
$0 < t \leqslant 1$	63	0.5	31.5
$1 < t \leqslant 3$	15	2	30
$3 < t \leqslant 7$	27	5	
$7 < t \leqslant 10$	24		
$10 < t \leqslant 15$	21		
Total	150		

The estimates of the subtotals are found by multiplying each midpoint by the corresponding frequency, for example $63 \times 0.5 = 31.5$
Each midpoint is the mean of the endpoints of the class, for example $\frac{3 + 7}{2} = 5$

    a   Copy and complete the table to find an estimate of the mean time spent exercising last week.

    b   Find the class in which the median lies.

Hint: find an estimate of the overall total by adding up the subtotals.

6 For the data in each of the tables, find

    i the modal class

    ii the class in which the median lies

    iii an estimate for the mean.

a

Length (cm)	$0 < l \leqslant 10$	$10 < l \leqslant 30$	$30 < l \leqslant 60$	$60 < l \leqslant 75$
Frequency	18	35	32	15

b

Time online (hours)	$0 < t \leqslant 1$	$1 < t \leqslant 3$	$3 < t \leqslant 5$	$5 < t \leqslant 10$
Frequency	27	10	35	22

7 The mean height of a class of 30 students is 162 cm. A student who is 180 cm tall leaves the class. Find the new mean height of the class.

Might the other averages have changed? How can you tell?

8 The mean of five numbers is 16. When a sixth number is added, the mean increases to 17. Find the sixth number.

Challenges

1 Assume that the data is distributed evenly across each of the class intervals in Practice questions **4**, **5** and **6**. How could you use proportional reasoning to work out an estimate of the median?

2 Describe how you could test these hypotheses:

"The words used in the local newspaper are shorter on average than the words used in a national newspaper."

"Salt and vinegar crisps are the easiest to identify in a blind tasting."

"Goals are more likely to be scored in the last five minutes of a football match than in any other five-minute period."

What data should you collect? How would you present your findings? What charts or measures would you use?

3 Investigate the relationship between the amount of time a person spends online and the amount of time they spend sleeping. How is this investigation different from those suggested in Challenges question **2**?

Reminders and links

In Books 1 and 2 you learned about **linear sequences** (also called arithmetic sequences) where the difference between successive terms is constant.

17   21   25   29   …

+ 4   + 4   + 4

You also explored a variety of non-linear sequences such as:

- **geometric sequences** – where each term is found by multiplying the previous term by a constant multiplier

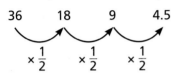

36   18   9   4.5   …

$\times \frac{1}{2}$ $\times \frac{1}{2}$ $\times \frac{1}{2}$

- **Fibonacci sequences** – where each term is the sum of two previous terms

4, 9, 13, 22, 35 …

4 + 9   9 + 13   13 + 22

- sequences formed from shapes, including **triangular numbers**, **square numbers** and **cube numbers**.

You learned how to generate sequences and to describe rules for sequences, using words and algebra.

Moving into Key Stage 4

- You will continue to work with all these types of sequences and will extend your knowledge to working with **quadratic sequences**, where the algebraic rules involve a term in n^2

$n^2 + 3n + 4$ When $n = 1$, $1^2 + 3 \times 1 + 4 = 8$

$n = 2$, $2^2 + 3 \times 2 + 4 = 14$

$n = 3$, $3^2 + 3 \times 3 + 4 = 22$

$n = 4$, $4^2 + 3 \times 4 + 4 = 32$

- You will look at the rules for generating geometric sequences and sequences involving surds, such as $\sqrt{2}$

- In addition, you will learn about **recurrence relations** such as $x_{n+1} = \sqrt{\dfrac{5}{x_n} + 3}$ and will use sequences of numbers to solve complex equations.

In this chapter, you will use and find algebraic rules for sequences.

You may already be familiar with some of this content if you completed all the Higher ⒽⓈ steps in Book 2

A **term-to-term** rule explains how one term in a sequence is connected to the next term.

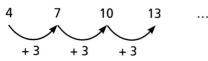

The term-to-term rule is "add three to the previous term".

A **position-to-term** rule explains how a term in a sequence is connected to its position in the sequence.

The position is usually denoted n, and the rule called the nth term rule.

Position (n)	1	2	3	4
Term	4	7	10	13

For a linear sequence, the rule will involve multiplying the position by the constant difference, in this case, $3 \times n = 3n$

You can work out the nth term rule by comparing $3n$ with the sequence itself.

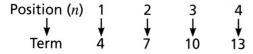

You need to add 1 to each value of $3n$ to get the corresponding term of the sequence.

The rule for the nth term of the sequence is $3n + 1$

Check: 4th term $= 3 \times 4 + 1 = 12 + 1 = 13$ ✓

Example 1

a Find the rule for the nth term of the sequence 10, 14, 18, 22...

b Find the 100th term of the sequence.

c Show that 836 is not a term in the sequence.

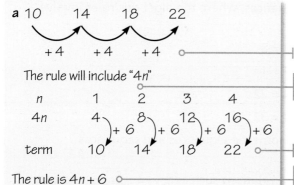

Find the common difference.

This is the coefficient of n in the rule for the sequence.

Compare the sequence with $4n$

As each term in the sequence is 6 greater than $4n$, the rule is $4n + 6$

b When $n = 100$

$4n + 6 = 4 \times 100 + 6$

$= 400 + 6$

$= 406$

Substitute $n = 100$ into the rule you found in part **a**

c If 836 is in the sequence, then

$4n + 6 = 836$

$4n = 830$

$n = 207.5$

The position of a term, n, must be an integer, so 836 is not a term in the sequence.

Form an equation using your rule.

Solve the equation to find the value of n

State your conclusion clearly.

Practice 17.2A

1 By finding the first four terms of each sequence, verify that the sequences given by the rules $5n + 3$, $5n - 1$, $5n + 11$ all have a common difference of 5 between successive terms.

2 **a** Which **two** of these sequences will have rules of the form $5n + k$ or $5n - k$?

A (7, 12, 17, 22...) B (8, 13, 19, 26...) C (1, 6, 11, 16...) D (5, 9, 13, 17...)

b Find the rule for the nth term of each of the two sequences you selected in part **a**

3 Find the rule for the nth term of each of these sequences.

a 5, 8, 11, 14...   **b** 3, 7, 11, 15...   **c** 6, 12, 18, 24...

d 5, 12, 19, 26...   **e** 10, 19, 28, 37...   **f** 7, 17, 27, 37...

g 12, 13, 14, 15...

4 Bobbie thinks the sequences that start 10, 13, 16, 19... and 13, 16, 19, 22... will have the same rule.

Find the rules for the nth term of the two sequences and show that Bobbie is wrong.

5 **a** What's the same and what's different about sequence A and sequence B?

A (7, 13, 19, 25...)   B (25, 19, 13, 7 ...)

b Verify that the rule for sequence B is $31 - 6n$

c Find the rule for the nth term of these sequences.

i 50, 40, 30, 20...

ii 75, 73, 71, 69...

iii 58, 55, 52, 49...

6 Which term of the sequence 7, 11, 15, 19... is equal to 875?

7 **a** Find the first term in the sequence 35, 41, 47, 53... that is greater than 1000

b How many terms in the sequence 230, 222, 214, 206... are positive?

Challenges

1 Here is a sequence of shapes made from matchsticks.

a Find the rule for the number of matchsticks needed to make the nth shape in the sequence.

b Explain how your rule is related to the shapes.

c What would the rule be if, instead of pentagons, the shapes in the sequence were nonagons?

d What would the rule be if the shapes were p-sided polygons?

e Investigate other shapes, such as the one shown below. Can you generalise?

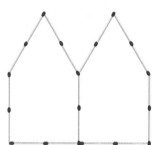

2 Investigate the sequences formed using these rules.

"Start at 8. The term-to-term rule is double the previous term and add 1"

"Start at 8. The term-to-term rule is halve the previous term and add 1"

a What's the same and what's different about these sequences?

b What happens if you vary the starting numbers and the instructions?

3 Investigate Flo's conjecture.

If the first term of a linear sequence is a and the common difference between the terms is d, then the nth term can be represented by the expression $a + (n - 1)d$

4 Show that the rule ar^n, where a and r are constants and n is the position of a term in the sequence, produces a geometric sequence. Investigate with positive and negative values of a and r

5 Here are the first and second differences of the sequence given by the rule n^2

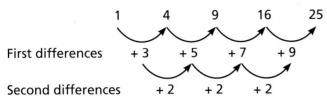

1   4   9   16   25

First differences   + 3   + 5   + 7   + 9

Second differences   + 2   + 2   + 2

a Find the first and second differences of the rules given by

   i $n^2 + 3n$

   ii $n^2 + 3n + 1$

   iii $n^2 + 3n - 1$

   What's the same and what's different? Investigate the sequences given by the rules $n^2 + an + b$ for other values of a and b, both positive and negative.

b By comparing them with the sequence given by n^2, find the rule for the nth term of each of these sequences.

   i 4, 13, 24, 37, 52

   ii −4, 1, 8, 17, 28

   iii −9, −16, −21, −24, −25

6 Here are the first few rows of Pascal's triangle, which is named after the French mathematician Blaise Pascal (1623–1662).

```
                    1
                 1     1
              1     2     1
           1     3     3     1
        1     4     6     4     1
     1     5    10    10     5     1
  1     6    15    20    15     6     1
```

a Write the next row of the triangle.

b What sequences can you see? Can you find any linear sequences or the triangle numbers? Can you make any predictions?

Reminders and links

In Chapter 12.4 of this book, you explored similar triangles and looked at the ratios of the lengths of the sides in right-angled triangles.

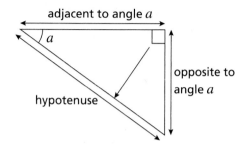

adjacent to angle a

a

opposite to angle a

hypotenuse

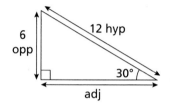

6 opp

12 hyp

30°

adj

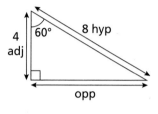

4 adj

60°

8 hyp

opp

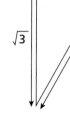

60°

$\sqrt{3}$

2

$$\sin 30° = \frac{6}{12} = \frac{1}{2}$$

$$\cos 60° = \frac{4}{8} = \frac{1}{2}$$

$$\tan 60° = \frac{\sqrt{3}}{1} = \sqrt{3}$$

Moving into Key Stage 4

You will extend your knowledge of the connections between the sides and angles in right-angled triangles and you will use this knowledge to solve problems.

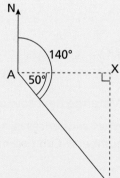

N

140°

A 50°

X

B

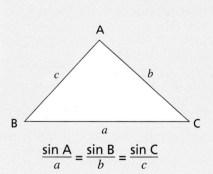

A

c

b

B

a

C

You will also learn how to use the trigonometric ratios in triangles that do not have a right angle.

$$\frac{\sin A}{a} = \frac{\sin B}{b} = \frac{\sin C}{c}$$

You already know that all right-angled triangles with an angle of 30° are similar.

The same is true for all right-angled triangles with an angle of 25°

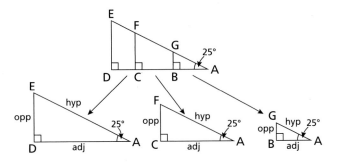

$$\sin 25° = \frac{DE}{AE} = \frac{CF}{AF} = \frac{BG}{AG}$$

$$\cos 25° = \frac{AD}{AE} = \frac{AC}{AF} = \frac{AB}{AG}$$

$$\tan 25° = \frac{DE}{AD} = \frac{CF}{AC} = \frac{BG}{AB}$$

This is also true for any other angle, x

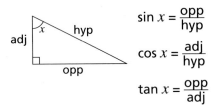

$$\sin x = \frac{opp}{hyp}$$

$$\cos x = \frac{adj}{hyp}$$

$$\tan x = \frac{opp}{adj}$$

For any given angle, the ratios of the lengths of the sides in the triangles are constant. The values are stored in your calculator. You can use these ratios to work out missing sides and angles in right-angled triangles.

You use this key with the sin, cos, tan keys to find the angle when you know the ratio.

sine

tangent

cosine

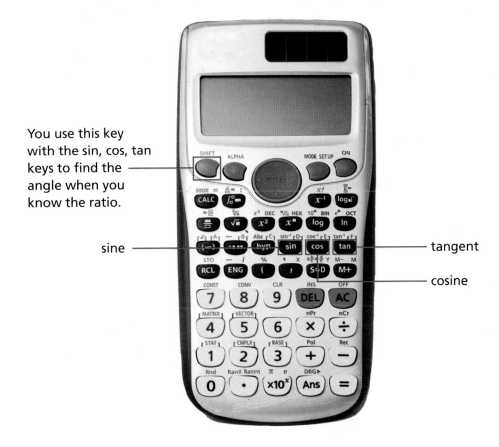

Example 1

Work out the length of side AB

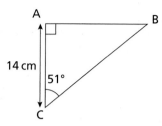

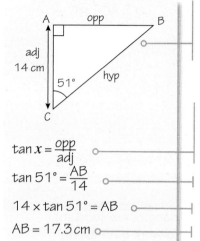

First, label the sides of the triangle opposite (opp), adjacent (adj) and hypotenuse (hyp) using the positions of the right angle and the given angle to decide which is which.

$\tan x = \dfrac{opp}{adj}$

$\tan 51° = \dfrac{AB}{14}$

Decide which ratio to use and write it down. Here, you know the adjacent side and want to find length of the opposite side, so you need to use the ratio that features both of these sides – the tangent ratio.

Substitute the information you know.

$14 \times \tan 51° = AB$

Rearrange the equation to find AB

$AB = 17.3\,cm$

Use your calculator to find the answer. Include the units.

Example 2

Work out the size of angle PQR

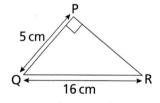

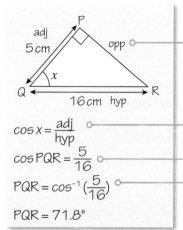

First, label the required angle. Then label the sides of the triangle using the positions of the right angle and angle PQR to decide which side is which.

$\cos x = \dfrac{adj}{hyp}$

$\cos PQR = \dfrac{5}{16}$

Decide which ratio to use and write it down. Here, you know the adjacent side and the hypotenuse, so you need to use the ratio that features both these sides – the cosine ratio.

Substitute the information you know.

$PQR = \cos^{-1}\left(\dfrac{5}{16}\right)$

If you know a ratio (sin, cos, or tan) you use the **inverse function** on your calculator to find the angle. These are denoted $\sin^{-1}$, $\cos^{-1}$ and $\tan^{-1}$

$PQR = 71.8°$

Practice 17.3A

Unless otherwise stated, give the answers in this exercise correct to 3 significant figures.

1 **a** Check that your calculator is in the correct mode (degrees mode) by working out the value of sin 30°. You should get the answer $\frac{1}{2}$ or 0.5. If you don't, investigate how to change to the correct mode on your calculator and then try again.

b Find

i sin 40°   **ii** cos 48°   **iii** tan 62°

iv sin⁻¹ 0.7   **v** cos⁻¹ 0.7   **vi** tan⁻¹ 0.7

c Find the sine, cosine and tangent of 70°

d **i** The sine of an angle is 0.4. What is the angle?

ii The cosine of an angle is 0.4. What is the angle?

iii The tangent of an angle is 0.4. What is the angle?

2 Copy the triangles and label the hypotenuse (hyp), the side opposite the given angle (opp) and the side adjacent to the given angle (adj).

a
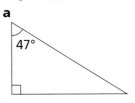
47°

b
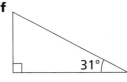
34°

c
72°

d
58°

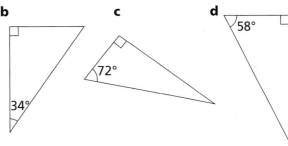

e
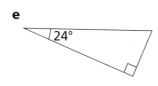
24°

f
31°

g
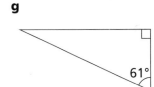
61°

3 Solve these equations.

a $\frac{t}{9} = 1.2$ **b** $8 = \frac{g}{0.6}$ **c** $0.4 = \frac{12}{j}$

4 Find the length of each side labelled x in these triangles.

Start by sketching each triangle and labelling the sides opp, adj and hyp to help you to select the correct trigonometric ratio.

a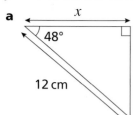

b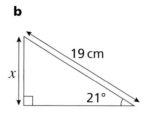

c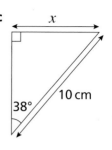

d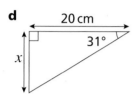

e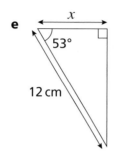

f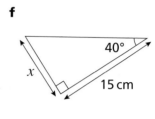

5 Find the size of each angle labelled x in these triangles.

Start by sketching each triangle and labelling the sides to help you to select the correct trigonometric ratio to use.

a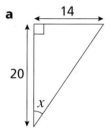

b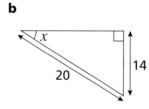

c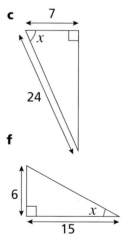

d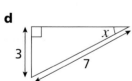

e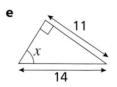

f

6 Find the length of the side labelled x

Hint: in part **a** you will need to rearrange the formula to make the denominator the subject.

a by using the 54° angle

b by using the unmarked angle in the triangle.

7 Work out size of angle ACD

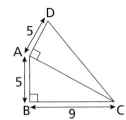

8 A rectangle measures 8 cm by 5 cm. Find the size of the angle between a diagonal and one of the longer sides of the rectangle.

9 An isosceles triangle has sides of length 10 cm, 10 cm and 8 cm. Find the area of the triangle using

    a Pythagoras' theorem   **b** trigonometry.

10 WXYZ is a rhombus with sides of length 10 cm. Angle WXY is 80°

Find the lengths of the diagonals of rhombus WXYZ

Challenges

1 **a** Use sketches of right-angled triangles to explain why the sine and cosine of an angle must be less than 1, but the tangent of an angle can be more than 1

    b Use a sketch to determine $\tan^{-1} 1$

2 **a** Find

    i $\sin 20°$ **ii** $\cos 70°$ **iii** $\cos 40°$ **iv** $\sin 50°$ **v** $\sin 10°$ **vi** $\cos 80°$

    Give your answers to 3 significant figures.

    b What do you notice about your answers to part **a**? Generalise your result and use a sketch of a right-angled triangle to explain why it is true.

    c Use the formulae for sine, cosine and tangent to show that Jackson's conjecture is true.

> If you divide the sine of an angle by the cosine of an angle, the result is the tangent of that angle.

3 Use the fact that $\sin 60° = \dfrac{\sqrt{3}}{2}$ to prove that the area of an equilateral triangle with sides of length a is $\dfrac{\sqrt{3}a^2}{4}$

4 Another way of defining the sine and cosine of an angle is as the coordinates of a point on a circle of radius 1 (this is called a unit circle).

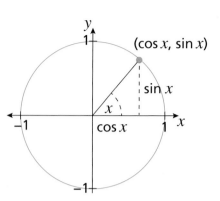

Use the unit circle to explore the sine and cosine of angles greater than 90°. In which quadrants will these ratios be positive and in which will they be negative? Extend your investigation to include the tangent. (Recall from question **2** that $\tan x = \frac{\sin x}{\cos x}$)

5 The picture shows a student using a clinometer to find the height of a tree.

Investigate how to make your own clinometer. Use it to estimate the heights of tall objects such as buildings or trees around your home or school.

Glossary

2-D shape – a flat shape with two dimensions such as length and width

3-D shape – a shape with three dimensions: length, width and height

Acute angle – an angle less than 90°

Adjacent sides/angles – sides or angles that are next to each other

Alternate angles – a pair of angles between a pair of parallel lines on opposite sides of a transversal

Annual – covers a period of one year

Area – the amount of space inside a 2-D shape

Axis (plural: **axes**) – a reference line on a graph

Balance – an amount of money in an account

Best buy (or **best value**) – the item which is cheapest when equal-sized amounts of different items are compared

Bill – shows how much money is owed for goods or services

Binomial – expression with two terms

Bisector – a line that divides something into two equal parts

Capacity – how much space a 3-D shape, or container, holds

Centre of enlargement – point from which an enlargement is made

Co-interior angles – a pair of angles between a pair of parallel lines on the same side of a transversal

Compound shape – also known as a composite shape, this is a shape made up of two or more other shapes

Congruent – exactly the same size and shape, but possibly in a different orientation

Conjecture – a statement that might be true that has not yet been proved

Constant – not changing

Construct – draw accurately using a ruler and compasses

Conversion graph – a graph used to change from one unit to another

Convert – change from one form to another, for example a percentage to a decimal

Coordinate – an ordered pair used to describe the position of a point

Corresponding angles – a pair of angles in matching positions compared with a transversal

Counterexample – an example that disproves a statement

Credit – an amount of money paid into an account

Cross-section – the shape you get when you slice a prism parallel to its base

Currency – the system of money used in a particular country

Debit – an amount of money taken out of an account

Decimal – a number with digits to the right of the decimal point

Denominator – the bottom number in a fraction; it shows how many equal parts one whole has been divided into

Density – a measure of mass per unit of volume

Deposit – amount of money paid into a bank account

Depreciate – reduce or decrease in value

Dimensions – measurements such as the length, width and height of something

Direct proportion – two quantities are in direct proportion when, as one increases or decreases, the other increases or decreases at the same rate

Directed numbers – numbers that can be negative or positive

Divide in a ratio – share a quantity into two or more parts so that the shares are in a given ratio

Edge – a line segment joining two vertices of a 3-D shape; it is where two faces of a 3-D shape meet

Enlargement – making a shape bigger, or smaller

Equation – a statement with an equal sign, which states that two expressions are equal in value

Equidistant – at the same distance from

Equivalent – numbers or expressions that are written differently but are always equal in value

Error interval – the range of values a number could have taken before being rounded

Estimate – an approximate answer or to give an approximate answer

Exchange rate – the value of a currency compared to another

Expand – multiply to remove brackets from an expression

Expected outcomes – an estimate of how many times a possible outcome will occur

Exterior angle – an angle between the side of a shape and a line extended from the adjacent side

Face – a flat surface of a 3-D shape

Factor – a positive integer that divides exactly into another positive integer

Factorise – find the factors you need to multiply to make an expression

Fair – equally likely to land on any possible outcome

Formula (plural: **formulae**) – a rule connecting variables written with mathematical symbols

Fractional scale factor – a scale factor that is a fraction, not an integer

Front/side elevation – when an object is viewed from the front or side

Give a reason – state the mathematical rule(s) you have used, not just the calculations you have done

Gradient – the steepness of a line

Graphical – using a graph

Hypotenuse – the side opposite the right angle in a right-angled triangle

Income tax – a tax that is payable on personal income such as wages and salary

Independent events – not affected by other events

Inequality – a comparison between two quantities that are not equal to each other

Integer – a whole number

Intercept – the point at which a graph crosses, or intersects, a coordinate axis

Interest – a percentage fee paid when borrowing money or a percentage earned when you deposit money into a savings account

Interior angle – an angle on the inside of a shape

Inverse proportion – if two quantities are in inverse proportion, when one quantity increases, the other decreases at the same rate

Irrational number – a number that cannot be written in the form $\frac{a}{b}$ where a and b are integers

Irregular – a shape that has unequal sides and unequal angles

Isometric drawing – a method of drawing 3-D shapes in two dimensions using special dotty paper

Isosceles – having two sides the same length

Line of symmetry – a line that cuts a shape exactly in half

Line segment – part of a line that connects two points

Litre – 1000 cm³

Locus (plural: **loci**) – a set of points that describe a property

Loss – if you buy something and then sell it for a smaller amount,
loss = amount paid – amount received

Mass – the amount of matter that makes up an object

Midpoint – the point halfway between two others

Multiple – the result of multiplying a number by a positive integer

Multiplier – a number you multiply by

Negative numbers – numbers less than zero

Net – a 2-D shape that can be folded to make a 3-D shape

Numerator – the top number in a fraction that shows the number of parts

Obtuse angle – an angle more than 90° but less than 180°

Opposite sides/angles – sides or angles that are not next to each other

Orientation – the position of an object based on the direction it is facing

Original value – a value before a change takes place

Overtime – the time worked in addition to a person's normal working hours

Parabola – a type of curve that is approximately U-shaped and has a line of symmetry

Parallel – always the same distance apart and never meeting

Per annum – means "per year"

Perpendicular – at right angles to

Piece-wise graph – a graph that consists of more than one straight line

Plan view – when an object is viewed from above

Point of intersection – the point where two graphs cross each other

Power (or **exponent**) – this is written as a small number to the right and above the base number, indicating how many times to use the number in a multiplication. For example, the 5 in 2^5

Prime factor decomposition – writing numbers as a product of their prime factors

Prime number – a positive integer with exactly two factors, 1 and itself

Principal – an initial amount invested

Prism – a solid shape with polygons at its ends and flat surfaces

Probability – how likely an event is to occur

Profit – if you buy something and then sell it for a higher amount,
profit = amount received – amount paid

Proof – an argument that shows that a statement is true

Prove – to show that something is always true

Quadratic – an expression is quadratic when the highest power of the variable is 2

Rate – a comparison between two quantities

Ratio – a ratio compares the sizes of two or more values

Rational number – a number that can be written in the form $\frac{a}{b}$ where a and b are integers

Real number – all positive and negative numbers including decimals and fractions

Reciprocal – the result of dividing 1 by a given number. The product of a number and its reciprocal is always 1

Reflection – a transformation resulting in a mirror image

Regular – a shape that has equal sides and equal angles

Regular polygon – a polygon whose sides are all equal in length and whose angles are all equal in size

Relative frequency – the number of times an event occurs divided by the number of trials

Repeated percentage change – when an amount is changed by one percentage followed by another

Replacement – putting back – when an item is replaced, the probabilities do not change; when an item is not replaced, the probabilities do change

Reverse percentage – a problem where you work out the original value

Right angle – an angle of 90°

Rotation – turn a shape around a fixed point called the centre of rotation

Rotational symmetry – when a shape still looks the same after turning

Satisfy – make an equation or inequality true

Scale – the ratio of the length in a drawing or a model to the actual object

Scale factor – how much a shape has been enlarged by

Significant figure – the most important digits in a number that give you an idea of its size

Similar – two shapes are similar if their corresponding sides are in the same ratio

Simultaneous – at the same time

Sketch – a rough drawing

Solution – a value you can substitute in place of the unknown in an equation or inequality to make it true

Solve – find a value that makes an equation true or find a set of values that make an inequality true

Speed – the rate at which an object is moving

Square root – a square root of a number is a value that, when multiplied by itself, gives the number

Standard form – a number written in the form $A \times 10^n$ where A is at least 1 and less than 10, and n is an integer

Subject – the variable in a formula that is expressed in terms of the other variables

Surd – a root that cannot be written as an integer

Surface area – the sum of the areas of all the faces of a 3-D shape

Table of values – this lists pairs of values that represent a relationship between two variables and that can be used to plot a graph

Tax allowance – the amount that you can earn before being taxed

Transformation – a way of changing the size or position of a shape

Translation – this moves a shape up, down or from side to side but it does not change its appearance in any other way

Transversal – a line that crosses at least two other lines

Tree diagram – a way of recording possible outcomes that can be used to find probabilities

Trigonometry – the study of lengths and angles in triangles

Unit cost/price – the cost or price of 1 item or 1 unit of an item

Unitary method – a technique for solving problems by first finding the value of one unit

Variable – a numerical quantity that might change, often denoted by a letter, for example x or t

VAT (value-added tax) – a tax that is added to the price of some goods and services

Vector – describes movement from one point to another

Vertex (plural: **vertices**) – a point where two line segments meet; a corner of a shape

Volume – the amount of space taken up by a 3-D shape

Answers

Block 1 Straight line graphs

Chapter 1.1

Are you ready?

1 a i They all have a 3 in the x position
  ii In each pair of the coordinates, the x-value is equal to 3
  b i They all have −7 in the y position
  ii In each pair of the coordinates, the y-value is equal to −7

2 a Any three pairs of coordinates of the form $(-1, a)$
  b Any three pairs of coordinates of the form $(b, 4)$

3 b three lots of x plus five
  c five lots of x subtract four
  d negative x plus one

4 a 11 **b** 20 **c** 21 **d** −4

Practice 1.1A

1 a Correct coordinate grid
  b i Any three pairs of coordinates of the form $(a, 3)$
  ii

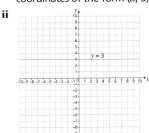

  iii For example (9, 3); the y-value is equal to 3
  c

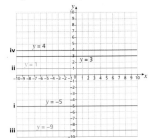

  d Each line of the form $y = a$ is a horizontal line through point $(a, 0)$ on the y-axis.

2 a i Any three pairs of coordinates of the form $(3, b)$
  ii

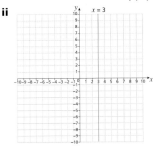

  iii For example (3, 9); the x-value is equal to 3
  b

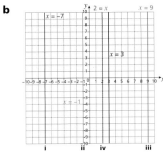

  c Each line of the form $x = b$ is a vertical line through point $(b, 0)$ on the x-axis.

3

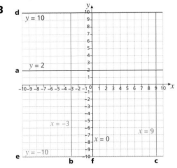

4 a $x = 18$ **b** $y = -12$
  c $x = -100$ **d** $y = 83$

5 a i Any pair of coordinates of the form (a, a)
  ii

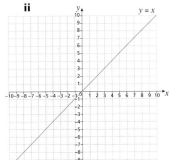

  iii For example $(-7, -7)$; the y-value is equal to the x-value
  b

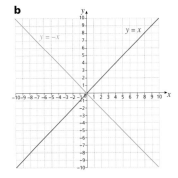

What do you think?

1 No; such a line is parallel to the y-axis

2 a $x = -17$ **b** $y = 12.4$ **c** $y = 4$
  d $y = -7$ **e** $y = x$

3 No; it depends on the scales used on the two axes

Practice 1.1B

1 a $y = 3x$ **b** $y = 3 + x$
  c $y = -3x$ **d** $y = x - 3$

2 a i At each point on the line $y = 3x$ the y-value is equal to 3 times the x-value
  ii

x	−2	−1	0	1	2
y	−6	−3	0	3	6

  iii

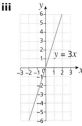

  b i At each point on the line $y = -3x$ the y-value is equal to −3 times the x-value
  ii

x	−2	−1	0	1	2
y	6	3	0	−3	−6

  iii

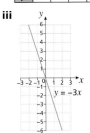

  c i At each point on the line $y = x + 3$ the y-value is equal to three more than the x-value

ii

x	-2	-1	0	1	2
y	1	2	3	4	5

iii

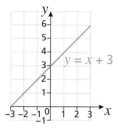

$y = x + 3$

d i At each point on the line $y = x - 3$ the y-value is equal to three less than the x-value

ii

x	-2	-1	0	1	2
y	-5	-4	-3	-2	-1

iii

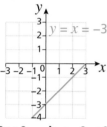

$y = x = -3$

3 a $y = 3x + 2$ **b** $y = 2x + 3$
c $y = 3(x + 2)$ **d** $y = \frac{x + 3}{2}$

4 a i The y-value is equal to two lots of the x-value plus three
ii The y-value is equal to three lots of the x-value plus two
iii The y-value is equal to three lots of two more than the x-value
iv The y-value is equal to half of three more than the x-value

b i

x	-2	-1	0	1	2
y	-1	1	3	5	7

ii

x	-2	-1	0	1	2
y	-4	-1	2	5	8

iii

x	-2	-1	0	1	2
y	0	3	6	9	12

iv

x	-2	-1	0	1	2
y	0.5	1	1.5	2	2.5

c

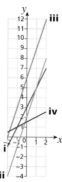

iii
iv
i
ii

5 a i The y-value is equal to negative two lots of the x-value plus three
ii The y-value is equal to negative three lots of the x-value plus two
iii The y-value is equal to negative three lots of two more than the x-value
iv The y-value is equal to half of three more than the negative of the x-value

b i

x	-2	-1	0	1	2
y	7	5	3	1	-1

ii

x	-2	-1	0	1	2
y	8	5	2	-1	-4

iii

x	-2	-1	0	1	2
y	12	9	6	3	0

iv

x	-2	-1	0	1	2
y	2.5	2	1.5	1	0.5

c

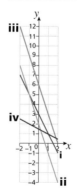

iii
iv
i
ii

6 The numbers and operations in the equations are the same but the coefficients of x are different. In all parts of question 5 the coefficient of x is negative, so the lines slope downwards to the right. In all parts of question 4 the coefficient of x is positive, so the lines slope upwards to the right. All the lines in question 4 have a positive gradient whereas all the lines in question 5 have a negative gradient.

7 i
a $y = x + 8$

x	-2	-1	0	1	2
y	6	7	8	9	10

b $y = x - 2$

x	-2	-1	0	1	2
y	-4	-3	-2	-1	0

c $y = 4x$

x	-2	-1	0	1	2
y	-8	-4	0	4	8

d $y = \frac{x}{2}$

x	-2	-1	0	1	2
y	-1	-0.5	0	0.5	1

ii

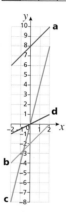

8 i
a $y = -5x$

x	-2	-1	0	1	2
y	10	5	0	-5	-10

b $y = 2x + 5$

x	-2	-1	0	1	2
y	1	3	5	7	9

c $y = -1 + 3x$

x	-2	-1	0	1	2
y	-7	-4	-1	2	5

d $y = 4x - 3$

x	-2	-1	0	1	2
y	-11	-7	-3	1	5

ii

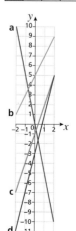

9 i

a $y = \dfrac{5 - x}{4}$

x	−2	−1	0	1	2
y	1.75	1.5	1.25	1	0.75

b $y = 2 - 6x$

x	−2	−1	0	1	2
y	14	8	2	−4	−10

c $y = 3(x + 1)$

x	−2	−1	0	1	2
y	−3	0	3	6	9

d $y = 3(2x + 1)$

x	−2	−1	0	1	2
y	−9	−3	3	9	15

ii

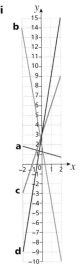

What do you think?

1 Zach can't be correct as his sketch has a negative gradient. Flo can't be correct as her sketch goes through the origin.

2 Add 15 to the previous term

3 a The x-values aren't consecutive
  b It will work when the x-values are consecutive

Consolidate

1 a $x = 1$ **b** $y = -4$
  c $y = x$ **d** $y = -x$

2 a i $y = 2x$

x	−2	−1	0	1	2
y	−4	−2	0	2	4

ii $y = 4x$

x	−2	−1	0	1	2
y	−8	−4	0	4	8

iii $y = x + 2$

x	−2	−1	0	1	2
y	0	1	2	3	4

iv $y = x + 4$

x	−2	−1	0	1	2
y	2	3	4	5	6

b

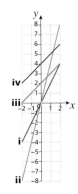

c Lines **i** and **ii** go through the origin, whereas lines **iii** and **iv** do not. Lines **iii** and **iv** have the same gradient, where as lines **i** and **ii** do not.

3 a i $y = 2x - 1$

x	−2	−1	0	1	2
y	−5	−3	−1	1	3

ii $y = 2x + 2$

x	−2	−1	0	1	2
y	−2	0	2	4	6

b

c The lines are parallel but go through different points on the y-axis

Stretch

1 The line doesn't go through the origin

2 No; $y = 7x + 3$ goes through (0, 3) but line this goes through (0, −5)

3 $(17, -\frac{34}{2})$ and $(7 - q, q - 7)$

4 (−19.8, 17)

5 Yes; $7 = x$ and $-2 = y$ meet at (7, −2) The point (7, −2) is on the line $y = 3 - \frac{5x}{7}$; substituting $x = 7$ into the equation gives $3 - \frac{5 \times 7}{7} = 3 - 5 = -2$ so $y = -2$

6 a i For example (3, 2)
    ii For example (3, 1)
  b (−1, 2), (6, 2), (6, −7) and (−5.5, −7)
  c 83.25 square units

Chapter 1.2

Are you ready?

1

x	−2	−1	0	1	2
y	−9	−5	−1	3	7

2

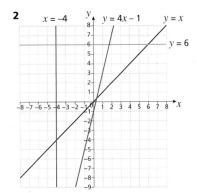

3 a i 4, 7, 10, 13, 16
    ii Add 3 to the previous term
  b i −1, 2, 5, 8, 11
    ii Add 3 to the previous term
  c They have the same term-to-term rule because both of the rules involve $3n$

Practice 1.2A

1 a

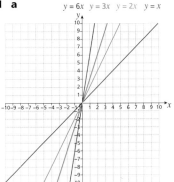

b They all go through the origin but have a different steepness
  c The greater the gradient, the greater the coefficient of x

2 a

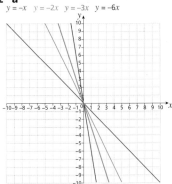

b They are all sloping downwards through the origin and have a different steepness
  c The graph slopes downwards from left to right

3 **a** Positive
b Negative
c Negative
d Positive

4 **a**

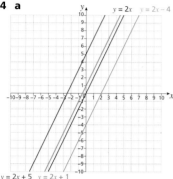

$y = 2x + 5$ $y = 2x + 1$

b **i** The lines don't meet and
will never meet
ii They have the same
coefficient of x
c Any equation of the form
$y = 2x + a$

5 **a**

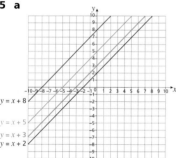

$y = x + 8$
$y = x + 5$
$y = x + 3$
$y = x + 2$

b The lines are all parallel
through different points
c The line crosses the y-axis
where the y-value is positive
(above the x-axis)

6 **a**

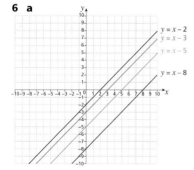

$y = x - 2$
$y = x - 3$
$y = x - 5$
$y = x - 8$

b The lines are all parallel
through different points
c The line crosses the y-axis
where the y-value is negative
(below the x-axis)

7 **a** Positive   **b** Negative
c Negative   **d** Positive

8 **a**

$y = 3 - x$ $y = 2x + 3$ $y = 3 + 4x$ $y = 5x + 3$

b **i** The lines all cross the y-axis
at the same point
ii The equations all have a
"+3" term in their
equations
c Any equation of the form
$y = ax + 3$

What do you think?

1 No; the 2 in $y = 2 + 3x$ isn't the
gradient, it is the y-intercept
2 **a** Any equation with
x coefficient greater than 3
b Any equation with
x coefficient 3
c Any equation with
x coefficient less than 3 but
greater than 0
d Any equation with negative
x coefficient
3 **a** A; it crosses the other lines
b A is of the form $y = ax + 9$
(where a is greater than 5),
C is $y = 5x$ and D is of the form
$y = 5x + b$ (where b is negative)
4 $b = -7$

Practice 1.2B

1 **a** **i** 5   **ii** 2   **iii** −4
iv −3   **v** $\frac{3}{5}$ **vi** $-\frac{1}{3}$
vii 7   **viii** −100
b **i** (0, 3) **ii** (0, −7) **iii** (0, 1)
iv (0, −9) **v** (0, 0) **vi** (0, 8)
vii (0, −11) **viii** (0, 134)
2 **a** **i** 10   **ii** −10
iii 10   **iv** −10
b **i** (0, 19)   **ii** (0, 19)
iii (0, −19)   **iv** (0, −19)
c **i** and **iii**, and **ii** and **iv** have
the same gradient as each
other but different
y-intercepts; **i** and **ii**, and **iii**
and **iv**, have the same
y-intercept as each other but
different gradients
3 **a** **i** 3   **ii** −2
iii −4   **iv** 3
v $\frac{3}{5}$ **vi** $-\frac{2}{3}$
vii −11   **viii** −134
b **i** (0, 5)   **ii** (0, 7)
iii (0, 0.5)   **iv** (0, 19)
v (0, 11)   **vi** (0, 54)
vii (0, 7)   **viii** (0, −100)

4 **a** **i** and **vii**; **ii** and **vi**; **iii** and **viii**;
iv and **v**
b **i** and **vi**; **ii** and **v**; **iii** and **iv**;
vii and **viii**
c Parallel lines have the same
gradient, those with the same
y-intercept have the same
value for c
5 **a** Any three equations of the
form $y = 23x + a$
b Any three equations of the
form $y = a - 7x$
c Any three equations of the
form $y = ax + 9$
d Any three equations of the
form $y = ax + 1$
6 **a** Not parallel; gradients are
different
b Parallel; both have gradient −1
c Parallel; both have gradient $\frac{2}{3}$
d Not parallel; different
gradients
e Not parallel; different
gradients
f Parallel; both have gradient 2
7 **a** Yes; at (0, 5)
b Yes; at (0, 0.5)
c No; different intercepts
d Yes; at (0, 11)
e No; different intercepts
f Yes; at (0, −7)
8 **a** $y = 7x + 2$
b $y = 4x - 9$
c $y = -5x$
d $y = 11 - x$
9 $y = 72x - 5$
10 $y = 10x - 21$

What do you think?

1 **a** Jackson has read the numbers
in order rather than reading
the gradient as the coefficient
of x- and the y-intercept as
the constant
b Gradient is 2 and y-intercept is 5
2 No; one has gradient −3 and one
has gradient +3
3 The gradient is just $\frac{119}{21}$; you don't
include the x
4 Gradient $\frac{4}{5}$; y-intercept 5

Consolidate

1 **a** 8   **b** 1   **c** −6
2 **a** (0, 17) **b** (0, $\frac{2}{3}$) **c** (0, −12.5)
3 Yes, they have the same gradient
and y-intercept so they are the
same line
4 **a** Any three equations with
gradient greater than 2
b Any three equations of the
form $y = ax + 4$
c Any three equations of the
form $y = 7x + b$

Stretch

1 Marta; they're the same line so
they're not parallel, they are the
same line
2 **a** $y = 19x - 13$
b $y = -19x - 13$
3 $y = \frac{5}{4}x + 34$

4 a $a = 8$
b $j = -\frac{1}{2}$, $k = 12$
5 a i The y-intercept is negative
ii This just means that a in the original equation is negative
b

6 a There is more than one line parallel to $y = 12x + 7$
b Yes; there are an infinite number of lines parallel to $y = 12x + 7$

Chapter 1.3

Are you ready?

1 a 140 **b** 280 **c** 14
d 42 **e** $70 - 5y$
2 a -4 **b** $(0, 9)$
3 $y = 7x - 3$
4 £93.75

Practice 1.3A

1 a $y = 9x + 10$ **b** $y = 2x + 7$
c $y = 4x + 5$ **d** $y = 4x + 1$
e $y = 2x + 18$ **f** $y = 4x + 8$
g $y = 11x + 5$ **h** $y = 2x + 2.5$
i $y = 3.5x + 4$ **j** $y = 6x + 1.5$
k $y = 0.5x + 1.5$ **l** $y = \frac{2}{3}x + \frac{1}{3}$
2 a i 9 **ii** $(0, 10)$
b i 2 **ii** $(0, 7)$
c i 4 **ii** $(0, 5)$
d i 4 **ii** $(0, 1)$
e i 2 **ii** $(0, 18)$
f i 4 **ii** $(0, 8)$
g i 11 **ii** $(0, 5)$
h i 2 **ii** $(0, 2.5)$
i i 3.5 **ii** $(0, 4)$
j i 6 **ii** $(0, 1.5)$
k i 0.5 **ii** $(0, 1.5)$
l i $\frac{2}{3}$ **ii** $(0, \frac{1}{3})$
3 The equation isn't in the form $y = mx + c$
4 a $y = 2x - 9$ **b** $y = 8x + 11$
c $y = -x + 1$ **d** $y = 3x + 11$
e $y = 4x - 12$ **f** $y = 10x + \frac{1}{2}$
g $y = 5x + 1$ **h** $y = 11x - 7$
5 a i 2 **ii** $(0, -9)$
b i 8 **ii** $(0, 11)$
c i -1 **ii** $(0, 1)$
d i 3 **ii** $(0, 11)$
e i 4 **ii** $(0, -12)$
f i 10 **ii** $(0, \frac{1}{2})$
g i 5 **ii** $(0, 1)$
h i 11 **ii** $(0, -7)$
6 Equation of any line with these gradients
a 3 **b** 6 **c** -11
d 2 **e** 1 **f** -0.5
g -1 **h** 4

7 Equation of any line with these y-intercepts
a 0 **b** 4 **c** 3
d 7 **e** 1 **f** 17
g 2.5 **h** 27.5

What do you think?

1 a i $4 \times 9 = 12 \times (2 + 1)$ and $9 - 3 = 3 \times 2$
ii $4 \times 0 = 12 \times (-1 + 1)$ and $0 - 3 = 3 \times -1$
b They are the same line
2 a $121 + 33x = 11y$
b $20y = 220 - 20x$
c $y - \frac{x}{3} = \frac{11}{3}$
d $\frac{y}{8} = 1.375 - \frac{3}{8}x$
3 a i 20 **ii** $(0, -28)$
b i 4.5 **ii** $(0, 3)$
c i $\frac{5}{3}$ **ii** $(0, \frac{35}{3})$
d i $\frac{32}{35}$ **ii** $(0, -\frac{24}{5})$

Practice 1.3B

1 a $y = x - 2$ **b** $y = -2x - 2$
c $y = 3x - 3$ **d** $y = -\frac{1}{2}x + 2$
2 a 2 **b** -3
c 4 **d** $-\frac{1}{2}$
3 a $y = 2x$ **b** $y = -3x$
c $y = 4x$ **d** $y = -\frac{1}{2}x$
4 a $y = 2x - 2$ **b** $y = -3x + 4$
c $y = 4x - 3$ **d** $y = -\frac{1}{2}x + 1$
5 a $y = 3x + 9$ **b** $y = -4x - 7$
c $y = \frac{1}{4}x - 3$ **d** $y = -2x + 10$
6 a i £35 **ii** £5
b i $y = 5x$ **ii** The gradient is the cost per book, and the y-intercept (0) shows that the cost of buying zero books is nothing
7 a i 3 **ii** 6
b 60p
c The gradient is the cost per mile and the y-intercept is the fixed charge
8 a i 0.4 kg **ii** 0.7 kg
b i $y = 0.4x + 0.7$
ii The gradient is the mass of a tin and the y-intercept is the mass of the empty box

What do you think?

1 a The gradient is negative as the line slopes downwards to the right
b Emily calculated change in x divided by change in y
2 Beca hasn't noticed that the scale on the y-axis is different from the scale on the x-axis. (The correct equation is $y = x$)
3 $y = \frac{1}{3}x - 7$

Consolidate

1 a $y = 2x + 3$ **b** $y = 4x - 3$
c $y = 6x + 7$
2 $y = 6x + 4$
3 $y = 20x + 4$
4 a Gradient 2; y-Intercept $(0, 5)$
b Gradient 3; y-Intercept $(0, -1)$
c Gradient 5; y-Intercept $(0, 3)$
5 a $y = x - 1$ **b** $y = 2x - 3$
c $y = -2x + 3$ **d** $y = 0.5x + 1$

Stretch

1 True
2 a $y = \frac{1}{4}x - 1$ **b** $y = \frac{1}{2}x - 2$
c $y = \frac{1}{2}x - 1$ **d** $y = \frac{1}{8}x - 1$
3 $y = \frac{5}{3}x - 30$
4 a Correct line **b** $y = -2x - 1$
5 a It is a linear relationship, increasing by £20 per hour
b i $y = 20x + 65$
ii The gradient is the cost per hour and the y-intercept is the set-up fee
6 a $a = 36$
b $b = -3$
c $(0, \frac{5}{8})$

Chapter 1.4

Are you ready?

1 a 1 cm **b** 16 cm
c 2 cm **d** 12 cm
e 6 cm **f** 9.6 cm
2 a 2 **b** 3
c 9 **d** $\frac{1}{11}$
e $\frac{1}{100}$ **f** 5
g $\frac{5}{2}$ **h** $\frac{4}{3}$
3 a $\frac{1}{3}$ **b** $\frac{1}{7}$
c 4 **d** 10
e $\frac{2}{3}$ **f** $\frac{8}{7}$

Practice 1.4A

1 a $2 \times -\frac{1}{2} = -1$ **b** $\frac{1}{3} \times -3 = -1$
c $4 \times -\frac{1}{4} = -1$ **d** $\frac{2}{3} \times -\frac{3}{2} = -1$
2 a -2 **b** -3
c -9 **d** $\frac{1}{11}$
e $\frac{1}{100}$ **f** -5
g $-\frac{5}{2}$ **h** $\frac{4}{3}$
3 a $-\frac{1}{3}$ **b** $-\frac{1}{7}$
c 4 **d** 10
e $-\frac{2}{3}$ **f** $\frac{8}{7}$
4 a $y = -\frac{1}{3}x$ **b** $y = -\frac{1}{7}x$
c $y = 4x$ **d** $y = 10x$
e $y = -\frac{2}{3}x$ **f** $y = \frac{8}{7}x$
g $y = -\frac{1}{9}x$ **h** $y = x$
5 $5y = 10 - x$ can be rearranged to give $y = 2 - \frac{1}{5}x$; $5 \times -\frac{1}{5} = -1$
6 Equation of any line with these gradients
a $-\frac{4}{3}$ **b** $\frac{1}{7}$ **c** -3
d $-\frac{1}{12}$ **e** $\frac{1}{2}$ **f** $-\frac{1}{11}$
g $\frac{7}{5}$ **h** -10
7 $y = -\frac{1}{8}x - 1$
8 $y = 4x + 28$

What do you think?

1 Zach has given the reciprocal, not the negative reciprocal
2 The gradient of A must be 0.6 because A and B are parallel. The gradient of C must be $-\frac{5}{3}$ because B and C are perpendicular. $0.6 \times -\frac{5}{3} = -1$, so A and C are perpendicular.
3 a i Equation of any line with gradient $-\frac{4}{3}$
ii $y = -\frac{4}{3}x + 18$

b There is an infinite number of lines perpendicular to $y = 0.75x + 5$ but only one of them passes through $(0, 18)$

Practice 1.4B

1 a 30 minutes

b

Number of bakers	1	2	5	10	15	20	30
Time taken (minutes)	60	30	12	6	4	3	2

c

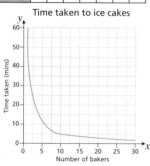
Time taken to ice cakes

d Assumption: each baker works at the same rate

2 a i 18 hours
   ii 6 hours

b

Number of painters	1	2	3	5	6	9	18
Time taken (hours)	18	9	6	3.6	3	2	1

c

Time taken to paint house

d Assumption: each painter works at the same rate

3 a 3 hours

b

Number of machines	1	2	3	4	5	10	20
Time taken (hours)	15	7.5	5	3.75	3	1.5	0.75

c

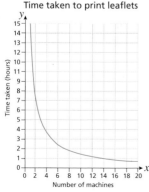
Time taken to print leaflets

4 a

Time taken to build a shopping centre

b 45

What do you think?

1 If there are no bakers/painters/machines the job will not get done. You cannot divide by 0

2 a Points are plotted correctly
   b Line of best fit should be a smooth curve, not a straight line

3 The baking time will be the same for each batch. The number of ovens and the baking time are not inversely proportional.

Consolidate

1 a $-\frac{1}{5}$ **b** -3

c $\frac{1}{7}$ **d** $\frac{5}{4}$

2 a $y = -\frac{1}{5}x$ **b** $y = \frac{1}{3}x$

c $y = -\frac{1}{8}x$ **d** $y = \frac{1}{6}x$

e $y = -\frac{1}{2}x$ **f** $y = \frac{1}{4}x$

3 a $y = -\frac{1}{2}x + 1$ **b** $y = \frac{1}{5}x + 3$

c $y = 2x - 4$

4 a Less

b i 14 hours   **ii** 3.5 hours

c They all work at the same rate

Stretch

1 $-\frac{4}{3a}$

2 $\frac{xy}{7}$

3 $p = -20$

4 $y = -\frac{3}{2}x + 16$

5 $y = -2x + 20$

Check my understanding

1 a

x	-2	-1	0	1	2
y	-11	-8	-5	-2	1

b

2 a 2   **b** $(0, 7)$
3 a $\frac{15}{8}$ **b** $(0, 1.5)$
4 $y = -4x + 6$
5 a £4   **b** £1.50

6 4 hours
7 $y = \frac{1}{3}x + 4$

Block 2 Forming and solving equations

Chapter 2.1

Are you ready?

1 a 4 multiplied by n
   b n divided by 4
   c n multiplied by itself
   d 4 more than n multiplied by 3
   e the product of 3 and four more than n

2 a 80   **b** 5   **c** 2   **d** 30
3 a 30   **b** 120   **c** 18   **d** 62
4 a -9 **b** $+7$ **c** $\times 3$ **d** $\div 9$
5 a $3x + 12$ **b** $5y - 35$
   c $14 + 7z$ **d** $20 - 6x$

Practice 2.1A

1 a $x = 30$ **b** $y = 4.5$ **c** $z = 22$
   d $q = 22$ **e** $d = -3$ **f** $g = -4$
   g $b = -50$ **h** $h = 28$

2 a $n > 7$ **b** $p < 20$ **c** $f \geqslant 5$
   d $x > 3$ **e** $d > 13$ **f** $d > 1$
   g $m > 10$ **h** $m > -10$

3 a Students' own answers
   b i $c = 7$ **ii** $n = 5.5$ **iii** $k = 14$
      iv $k = 5$ **v** $q = -0.6$
   c C $(20 + 4b = 7)$; the others are all the same equation written differently

4 a Students' own answers
   b Negatives are more difficult to show
   c i $k = 65$ **ii** $p = 4$
      iii $t = -9$ **iv** $a = 63$
      v $n = -6$
   d The operations are the same but are in a different order

5 a $m = 12$ **b** $m > 12$
   c Discuss as a class

6 a $t < 3$ **b** $t \leqslant 9$ **c** $t > 38$
   d $t > 9$ **e** $t < 4$

7 B, C and F

8 a $36 - 3 = 33$, $33 \div 4 = 8.25$
       $8.25 \neq 6$
   b He should have multiplied by 4 first
   c $a = 27$

9 a i $n = 8$ **ii** $n = 17$ **iii** $n = 4\frac{1}{4}$
      iv $n = 8\frac{1}{2}$ **v** $n = -11\frac{1}{2}$
   b i $p > -20$ **ii** $p < \frac{4}{5}$ **iii** $p \geqslant 4$
      iv $p \leqslant -16$ **v** $p < -3\frac{1}{2}$

What do you think?

1 Compare answers as a class, noting e.g. that the equation $x^2 = 25$ has two solutions

2 a An open circle means that the value is not included in the solution set; a filled circle means that it is included
   b A: $x > -1$ B: $x \leqslant 1$
      C: $x \geqslant -1$ D: $x < 1$
   c Compare answers as a class

3 a i x is any number between -2 and 5 inclusive

ii For the second inequality –2 and 5 are not included in the solution set

b 8 and 6

c Compare answers as a class

4 a $a = 30$ **b** $a = 27$ **c** $a = 13.5$
d $a = 16.5$ **e** $a = 3.3$
Discuss methods as a class

Practice 2.1B

1 a Students' own answers
b i $a = 8$ **ii** $b = 3$
c 56 is divisible by 7 so Method 2 is easier for the first equation. 30 is not divisible by 7 so Method 1 is easier for the second equation.

2 a $p = 27.5$ **b** $p = 25$ **c** $p = 12.5$
d $p = 6.25$ **e** $p = 8$ **f** $p = 10.5$
g $p = 5.25$

3 a $x = 15$ **b** $x = 12.5$
c $x = 10$ **d** $x = 5$

4 a $p = 4$ **b** $p < 4$
c Discuss as a class

5 a $x < -1$ **b** $x < 5$ **c** $x < 1$
d $x < 5$ **e** $x > -5$ **f** $x \geqslant -1\frac{1}{3}$
g $x \leqslant 0$

What do you think?

1 a A: $4a + 15 = 7$ B: no error
C: $8a + 26 = 10$ D: $4a + 8 = 0$
b Compare answers as a class

2 a length $= x + 4$, perimeter $=$ $2(\text{length} + \text{width}) = 2(x + 4 + x)$ $= 2(2x + 4)$; perimeter $= 38$, so $2(2x + 4) = 38$, which gives $x = 7.5\,\text{cm}$
b $86.25\,\text{cm}^2$

Consolidate

1 a $x = 3$ **b** $x = 2.5$ **c** $x = 3.5$
d $x = 1$ **e** $x = 5$

2 a $x < 3$ **b** $x < 2.5$ **c** $x \geqslant 3.5$
d $x < 1$ **e** $x \leqslant 5$

3 a $c = 7$ **b** $x < 15$ **c** $y < 15$
d $t = 30$ **e** $p \geqslant 79$ **f** $m < 3.4$
g $g = 0.5$ **h** $h < 7$

Stretch

1 a There are infinitely many solutions; compare answers as a class
b

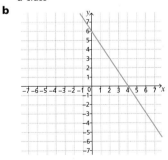

Each point on the graph represents a solution to the equation

c i Draw the graph $y = x$ and find the point where the two lines meet

---

ii Replace x (or y) in the equation to get $5y = 12$ (or $5x = 12$)
The algebraic method will be more accurate as the solution is not an integer and so will be hard to read from the graph

2 Discuss approaches as a class; equations with unknowns on both sides are covered in Chapter 2.3

3 a $x = 4, x = -4$ **b** $y = 5, y = -5$
c $z = 4, z = -4$ **d** $m = \frac{3}{2}, m = -\frac{3}{2}$
e $m = -1, m = -5$ **f** $m = \frac{1}{4}, m = -\frac{7}{4}$

Chapter 2.2

Are you ready?

1 a $>$ **b** $>$ **c** $<$ **d** $<$
2 $-10, -6, -4, -3, -1.5, 7, 10, 12$
3 a $x = 7$ **b** $y < 12$
4 a $a = -3$ **b** $b > -1$

Practice 2.2A

1 Discuss as a class
2 a $x = 18$ **b** $x = 9$ **c** $x = 4$
d $x = 2$ **e** $x = 4$ **f** $x = 1.2$
3 A, B and C are true
4 a $y > 10$ **b** $y < -10$ **c** $y > 90$
d $y < -90$ **e** $y > 27$ **f** $y > 33$
g $y < -27$
5 a $y > 4$ **b** $y > 7$ **c** $y < -4$
6 a $x > -\frac{7}{3}$ **b** $t \leqslant 6$ **c** $p < -30$
d $q < 18$ **e** $g \geqslant -\frac{3}{4}$ **f** $p \geqslant 3$
g $p \geqslant 21$
7 a i $x = 6, x = -\frac{2}{3}$
ii Discuss methods as a class
b i $p \leqslant 5.5$ **ii** $p \geqslant -2.5$

What do you think?

1 a The sign of the 4 is different
b $x = -\frac{5}{3}, x = 1$
2 a $2x - 10 > 15, x > 12.5$
b $10 - 2x > 15, x < -2.5$
c $10 - \frac{x}{2} \leqslant 8, x \geqslant 4$
d $10 - 2(x - 4) > 7, x < 5.5$

Consolidate

1 a $2y$ on second line should be $-2y$
b Second line should be $3x + 21 = 18$; then the equation is repeated, but with a copying error
c Answer should be $x < 2$
2 a $t < -3$ **b** $t > 3$ **c** $t > -3$
d $t < 3$ **e** $t > -20$ **f** $t > 60$
g $t \geqslant 17$ **h** $t \leqslant -3$
3 a i $x = 7$ **ii** $x = -7$
b i $x > 7$ **ii** $x < -7$

Stretch

1 a Both 4^2 and $(-4)^2$ are 16
b $10, -10, 5, -5, 5.1, -5.1$
c Discuss as a class
d i $x > 3, x < -3$
ii $x > 10, x < -10$
iii $x > 12, x < -12$
iv $x > 7, x < -7$

---

2 a x can take any value between –6 and 6
b i $-4 < x < 4$ **ii** $-7 < x < 7$
iii $-4 \leqslant x \leqslant 4$ **iv** $-8 < x < 8$
c B and D
3 Compare answers as a class

Chapter 2.3

Are you ready?

1 a $x = 3$ **b** $x = 20$
c $x = 40$ **d** $x = -20$
2 a $y = 11$ **b** $y = 16$
c $y = 8$ **d** $y = -8$
3 They are the same equation
4 a $t > 17$ **b** $t > 8.5$ **c** $t > -3$
d $t < 3$ **e** $t < 1.5$
5 a $4x$ **b** $-4x$ **c** $4x$ **d** $-10x$

Practice 2.3A

1 $a = 4$
2 a $5b + 2 = 3b + 14$
b $2b + 2 = 14$
c $b = 6$
d Both sides are equal to 32
3 a i

x	x	x
x	20	

ii Subtract x from each side
iii $x = 10$
iv Left-hand side: $3 \times 10 = 30$
Right-hand side: $10 + 20 = 30$
b i

q	q	16		
q	q	q	q	q	q

ii Subtract $2q$ from each side
iii $q = 4$
$16 + 2 \times 4 = 24$
$6 \times 4 = 24$
4 a Abdullah's method is more sensible as Benji's still leaves terms in x on both sides
b $x = 3$
5 a i Subtract $3g$ from each side
ii Subtract $3m$ from each side
iii Subtract t from each side
b i $g = 1$ **ii** $m = 8$ **iii** $t = 7$
6 a $x = 3$ **b** $x = 7$
c Discuss as a class
7 a $n = 5$ **b** $c = 3$ **c** $h = -9$
8 First step gives $-6 = 2 + 2x$
This gives $x = -4$
9 a $m = 1$ **b** $t = \frac{11}{3}$ **c** $h = \frac{1}{4}$
d $w = \frac{16}{3}$ **e** $p = \frac{8}{3}$ **f** $p = -\frac{8}{3}$
g $k = -\frac{16}{5}$ **h** $t = \frac{8}{5}$

What do you think?

1 Compare answers as a class
2 a i Jakub's, as then there will be a positive coefficient of t on one side of the equation. Chloe's method gives a negative coefficient of t
ii $t = 1$
b i $t = 1\frac{2}{3}$ **ii** $m = -2$ **iii** $x = 3$
iv $g = 4$ **v** $v = 1$ **vi** $f = -0.2$

3 a $2t + 12 = 3t$, $t = 12$
b $2t + 12 = t$, $t = -12$
c $2(t + 12) = 4t$, $t = 12$
d Compare answers as a class

Practice 2.3B

1 a Compare answers as a class
b i $x > 6$ **ii** $2 \leqslant 2x$, $x \geqslant 1$
iii $4x + 3 \geqslant 19$, $4x \geqslant 16$, $x \geqslant 4$
2 a i Subtract x from each side
ii Subtract $2x$ from each side
iii Subtract $3x$ from each side
iv Subtract $4x$ from each side
b i $x > 3$ **ii** $x > 4$ **iii** $x > 6$
iv $x > 12$
3 a $x > 4$ **b** $x \leqslant 4$
c $x \geqslant 4$ **d** $x > 4$
4 a $y > 5$ **b** $y > 6$
5 a $w > 3$ **b** $w > 5$
c $w > -5$ **d** $w > -3$
e $q \leqslant -17$ **f** $m < 10$
g $y \leqslant -\frac{8}{3}$ **h** $p < -5\frac{1}{2}$
6 a 1 **b** 2 **c** 5 **d** -9

What do you think?

1 a i Because this results in there being a positive y term on only one side
ii $5y > 25$, $y > 5$
b i $a < 4$ **ii** $b \geqslant 2$ **iii** $c > \frac{1}{2}$
2 a i Faith's, as the only term in y will have a positive coefficient (although Ed's method will work too)
ii $y > 2$
b When multiplying by a negative number the inequality sign is reversed. Seb hasn't done this.
c i $d \geqslant -2.5$ **ii** $d > 4$
iii $d > -1$

Consolidate

1 a $4a + 3 = 2a + 11$
b $2a + 3 = 11$ **c** $a = 4$
d Both sides are 19
2 a Compare answers as a class
b i $a = 2$ **ii** $m = 4$
iii $t = 3$ **iv** $f = 2$

3 a $a = 2$ **b** $a = -2$
c $a = 5\frac{1}{2}$ **d** $a = \frac{5}{8}$
4 a $v = 12$ **b** $v = -5$ **c** $b = 1$
d $b = -5$ **e** $b = -1$
5 a $x > 1$ **b** $x < 1$ **c** $x > 2\frac{1}{2}$
d $x < -1$ **e** $x < 3$ **f** $x < -1$

Stretch

1 a Discuss as a class; the solution is $x = 8$
b i $x = 24$ **ii** $x = 40$ **iii** $x = 16$
c i $y < -6$ **ii** $y \leqslant 30$ **iii** $y < 72$
2 a $x = -13$ **b** $x = -14$ **c** $x = \frac{1}{3}$
d No solution **e** $x = 5$
3 a $t < -\frac{1}{2}$ **b** $t \geqslant \frac{11}{8}$
c $t \geqslant \frac{11}{8}$ **d** $t > -\frac{1}{2}$
4 a 2, 3, 4 and 5 **b** $1 < x < 4$
c Because $3x + 1$ cannot be less than $3x$

---

5 Only 0 can be both added to and subtracted from 8 to give the same result, so $7x = 0$, which means $x = 0$

Chapter 2.4

Are you ready?

1 a e.g. sum of angles on a straight line, sum of co-interior angles, sum of angles in a triangle
b e.g. sum of angles in a quadrilateral
c e.g. corresponding, alternate, vertically opposite
2 a $2n + 40$ **b** $2n - 40$ **c** $\frac{n}{2} + 40$
d $\frac{n}{2} - 40$ **e** $40 - 2n$ **f** $2(n + 40)$
g $\frac{n - 40}{2}$
3 a $g = 7$ **b** $p = 23.5$ **c** $q = 11$
d $m = 60$ **e** $n = 45$ **f** $a = 4$
4 0.16

Practice 2.4A

1 a 180°
b i 100° **ii** 33° and 66° **iii** 37°
2 a Zach: e.g. $6a = 504$, $a = 84$
Faith: $b + 18 = 161$, $b = 143$
Flo: $\frac{c}{4} - 5 = 30$, $c = 140$
b Compare answers as a class
3 a C ($3x + 18 = 99$) **b** 27p
4 a $2a + 54 = 100$ **b** £28
c £402
5 a $12m + 59 < 150$ **b** 7 miles
6 a $x = 24$ **b** $y = 29$
c $z = 15$ **d** $p = 6$, $q = 34$
e $t = 17$, $w = 29$
7 a $45 + 12x < 120$ **b** 6 T-shirts
8 a $x = 0.1$ **b** 0.6
9 $x < 4$

What do you think?

1 a $5(x + 3) + 10x = 180$, $x = 11$
b $p + p + 5 < 60$; the price of a saucepan is less than £27.50
2 a 110
b i The number is at least 12
ii The number is less than 12
c You can't have 22.5 mints in a packet, but you can have 33
3 a $\frac{4}{21}$
b e.g. P(score a 1 or a 6), P(score is a factor of 4)

Practice 2.4B

1 a 680
b i $45w = 720$ **ii** $w = 16$ cm
2 a 65 **b** 23rd
c $304 - 17$ is not divisible by 4
d 1001, when $n = 246$
3 a 6 or -6 **b** 17 or -17
c 4 or -4
d You can't take the square root of a negative number
4 a C ($C = 0.9m + 2.8$) **b** £13.60
c 8 miles
5 a $A = \frac{1}{2}bh$
b i 1.6 cm **ii** 4 cm
6 a $s = 500$ **b** $t = 5$ **c** $v = -3$

What do you think?

1 Compare answers as a class
2 a i 3 **ii** $\frac{3}{8}$
b $p = 9$

Consolidate

1 36°, 72°, 108° and 144°
2 9
3 12
4 30
5 9
6 a $P = 148$ **b** $c = 5$ **c** $d = 4.8$
7 12

Stretch

1 a 88 **b** -41
c The sum is three times the middle number
2 a 6 **b** 14
3 a $D = -35$ **b** $a = 3$
c Compare answers as a class

Chapter 2.5

Are you ready?

1 a $+ 10$ **b** $\div 12$ **c** $\times 10$ **d** $- 15$
2 a $A = 25.46$ **b** $b = 4.5$
c $h = 48$
3 a $x = 40$ **b** $x = 52$ **c** $x = 16$
4 a 8 (or -8) **b** 4096

Practice 2.5A

1 a P **b** l **c** F **d** a
e m **f** s **g** I **h** t
2 Discuss as a class
3 a $y = x - 3$ **b** $y = x + 3$
c $y = \frac{x}{3}$ **d** $y = 3x$
4 a $t = \frac{p - 7}{3}$ **b** $t = \frac{p + 7}{3}$
c $t = \frac{p - a}{3}$ **d** $t = \frac{p + a}{3}$
e $t = 2(p - 7)$ **f** $t = 2(p + 7)$
g $t = 2(p - a)$ **h** $t = 2(p + a)$
5 a 12.7 cm **b** 12.7 cm
6 a $m = \frac{t - 3n}{2}$ **b** $n = \frac{t - 2m}{3}$
7 D ($b = \frac{a}{2c}$)

What do you think?

1 a First line should be $a - b = -2c$
b $2c = b - a$, $c = \frac{b - a}{2}$
c Discuss as a class
d i $c = x - y$ **ii** $c = y - x$
iii $c = \frac{y - x}{4}$ **iv** $c = 2(y - x)$
2 The inverse of "subtract from 10" is "subtract from 10" (This is called "self-inverse".) So $x = \frac{10 - t}{3}$
3 a $y = 10 - x$
b Negative, as the coefficient of x is -1
c i -4 **ii** -2 **iii** 2
d Discuss as a class
4 Discuss as a class

Practice 2.5B

1 Students' own answer
2 a $q = \frac{p - 4}{2}$ **b** $q = \sqrt{p - 4}$
c $q = \sqrt{\frac{p - 4}{2}}$
3 b $w = \frac{P - 2l}{2}$ and $w = \frac{P}{2} - l$

4 Possible answers include

a $t = \frac{x}{3} - 4$　　**b** $t = \frac{x-12}{6}$

c $t = \frac{x+12}{6}$　　**d** $t = \frac{x-12}{15}$

e $t = \frac{12-x}{15}$　　**f** $t = \sqrt{\frac{x}{3} - 4}$

5 Possible answers include

a $x = 5y$　　**b** $x = 5(y-1)$

c $x = 5y - 1$　　**d** $x = y - \frac{1}{5}$

e $x = \frac{5y+1}{3}$　　**f** $x = \sqrt{\frac{5y+1}{3}}$

g $x = \frac{1-5y}{3}$　　**h** $x = \sqrt{\frac{5y+1}{3}}$

i $x = \sqrt{\frac{5y+1}{3}}$　　**j** $x = \frac{(5y-3)}{2} - 1$

6 a B $(a = \frac{y^2}{3})$, check by substitution

b i $a = \frac{y^2}{16}$　　**ii** $a = (y-4)^2$

iii $a = y^2 - 4$　　**iv** $a = y^2 + 4$

v $a = 4 - y^2$　　**vi** $a = 4y^2$

vii $a = 4(y-1)^2$ **viii** $a = 4(y^2-1)$

ix $a = \sqrt{4(y^2-1)}$ **x** $a = \sqrt{y^2 - \frac{1}{4}}$

What do you think?

1 a $I = \frac{V}{R}$　　**b** $R = \frac{V}{I}$

2 a $F = AP$　　**b** $A = \frac{F}{P}$

3 a i $m = \frac{2K}{v^2}$　**ii** $v = \sqrt{\frac{2K}{m}}$

b i $\frac{K}{mv^2}$　　**ii** $\frac{mv^2}{K}$

4 a $f = \frac{u+v}{uv}$　　**b** $v = \frac{u-f}{fu}$

Consolidate

1 a $a = 4b$, $b = \frac{a}{4}$

b $x = y + 10$, $y = x - 10$

c $m = \frac{1}{2}t$, $t = 2m$

2 a $y = \frac{p}{6}$　　**b** $y = 6q$

c $y = m - 6$　　**d** $y = n + 6$

3 a $c = 4d + 12$　　**b** $d = c - \frac{c-12}{4}$

4 a $d = \frac{c}{3} + 4$　　**b** $d = \frac{c+12}{3}$

5 a $t = \frac{y}{2}$　　**b** $t = y + 2$

c $t = 2y$　　**d** $t = y - 2$

e $t = \frac{y-5}{2}$　　**f** $t = 2(y-5)$

g $t = 2(y+5)$　　**h** $t = \frac{y+5}{2}$

6 a $y = -3x + 7$　　**b** $y = 2x - 4$

c $y = \frac{1}{2}x - 5$　　**d** $y = \frac{3}{5}x + 2$

7 a $m = \frac{E}{c^2}$　　**b** $c = \sqrt{\frac{E}{m}}$

Stretch

1 a i $F = 2C + 30$ **ii** $C = \frac{F-30}{2}$

b $C = \frac{5}{9}(F - 32)$

c i $C = 10$　　**ii** $F = 50$

2 a $a = \frac{v-u}{t}$

b $at = v - u$, $v = u + at$

c i $a = \frac{v^2-u^2}{2s}$ **ii** $s = \frac{v^2-u^2}{2a}$

iii $u = \sqrt{v^2 - 2as}$

d $s = \frac{1}{2}(u + u + at)t$, $s = \frac{1}{2}(2u + at)t$,

$s = \frac{1}{2}(2ut + at^2)$, $s = ut + \frac{1}{2}at^2$

e t cannot be negative, but s, u, v and a can be

---

3 a One of the variables, c, appears twice

b $c = \frac{q-p}{7}$　　**c** $t = \frac{7x-5y}{3}$

Check my understanding

1 $4y = 18$, $2y - 7 = 2$ and $3(y - \frac{1}{2}) = 12$

2 a $p > 9$　　**b** $q \leqslant -2.5$

c $t < -\frac{3}{4}$

3 $2n - 3 = 18 - n$; $n = 7$

4 Yes; $6n - 3 = 879$ has solution $n = 147$, so it is the 147th term

5 a $P = 6a + 2b$　　**b** $b = 4$

c $a = \frac{P-2b}{6}$ or $a = \frac{P}{6} - \frac{1}{3}b$

6 a $x = \frac{m}{5}$　　**b** $x = \sqrt{n}$

c $x = \sqrt{\frac{q}{5}}$　　**d** $x = \frac{(10-g)^2}{49}$

e $x = \frac{\sqrt{2p - 7V}}{5}$

Block 3 Testing conjectures

Chapter 3.1

Are you ready?

1 a 4　　**b** 8　　**c** 12

d 16　　**e** 20

2 a Even numbers

b Final digit is 0, 2, 4, 6 or 8

3

×	2	4	6
3	6	12	18
5	10	20	30
8	16	32	48

×	2	5	7
3	6	15	21
4	8	20	28
9	18	45	63

Practice 3.1A

1 a $3 \times 4 = 12$ (or $4 \times 3 = 12$)

$2 \times 6 = 12$ (or $6 \times 2 = 12$)

$1 \times 12 = 12$ (or $12 \times 1 = 12$)

b 1, 2, 3, 4, 6, 12

2 a 5 by 3 and 1 by 15

b 1, 3, 5, 15

c i 1, 2, 4, 5, 10, 20

ii 1, 2, 3, 4, 6, 8, 12, 24

iii 1, 2, 5, 6, 10, 15, 30

iv 1, 3, 9, 27

3 a i 16, 18, 20, 22, 24

ii 15, 18, 21, 24

iii 16, 20, 24

iv 18, 24

v 16, 24

b Compare answers as a class

4 a 8 by 2, 4 by 4, 16 by 1

b 17 by 1

c 17 has only two factors and 16 has more

d 7, 13, 19, 29, 31

5 a If dividing results in an integer

b i 90, 93, 96, 99

ii 91, 98

c 97

6 a

$$\begin{array}{ccc} & 24 & \\ 4 & & 6 \\ ②\quad② & ② & ③ \end{array}$$

b $24 = 2^3 \times 3$

---

c i $2 \times 3 \times 5$

ii $2^3 \times 5$　　**iii** $2 \times 3 \times 11$

iv $2^2 \times 7$　　**v** 3^4

vi $2^3 \times 3 \times 5$

7 a 2×13 **b** 2×19 **c** 2×23

d 5×11 **e** 5×13 **f** 3×31

8 a $2^2 \times 3^3$　　**b** $2^3 \times 3^3$

c $2^4 \times 3^3 \times 5$

9 b i $2^2 \times 5^3$

ii $2^2 \times 3^2 \times 7$

iii $2^3 \times 3 \times 11$

10 a 3　　**b** 3600

What do you think?

1 Marta is correct

2 a $2 \times 43 \times 47$

b $2 \times 5 \times 43 \times 47$

c $2 \times 3 \times 5 \times 43 \times 47$

3 Compare answers as a class

4 a 40%　　**b** 25%

c 16.8%

d An increasing proportion of the large numbers are multiples of lower integers and so are not prime

Consolidate

1 a $3 \times 6 = 18$ or $6 \times 3 = 18$

b 2 by 9 and 18 by 1

2 a 1, 2, 5, 10

b 1, 2, 4, 7, 14, 28

c 1, 2, 4, 5, 8, 10, 20, 40

3 a i 27, 30, 33 **ii** 28, 32

b i 30　　**ii** 29 and 31

4 a 11, 13, 17, 19 **b** 37, 41, 43

5 3 and 15

6 a 3×11 **b** 2×17 **c** 5×7

d 2×23 **e** 3×17 **f** 5×17

7 a 3×5^2　　**b** $2^4 \times 3$

c $2^2 \times 3 \times 5$ **d** $2^3 \times 3 \times 5^2$

e $2^5 \times 3$　　**f** $3^2 \times 5^2$

Stretch

1 a i e.g. 3 and 6

ii e.g. 2 and 4

b True for odd numbers

2 a 43, 47, 53, 61

b They are all prime

c $41^2 + 41 + 41$ will have a factor of 41

d Compare answers as a class

3 a 60, 72, 84, 90 and 96

b Compare answers as a class

4 a $5^2 \times 7^2 = (5 \times 7)^2$

b 1764, 729, 3969 and 1 000 000

c 729 and 1 000 000

d Compare answers as a class

Chapter 3.2

Are you ready?

1 a 5, 10, 15, 20, 25, 30

b Final digit is either 5 or 0

2 a 1, 5, 7, 35　　**b** 1, 5, 11, 55

c 1, 2, 5, 7, 10, 14, 35, 70

3 a 3　　**b** (0, −5)

4 a $\frac{1}{5}$　　**b** $\frac{1}{5}$　　**c** $\frac{3}{5}$

Practice 3.2A

1 a False　**b** False　**c** True

d False　**e** False

2 a False　**b** True　**c** False

d True　**e** True　**f** False

g True

3 a Because $846 \div 2 > 800 \div 2$
　　which is 400
b i True　**ii** True　**iii** False
　　iv False
4 a False　**b** True　**c** True
　　d True
5 a False　**b** False　**c** True
6 False, $91 = 7 \times 13$
7 a False　**b** True　**c** True
　　d True
8 No, $h = 8$
9 a False, the gradient is -5
　　b True
　　c False, the gradient is 3
　　d False, the y-intercept is $(0, \frac{1}{2})$
　　e False　**f** True

What do you think?
1 a 1, 4, 9, 16, 25, 36, 49
　　b Square numbers
2 a True　**b** False　**c** False

Practice 3.2B
1 a e.g. true for 3, false for 2
　　b e.g. 2 (even, is a factor of 2)
　　　and 1 (odd, is a factor of 2)
　　c e.g. 16 for true and 14 for false
　　d e.g. $1 + 2 + 3$ for true and
　　　$2 + 3 + 4$ for false
2 a Sometimes　**b** Sometimes
　　c Always
3 a 5, 9, 13
　　b i Never　　**ii** Always
　　iii Always　**iv** Always
　　v Sometimes
4 a Sometimes　**b** Always
　　c Sometimes　**d** Sometimes
　　e Sometimes　**f** Sometimes
　　g Never　　**h** Always
5 a Always　　**b** Sometimes
　　c Sometimes　**d** Never
　　e Sometimes
6 a Sometimes　**b** Never
　　c Sometimes　**d** Sometimes
　　e Sometimes　**f** Sometimes
　　g Always

What do you think?
1 a Always　　**b** Always
　　c Sometimes　**d** Sometimes
2 a True for e.g. 1, 1, 9, 10 but
　　　false for e.g. 1, 2, 2, 2
　　b Compare answers as a class
3 a cm^2 are not the same as cm
　　b Sometimes
4 The area is always reduced; the
　　perimeter is sometimes reduced

Consolidate
1 a True　**b** True　**c** True
　　d False　**e** False
2 a False　**b** False　**c** True
　　d True　**e** False　**f** False
3 Sometimes
4 a Sometimes　**b** Sometimes
　　c Sometimes　**d** Sometimes
　　e Never

Stretch
1 a True　**b** True　**c** False
　　d True　**e** True
2 Put $x = 0$ and $y = 0$ and rearrange

---

3 a Always　　**b** Sometimes
　　c Always

Chapter 3.3

Are you ready?
1 Odd: 371, 4005, 1234567;
　　Even: 18, 56, 90, 40008 and
　　123456. Look at the final digit;
　　if it is 0, 2, 4, 6 or 8, then the
　　number is even.
2 a $4a + 28$　　**b** $24 - 8b$
　　c $x^2 - 5x$　　**d** $6a + 15b$
3 a $5(3b + 2c)$　**b** $6(2 - t)$
　　c $p(p + 12)$
4 a -2　　　**b** -8
　　c -6　　　**d** 2
　　e 2

Practice 3.3A
1 a Both are 320
　　b Discuss as a class
　　c e.g. both are $2 \times 40 \times 4$
2 a Both are 1200
　　b Both are 300
　　c Both are 10
3 a $\frac{4}{5} = \frac{8}{10} = 0.8$
　　b $\frac{9}{20} = \frac{45}{100} = 0.45$
4 Compare answers as a class
5 a $32 > 30$
　　b i Both are 54
　　ii Both are 15
　　iii Both are 45
　　c e.g. "$a\%$ of $b = b\%$ of a"
6 a $2 \times (2a + a + 5) \equiv 2 \times (3a + 5)$
　　　$\equiv 6a + 10$
　　b $2a(a + 5) \equiv 2a^2 + 10$
7 a $3(p + 2q) + 2(p + 3q) \equiv$
　　　$3p + 6q + 2p + 6q \equiv 5p + 12q$
　　b $3(p + 2q) - 2(p + 3q) \equiv$
　　　$3p + 6q - 2p - 6q \equiv p$
　　c Both sides expand to $36p + 18q$
8 Compare answers as a class
9 a e.g. $96 > 72$　**b** Both are 64
　　c $3x > 2.5x$

What do you think?
1 e.g. $21x^2 > 20x^2$
2 They factorise to 13×9, 11×17,
　　13×23, 29×29 and 31×41
3 a $2304 = 2^8 \times 3^2 = (2^4 \times 3)^2$
　　　but 8 and 2 are not divisible
　　　by 3 so it is not a cube number
　　b $8000 = 2^6 \times 5^3 = (2^2 \times 5)^3$ but 3
　　　is not divisible by 2 so it is not
　　　a square number
　　c $46656 = 2^6 \times 3^6 (2^3 \times 3^3)^2 =$
　　　$(2^2 \times 3^2)^3$ so it is both a square
　　　and a cube number
4 a e.g. $100 \times 1.5 \times 0.5 = 75 \neq 100$
　　b e.g. $100 \times \frac{4}{3} \times \frac{3}{4} = 100$
　　c Compare answers as a class

Practice 3.3B
1 e.g. $3 + 5 = 8$ and 8 is not odd
2 a There are 2 equal rows
　　b 1 more than 2 equal rows
　　c Discuss as a class
3 All the conjectures are true.
　　Compare representations used as
　　a class

---

4 a He does not consider all the
　　　multiples of 3
　　b–e Compare answers as a class
5 a Even
　　b e.g. the product of two even
　　　numbers is always even
　　c e.g. the product can be shown
　　　as 4 equal rectangles, so it is a
　　　multiple of 4
6 a True　　　　**b** False

What do you think?
1 Compare answers as a class, e.g.
　　"The sum of an odd set of odd
　　numbers is odd but the sum of an
　　even set of odd numbers is even."
2 a, b Compare answers as a class
　　c True
　　d Compare answers as a class

Consolidate
1 a Both are 1000
　　b Compare answers as a class
2 a Both are 1800
　　b Both are 180
　　c Both are 8
3 a $\frac{3}{5} = \frac{6}{10} = 0.6$
　　b
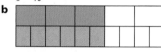
　　c $\frac{0.6}{5\overline{)3.^30}}$
4 Both are 28
5 a Both are $8x + 12y$
　　b They are equal as $4(2x + 3y) \equiv$
　　　$2 \times 2 \times (2x + 3y) \equiv 2(4x + 6y)$
6 a Areas are all $64p^2$
　　b Rectangle $= 40p$,
　　　parallelogram $= 70p$,
　　　square $= 32p$
7 a, b and **d** are true

Stretch
1 a e.g.
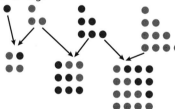
　　c After $T(3)$, one of n or $n + 1$ is
　　　even and greater than 2 so $\frac{n}{2}$
　　　or $\frac{(n + 1)}{2}$ is a factor
2 b $10a + b = 9a + (a + b)$. If the
　　　digit sum $a + b$ is divisible by 9
　　　then $9a + (a + b) \equiv 9a + 9k$ for
　　　an integer k, which is the same
　　　as $9(a + k)$, so the number is
　　　divisible by 9
　　c Use $100a + 10b + c$ and
　　　proceed as in **a**
　　d Yes – discuss as a class
3 a If a number is divisible by 9,
　　　then the digit sum of the
　　　number is divisible by 9
　　b Yes
4 Compare answers as a class

Chapter 3.4

Are you ready?

1 a $4a + 16$ **b** $a^2 + 4a$
  c $4a - 16$ **d** $a^2 - 4a$
  e $ab + 3b$ **f** $3b + ab$
  g $3b + b^2$ **h** $3b - b^2$

2 a $7y$ **b** $-y$
  c y **d** $-7y$

3 x^2 and $2x^2$; $2x$, $7x$ and $-2x$;
  2, 7 and -2

4 a $15p + 26q$ **b** $15p - 14q$
  c $-9p + 26q$

5 a $2(a + b)$ **b** $3(p + q + 2)$
  c $4(m + 2n + 1)$

Practice 3.4A

1 a ab, $3b$ and 12
  b $ab + 4a + 3b + 12$
  c $1200 + 80 + 180 + 12 = 1472$

2 a $ab + 5a + 2b + 10$
  b $pq + 4p + 4q + 16$
  c $cd + c + 7d + 7$
  d $fg + 7f + 6g + 42$

3 a x^2, $3x$, x, 3
  b $x^2 + 4x + 3$
  c Students' own answers

4 a $x^2 + 6x + 8$ **b** $x^2 + 8x + 15$
  c $x^2 + 9x + 14$ **d** $x^2 + 8x + 15$
  e $x^2 + 6x + 8$ **f** $x^2 + 9x + 14$
  g $x^2 + 7x + 6$ **h** $x^2 + 7x + 6$

5 a and **e**, **b** and **d**, **c** and **f**, **g** and **h**
  Multiplication is commutative

6 a $x^2 + 8x + 16$
  b i $x^2 + 10x + 25$
    ii $x^2 + 6x + 9$
    iii $49 + 14x + x^2$
  c i 625   **ii** 1089
    iii 3249

7 a D   **b** $x^2 + 2x - 3$

8 a $5x - x$ simplifies to $4x$, not 5
  b $x^2 + 4x - 5$

9 a $x^2 + x - 6$
  b $x^2 - x - 6$
  c $y^2 + y - 20$
  d $p^2 + 2p - 15$
  e $x^2 - 7x + 10$
  f $g^2 - 8g + 12$
  g $t^2 + 10t + 25$
  h $t^2 - 10t + 25$
  i $20 + x - x^2$

What do you think?

1 a e.g.

	c	2
a		
b		
3		

  $ac + 2a + bc + 2b + 3c + 6$
  b No, all the terms are unlike
  c i $ac + 2a + bc + 2b - 3c - 6$
    ii $ac - 2a - bc + 2b + 3c - 6$
    iii $ac - 2a - bc + 2b - 3c + 6$
  d i $a^2 + 5a + ab + 2b + 6$
    ii $a^2 + 2ab + b^2 + 3a + 3b$
    iii $6a - a^2 - 3b + ab - 9$

2 a i $x^2 - 25$ **ii** $x^2 - 16$
    iii $x^2 - 4$ **iv** $x^2 - y^2$

b i $x^2 - 81$ **ii** $y^2 - 9$
  iii $p^2 - 36$ **iv** $36 - p^2$

c i $(x + 12)(x - 12)$
  ii $(53 + 47)(53 - 47)$
  iii $100 \times 6 = 600$
  iv $10 \times 4.6 = 46$

3 a C
  b Compare answers as a class

Practice 3.4B

1 a Odd   **b** Even
  c Even   **d** Odd
  e Even   **f** Odd

2 a i Even **ii** Odd **iii** Odd
  b Faith's claim is true; since p is
  even, $3p$ is always even so
  Jakub's claim is false

3 a $n + 1$ **b** $n - 2$ **c** $2n$
  d $n + 2$ **e** $4n$

4 a $6k = k \times 6$, so it is a multiple of 6
  b i $5k$ **ii** $8k$ **iii** $7k - 1$
  c $6k = 3 \times 2k$, so it is a multiple
  of 3
  d $4k$ and $10k - 2$ are definitely
  even, $6k + 1$ is definitely odd,
  the others could be even or
  odd

5 a Darius' approach works for the
  sum of any two odd numbers,
  which could be different
  b $2(m + n + 1)$

6 a $2a + 2b + 2c \equiv 2(a + b + c)$
  b $5k + 5m \equiv 5(k + m)$
  c $n + n + 1 \equiv 2n + 1$
  $2n$ is even, so $2n + 1$ is odd
  d $n + n + 1 + n + 2 \equiv 3n + 3 \equiv$
  $3(n + 1)$

7 a e.g. $2 + 10 = 12$
  b e.g. $m = 6$
  c e.g. $x = 3$

8 a True; $4n + 2 \equiv 2(2n + 1)$, so it is
  a multiple of 2
  b Compare answers as a class

9 a Even
  b $2m \times 2n \equiv 4mn \equiv 2 \times 2mn$,
  so it is a multiple of 2
  c $2m \times 2n \equiv 4mn \equiv 4 \times mn$, so it
  is a multiple of 4
  d e.g. always a multiple of 9
  e Always odd e.g. $(2m + 1)(2n + 1)$
  $\equiv 4mn + 2m + 2n + 1 \equiv$
  $2(2mn + m + n) + 1$, so it is one
  more than a multiple of 2

10 Compare answers as a class

What do you think?

1 a $2n \times 2n$ should be $4n^2$
  b $4n^2 \equiv 2 \times 2n^2$, so it is a multiple
  of 2
  c Always
  d $(2m + 1)(2m + 1) \equiv 4m^2 + 4m + 1$
  $\equiv 2 \times 2(m^2 + m) + 1$, so it is one
  more than a multiple of 2, so
  it is odd. It is also one more
  than a multiple of 4

2 $2m + 1 + 2n + 1 + 2k + 1 \equiv$
  $2m + 2n + 2k + 3$
  $\equiv 2m + 2n + 2k + 2 + 1 \equiv$
  $2(m + n + k + 1) + 1$, so odd
  $2m + 1 + 2n + 1 + 2k + 1 + 2q + 1$
  $\equiv 2m + 2n + 2k + 2q + 4 \equiv$
  $2(m + n + k + q + 2)$, so even

3 a Total is $n + n + 1 + n + 2 \equiv$
  $3n + 3 \equiv 3(n + 1)$, so the mean
  is $3(n + 1) \div 3 = n + 1$, which is
  the middle number
  b Compare answers as a class

Consolidate

1 a ab, $2a$, 10
  b $ab + 2a + 5b + 10$
  c $600 + 40 + 150 + 10 = 800$

2 a $ab + 4a + 3b + 12$
  b $cd + 3c + 6d + 18$
  c $fg + 2g + 7f + 14$

3 a i $xy + 5x + 4y + 20$
    ii $x^2 + 9x + 20$. This has been
    simplified but **i** cannot be
  b i $y^2 + 13y + 30$
    ii $p^2 + 10p + 24$
    iii $m^2 + 13m + 36$
    iv $n^2 + 2n + 1$

4 a i $t^2 + 7t + 12$
    ii $t^2 - t - 12$
    iii $t^2 + t - 12$
    iv $t^2 - 7t + 12$
  b i $p^2 - p - 30$
    ii $q^2 - 8q + 12$
    iii $g^2 + 3g - 28$
    iv $f^2 + 6f + 9$

5 a $q - 1$ **b** $q + 2$ **c** $3q$
  d $q - 2$ **e** $q + 3$

6 Odd, $2m + 2n + 1 \equiv 2(m + n) + 1$,
  so it is one more than a multiple
  of 2, so odd

7 a $3k$ **b** $5k$
  c Only sometimes true, e.g.
  $3 + 5 = 8$, but $6 + 5 = 11$
  d $6k + 9m \equiv 3(2k + 3m)$, so it is a
  multiple of 3

8 a $10k \equiv 5k \times 2$, so even
  b $10k \equiv 2k \times 5$, so it is a multiple
  of 5

9 e.g. always a multiple of 9,
  $3k \times 3m \equiv 9km$

Stretch

1 b $x^3 + 6x^2 + 11x + 6$
  d i $x^3 + 3x2 + 3x + 1$
    ii $x^3 - 3x^2 + 3x - 1$
    iii $x^3 + 6x^2 + 12x + 8$
    iv $x^3 - 6x^2 + 12x - 8$
  e $x^3 + 9x^2 + 27x + 27$,
  $x^3 - 9x^2 + 27x - 27$

2 a Compare answers as a class
  b i $(x + 1)^2 - x(x + 2) \equiv$
    $x^2 + 2x + 1 - (x^2 + 2x) \equiv 1$
    ii $(x - 1)^2 - x(x - 2) \equiv$
    $x^2 - 2x + 1 - (x^2 - 2x) \equiv 1$
    It is easier to deal with
    positive coefficients

3 Compare answers as a class

Chapter 3.5

Are you ready?

1 a $n + 1$ **b** $n + 10$
  c $n - 11$

2 a $x^2 + 2x$ **b** $y^2 + 10y$
  c $t^2 - 4t$ **d** $p^2 + 13p + 30$

3 a $4a + 22$ **b** $4b + 36$
  c $5c$

Practice 3.5A

1 **b** 99
  c It doesn't fit on the grid
  d $x + 2$ **e** $3x + 3$
  f 240
2 **a** 109   **b** 161
  c $x + 10$, $x + 20$ **d** $4x + 49$
  e 289
  f e.g. $2(2x + 24) + 1$ is one more than a multiple of 2
  g 73
3 **a** $6x + 66$
  b No, $6x + 66 = 428$ does not have an integer solution
  c 23
  d $6x + 66 = 6 \times (x + 11)$
4 **a** 30
  b $(x + 3)(x + 10) - x(x + 13) \equiv$
$x^2 + 13x + 30) - (x^2 + 13x) \equiv 30$
  c $(x + 1)(x + 30) - x(x + 31) \equiv$
$x^2 + 31x + 30 - (x^2 + 31x) \equiv 30$

What do you think?

1 Compare answers as a class
2 Compare answers as a class

Consolidate

1 **b** 69
  c It doesn't fit on the grid
  d $x + 10$ **e** $3x + 30$
  f 180
2 **a** 89   **b** 337
  c $x + 10$, $x + 11$ **d** $4x + 41$
  e 365
  f Always odd; e.g. $2 \times (2x + 20) + 1$
  gi–iii Compare answers as a class

Stretch

1 **a** The total on the 10 by 10 grid is 5 greater than the total on the 9 by 9 grid
  b Compare answers as a class. For grid size g an expression for the total might be $5x + 5g + 2$
  c Compare answers as a class. You could work in groups to present your findings.
2 Compare answers as a class. You could work in groups to present your findings.

Check my understanding

1 **a** $80 = 10 \times 8$, so it is a multiple of 8. 80 divides into 800 with no remainder, so it is a factor of 800
  b $2^4 \times 5$
2 **a** False: e.g. $11 \times 7 = 77$ is in both
  b True: $3\,km = 3\,000\,000\,mm$ which is greater than $300\,000\,mm$
3 **a** e.g. true for $k = 1, 2, 3, 5, 6, 7…$ but false for $k = 4, 8, 14…$
  b e.g. both can be written $4 \times 4 \times 10 \times 25$
  c $9m \times 4n \equiv 36\ mn \equiv 12 \times 3mn$
4 **a** **i** $x^2 - 3x - 40$
    ii $xy - 8x + 5y - 40$
    iii $p^2 - 2p + 1$
  b $100^2 - 200 + 1 =$
$10\,000 - 200 + 1 = 9801$

---

5 **a** 75
  b True. $B_x = x + x + 10 + x + 11 + x + 12 + x + 21 = 5x + 54$. When x is even, $5x$ is even so $5x + 54$ is even. When x is odd, $5x$ is odd so $5x + 54$ is odd.

Block 4 Three-dimensional shapes

Chapter 4.1

Are you ready?

1 **a** Square
  b Regular pentagon
  c Parallelogram **d** Circle
2 **a** right-angled triangle
  b scalene triangle
  c isosceles triangle
  d equilateral triangle
3 A and C are hexagons
4 **a** cube **b** cone **c** cylinder
  d square-based pyramid

Practice 4.1A

1 Rectangle or square
2 Regular pentagon
3 No, it could also be a rhombus
4 **a** and **iv** or **i** **b** and **i**
  c and **iii**   **d** and **ii**
5 **a** Similarities include: two pairs of parallel sides, opposite angles equal, sum of interior angles is 360°, opposite sides equal in length
  b Differences include: rhombus has all sides equal in length, in a rhombus the diagonals meet at 90°, a rhombus has reflection symmetry
6 The other two angles would also be 110° and 60° and the sum of all four would be 340° and not 360°
7 **a** Tetrahedron **b** Cuboid
  c Triangular prism
  d Sphere
8 **a** 6 faces, 12 edges, 8 vertices
  b 5 faces, 9 edges, 6 vertices
  c 5 faces, 8 edges, 5 vertices
  d 7 faces, 12 edges, 7 vertices
9 Triangular prism

What do you think?

1 Zach and Beca are both right. A square is a rectangle with sides of equal length and a rectangle is a parallelogram with equal angles.
2

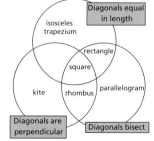

3 **a** 4 faces, 6 edges and 4 vertices
  b Benji is wrong. Some edges, faces and vertices will overlap.

Practice 4.1B

1 **a** Hexagonal prism
  b Cuboid
  c Cylinder
  d Triangular prism
2 A and B
3 **a**

Name	Number of faces	Number of edges	Number of vertices
cuboid	6	12	8
triangular prism	5	9	6
hexagonal prism	8	18	12
pentagonal prism	7	15	10

  b Faith is correct
  c **i** and **ii** Yes **d** $F + V = E + 2$
4 16 edges
5 Because $4 + 4 \neq 7 + 2$
6 Octagon

What do you think?

1 No, Marta is not right. The triangles are not identical in size.
2 Yes, it's a prism with a rectangular cross-section.
3 Prisms have rectangular faces but a cylinder has a curved surface instead. Cylinders do have a uniform cross-section.
4 $n + 2$ faces, $3n$ edges and $2n$ vertices

Consolidate

1 **a** Regular pentagon
  b Kite
  c Right-angled triangle
  d Trapezium
2 Rhombus or square
3 Isosceles triangle (could be right-angled at the same time)
4 A is true
5 **a** Similarities include: diagonals meeting at 90°, sum of interior angles is 360°, adjacent sides are equal in length
  b Differences include: number of lines of reflective symmetry, a rhombus has parallel sides, in a rhombus both pairs of opposite angles are equal
6 **a** Cube **b** Cylinder
  c Cone **d** Hexagonal prism
7 **a** 8 faces, 18 edges and 12 vertices
  b 4 faces, 6 edges and 4 vertices
  c 6 faces, 12 edges and 8 vertices
  d 5 faces, 9 edges and 6 vertices
8 Examples include a cube, cuboid or pentagonal pyramid

Stretch

1 Examples include a cylinder or sphere
2 Abdullah is wrong; a cone has no edges. An edge is where two flat faces meet. A cone has one face and one curved surface.
3 **a** Examples include a pentagonal prism or hexagonal-based pyramid

b Pentagonal prism: 15 edges and 10 vertices
Hexagonal-based pyramid: 12 edges and 7 vertices.

4 a

Shape	i	ii	iii	iv
Faces	6	7	8	5
Edges	10	12	14	8
Vertices	6	7	8	5

b The number of faces and vertices is always the same
c $n + 1$ faces, $2n$ edges, $n + 1$ vertices

5 All of the platonic solids have faces which are regular polygons

Chapter 4.2

Are you ready?

1 6
2 a 5
b Two triangles and three rectangles
3 a Check accuracy with a partner
b 12 cm²
4 a Check accuracy with a partner
b 5 cm

Practice 4.2A

1 For example:

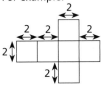

2 For example:

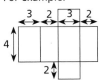

3 3 cm by 6 cm by 1 cm
4 a i FG   **ii** DE
b B and H
5 For example:

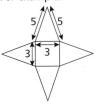

6 A and D
7 One of the squares needs to be on the other side – it currently has two bases
8 a 5   **b** 6   **c** 4
9 a Cuboid
b Triangular prism
c Tetrahedron
d Square-based pyramid

What do you think?

1 Check answers as a class. There are 11 unique nets for a cube.

2 For example:

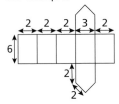

3 a For example:

b The circumference of the circle.
4 Discuss as a class

Practice 4.2B

1

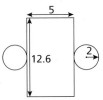

2 Compare answers as a class
3 In order: plan, front, side (front and side are interchangeable)
i

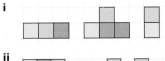

ii

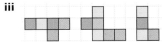

iii

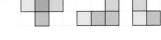

4 a

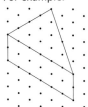

b

5 a

b

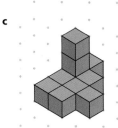

c

6 a 9   **b** 16   **c** n^2
d Check answer with a partner

What do you think?

1 Compare answers as a class
2 For example:

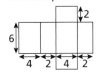

3 Abdullah is wrong as some of the straight lines will meet to create one edge

Consolidate

1 For example:

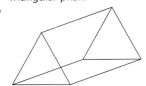

2 a Triangular prism
b

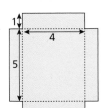

3

4 Possible answers include:

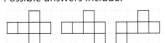

5 For example:

6 Compare answers as a class

7 a

b

c

8 In order: plan, front, side (front and side are interchangeable)

i

ii

iii

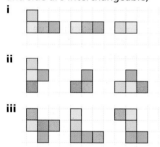

Stretch

1 a Both a cone and pyramid would look the same if viewed from any of the sides

b i ⬤ **ii** ⬡

c A triangular prism

2 D

3 a For example:

b For example:

4 A and D

Chapter 4.3

Are you ready?

1 a $16\,cm^2$ **b** $10\,cm^2$
   c $7.5\,cm^2$ **d** $10\,cm^2$
2 a diameter
   b circumference
   c radius
3 a 24.35   **b** 5.66   **c** 8.00
   d 0.06   **e** 104.10
4 a $x = 12$ **b** $x = 2.3$ **c** $x = 46$
   d $x = 5$ or $x = -5$
   e $x = 0.6$ or $x = -6$
5 a 100   **b** 30   **c** 3.42   **d** 7.6

Practice 4.3A

1 a $36\,cm^2$ **b** $24\,cm^2$
   c $392\,mm^2$ **d** $39\,cm^2$
   e $7.44\,m^2$ **f** $58.5\,cm^2$
   g $336\,cm^2$ **h** $24\,cm^2$
   i $3520\,mm^2$ **j** $78\,cm^2$
   k $2.25\,m^2$ **l** $0.715\,m^2$
   The shapes are squares, rectangles, triangles and trapeziums.
2 a $28.3\,cm^2$ **b** $1256.6\,mm^2$
   c $254.5\,cm^2$ **d** $25.1\,cm^2$
3 a $28\,cm^2$ **b** $54\,cm^2$
   c $2080\,mm^2$ **d** $21\,m^2$
   e $132.5\,cm^2$ **f** $39.6\,cm^2$
4 Faith has used the slant height instead of the height that is perpendicular to the base. The correct answer is $60\,cm^2$
5 a 7 cm   **b** 6 mm
   c 5 cm   **d** 5.5 cm
6 Answer: $179\,cm^2$. Check methods with a partner
7 $x = 4$
8 12 cm
9 a $d^2\,cm^2$ **b** $8e\,cm^2$
   c $9f\,cm^2$ **d** $36gh\,cm^2$
10 Five boxes, costing £65

What do you think?

1 The square
2 a $12.4\,cm^2$ **b** $30.9\,cm^2$
3 $1560\,cm^2$
4 $146.1\,cm^2$
5 Check answers with a partner

Practice 4.3B

1 a i $10\,cm^2$ **ii** $16\,cm^2$ **iii** $40\,cm^2$
   b $132\,cm^2$
2 a $152\,cm^2$ **b** $110\,cm^2$
   c $1480\,mm^2$ **d** $9.36\,m^2$
3 A has the larger surface area
   A: total surface area
   $= 2(3 \times 8 + 3 \times 5 + 8 \times 5)$
   $= 158\,cm^2$
   B: total surface area
   $= 2(3 \times 10 + 3 \times 10 + 3 \times 3)$
   $= 138\,cm^2$

4 Ed is wrong because some of the faces will be hidden and will no longer contribute to the overall surface area
5 $459\,cm^2$
6 $6x^2\,cm^2$
7 a $25\,cm^2$ **b** 5 cm
8 He needs two tins, which will cost £16

What do you think?

1 a $38x^2\,cm^2$
   b $(2pq + 2pr + 2qr)\,cm^2$ or $2(pq + pr + qr)\,cm^2$
2 $x = 8$
3 Minimum surface area arrangement is a cube ($2 \times 2 \times 2$)

Consolidate

1 a $128\,mm^2$ **b** $56\,cm^2$
   c $60.5\,cm^2$ **d** $137.5\,cm^2$
   e $30\,cm^2$ **f** $0.84\,cm^2$
2 a $169\pi\,cm^2$ **b** $7.84\pi\,mm^2$
   c $36\pi\,cm^2$ **d** $9\pi\,cm^2$
3 a $40\,cm^2$ **b** $5.51\,cm^2$
   c $86.5\,cm^2$
4 a 12 cm   **b** 4 cm
   c 7 cm   **d** 1.5 cm
5 He has used the formula incorrectly; it should be the sum of the parallel sides (6.5 + 9.5) but he has multiplied instead. Answer should be $24\,cm^2$
6 $54\,cm^2$
7 a $242\,cm^2$ **b** $83\,cm^2$
   c $108.1\,m^2$ **d** $250\,cm^2$
8 a 48 tiles
   b £64.95

Stretch

1 $x = 5\,cm$
2 $1.8\,cm^2$
3 a $x = 4$ **b** $x = 1.5$
4 Compare answers with a partner

Chapter 4.4

Are you ready?

1 a $81\,cm^2$ **b** $120\,cm^2$
   c $54\,cm^2$ **d** $71.25\,cm^2$
2 a 56.5 cm   **b** 52.8 mm
   c 44.0 cm   **d** 10.7 m
3 a $36\pi\,cm^2$ **b** $6.25\pi\,m^2$
   c $49\pi\,m^2$ **d** $56.25\pi\,cm^2$
4 a 3.3   **b** 0.4
   c 0.1   **d** 29.0
5 a Cube
   b Triangular prism
   c Cylinder

Practice 4.4A

1 a For example:

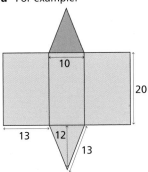

b 840 cm²
2 a 468 cm²　　**b** 108 cm²
　　c 744 mm²
3 a 210 cm²
　　b Marta is wrong. Doubling the length will only double the areas of the rectangular faces but the triangular faces will have the same area.
4 Abdullah has mixed his units. He needs to convert the lengths so that they are all in cm or mm.
5 17.1%
6 $x = 11$

What do you think?

1 5:8
2 Cutting the cube in half will mean that there are new extra surfaces as well as the faces of the original cube, so Flo is wrong. The surface area of each prism will be bigger than half of the surface area of the cube.
3 211.2 cm²

Practice 4.4B

1 a For example:

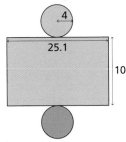

b 351.9 cm²
2 a 942.5 cm²　　**b** 1357.2 mm²
3 a 48π cm²　　**b** 128π cm²
4 462 cm²
5 Shape A has the larger surface area. Curved area is the same for both cylinders. Cylinder A has a larger top and base
6 a 2563.5 cm²　　**b** 6.25%
7 78.4 cm²
8 $h = 12$

What do you think?

1 $x = 9$
2 668.4 cm²
3 357.9 cm²

Consolidate

1 a 175.9 cm²　　**b** 520 mm²
　　c 696 cm²　　**d** 622 cm²
2 a 528π mm²　　**b** 768π cm²
3 79.4%
4 Six tins
5 $x = 8.56$
6 $h = 20$ cm

Stretch

1 444 cm²
2 599.1 cm²
3 3.34 cm and 4.73 cm

Chapter 4.5

Are you ready?

1 a 55 cm²　　**b** 24 cm²
　　c 38.5 cm²　　**c** 96 cm²
2 a 2.3　　**b** 29.0
　　c 106.0　　**d** 1.0
3 a 14　　**b** 16
　　c 4　　**d** 36
　　e 48
4 a $x = 1.4$
　　b $x = 3.0$ or $x = -3.0$
　　c $x = 3.5$ or $x = -3.5$
　　d $x = 2.2$
　　e $x = 4.6$ or $x = -4.6$

Practice 4.5A

1 a 24 cm³　**b** 48 cm³
　　c 42 cm³
2 a 84 cm³　**b** 24 cm³
　　c 1728 mm³
3 $x = 12$
4 302.5 cm³
5 240 litres
6 a $y = 9$　　　　**b** $y = 3$
　　c $y = 14$　　　**d** $y = 9$
7 Multiplication is commutative, so it does not matter in which order the length, width and height are multiplied
8 2016
9 a 125 cm³　　**b** 33.1%
　　c Because each length is increased by 10%, so this is the same as the volume being increased by 10% in three successive calculations

What do you think?

1 512 cm³
2 10 hours and 40 minutes
3 3.2 cm

Practice 4.5B

1 a 180 mm³　　**b** 3600 cm³
　　c 612 cm³　　**d** 37 500 mm³
2 a 40π cm³　　**b** 8960π mm³
　　c 64.8π cm³　　**d** 1134π cm³
3 $x = 9.86$
4 102.65 cm³
5 6 full glasses
6 No, he needs 3.5 litres
7 a $y = 16$　　　**b** $y = 4$
　　c $y = 12.5$
8 2610 cm³

What do you think?

1 10.3 cm
2 21.5%

3 6.97 cm

Practice 4.5C

1 a 2145 cm³　　**b** 150 cm³
　　c 3.0 m³　　　**d** 56.5 mm³
2 19 905 mm³
3 No, Chloe is wrong. The pyramid has volume 24 cm³ and the cone has volume 18.8 cm³
4 47.6%
5 a $x = 8$　　**b** $x = 15$　　**c** $x = 6.5$

What do you think?

1 4.16 cm
2 $r = 3$ cm

Consolidate

1 a 30 cm³　　**b** 128 cm³
　　c 40.5 m³
2 a 5632 cm³　　**b** 4.48 cm³
　　c 520 mm³
3 a 72π m³　　**b** 5070π mm³
　　c 1.568π cm³
4 a $x = 50$　　**b** $x = 9$
　　c $x = 3.5$
5 No it will not. The cylinder has volume 864 cm³ and the cuboid has volume 504 cm³
6 $x = 13.8$
7 78.5%
8 Pyramid (5.13 cm³), cuboid (7.84 cm³), cone (32.57 cm³), sphere (38.79 cm³)

Stretch

1 13.8 cm
2 17:9
3 11.2 cm
4 94.5 litres
5 a The radius of the cone is not constant therefore the cone is less than half full
　　b $\frac{1}{8}$
　　c 301.6 ml

Check my understanding

1 6 faces, 12 edges and 8 vertices
2 B only; uniform cross-section
3 a i Cuboid　　**ii** 162 mm²
　　iii 126 mm³
　　b i Cube　　　**ii** 384 cm²
　　iii 512 cm³
　　c i Cylinder　　**ii** 180.6 cm²
　　iii 176.7 cm³
　　d i Triangular prism
　　ii 336 cm²　　**iii** 288 cm³

Block 5 Constructions and congruency

Chapter 5.1

Are you ready?

1 a 30°　　**b** 65°　　**c** 45°
　　d 15°　　**e** 35°
2 a ∠ABE, ∠EBC, ∠AEB, ∠DEB, ∠BAC
　　b ∠ABC, ∠AED
3 ∠ABC, ∠AED, ∠DEA
4 a 65° is acute, but the diagram shows an angle more than 90°

b 103° is obtuse but the diagram shows an angle less than 90°

Practice 5.1A

1 Check accuracy with a partner
2 **a** Compare answers as a class.
   b i 55° **ii** 140° **iii** 98° **iv** 62°
3 **a** 180 + 50 = 360 − 130 = 230
   b and **c** Check accuracy with a partner
4 **a** About 63° and 117°
   b Both pairs of opposite angles, as opposite angles in a parallelogram are equal
   c All four pairs of adjacent angles, as co-interior angles add up to 180°
5 **a** 8 cm by 5 cm
   b i 16 m by 10 m
   ii 4 m by 2.5 m
   iii 40 m by 25 m
   iv 8 mm by 5 mm
6 600 m
7 **a** 500 000 cm, 5000 m, 5 km
   b 90 km **c** 42.4 cm
8 **a i** 82.8 cm **ii** 8.28 cm
   iii 165.6 cm **iv** 331.2 cm
   b i About 1 : 800
   ii About 1 : 1600
   iii About 1 : 2000
9 Compare answers as a class

What do you think?

1 So that you can easily draw angles both clockwise and anticlockwise
2 Discuss as a class
3 **a** 50 000 m²
   b 500 cm² or 0.05 m²
   c Both dimensions have been scaled by $\frac{1}{1000}$, so the area is scaled by $\frac{1}{1000} \times \frac{1}{1000} = \frac{1}{1000000}$

Practice 5.1B

1 Compare accuracy with a partner
2 **a–c** Compare explanation and accuracy with a partner
   d The sides will not meet
3 Compare explanation and accuracy with a partner
4 A = angle, S = side; the number of each letter indicates the number of angles/sides needed when constructing. For example, if there are two sides given, then the angle between them is needed (SAS).
5 **a** A sketch shows the general shape and lengths of the sides, but is not drawn to scale
   b Compare accuracy with a partner
6 **a–d** Compare accuracy with a partner
   e ∠XZY = 80°, ∠FDE = ∠DFE = 55°
7 Compare accuracy with a partner

What do you think?

1 **a** It depends on the type of triangle (equilateral, isosceles or scalene) and what information is given. Compare findings as a class
   b Given three angles there are an infinite number of possibilities
2 $x > y + z$, $y > x + z$, $z > x + y$
3 **a** Compare accuracy with a partner
   b 120°
   c Discuss answers as a class (all the angles can be found)

Consolidate

1 Compare accuracy with a partner
2 **b i** 95° **ii** 160° **iii** 70°
   iv 42° **v** 345°
3 **a** 1.2 m **b** 120 m
   c 12 m **d** 30 m
   e 30 m **f** 6 km
4 **a** 5 **b** 1 : 500 000
5 Compare accuracy with a partner

Stretch

1 **a** 360 − 40 = 320°
   b i A = 050° **ii** C = 140°
   iii D = 230°
2 **a–d** Check accuracy with a partner
   e i 305° **ii** 060° **iii** 015° **iv** 244°
   f 180° more or less than the original bearing
   g If $x < 180°$, then $(x + 180)°$; if $x > 180°$, then $(x − 180)°$
3 Compare answers as a class

Chapter 5.2

Are you ready?

1–3 Check accuracy with a partner

Practice 5.2A

1 Check accuracy with a partner
2 Check accuracy of circles with a partner
3 **a–c**

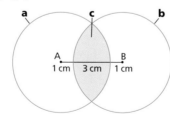

   d D
4 **a–c**

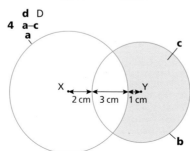

5

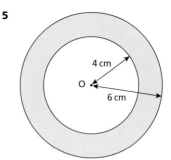

6 **a**

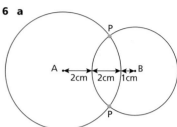

   b AB = 5 cm, AP = 4 cm, BP = 3 cm; a 3 : 4 : 5 triangle has a right angle
7 **a i**

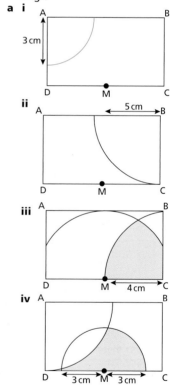

   b Compare answers as a class

What do you think?

1 **a** Diagram A includes regions that are beyond the lawn

b i

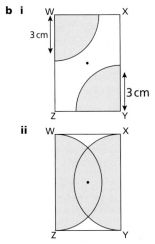

ii

2 Compare answers as a class

Practice 5.2B
1 Compare answers as a class
2 a Discuss as a class
b i

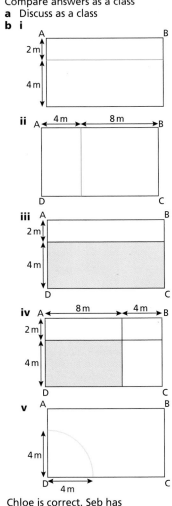

ii

iii

iv

v

3 Chloe is correct. Seb has forgotten that the locus from the end points will be curved.

4

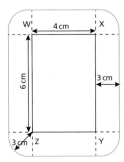

5

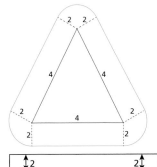

6

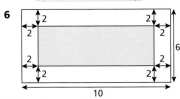

What do you think?
1 a

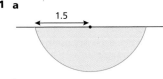

b

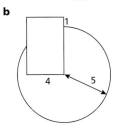

2 Compare answers as a class

Consolidate
1 Check accuracy of circles with a partner
2

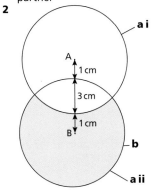

3 a i

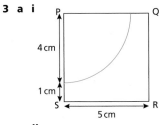

ii

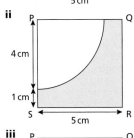

iii

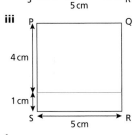

iv

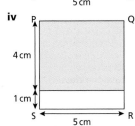

v

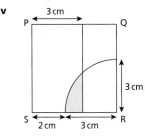

b

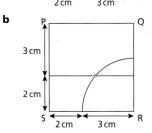

4 a

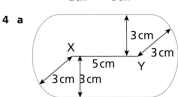

b

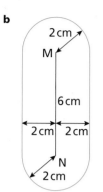

5 a

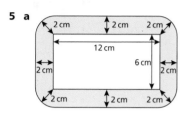

b

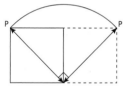

Stretch

1 and **2** Compare answers as a class

3

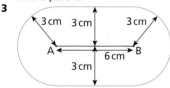

Chapter 5.3

Are you ready?

1 PQ and RS, WX and YZ
2 Check the accuracy of the circle with a partner
3

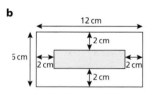

Practice 5.3A

1–3 Check results with a partner

What do you think?

You only need one pair of arcs on each side of the line; two pairs of arcs on the same "side" would also work

Practice 5.3B

1–6 Check accuracy with a partner

What do you think?

1 Check accuracy with a partner
2 b i Inside the triangle
    ii Outside the triangle
3 Check accuracy with a partner

Consolidate

1–5 Check accuracy with a partner

Stretch

1 Compare answers as a class
2 a Parallel to l_1 and l_2
   b A straight line parallel to l_1 and l_2 and halfway between them
3 a Check accuracy with a partner
   b $y = 2x - 1$
   c $(0, -1)$
   d $y = -\frac{1}{2}x - 1$
   e The product of the gradients is -1

Chapter 5.4

Are you ready?

1 and **2** Check accuracy with a partner
3

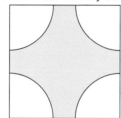

Practice 5.4A

1–5 Check accuracy using a protractor or with a partner
6 a–c Check accuracy with a partner
   d You only need two bisectors, as the third bisector will also meet at the same point. The third bisector is a useful confirmation
7 Check accuracy using a protractor or with a partner
8 a Check accuracy of the bisector of angle BAC with a partner
   b This locus would be the perpendicular bisector of the line segment joining points B and C. Check the accuracy of this perpendicular bisector with a partner.

What do you think?

1 a Discuss as a class
   b

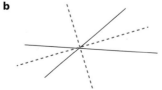

2 The arcs on the arms of the angle have different radii

3 a 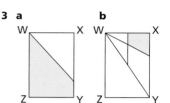 **b**

   c Compare answers as a class

Consolidate

1 Check accuracy using a protractor or with a partner
2 a No
   b i No   **ii** Yes
     iii Yes for the bisectors of the two unequal angles in the kite, no for the bisectors of the two equal angles
3 Check accuracy with a partner

Stretch

1 Check accuracy with a partner. Compare methods, for example $75° = 90° - (60°$ bisected twice$)$
2 The theorem is correct

Chapter 5.5

Are you ready?

1 Compare accuracy with a partner
2 a Isosceles
   b Isosceles right-angled
   c Equilateral
   d Scalene right-angled
   e Isosceles
3 a 180°   **b** 360°

Practice 5.5A

1 Answers include

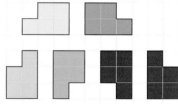

2 Compare answers as a class
3 C
4 The hatch marks and symbols on the angles show that all the corresponding sides and all the corresponding angles are equal
5 Compare answers as a class
6 Marta is correct but Seb is not
7

8 QR = 4 cm   BC = 5 cm
   △ADC = 90°   △PSR = 74°

What do you think?

1 a Need more information
   b Definitely congruent
   c Need more information
   d Need more information
2 a Always congruent
   b May be congruent
   c Always congruent
   d Always congruent
   e May be congruent

Practice 5.5B

1 **a** △ABC and △FED
   b △PQR and △YXZ
   c △HIJ and △MKL
2 **a** △PQR and △EDF
   b △XYZ and △ACB
   c △PQR and △PSR
3 **a** △ABC and △UVT
   b △DEF and △XZY
   c △PQR and △LKJ
4 **a** △ABC and △GKH (SSS)
   b △LMN and △RQP (AAS)
   c △XYZ and △TUS (SAS)
   d △DEF and △HGF (SAS)
5 **a** Vertically opposite angles are equal
   b Yes (SAS)
6 **a** QR = RS (given), △PQR and △PSR are congruent (SSS)
   b AB = CD (given), BC = DA (given), AC = AC (diagonal in common), △ABC and △CDA are congruent (SSS)
   c Yes. SSS, as diagonal is in common and all other sides are equal (given)
7 Use the fact that angles in a triangle add up to 180°. △ABC and △EDF
8 Discuss as a class

What do you think?

1 Discuss as a class
2 **a** SSS   **b** SAS   **c** RHS
   d AAS   **e** RHS   **f** SAS

Consolidate

1 **a** Yes   **b** Yes   **c** Yes
   d No   **e** No   **f** Yes
2 Compare answers as a class
3 All the statements are true
4 **a** 4cm   **b** 240cm²
5 Yes, AAS
6 **a** △DEF and △DGF
   b △PQS and △RSQ
   c △WYZ and △WXZ
   d △JKM and △LKM

Stretch

1 **a** Faith
   b

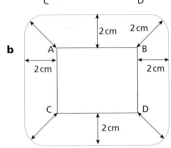

2 **a** Compare answers as a class
   b 4
3 Compare answers as a class
4 Discuss as a class

Check my understanding

1 **a** Check accuracy with a partner
   b 143°
2 A circle centre P radius 3.5cm

---

3 **a**

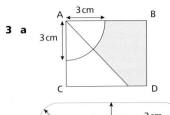

   b

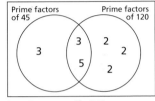

   c

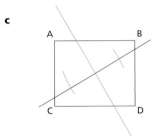

4 Check accuracy with a partner
5 Check answer with a partner
6 Yes, AAS

Block 6 Numbers

Chapter 6.1

Are you ready?

1 **a** 20, 40, 60, 80, 100
   b 16, 32, 48, 64, 80
   c 32, 64, 96, 128, 160
   d 48, 96, 144, 192, 240
   e 17, 34, 51, 68, 85
2 **a** 1, 2, 4, 5, 10, 20
   b 1, 2, 4, 8, 16
   c 1, 2, 4, 8, 16, 32
   d 1, 2, 3, 4, 6, 8, 12, 16, 24, 48
   e 1, 17
3 24, 30, 36, 42, 48

Practice 6.1A

1 10, −3, 117
2 **a** Any four integers between 5 and 15, e.g. 6, 9, 13, 14
   b Any four non-integers between 2 and 10, e.g. 2.7, 3.1, 9.9, 8.5
   c Any four values between 3 and 4, e.g. 3.01, 3.25, 3.81, 3.999
3 $\frac{7}{2}$, 1.5, 11
4 **a** $\sqrt{2}$, $\sqrt{3}$, $\sqrt{5}$, $3\sqrt{7}$, $-\sqrt{3}$
   b All of them
5 **a** 0.7 = 7 tenths = $\frac{7}{10}$
   b No; 0.7777... = $\frac{7}{9}$
6 No; 1.76503 = $\frac{176\,503}{100\,000}$ and $0.\dot{6} = \frac{2}{3}$

What do you think?

1 **a** Discuss as a class
   b Faith, because $\frac{22}{7}$ is only approximate
   c Beca, because it's 5
2 The conjecture is false as $\sqrt{9}$ is 3 so $2\sqrt{9}$ is 2 × 3 = 6

Practice 6.1B

1 **a** 1, 2, 4, 5, 10, 20
   b 1, 2, 5, 10, 25, 50
   c 1, 2, 5, 10
   d 10
2 **a** 4, 8, 12, 16, 20, 24, 28, 32, 36, 40
   b 6, 12, 18, 24, 30, 36, 42, 48, 54, 60
   c 12, 24, 36   **d** 12
3 **a** **i** 45 = 3 × 3 × 5
      ii 120 = 2 × 2 × 2 × 3 × 5
   b

Prime factors of 45		Prime factors of 120
3	3 5	2 2 2

   c **i** 15   **ii** 360
4 **a** 30   **b** 42   **c** 20   **d** 48
5 **a** 200   **b** 72   **c** 300   **d** 2112
6 **a** 120   **b** 13 566 000
   c A, because 2 × 5 × 5 × 7 > 17 × 19
7 **a** 150   **b** 693 000
8 **a** **i** The LCM of 12 and 6 is 12
      ii The LCM of 12 and 18 is 36
   b Yes, e.g. 36
9 11:12 a.m.

What do you think?

1 Beca is correct and Abdullah is not. 1 is the highest common factor of two numbers when they have no prime factors in common, e.g. 2 and 3, 10 and 21 and so on
2 **a** It is 1 because each pair of numbers have no prime factors in common
   b X is not a multiple of 11
   c Y is a multiple of 11
3 Any pair of numbers such that the HCF is 18, e.g. 18 and 36 or 36 and 54

Practice 6.1C

1 Surds: $\sqrt{24}$ $\sqrt{2}$ $\sqrt{8}$ $\sqrt{130}$
   Not surds: $\sqrt[3]{8}$ $\sqrt{25}$ $\sqrt{1}$ $\sqrt{169}$
2 **a** **i** 2   **ii** 10
   b **i** 2 × 10 = 20 **ii** 10 ÷ 2 = 5
3 **a** **i** 3   **ii** 4
   b 3 + 4 ≠ 5 and 4 − 3 = 1 and $\sqrt{7} \neq 1$
4 **a** $\sqrt{15}$ **b** $\sqrt{77}$ **c** $\sqrt{102}$ **d** $\sqrt{315}$
5 **a** $\sqrt{7}$ **b** $\sqrt{13}$ **c** $\sqrt{3}$ **d** $\sqrt{\frac{1}{7}}$
6 **a** $\sqrt{4} \times \sqrt{2}$ **b** $\sqrt{25} \times \sqrt{2}$
   c $\sqrt{9} \times \sqrt{3}$ **d** $\sqrt{100} \times \sqrt{11}$
   e $\sqrt{49} \times \sqrt{10}$ **f** $\sqrt{16} \times \sqrt{3}$
   Multiple answers are available where there is more than one square factor

7 a $2\sqrt{2}$ **b** $5\sqrt{2}$
  c $3\sqrt{3}$ **d** $10\sqrt{11}$
  e $7\sqrt{10}$ **f** $4\sqrt{3}$
  Multiple answers are available where there is more than one square factor. Full simplification should be encouraged

What do you think?

1 a Samira has identified a square factor of 200
  b She could have used the greatest square factor, i.e. $\sqrt{100}$
2 $\frac{18y}{2y} = 9$ and $\sqrt{9} = 3$
3 a $9\sqrt{3} = \sqrt{81} \times \sqrt{3} = \sqrt{243}$
  b $6\sqrt{5} = \sqrt{36} \times \sqrt{5} = \sqrt{180}$

Consolidate

1 a 1, 2, 3, 4, 6, 12
  b 1, 2, 3, 5, 6, 10, 15, 30
  c 1, 2, 4, 5, 10, 20, 25, 50, 100
2 a 1, 2, 3, 6   **b** 1, 2, 4
  c 1, 2, 5, 10
3 a i $2 \times 2 \times 2 \times 2 \times 5$
    ii $2 \times 2 \times 2 \times 7$
  b i 560   **ii** 8

Stretch

1 a Never true
  b Always true
  c Always true
  d Sometimes true
  e Sometimes true
  f Always true
2 a 35 (HCF), 980 (LCM)
  b $2 \times 2 \times 5 = 20$ but neither 7×7 nor 5×7 give 20
  c X = 140 and Y = 245
3 a e.g. $\sqrt{32} = 4\sqrt{2}$
  b e.g. $\sqrt{11}$
4 a $x = 6$ **b** $x = 10$ **c** $x = 637$

Chapter 6.2

Are you ready?

1 a 600   **b** 10 000   **c** 0.07
  d 7   **e** 100
2 a 1000   **b** 3000   **c** 400
  d 60   **e** 43 000
3 1200, 950, 1100

Practice 6.2A

1 a i 900   **ii** 300
  b i 1200   **ii** 600
   iii 270 000   **iv** 3
2 a 6000   **b** 8700   **c** 700
  d 100   **e** 1200   **f** 4.8
  g 61   **h** 0.24   **i** 5
  j 0
3 a 6020   **b** 435   **c** 2100
  d 80   **e** 480   **f** 15
4 £360
5 a 140 m²   **b** 800 mm²
  c 1000 km²   **d** 150 cm²
  e 400 cm²   **f** 26 cm²
6 a £50   **b** £1000
  c £25 000
7 a 310   **b** 290
  c 3000   **d** 30
  e 90 100   **f** 96 100
8 $\frac{16}{3}$ kg

What do you think?

1 a 14
  b Overestimate; both the numerators were rounded up and the denominator was rounded down (which will also increase the estimate)
2 a e.g. $\frac{250 \times 45}{2.4}$
  b e.g. $\frac{349.9 \times 54}{1.6}$
3 a Chloe has rounded to 1 significant figure correctly
  b She rounded again mid calculation
4 Ed's estimate will be more accurate

Practice 6.2B

1 a 4.4, 4.403, 3.91, 3.51, 4.06
  b $3.5 \leqslant x < 4.5$
2 a Any three correct values
  b 88°
3 a $12.35 \leqslant y < 12.45$
  b i $59.5 \leqslant p < 60.5$
   ii $24.65 \leqslant q < 24.75$
   iii $5.605 \leqslant r < 5.615$
  c Nearest integer $84.5 \leqslant x < 85.5$
   1 d.p. $84.95 \leqslant x < 85.05$
4 a $89.5 \leqslant x < 90.5$
  b $85 \leqslant x < 95$
5 a 39.5 m
  b $39.5 \leqslant l < 40.5$
6 $4999.5 \leqslant a < 5000.5$
  $4995 \leqslant b < 5005$
  $4950 \leqslant c < 5050$
  $4500 \leqslant d < 5500$
7 $9500 \leqslant n < 15 000$
  If it is the area of a field it can be given to any number of decimal places, e.g. 14 999.999 999
  If it is a number of people, then the maximum value is 14 999 as it can only be an integer
8 a $250 \leqslant n < 350$
  b $295 \leqslant n < 305$
  c $299.5 \leqslant n < 300.5$

What do you think?

1 a 81.9999999999 cm or 82 cm
  b 80 cm
2 a $45 \leqslant p < 55$
  b $47.5 \leqslant q < 52.5$
  c $49 \leqslant r < 51$
  d $49.75 \leqslant s < 50.25$
  e $50 - \frac{x}{2} \leqslant t < 50 + \frac{x}{2}$

Consolidate

1 a 200   **b** 10
  c 60   **d** 0.0008
  e 40 000
2 £25
3 a 1000   **b** 2700
  c 50   **d** 80 000
  e 100   **f** 1000
4 2400   Overestimate as both numbers were rounded up
5 a 7   **b** 4000   **c** 40

Stretch

1 a 50π cm²   **b** 150 cm²
  c a because pi is more accurate than 3

2 a The length can be 0.5 cm greater than or less than 20 cm
  b i $19.5 \leqslant l < 20.5$
   ii $78 \leqslant p < 82$
   iii $380.25 \leqslant a < 420.25$
3 a $8.44 \leqslant s < 8.51$
  b $8.43 \leqslant s < 8.52$
  c $8.40 \leqslant s < 8.55$
4 a $20 \times 40 = 800$
  b 17.45 is 2.55 less than 20 and 42.55 is 2.55 more than 40 You are "adding on" more (2.55×42.55) than you are "taking off" (2.55×17.45)

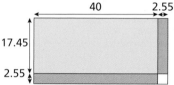

5 This conjecture is sometimes true, e.g. if 100 has been obtained by rounding to the nearest integer, then $99.5 \leqslant n < 100.5$ and $100 - 99.5 = 100.5 - 100$
If 100 has been obtained by rounding to 1 significant figure, then $95 \leqslant n < 150$ and $100 - 95 \neq 150 - 100$

Chapter 6.3

Are you ready?

1 a 497   **b** 2856   **c** 743
  d 23 491   **e** 416   **f** 912
  g 4025   **h** 129 600 **i** 36
  j 153   **k** 798   **l** 266
2 a 3   **b** −3   **c** 5
  d −7
3 Any three correct calculations

Practice 6.3A

1 £216
2 a 9   **b** 7
3 7 weeks
4 £18 620
5 200 ml
6 31

What do you think?

1 124   680   165
  4 200 000   120   £145
  52 000   5   347 000
2 a Both give information about the cost of items, but the first gives the cost of an item and the difference in cost while the second gives the difference in cost and the total cost
  b Scooter and bike cost £121, skateboard costs £47 and football costs £19
3 5 nights (M to Th, plus either Su or F)

Practice 6.3B

1 a Charlie has used incorrect place value
  b 14.07
2 a 3.84   **b** 25.6 kg
  c 44p   **d** 8.7

3 Divide the answer by 100 because
$17 \times 39 = 1.7 \times 10 \times 3.9 \times 10$
$= 1.7 \times 3.9 \times 100$

4 a 72 **b** 30.34 **c** 32.9
d 2.275

5 a 0.9 **b** 0.6 **c** 9
d 6 **e** 2.4 **f** 0.24

6 a 50.4 **b** 5.04 **c** 1008
d 100.8 **e** 5.04 **f** 0.504
g 36 **h** 360 **i** 140
j 1.4

7 a £17.10 **b** £4.75

8 £13.76

9 9.6 kg

10 2.25 + 4.8 < 8 so yes

What do you think?

1 a 180 **b** 18 **c** 1.8
d 0.18 **e** 0.018

2 a Any correct method
b Compare answers with a partner

3 Discuss methods as a class

Practice 6.3C

1 a Faith is confusing the rules for calculating with negative numbers
b −8

2 a −10 **b** −3 **c** −11 **d** −110

3 Rob has made 5 and −3 using counters. There are three zero pairs which cancel out to leave two +1 counters, so the answer is 2

4 a 4 **b** 4 **c** −4
d −4 **e** 6 **f** −3

5 a 7, −7 **b** −5, 5 **c** 3, −3
d −6, −12

6

	−2	−3	−5	4	5	8
−2	4	6	10	−8	−10	−16
−3	6	9	15	−12	−15	−24
−5	10	15	25	−20	−25	−40
4	−8	−12	−20	16	20	32
5	−10	−15	−25	20	25	40
8	−16	−24	−40	32	40	64

7 a 21 **b** −10 **c** −10
d 10 **e** −6 **f** −5
g 9 **h** 30 **i** −30
j −1 **k** −2 **l** −0.5

What do you think?

1 a The product of two negative numbers is positive
b Lydia should have used brackets, i.e. entered $(-5)^2$
c $-6^2 = -1 \times 6^2 = -36$ whereas $(-6)^2 = -6 \times -6 = 36$

2 a Negative **b** Negative
c Negative **d** Positive
e Negative **f** Positive

3 Compare answers with a partner

Consolidate

1 a 680 **b** 473
2 16 cm
3 1071

---

4 744
5 39
6 a i 2 **ii** 3 **iii** −5
b i 2 **ii** 3 **iii** −5
7 a 11 **b** −11 **c** −11
d 7 **e** −18 **f** 18
g −4.5 **h** 4.5

Stretch

1 140 m
2 a 155 cm
b 135 cm, 159.5 cm, 159.5 cm, 166 cm or 135 cm, 153 cm, 166 cm, 166 cm
3 105
4 9.95 litres
5 £2.96
6 3:07 p.m.
7 Compare answers with a partner
8 a e.g. a contestant could get twice as many incorrect as correct
b No

Chapter 6.4

Are you ready?

1 a $\frac{3}{4}$ **b** $\frac{4}{7}$ **c** $\frac{3}{10}$ **d** $\frac{1}{2}$

2 $2\frac{1}{3}$ $\frac{7}{3}$

3 a $\frac{3}{2}$ **b** $\frac{9}{5}$ **c** $\frac{11}{3}$

4 a $1\frac{5}{6}$ **b** $3\frac{3}{4}$ **c** $8\frac{1}{3}$

5 a 1 **b** 1 **c** $\frac{5}{6}$

Practice 6.4A

1 a Set A $\frac{7}{12}$ $\frac{11}{15}$ $\frac{1}{18}$
Set B $\frac{7}{10}$ $\frac{4}{12}$ $(\frac{1}{3})$ $\frac{2}{9}$
Set C $\frac{5}{12}$ $\frac{17}{18}$ $\frac{15}{30}$ $(\frac{1}{2})$

b In set A, the common denominator is the product of the denominators in the question. In set B, the common denominator is one of the denominators from the question. In set C, the common denominator is a common multiple of the denominators that is less than their product

c Any correct questions that fit the descriptions in part **c**

2 a $\frac{13}{15}$ **b** $\frac{7}{20}$
c $\frac{13}{14}$ **d** $\frac{7}{10}$
e $\frac{1}{5}$ **f** $\frac{11}{20}$
g $\frac{25}{12}$ or $2\frac{1}{12}$ **h** $\frac{43}{36}$ or $1\frac{7}{36}$
i $\frac{1}{16}$ **j** $\frac{23}{18}$ or $1\frac{5}{18}$
k $\frac{1}{16}$ **l** $\frac{219}{100}$ or $2\frac{19}{100}$

3 There are three whole ones altogether, then
$\frac{1}{3} + \frac{5}{6} = \frac{2}{6} + \frac{5}{6} = \frac{7}{6} = 1\frac{1}{6}$
$3 + 1\frac{1}{6} = 4\frac{1}{6}$

4 a $2\frac{9}{10}$ **b** $1\frac{2}{9}$ **c** $4\frac{2}{3}$
d $4\frac{7}{10}$ **e** $6\frac{7}{12}$ **f** $5\frac{2}{15}$
g $12\frac{4}{5}$ **h** 9 **i** $218\frac{7}{12}$
j $15\frac{19}{36}$ **k** $9\frac{1}{12}$ **l** $14\frac{29}{30}$

5 a 1 **b** $\frac{5}{9}$ **c** $4\frac{1}{4}$
d $1\frac{23}{30}$ **e** $5\frac{7}{8}$ **f** $1\frac{3}{5}$

What do you think?

1 a Discuss as a class
b Method 1 because it avoids long multiplication

2 a Advantages: all questions can be answered with the same approach
Disadvantages: it risks numerical error, it is harder, takes longer and will need simplifying
b Recognise that 36 is double 18, so use 36 as a common denominator

Practice 6.4B

1 a i $\frac{1}{5}$ **ii** $\frac{2}{5}$ **iii** $\frac{3}{5}$
iv $\frac{4}{5}$ **v** $\frac{5}{5}$ **vi** $\frac{6}{5}$
vii $\frac{7}{5}$

b i $\frac{4}{3}$ **ii** $\frac{6}{4}$ **iii** $\frac{10}{9}$
iv $\frac{4}{9}$ **v** $\frac{6}{7}$ **vi** $\frac{18}{10}$
vii $\frac{22}{20}$

2 a The square is split into four columns of equal size so three columns represents $\frac{3}{4}$
The square is split into three rows of equal size so two rows represents $\frac{2}{3}$
The overlap of $\frac{3}{4}$ and $\frac{2}{3}$ is 6 cells which represents $\frac{6}{12}$ or $\frac{1}{2}$

b i

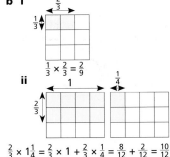

$\frac{1}{3} \times \frac{2}{3} = \frac{2}{9}$

ii

$\frac{2}{3} \times 1\frac{1}{4} = \frac{2}{3} \times 1 + \frac{2}{3} \times \frac{1}{4} = \frac{8}{12} + \frac{2}{12} = \frac{10}{12}$

3 a $\frac{2}{15}$ **b** $\frac{3}{20}$ **c** $\frac{3}{14}$
d $\frac{3}{25}$ **e** $\frac{32}{45}$ **f** $\frac{9}{7}$
g $\frac{7}{10}$ **h** $\frac{11}{6}$ **i** 3
j $\frac{85}{11}$ **k** $\frac{299}{42}$ **l** 1

4 a $\frac{1}{24}$ **b** $\frac{7}{20}$ **c** 1

5 a 3 **b** 5 **c** 5
d $\frac{1}{2}$ **e** e.g. 9, 4 **f** 5

6 a $\frac{3}{2}$ **b** $\frac{21}{4}$ **c** $\frac{14}{15}$
d $\frac{18}{25}$ **e** $\frac{20}{21}$

7 a The first is how many fifths are in 2 and the second is how many 2s are in one-fifth
b i 6 **ii** $\frac{15}{2}$ **iii** $\frac{40}{3}$
iv $\frac{1}{6}$ **v** $\frac{2}{15}$ **vi** $\frac{3}{40}$

8 a $\frac{5}{2}$ **b** $\frac{52}{5}$ **c** $\frac{36}{25}$ **d** $\frac{34}{9}$
e $\frac{5}{4}$ **f** $\frac{11}{6}$

What do you think?

1 a Several possible solutions as long as the second number is half the first

b One possible solution, 5

c Several possible solutions of the form $\frac{a}{b}$ such that 10b is double 7a

d Several possible solutions as long as the product of the two numbers is 144

2 Cancel common factors first:

$\frac{121}{180} \times \frac{36}{55} =$

$\frac{11 \times 11}{36 \times 5} \times \frac{36}{5 \times 11} =$

$\frac{11 \times \cancel{11} \times \cancel{36}}{5 \times 5 \times \cancel{11} \times \cancel{36}} = \frac{11}{5 \times 5} = \frac{11}{25}$

Practice 6.4C

1 **a** $7\frac{29}{60}$ cm **b** $7\frac{19}{20}$ cm

2

$2\frac{3}{4}$

$1\frac{1}{5}$ $1\frac{11}{27}$

$\frac{1}{2}$ $\frac{7}{10}$ $\frac{17}{20}$

3 **a** $24\frac{7}{10}$ cm^2 **b** $20\frac{7}{10}$ cm^2

4 $5\frac{23}{24}$ cm^2

5 $1\frac{7}{20}$ litres

6 £22.95

7 $\frac{17}{40}$

What do you think?

1 **a** **i** $\frac{11}{10}$ **ii** $\frac{25}{2}$

b $x = -2\frac{7}{20}$

2 $\frac{7}{12}$ cm^2

3 11, 12, 13, 14, 15, 16 or 17

Consolidate

1 **a** 24 **b** $\frac{19}{24}$

2 $\frac{1}{2} + \frac{1}{5} = \frac{5}{10} + \frac{2}{10} = \frac{7}{10}$

3 **a** $\frac{1}{3}$ **b** $\frac{17}{30}$ **c** $\frac{13}{24}$

d $\frac{1}{21}$ **e** $\frac{1}{18}$ **f** $\frac{7}{12}$

4 $\frac{7}{12}$

5 **a** $\frac{8}{9}$ **b** $\frac{5}{7}$ **c** $\frac{8}{9}$

d $\frac{4}{7}$

6 **a** $\frac{2}{15}$ **b** $\frac{1}{4}$ **c** $\frac{21}{80}$

d $\frac{1}{14}$

7 **a** $-\frac{14}{27}$ **b** $\frac{1}{2}$ **c** $\frac{1}{225}$

8 **a** 20 **b** 9 **c** $\frac{12}{5}$

d 6 **e** $\frac{50}{7}$

9 $\frac{17}{60}$ litre

Stretch

1 **a** **i** $\frac{4}{5}$ **ii** $\frac{13}{10}$

iii $\frac{4}{5} \times \frac{13}{10} = \frac{52}{50} = \frac{26}{25}$

b **i** $\frac{7}{20}$ **ii** $\frac{93}{10000}$

iii $\frac{3}{4}$

2 $\frac{980}{1377}$

3 **a** $2\frac{1}{4} + 2\frac{3}{20} = 4\frac{4}{10} = 4\frac{2}{5}$

b $1\frac{3}{10} + 1\frac{7}{8} = 3\frac{7}{40}$

c $2\frac{1}{4} - 1\frac{3}{10} = \frac{19}{20}$

d $2\frac{1}{4} - 2\frac{3}{20} = \frac{1}{10}$

4 $\frac{1}{4}$ cm

---

5 **a** $\frac{3}{10}$ **b** $\frac{2}{7}$ **c** $\frac{7}{30}$ **d** $\frac{44}{35}$

6 **a** $\frac{3x^2}{50y}$ mm^2 **b** $\frac{17x}{10}$ mm

7 $\frac{21p}{20} + \frac{3}{4}$

8 $-\frac{2g}{9} \times \frac{9}{2g} = -1$ therefore the lines are perpendicular

Chapter 6.5

Are you ready?

1 **a** 50 400 **b** 670 000
c 23 005 800 **d** 9600
e 0.405 **f** 0.697
g 2.88 **h** 0.9996

2 **a** **i** 1000 **ii** 10 000
iii 100 000 **iv** 10 000 000
b **i** 50 000 **ii** 5000
iii 5 **iv** 500 000

3 3.7, 4, 9.999, 1

4 **a** 100 **b** 1000
c 10 000 **d** 1 000 000
e 1 **f** 0.1

Practice 6.5A

1 **a** 3×10^8 **b** 3.7×10^8
c 3.77×10^8

2 **a** 8×10^{-7} **b** 8×10^{-6}
c 8.1×10^{-7}

3 **a** 6.1×10^4 **b** 7.41×10^7
c 9.03×10^7 **d** 5.904×10^{12}
e 3×10^6 **f** 4.58×10^5
g 5.7×10^9

4 **a** 8×10^{-5} **b** 2.6×10^{-6}
c 5.07×10^{-5} **d** 5×10^{-2}
e 4.009×10^{-4} **f** 7.133×10^{-9}
g 4×10^{-2} **h** 7×10^{-3}

5 The mass of Earth is 6×10^{24} kg
On average, YouTube is watched for a total of 3×10^9 hours each month.

6 **a** > **b** > **c** > **d** <

7 **a** 1.2×10^4 **b** 3.4×10^{10}
c 7×10^4 **d** 3.2×10^5
e 7.35×10^{13} **f** 7.02×10^5

8 **a** 3.5×10^5 **b** 1.67×10^5
c 3.15×10^3 **d** 1.92×10^4
e 7.486×10^5 **f** 3.8419×10^7
g 4.4×10^{-3} **h** 3.368×10^{-1}
i 2.2164×10^4

9 **a** 1.5×10^8 **b** 1.5×10^9
c 8.4×10^3 **d** 1.17×10^4
e 5.58×10^{12} **f** 1.98×10^5

10 **a** 2×10^3 **b** 4×10^2
c 1.2×10^2 **d** 1.5×10^3
e 1.1×10^1 **f** 6×10^1

11 2×10^3 grams

12 Sixty million

13 1.514×10^8 km

What do you think?

1 1.25×10^7

2 **a** 6.72×10^5 km
b 2.4528×10^8 km

3 6.638×10^7

Consolidate

1 3×10^{12}

2 **a** 29 000 **b** 2090
c 2 900 000 **d** 20 900

3 **a** 2×10^3 **b** 2.1×10^3
c 7.1×10^4 **d** 8.04×10^4
e 3.2×10^5 **f** 5.6×10^6

---

g 4×10^{-5} **h** 1×10^{-4}
i 8.7×10^{-5} **j** 9.43×10^{-7}
k 7.02×10^{-5}

4 **a** 7×10^6 **b** 8.1×10^8
c 8×10^{-3} **d** 1.5×10^7
e 3.77×10^5 **f** 3.05×10^6
g 2.44×10^{-2} **h** 4.293×10^{-2}

5 **a** 5.6×10^5 **b** 4.73×10^7
c 3.19×10^{-3} **d** 4.189×10^{-3}
e 5.708×10^5 **f** 2.7275×10^7

6 **a** 8×10^4 **b** 9×10^{-3}
c 9.24×10^7 **d** 8×10^6
e 9×10^{-4} **f** 6.8×10^8

7 **a** 4×10^4 **b** 2×10^{-3}
c 9.24×10^4 **d** 2×10^4
e 4×10^1 **f** 1.7×10^2

Stretch

1 30%

2 **a** 7.5×10^9 **b** 6.25×10^8
c 7.466×10^9 **d** 2.55×10^{17}
e 3.06×10^{18}

3 Large: 4.2×10^6 g
Small: 1.8×10^6 g
Total: 6×10^6 g

4 2×10^x

5 $y = w + \frac{x}{100}$

6 **a** **i** $xy \times 10^{a+b}$
ii $xy \times 10^{a+b+1}$
b Then A and B would not be written in standard form

7 **a** $p = 5$ and $q = 4$
b $n = 4$ **c** $x = 6$ and $y = 5$

8 3.84×10^{38} mm^2

Check my understanding

1 **a** -16 **b** -6 **c** -80 **d** 11

2 **a** 1.45×10^8 **b** 6.07×10^{-6}
c 0.000 047

3 £985

4 £138.60

5 6:39 a.m.

6 $\frac{38}{25}$

7 $5\sqrt{2}$

Block 7 Percentages

Chapter 7.1

Are you ready?

1 **a** 30% **b** $\frac{3}{10}$

2 **a** 0.55 **b** $\frac{11}{20}$

3 **a** 46 **b** 45.6 **c** 45.58

4 45

5 £96

Practice 7.1A

1 **a** **i** $\frac{43}{100}$ **ii** 43% **iii** 0.43

b **i** $\frac{17}{100}$ **ii** 17% **iii** 0.17

c **i** $\frac{14}{25}$ **ii** 56% **iii** 0.56

d **i** $\frac{3}{10}$ **ii** 30% **iii** 0.3

2 **a** **i** $\frac{1}{5}$, 0.2 **ii** $\frac{21}{50}$, 0.42

iii $\frac{7}{40}$, 0.175 **iv** $\frac{7}{5}$ or $1\frac{2}{5}$, 1.4

v $\frac{26}{25}$ or $1\frac{1}{25}$, 1.04

b **i** 0.3, 30% **ii** 0.8, 80%
iii 0.857 (3 s.f.), 85.7%
iv 1.125, 112.5%
v 0.3, 33.3% (3 s.f.)

c **i** $\frac{7}{10}$, 70%   **ii** $\frac{63}{100}$, 63%

   iii $\frac{1}{20}$, 5%   **iv** $\frac{33}{25}$, 132%

   v $\frac{182}{125}$, 145.6%

3 No. Chloe is wrong as $\frac{35}{40}$ is 87.5% but $\frac{52}{60}$ is only 86.7%

4 **a** True **b** False **c** False **d** True

5 **a** $\frac{1}{50}$, 0.05, 15%, $\frac{1}{5}$, $\frac{15}{10}$, 1.55

   b 0.06, 0.106, 16%, $\frac{1}{6}$, 106%, 1.6

6 **a** =   **b** >   **c** <   **d** =

7 **a** 7   **b** 10.5 **c** 30.8 **d** 84

8 38.5%

What do you think?

1 36%
2 240
3 30%
4 75%
5 20%

Practice 7.1B

1 **a** 75%   **b** 0.75   **c** £45
2 **a** 103% **b** 1.03   **c** £252.35
3 **a** 0.8   **b** 0.65   **c** 0.95
   d 0.875 **e** 0.985
4 **a** 1.15 **b** 1.28   **c** 1.03
   d 1.175 **e** 1.024
5 **a** 2% decrease
   b 5% increase
   c 33% increase
   d 66% decrease
   e 23.5% decrease
   f 23.5% increase
6 £1008
7 502 029
8 **a** £19 360
   b Correct value: £17 036.80
   Mistake: in the second year you subtract 12% of the new value not 12% of £22 000
9 **a** 21 cm²
   b 18.48 cm²
10 B (Buy one get one half price)

What do you think?

1 **a** £10.08
   b Discount is 20% of 14 and then 10% of what's left (not 10% of 14)
2 **a** 43.75%
   b Discuss answers as a class
3 C (0.24 × 63) and E (63 ÷ 100 × 24)
4 Both the same
5 1.2 is the same as $\frac{6}{5}$

Practice 7.1C

1 **a** £15   **b** 20%
2 **a** £15   **b** 18.75%
3 24.6%
4 31.25%
5 20%
6 1st row: 24
   2nd row: 40.5
   3rd row: 1.25, 25% increase
   4th row: 0.8, 20% decrease
7 **a** 25%
   b 18.2%
   c His rent has increased by a higher percentage than his earnings
8 Ed (higher percentage change)

What do you think?

1 25%
2 **a** 180   **b** £295   **c** 145.8%
3 **a** Abdullah is not correct. Reducing 800 by 100% would leave £0
   b Yes, 2.5
   c No, because it would mean getting negative values

Consolidate

1 **a** 0.2, $\frac{1}{5}$ **b** 0.5, $\frac{1}{2}$
   c 0.075, $\frac{3}{40}$ **d** 0.125, $\frac{1}{8}$
   e 1.4, $\frac{7}{5}$
2 **a** 0.34, 34%
   b 0.6, 60%
   c 0.65, 65%
   d 0.125, 12.5%
   e 0.$\dot{3}$, 33.$\dot{3}$% (3 s.f.)
3 **a** $\frac{1}{5}$
   b 42%
   c 24.2%
4 **a** 0.22, $\frac{2}{5}$, $\frac{1}{2}$, 0.502, 0.52, 53%
   b $\frac{3}{10}$, 31%, 33%, $\frac{1}{3}$, 0.35, 3.3
5 English (88%)
6 450 g
7 £35.70
8 40%
9 29.2%
10 66.7%

Stretch

1 **a** 37.5%
   b 20%
   c Discuss answers as a class
2 After 11 bounces

Chapter 7.2

Are you ready?

1 **a** 0.3   **b** 0.56
   c 0.174 **d** 1.2
2 **a** 1.25   **b** 0.65
   c 0.58   **d** 1.045
   e 2.23   **f** 0.96
3 **a** $x = 40$
   b $x = 93.75$
   c $x = 145$ (3 s.f.)
   d $x = 22.3$ (3 s.f.)

Practice 7.2A

1 **a** 25 **b** 12.5 **c** 150 **d** 250
2 **a** 8   **b** 16   **c** 28 **d** 80
3 £800
4 **a** 120   **b** 84
5 £20
6 £1400
7 £940
8 600 g
9 £375
10 £62
11 **a** He has worked out 15% of the final amount but the increase would have been 15% of the original amount.
   b £1200
12 1st row: 54
   2nd row: decrease by 20%
   3rd row: 120
   4th row: 90
   5th row: 80

13 £110
14 £63.20
15 £360

What do you think?

1 £150
2 £650
3 Marta is incorrect. The increase is based on the initial value; the decrease is based on the new value.

Consolidate

1 **a** 12   **b** 48   **c** 66   **d** 120
2 827 mg (3 s.f.)
3 **a** £600 **b** £420 **c** £40 **d** £75
4 200 ml
5 £35.00
6 171.4 cm (1 d.p.)
7 £3125
8 £85
9 £45
10 **a** 32 mins   **b** 20%

Stretch

1 **a** 4%   **b** £23 000
2 Reducing a value by 20% leaves 80%, so finding 80% of an amount is the same as reducing it by 20%
3 **a** 24 cm²   **b** 4 cm
   c 25%; since the area has increased by 25% and the length has stayed the same, the width must have increased by 25%
4 This would mean there were 2307.69… people at the match last week but the number must be an integer
5 £350
6 **a** 100 cm   **b** 6 : 5

Chapter 7.3

Are you ready?

1 **a** 2.35   **b** 1400
   c 0.04   **d** 56.8
2 2nd row: £1 loss
   3rd row: £16
   4th row: £25
3 **a** 72   **b** 9   **c** 117 **d** 5.4
4 **a** 11.4   **b** 27.8
   c 5.7   **d** 385.6
5 **a** 10% increase
   b 34% increase
   c 2% decrease
   d 40% decrease
   e 4.5% increase
   f 2.5% decrease

Practice 7.3A

1 **a** 25%   **b** 144
2 **a** 29%   **b** 15%
   c 25%   **d** 60%
   e 30%
3 £960
4 £360
5 85%
6 120
7 420
8 £250
9 **a** 80%   **b** 25%   **c** 40%
10 £19
11 £2

What do you think?

1 Beca's idea
2 16%

Practice 7.3B

1 £20.24
2 4895
3 Ed is wrong. Flo scored a higher percentage.
4 a £135
b 55.1% (3 s.f.)
5 20%
6 £12.60
7 3 full cups
8 92.1 cm (3 s.f.)
9 B (55 × 1.12), C (0.12 × 55 + 55), D ($\frac{55}{100}$ × 112)
10 £920
11 32%
12 £11 708.93

What do you think?

1 £23
2 a £4800   **b** £600
c 1:7
3 a 60   **b** 80
c 17.1% (3 s.f.)

Consolidate

1 46%
2 a 18   **b** $\frac{19}{20}$
3 a 54   **b** 54
c They are the same
4 £42.24
5 35 min
6 35%
7 a £23.40   **b** £680
8 Seb
9 27.72 cm²

Stretch

1 £268.80
2 £465
3 27.7% (3 s.f.)
4 a Ed: £140, Seb: £350, Faith: £280
b 36.4% (3 s.f.)
5 2%

Chapter 7.4

Are you ready?

1 a 1.1   **b** 1.15   **c** 1.285
d 1.03   **e** 1.025
2 a 0.9   **b** 0.96   **c** 0.51
d 0.875   **e** 0.945
3 a 8   **b** 30   **c** 25.5
d 11.97   **e** 7.65
4 a 4.34   **b** 426.8   **c** 54.8
d 53 000   **e** 15.70

Practice 7.4A

1 a £300   **b** £150
c It doesn't go back to £200
2 £43.20
3 a 125 cm   **b** 15.625 cm
4 a 318
b 401 (nearest integer)
5 a £13 200   **b** £8995.43
c £4177.51
6 £1955.20
7 359

8 The first increase would have been 10% of the original rent and the second increase would have been 15% of the new (higher rent). Zach is wrong.
9 C (× 0.85⁴)
10 a 216 cm³   **b** 287.496 cm³
c 33.1%
d It would not change. 1.1³ = 1.331, which is a 33.1% increase and is independent of the side length.

What do you think?

1 Chloe is wrong. Each time it will be 10% of the new amount (which will be smaller) so it will never reach £0
2 After 10 weeks
3 Jakub is wrong. 10% of the original amount won't be the same as 10% of the new amount
4 a 44% decrease
b 49.5% increase

Consolidate

1 £364.50
2 158.65 cm
3 a 416   **b** 450
c 547
4 a £22 080   **b** £17 193.43
c £8823.99
5 £302 572.80
6 The first discount is 20% of the original amount and the second discount is 20% of the new (reduced) amount.
7 London = 9.45 million and New York City = 9.46 million so New York will have the bigger population

Stretch

1 270
2 26% increase
3 a £210   **b** 40% discount
4 72.8% increase
5 5 years

Check my understanding

1 a 0.4   **b** 40%
2 a $\frac{7}{20}$ **b** 0.35
3 A ($\frac{1}{3}$ of 60) and D (40% of 50)
4 £22.10
5 £127.50
6 12.5%
7 40%
8 £350
9 £217.35

Block 8 Maths and money

Chapter 8.1

Are you ready?

1 a £6.05   **b** £5.10
c £1.53   **d** £3.75
2 a 3.64   **b** 12.48
c 4.02   **d** 0.44

3 a −30   **b** 22
c −19   **d** 7
4 a £7.50   **b** £12.50

Practice 8.1A

1 a 6   **b** £8.25
c £17.28   **d** £2.72
2 Jackson
3 £7.25
4 £81.30
5 £42.56
6 Benji has converted incorrectly from pence to pounds. He has divided by 10 rather than 100
7 £29.27
8 Overdrawn means having a negative balance and owing the bank money
9 a £21.56
b Overdrawn by £18.44
10 a £192.01   **b** £20.00
c £192.01   **d** £147.01
e £450.00   **f** £586.82
g £559.16   **h** £559.16

What do you think?

1 £34.66
2 £4.20
3 142 units

Consolidate

1 a £20.80   **b** 42p
2 a £5.70
b No (£5.70 + £4.50 = £10.20)
3 £103.75
4 Kate should have worked out the difference between the meter readings. Correct bill is £51.53
5 −£23.43
6 a 53 miles   **b** £24.38

Stretch

1 £36.57
2 a The plan with no fixed cost (Red network). (150 × £0.02 + £10 = £13.00; 150 × £0.065 = £9.75)
b Approximately 223 minutes

Chapter 8.2

Are you ready?

1 a 21   **b** 27
c 135   **d** 21.6
2 a 1.1   **b** 1.03
c 0.8   **d** 1.045
3 a 605   **b** 627
c 671   **d** 566.5
4 a 221   **b** 265.2
c 336.6   **d** 282.2
5 a 1.331   **b** 1.216
c 13.339   **d** 1.277

Practice 8.2A

1 a £15
b i £315   **ii** £345
2 a £240   **b** £6960
3 £3400
4 a £180   **b** £150   **c** £225
5 a £585   **b** £614.25 **c** £51.19
6 £833.04
7 Both will be the same (£2000 × 0.04 = £80 and £4000 × 0.02 = £80)

What do you think?

1 $n = 5$
2 8%
3 $n = 320$, $x = 12.5\%$

Practice 8.2B

1 **a** £10 **b** £510
   c £10.20 **d** £520.20
2 **a** £2940.37 **b** £701.76
3 **a** Chloe **b** £15.25
4 £14 974.46
5 Option B
6 12 years
7 **a** Students' own values
   b No, because $1.06^{12} \approx 2$, so whatever the starting value, it will always be multiplied by 2 and hence double
8 C

What do you think?

1 **a** £1845 **b** £1694.14
   c It will take an extremely long time to pay off (approx. 11 years)
2 **a** $100 \times 1.075^{10} = £206.10$
   b No, because $1.075^{10} \approx 2$, so regardless of starting amount, the investment would double

Consolidate

1 £1500
2 Charlie: interest earned =
   $£3450 \times 1.04^7 - £3450 = £1089.96$
   Ali: interest earned =
   $£3450 \times 0.07 \times 4 = £966$;
   so Charlie will have earned more interest
3 **a** £3265.17 **b** 7 years
   c £3168, 9 years
4

Principal value	Interest rate	Interest type	After 2 years	After 5 years
£5000	3%	compound	£5304.50	£5796.37
£200	2%	simple	£208	£220
£1700	2.5%	compound	£1786.06	£1923.39
£23 000	8%	simple	£26 680	£32 200

5 No
6 Option A

Stretch

1 8 years
2 17 years
3 9%
4 $n = 5$
5 $n = £3200$ and $x = 12\%$

Chapter 8.3

Are you ready?

1 **a** 9 **b** 63.45
   c 58.2 **d** 2.56
2 **a** £35.69 **b** £1231
   c £14.26 **d** £27.83
3 **a** 30% increase
   b 20% decrease
   c 12% increase
   d 2.5% decrease
4 **a** £6.30 **b** £10.35
   c £3.60 **d** £22.05

Practice 8.3A

1 **a** £9.30 **b** £195.30
2 £4.20
3 £45.60
4 £25 200
5 £3.13
6 £234
7 £458.33
8 €2.20
9 £56.16

What do you think?

1 **a** The 5% will be of the £450 and the VAT will be of the new total, not of the £450
   b £567
2 £50 per chair
3 **a** £179.13
   b No, because multiplication is commutative so the order does not matter when using multipliers

Practice 8.3B

1 £351.50
2 **a** £402.50 **b** £661.25
3 38 hours
4 £10.80
5 **a** £1190 **b** £1438.33
6 **a** £49 550 **b** £12 320
7 **a** £22 880 **b** £2076
   c £1733.67 **d** 9.1%

What do you think?

1 20.4 hours per week
2 The job in France pays more: €32 594; the one in Italy pays €32 510

Consolidate

1 **a** £124 **b** £744
2 £176.40
3 £315
4 **a** £26 per hour **b** £676
5 £283.33
6 £64.29
7 **a** £3150 **b** £2091.67

Stretch

1 Jackson should accept the job in Egypt
2 The one is Switzerland costs approximately £95 less
3 Junaid would pay about £9 more

Chapter 8.4

Are you ready?

1 **a** £12.75 **b** £5.64
   c £17.55 **d** £679.70
2 **a** 7p **b** 23p
   c 22p **d** 20p
3 **a** 15 **b** 12 **c** 7 **d** 4

Practice 8.4A

1 55.56, 2778, £10, £20
2 **a** 345 euros **b** 21 750 yen
   c 19 380 rand **d** 76.04 dollars
   e £20.83 **f** £365.52
   g £19 607.84 **h** £3.55
3 50 750 yen
4 £117.39
5 Cheaper in the USA (£306.60)

6 Huda has £28.52, Flo has £27.93. Huda has more.
7 Cheapest in the USA
8 £23.41
9 **a** 630 yuan **b** £56

What do you think?

1 £4.40
2 Madrid is the better deal
3 £3379

Practice 8.4B

1 **a i** 36p **ii** 34p
   b B
2 **a i** £1.06 **ii** £1.05
   b The supermarket
3 300g for £7. Compare working as a class
4 B
5 C
6 School Stuff
7 Mario has worked out the grams per pence. So the better value would be the one that gives more grams per pence, which means that the 500 g is better value.
8 All methods work – discuss in pairs/groups

What do you think?

1 A
2 A
3 Price between £11.10 and £11.80

Consolidate

1 **a** $262.50 **b** £214.29
2 £13.61
3 Cheaper in Manchester, by 61p
4 A
5 B
6 C

Stretch

1 **a** $59.56
   b €1 = $1.19 or $1 = €0.84
2 B
3 B
4 **a** All cost £30
   b i No, this time all cost £24
      ii Yes, C would be best (even though offer B gives a free book)

Check my understanding

1 £2.53
2 **a** £150 **b** £3600
3 **a** £2008.17 **b** £401.20
4 £262.80
5 £379.17
6 £605
7 **a** $328.80 **b** £1021.90
8 B

Block 9 Deduction

Chapter 9.1

Are you ready?

1 **a** 316 **b** 56 **c** 111 **d** 141
2 **a** True **b** False **c** False **d** True
3 $a = 95°$ $b = 63°$ $c = 63°$ $d = 125°$
   $e = 55°$ $f = 143°$

Practice 9.1A

1 a q and u, r and v, p and t, s and w
 b s and u, r and t
 c r and u, s and t
2 There are other possible valid reasons for some parts. Compare answers with a partner
 a $a = 91°$ (alternate angles are equal)
 b $b = 47°$ (corresponding angles are equal)
 $c = 133°$ (angles on a straight line add up to 180°)
 c $d = 22°$ (co-interior angles add up to 180°)
 $e = 22°$ (alternate angles are equal)
 d $f = 115°$ (vertically opposite angles are equal)
 $g = 115°$ (corresponding angles are equal)
 e $h = 58°$ (angles on a straight line add up to 180°)
 $i = 58°$ (alternate angles are equal)
 f $j = 88°$ (corresponding angles are equal)
 $f = 88°$ (vertically opposite angles are equal)
 g $g = 60°$ (angles on a straight line add up to 180°)
 $m = 60°$ (co-interior angles add up to 180°)
 $n = 71°$ (alternate angles are equal)
3 a Right-angled **b** Scalene
 c Isosceles and right-angled
4 a $a = 50°$ **b** $b = 64°$ **c** $c = 125°$
 $d = 55°$ **d** $e = 254°$ $f = 12°$
 e $g = 130°$ $h = 34°$
5 Seb is correct about a
 Only b and 110° meet on a straight line and so b would be $180 - 110 = 70°$
6 a Yes, co-interior angles add up to 180°
 b No, the angles should add up to 180° but they do not
 c Yes, alternate angles are equal in parallel lines only
7 One pair of parallel sides
8 a $a = 62°$ $b = 118°$ $c = 62°$
 b $d = 106°$ $e = 158°$
9 Yes, Seb is right

What do you think?

1 $a = 101°$ $b = 51°$ $c = 101°$
2 AB, CD and GH are parallel
3 $x = 33$ $y = 30$

Consolidate

1 a e **b** g **c** h **d** c
2 There are other possible valid reasons for some parts. Compare answers with a partner
 a $a = 80°$ (co-interior angles add up to 180°)
 b $b = 52°$ (corresponding angles are equal)
 $c = 128°$ (co-interior angles add up to 180°)

c $d = 52°$ (angles on a straight line add up to 180°)
 $e = 52°$ (alternate angles are equal)
 d $f = 37°$ (corresponding angles are equal)
 $g = 37°$ (vertically opposite angles are equal)
 e $h = 99°$ (corresponding angles are equal)
3 Yes, they are parallel. The angle vertically opposite to 70° is also 70°. Then you have co-interior angles that sum to 180°, which is only true if AB and CD are parallel.
4 a 52° (angles on a straight line add up to 180°)
 b 52° (co-interior angles add up to 180°)
 c 128° (alternate angles are equal)
5 a $a = 51°$ $b = 94°$ $c = 35°$
 b $d = 82°$ $e = 67°$ $f = 31°$
6 a Isosceles
 b $a = 46°$ $b = 134°$ $c = 46°$

Stretch

1 a $a = 36°$ $b = 65°$ $c = 79°$
 b $d = 52°$ $e = 128°$
2 $a = 68°$ $b = 68°$
3 $a = 86°$ $b = 97°$
4 $x = 15$

Chapter 9.2

Are you ready?

1 a 315 **b** 55
 c 283 **d** 81
2 a $a = 48°$ **b** $b = 65°$
 c $c = 110°$ **d** $70°$
 d $e = 72°$
3 a Vertically opposite
 b Corresponding angles
 c Alternate angles
 d Co-interior angles

Practice 9.2A

1 a $\angle DEB = 50°$ (corresponding angles are equal), $\angle DEH = 130°$ (angles on a straight line add up to 180°)
 b $\angle RQS = 59°$ (corresponding angles are equal), $\angle RSQ = 88°$ (angles in a triangle add up to 180°), $\angle RSU = 92°$ (angles on a straight line add up to 180°)
2 There are a variety of ways of reaching the correct angle size. Compare answers with a partner or as a class.
 a $a = 59°$ **b** $b = 92°$ **c** $c = 53°$
 d $d = 60°$ **e** $e = 110°$ **f** $f = 83°$
 f $g = 89°$ $h = 107°$
3 a $a = 25°$ **b** $b = 45°$ **c** $c = 72°$
4 Ed is wrong when he writes that BEF is alternate to ABG. HEF is 81°
5 $a = 160°$ and $b = 20°$
6 $\angle ACD = \angle ADC = \angle DAC = 60°$ (angles in an equilateral triangle are all equal to 60°)
 $\angle CDB = 120°$ (angles on a straight line add up to 180°)
 $\angle BCD = 35°$ (angles in a triangle add up to 180°)

7 $\angle PRQ = 55°$ (angles on a straight line add up to 180°)
 $\angle RQP = 62.5°$ (angles in a triangle add up to 180° and two angles in an isosceles triangle are equal)
 $\angle CQR = 45°$ (angles in a triangle add up to 180°)
 $\angle BQP = 72.5°$ (angles on a straight line add up to 180°)
8 a They are on alternate sides of the transversal and outside the parallel lines
 b Compare answers with a partner or as a class

What do you think?

1

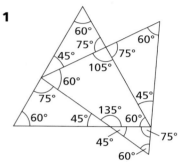

2 $y = 20°$
3 a $x = 130°$ **b** $x = 80°$

Consolidate

1 There are different methods and hence reasons. Compare answers with a partner
 a $a = 274°$ **b** $b = 74°$
 c $c = 70°$ **d** $d = 110°$ **e** $e = 70°$
2 There are different methods and hence reasons. Compare answers with a partner
 a $a = 138°$ **b** $b = 44°$ **c** $c = 55°$
 d $d = 103°$ **e** $e = 67°$
3 a $a = 124°$ **b** $b = 58°$ **c** $c = 36°$
4 a $\angle ADB = 75°$ (two angles in an isosceles triangle are equal)
 $\angle ABD = 30°$ (angles in a triangle add up to 180°)
 $\angle ADC = 105°$ (angles on a straight line add up to 180°)
 $\angle ACB = 37.5°$ (angles in a triangle add up to 180°)
 $\angle ABC = 30 + 37.5 = 67.5°$
 b $\angle EHF = 65°$ (two angles in an isosceles triangle are equal)
 $\angle FHG = 115°$ (angles on a straight line add up to 180°)
 $\angle HFG = 32.5°$ (angles in a triangle add up to 180°)
 $\angle EFG = 65 + 32.5 = 97.5°$
5 Flo is wrong. Angle b is not vertically opposite to 65°
6 a $x = 115°$, $y = 58°$
 b $x = 110°$ **c** $x = 55°$

Stretch

1 ∠BAC = 60° (vertically opposite
angles are equal)
∠ACD = 60° (corresponding
angles are equal)
∠BCF = 60° (angles on a straight
line add up to 180°)
∠ACB = 60° (angles on a straight
line add up to 180°)
∠ABC = 60° (alternate angles are
equal)
So all the angles of the triangle
ABC are equal to 60°. Therefore,
ABC is an equilateral triangle.

2 $x = 45°$

3 a 16. Any one angle is sufficient
 b Compare answers as a class

Chapter 9.3

Are you ready?

1 a 180° **b** 360°
 c equal **d** 360°
2 a 540° **b** 720°
 c 1080°
3 a $5x + 2$ **b** $6y - 13$
 c $2z + 31$
4 a $x = 77.5$ **b** $x = 114$
 c $x = 61$ **d** $x = 14$

Practice 9.3A

1 a Angles on a straight line add
 up to 180°
 b $3x + 66° = 180°$
 c $x = 38°$
2 a 540° **b** $y = 36°$
3 a $p = 28°$ **b** $q = 18°$
4 a $a = 54°$ **b** $b = 69°$
 c $c = 35°$ **d** $d = 13°$
 e $e = 23.5°$ **f** $f = 20°$
5 a $a = 33°$ **b** $b = 41°$
 c $c = 14°$ **d** $d = 30°$
 e $e = 37°$ **f** $f = 27.5°$
6 a $x = 30°, y = 42°$
 b $x = 27.5°, y = 60°$
7 $x = 35°$
8 a $x = 15°$ **b** $x = 10°$ **c** $x = 10°$
9 a $x = 6°$ **b** 44°

What do you think?

1 ABC and DEF are not parallel.
 Solving for x on the straight line:
 $5x + 6° + 16x + 6° = 180°$, gives
 $x = 8°$. ∠DEB = 20(8°) – 15° = 145°,
 ∠ABE = 5(8°) + 6° = 46°. 145° +
 46° ≠ 180°. If they were parallel,
 the co-interior angles would add
 up to 180°
2 18 sides
3 No. If it was, $25x - 5°$ would equal
 70°, giving $x = 3°$. Also, $12x + 68°$
 would equal 110°, but this gives
 $x = 3.5°$

Consolidate

1 a $f = 26.75°$ **b** $g = 15°$
 c $h = 8.8°$ **d** $i = 34°$
 e $j = 70°$ **f** $k = 43°$
2 a $a = 8°$ **b** $b = 15.5°$
 c $c = 12°$ **d** $d = 43°$
 e $e = 24°$ **f** $f = 10.5°$
3 a $u = 42°$ **b** $v = 30°$
4 $x = 11.4°$
5 a $x = 36°, y = 26\frac{2}{3}°$

b $x = 26°, y = 44°$
c $x = 7°, y = 48\frac{2}{3}°$

Stretch

1 a $x + y + z = 180°$; replace $x + y$
 with z; $z + z = 180°$, $2z = 180°$;
 hence $z = 90°$
 b $x + y + z = 180°$; replace $x + y$
 with $2z$; $2z + z = 180°$, $3z = 180°$;
 hence $z = 60°$
2 Solving $11a + 37 = 180°$ gives
 $a = 13°$. Substituting into the
 expressions for the angles gives
 angles of 44°, 68°, 68°. Since two
 angles are equal, it must be an
 isosceles triangle.
3 Chloe's conjecture is always true.
 Let the first angle be x, the next
 $x + a$ and the next $x + 2a$
 $x + x + a + x + 2a = 180°$ gives
 $3x + 3a = 180°$. Hence $x + a = 60°$
 (which is the middle angle).

Chapter 9.4

Are you ready?

1 a $x = 80°$ **b** $x = 115°$
 c $x = 54°, y = 126°$
 d $x = 74°, y = 128°$
2 a $a = 62°$ **b** $b = 89°$
 c $c = 64°$ **d** $d = 68°$
 $e = 80°$
3 a Corresponding angles are
 equal
 b Vertically opposite angles are
 equal
 c Alternate angles are equal
 d Angles on a straight line add
 up to 180°

Practice 9.4A

1 a False – a scalene triangle has
 no lines of symmetry
 b False – if it's also isosceles, it
 would have one line of
 symmetry
 c True
2 a Only true if the turns are in
 the same direction
 b False, for example 20° + 100° =
 120° which isn't reflex
 c True
3 a 60° (angles on a straight line
 add up to 180°)
 b 60° (angles in a triangle add
 up to 180°)
 c Yes, Marta is correct
4 ∠PQR = 75° (angles on a straight
 line add up to 180°)
 ∠RPQ = 75° (angles in a triangle
 add up to 180°)
 △PQR has angles of 75°, 75° and
 30°, so it must be isosceles
5 Any counterexample, such as an
 arrowhead
6 a Compare answers with a
 partner or discuss as a class
 b Chloe's conjecture is
 sometimes true, for example it
 is true for rectangles but not
 for trapezia
7 a Sometimes true (if the
 rectangle is a square)
 b False **c** True

d Sometimes true (if the
 parallelogram has sides of
 equal length)
8 ∠DBC = 25° (two angles in an
 isosceles triangle are equal)
 ∠BDC = 130° (angles in a triangle
 add up to 180°)
 ∠ADB = 50° (angles on a straight
 line add up to 180°)
 ∠DBA = 50° (two angles in an
 isosceles triangle are equal)
 ∠BAD = 80° (angles in a triangle
 add up to 180°)
9 Angles in a quadrilateral add up
 to 360°
 $14x + 10° = 360°, x = 25°$
 The angles are then: 70°, 110°,
 70° and 110° (in order)
 Each pair of opposite angles add
 up to 180°, so they are co-interior
 angles which only occur in
 parallel lines
10 ∠DCB = ∠DBC = 80° (angles in a
 triangle add up to 180°; two
 angles in an isosceles triangle are
 equal)
 ∠ABE = 80° (vertically opposite
 angles are equal)
 ∠BAE = 40° (angles in a triangle
 add up to 180°)
 40° is half of 80°, as required

What do you think?

1 Draw another parallel line going
 through B

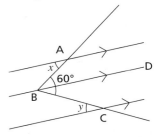

 ∠ABD = x (alternate angles are
 equal)
 ∠CBD = y (alternate angles are
 equal)
 Hence, $x + y = 60°$
2 ∠CAD = a (angles in isosceles
 triangle ACD)
 ∠ABC = a (angles in isosceles
 triangle BAD)
 In triangle BAD, angles are
 $a + (b + a) + a = 180°$
 $3a + b = 180°$
3 ∠BAC = y (angles in an isosceles
 triangle)
 ∠ACB = $180° - 2y$
 At C, angles on a straight line add
 up to 180°, so $180° - 2y + x = 180°$
 This gives $-2y + x = 0$; hence $x = 2y$
 Angle x is twice the size of angle y

Practice 9.4B

1 Rhombus. All sides are the same
 length
2 a Kite; AC = BC, OB = OA;
 adjacent pairs of sides are the
 same length

b OB = OA, ∠BOC = ∠AOC, OC is common to both. Hence the conditions for SAS are met. Triangles OBC and OAC are therefore congruent.

3 Amina is constructing a triangle with all sides the same length, that is, an equilateral triangle. Therefore, all angles will also be the same, 60°

4 a Check ideas with a partner
   b In the first, 17 > 4 + 6, so the two shorter sides can't meet. In the second, the angles do not add up to 180°. In the third, if two sides are equal, then two angles must also be equal. In the last, if all three angles are equal, then the sides will also be of equal length.

5 a Compare answers with a partner
   b The bisectors meet at the centre of the circle
   c Draw the perpendicular bisectors of AB and BC to find the centre of the circle. Then draw a circle going through the points A, B and C

What do you think?

1 a and **b** Check on diagram
   c D is the centre of the circle
   d The angles are 90°
   e Kite
   f They will always be true

2 The triangles have bases with equal length (DO = OB). The perpendicular from A to DB is the height of both triangles. Since both triangles have the same height and the same base, they will have the same area.

Consolidate

1 a False (for example, 20° + 20° = 40°, which is not obtuse)
   b True, since the interior angles of a regular hexagon are 120°
   c False, an equilateral triangle has angles of 60°

2 a Check ideas with a partner
   b Marta's conjecture is sometimes true, for example for a kite

3 a Check examples with a partner
   b Filipo's conjecture is not always true

4 a

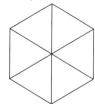

At the centre, 360° ÷ 6 = 60°
And at each vertex, 120° ÷ 2 = 60°
Therefore, they are all equilateral

b

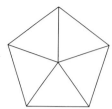

At centre, 360° ÷ 5 = 72° ≠ 60°
And at each vertex, 108° ÷ 2 = 54° ≠ 60°
Therefore, they are not equilateral

5 ∠ADB = 55° (angles at the base of an isosceles triangle are equal)
∠BDC = 125° (angles on a straight line add up to 180°)
∠BCD = $\frac{180° - 125°}{2}$ = 27.5°
(angles at the base of an isosceles triangle are equal)

6 Angles in a triangle add up to 180°
5x + 20° = 180°, x = 32° The angles are then 32°, 58° and 90°

Stretch

1 ∠BFG = q (alternate angles are equal)
∠DFE = 180° − (r + s) (angles in a triangle add up to 180°)
On a straight line at F:
q + p + 180 − (r + s) = 180°; hence q + p = r + s

2 Discuss as a class or with a partner

3 Jakub can join OACB to get a kite. AO and BO will be equal in length and AC and BC will be equal in length. He has constructed the line of symmetry of a kite. Faith can join ACBD to make a kite. BC and BD are equal in length and AC and AD are equal in length. She has used the fact that the diagonals of a kite meet at 90° to construct her kite.

Check my understanding

1 a a = 104° (angles in a triangle add up to 180°; angles on a straight line add up to 180°)
   b b = 55°, c = 70° (angles at the base of an isosceles triangle are equal; angles in a triangle add up to 180°)
   c d = 130°, e = 70° (co-interior angles add up to 180°)
   d f = 58°, g = 122° (corresponding angles are equal, co-interior angles add up to 180°)

2 (Many chains of reasoning are possible)
   a a = 88°   **b** b = 66°
   c c = 75°   **d** d = 78°

3 a False – in an equilateral triangle, the angles are less than 90°
   b Generally false – vertically opposite angles are equal, but if both were 90° then it would be true
   c False – for example, a scalene triangle has no lines of symmetry

Block 10 Rotations and translations

Chapter 10.1

Are you ready?

1
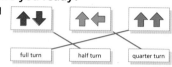

2 The base isn't the same length on the right of the reflection line as on the left

3 a i Regular pentagon
   ii Isosceles trapezium
   iii Isosceles triangle
   iv Equilateral triangle
   v Regular hexagon
   vi Hexagon
   b i, **iv** and **v**

Practice 10.1A

1 a 4   **b** 3   **c** 6
2 No, it has rotational symmetry of order 2
3 a 1   **b** 4   **c** 2
4 The order of rotational symmetry is equal to the number of sides if the shape is regular
5 Compare answers with a partner
6 a 1   **b** 3   **c** 3
7 a 5   **b** 5   **c** 8

What do you think?

1 Circle
2 Sometimes true

Practice 10.1B

1 a i 3   **ii** 5   **iii** 6
   b i 3   **ii** 5   **iii** 6
2 a i 0   **ii** 0   **iii** 0
   b i 1   **ii** 1   **iii** 1
3 In regular shapes, the order of rotational symmetry is equal to the number of lines of symmetry; this isn't the case in irregular shapes
4 Sometimes, for example the examples in question 2 don't have lines of symmetry but an isosceles triangle, which is an irregular shape, does
5 a No, it has rotational symmetry of order 4
   b 4

What do you think?

1 Compare answers with a partner
2 a Any correct shape, e.g. a non-isosceles trapezium
   b Any correct shape, e.g. the 'roundabout' road sign in Practice 10.1A

Consolidate

1 a i 4   **ii** 4
   b i 2   **ii** 2
   c i 0   **ii** 2
   d i 5   **ii** 5
2 a i 2   **ii** 2
   b i 1   **ii** 1
3 Compare answers with a partner

Stretch

1 a $\frac{7}{26}$ b 31%

An interesting discussion point is how font or handwriting can change the answer.

2 5

3 8

Chapter 10.2

Are you ready?

1 A is (2, 3), B is (0, 4), C is (−2, 1), D is (3, −2)

2 A and C

3 quarter turn = 90°
half turn = 180°
three-quarter turn = 270°
full turn = 360°

Practice 10.2A

1 a i

ii

iii

iv

b i

ii

iii

iv

c i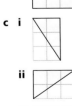

ii

iii

iv

d i

ii

iii

iv

iii and iv give the same result in each part; this is because a 90° anticlockwise turn is equivalent to a 270° clockwise turn

2 Benji has rotated the shape clockwise. Chloe's rotated shape isn't the same size as A. Seb has rotated the shape through 180°

3 a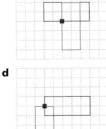

b

c

d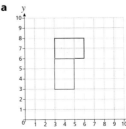

4 a

b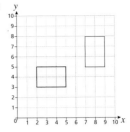

c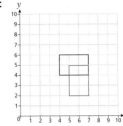

d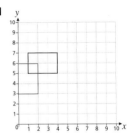

5 a

b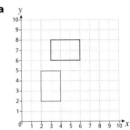

c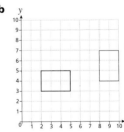

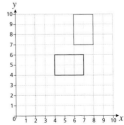

d

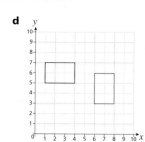

6 The orientation of the shapes is the same but their positions are different

7 a B is a rotation of 90° clockwise about (5, 2)

b B is a rotation of A 180° about (5, 5)

c Part **a** could be 270° anticlockwise and part **b** could be in either direction

8 a

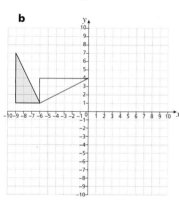

b

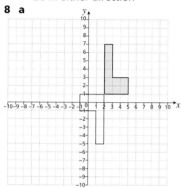

c

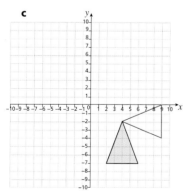

d

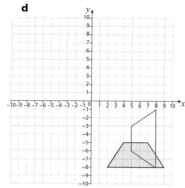

e

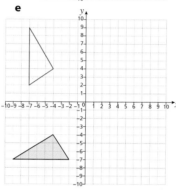

f

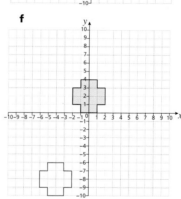

9 a A rotation through 180° about the origin

b A rotation through 90° anticlockwise about (−1.5, −2.5)

c A rotation through 180° about (−1, −1)

d A rotation through 90° clockwise about (−1.5, 2.5)

What do you think?

1 Either direction gives the same result

2 It would return to its starting position

3 Either 90° clockwise or 270° anticlockwise

4 Always true

Consolidate

1 a

b

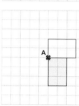

c

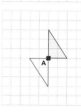

d

e

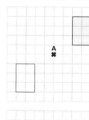

f

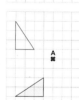

2 a

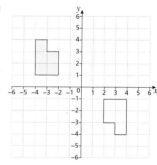

b

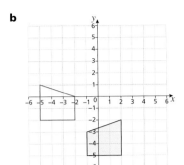

c

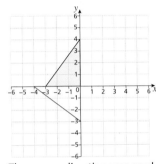

3 a The wrong direction was used
b The shape is in the wrong orientation
c The shape is the wrong size

Stretch

1 Rotation through 90° clockwise about (7, 3)
2 a The line segment corresponds to CD
b A becomes (−2, −4), B becomes (−3, −4), C becomes (−3, −1), D becomes (−2, −1)
3 Three: each of the vertices
4 Seb's conjecture is correct. If the centre of rotation isn't touching the original shape it can't be touching the rotated shape either.

Chapter 10.3

Are you ready?

1 a i From point P to point Q is 2 squares right and 3 squares up
ii From point Q to point R is 6 squares right and 2 squares up
iii From point R to point S is 9 squares left and 3 squares up
iv From point S to point P is 1 square right and 8 squares down
v From point Q to point P is 2 squares left and 3 squares down
b (5, 1)

Practice 10.3A

1

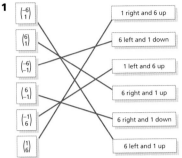

2 a Translated by vector $\binom{-2}{-3}$
b i She has counted 2 left and 3 down but then drawn the shape from the wrong vertex
ii Translated by vector $\binom{-4}{-3}$
3 a $\binom{2}{2}$ **b** $\binom{0}{-5}$ **c** $\binom{-4}{5}$
4 a

b

c

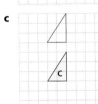

5 No, that is the translation from B to A. A to B is a translation by the vector $\binom{5}{3}$
6 a

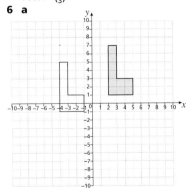

b

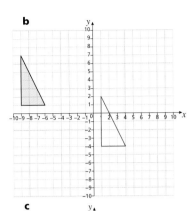

c

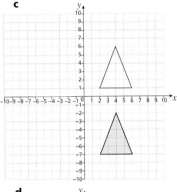

d

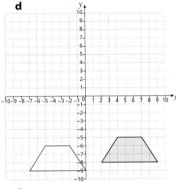

e

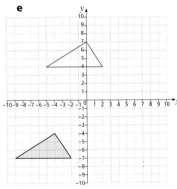

f

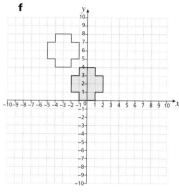

7 a Translation by the vector $\binom{5}{6}$

b Translation by the vector $\binom{-6}{0}$

c Translation by the vector $\binom{2}{0}$

d Translation by the vector $\binom{-6}{4}$

e Translation by the vector $\binom{1}{1}$

f Translation by the vector $\binom{0}{-1}$

8 a J is (2, 4), K is (7, 4), L is (7, 2), M is (2, 2)

b

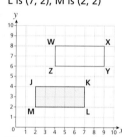

c W is (4, 8), X is (9, 8), Y is (9, 6), Z is (4, 6)

d The x-values are 2 more and the y-values are 4 more in WXYZ than in JKLM

e A is (6, 12), B is (6, 15), C is (9, 12), D is (9, 15)

What do you think?

1 a A, D and E

b Shapes that are translations are congruent and in the same orientation

2 a Right-angled

b D is (0, 0), E is (0, 19), F is (–5, 0)

Practice 10.3B

1 a, b, c

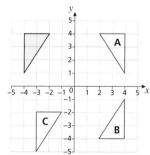

d It is the same triangle but it is in different orientations

2 Any correct answer with justification, for example rotation and reflection

3 Yes, they will end up in the same position

4

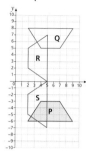

5 a

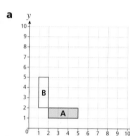

b Rotation 90° anticlockwise about (1.5, 1.5)

6 a Reflection in the x-axis or translation by $\binom{0}{-12}$

b For example, reflection in the y-axis then a 180° rotation about the origin

c Translation by the vector $\binom{10}{0}$, then a reflection in the x-axis and a reflection in the y-axis

What do you think?

1 Shape B is a reflection of A in the line $y = x$. Shape B is a rotation of A through 90° clockwise about (–2, –2)

2 a True   **b** True

Consolidate

1 a $\binom{-3}{4}$ **b** $\binom{5}{-7}$ **c** $\binom{0}{10}$

   d $\binom{-1}{-17}$ **e** $\binom{-15}{0}$

2 a Translation by the vector $\binom{5}{2}$

b Translation by the vector $\binom{-5}{-2}$

3 a Translation by the vector $\binom{5}{0}$

b Translation by the vector $\binom{-5}{-6}$

c Translation by the vector $\binom{-2}{-2}$

d Translation by the vector $\binom{7}{4}$

4 Rotation 180° about (–1, 1)

Stretch

1 a They will be the same

b They will be the same

c They will be the same

2 a They will be the same

b They will be the same

c They will be different

3 $a = –7$, $b = –4$

4 $y = x + 1$

Check my understanding

1 a i 4   **ii** 4

   b i 1   **ii** 1

2 a 9   **b** 9

3 a

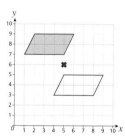

b

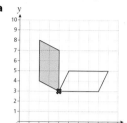

c

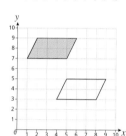

4 (–7, –3), (–2, –3), (–7, –5), (–2, –5)

Block 11 Pythagoras' theorem

Chapter 11.1

Are you ready?

1 a 36   **b** 64   **c** 2.25 **d** $\frac{4}{9}$

   e 36

2 a 5   **b** 10   **c** 13   **d** 1.5

   e $\frac{2}{3}$

3 False

4 a 25   **b** 10

Practice 11.1A

1 a Correct drawing

   b 5 cm   **c** BC   **d** No

2 a AC   **b** EF   **c** RQ

   d g **e** XZ   **f** AC

3 The hypotenuse in a right-angled triangle is always opposite the right angle

4 a ∠PQR   **b** ∠ACB   **c** ∠XYZ

What do you think?

1 Yes, the hypotenuse is always opposite the right angle

2 (7, 5) and (3, –2)

3 C

Practice 11.1B

1 $6^2 + 8^2 = 10^2$
2 $5^2 + 5^2 \neq 8^2$
3 Using Pythagoras' theorem:
   $7^2 + 24^2 = 25^2$ so the triangle is right-angled.
4 **a** No, $7^2 + 8^2 \neq 9^2$
   b Yes, $20^2 + 21^2 = 29^2$
   c Yes, $0.5^2 + 1.2^2 = 1.3^2$
   d Yes, $(\frac{3}{10})^2 + (\frac{2}{5})^2 = (\frac{1}{2})^2$
5 $12^2 + 35^2 = 37^2$ or $120^2 + 350^2 = 370^2$ or $0.12^2 + 0.35^2 = 0.37^2$

What do you think?

1 **a** Correct drawing
   b 9.9 cm
   c $7^2 + 7^2 \neq 100$
2 **a** $3^2 + 4^2 = 5^2$
   b Each is an enlargement of the original
   c Any correct triangles with dimensions $3p$, $4p$ and $5p$ where p is constant

Consolidate

1 **a** AB **b** EF **c** GH **d** JK
2 **a** $\angle$ABC **b** $\angle$DFE
   c $\angle$GIH **d** $\angle$JKL
3 **a** $8^2 + 15^2 = 17^2$
   b $0.6^2 + 0.11^2 = 0.61^2$
4 $23^2 + 28^2 \neq 31^2$

Stretch

1 **a** AB and BC are perpendicular
   b $5^2 + 12^2 = 13^2$
2 **a** $\frac{3^2\pi}{2} + \frac{4^2\pi}{2} = \frac{9\pi}{2} + \frac{16\pi}{2} = \frac{25\pi}{2}$
   $\frac{5^2\pi}{2} = \frac{25\pi}{2}$
   Therefore the area of semicircle A + the area of semicircle B = the area of semicircle C
   b $a^2 + b^2 = c^2$
   Multiplying both sides of the equation by
   $\frac{a^2\pi}{8} + \frac{b^2\pi}{8} = \frac{c^2\pi}{8}$
   These expressions are the areas of the semicircles
3 **a** They are sets of three numbers that are the sides of right-angled triangles
   b 6, 8 and 10 are double 3, 4 and 5, and 12, 16 and 20 are four times 3, 4 and 5
   c $(3x)^2 + (4x)^2 = 9x^2 + 16x^2 = 25x^2 = (5x)^2$

Chapter 11.2

Are you ready?

1 **a** 25 cm^2 **b** 49 cm^2 **c** 400 mm^2
2 **a** 3 cm **b** 10 mm **c** 8 m
3 **a** AC **b** XZ
4 $12^2 + 9^2 = 15^2$

Practice 11.2A

1 **a**
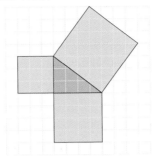

   b 9, 16 and 25 square units
   c $9 + 16 = 25$
2 Lydia has forgotten to find the square root of 169
3 **a** 10.8 cm **b** 13.0 mm
   c 18.2 m **d** 6.3 cm
4 42.9 cm (or 429 mm)
5 39.4 cm
6 11.2 cm
7 1 cm

What do you think?

1 **a** Yes **b** Yes **c** Yes
2 **a** 24.9 cm
   b It is an isosceles trapezium
   c Yes
3 Yes, 97 < 97.08

Practice 11.2B

1 **a** 13 is the hypotenuse but Bobbie has treated it like one of the shorter sides
   b 8.3 cm
2 **a** 9 cm **b** 6.2 cm
   c 173.1 mm **d** 4.3 m
3 33.5 inches
4 **a** 7.4 cm **b** 22.2 cm^2

What do you think?

1 No. In one of the triangles, 640 mm is the hypotenuse but in the other it is one of the shorter sides
2 **a** Compare diagrams with a partner
   b Yes, it will reach 4.14 metres up the wall
3 Discuss as a class

Consolidate

1 C
2 **a** 8.5 cm **b** 8 mm **c** 50 m
   d 0.8 m **e** 5.1 cm **f** 12.7 cm
   g 600.2 mm **h** 0.8 m **i** 10.5 cm

Stretch

1 **a** 14.1 cm **b** 16.6 cm
   c 53.2 cm **d** 156.1 cm^2
2 16.8 miles
3 1.0 m
4 **a** 10.6 cm **b** 112.5 cm^2
5 AC = 6 cm + 2 cm = 8 cm
   BC = 4 cm + 2 cm = 6 cm
   AB = 6 cm + 4 cm = 10 cm
   $8^2 + 6^2 = 10^2$

6
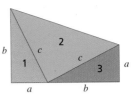

Area of triangle 1 = $\frac{1}{2}ab$
Area of triangle 2 = $\frac{1}{2}c^2$
Area of triangle 3 = $\frac{1}{2}ab$

Area of entire trapezium
   $= \frac{1}{2}(a + b)(a + b)$
   $= \frac{1}{2}(a^2 + 2ab + b^2)$
Area of triangle 1 + Area of triangle 2 + Area of triangle 3 = Area of entire trapezium
$\frac{1}{2}ab + \frac{1}{2}c^2 + \frac{1}{2}ab = \frac{1}{2}(a^2 + 2ab + b^2)$
$\frac{1}{2}c^2 + ab = \frac{1}{2}a^2 + ab + \frac{1}{2}b^2$
$\frac{1}{2}c^2 = \frac{1}{2}a^2 + \frac{1}{2}b^2$
$c^2 = a^2 + b^2$

Chapter 11.3

Are you ready?

1 **a** AB = 17 cm **b** EF = 7.4 mm
   c GH = 61.6 mm
2 **a** AC = 12.7 cm **b** DF = 7.02 mm
   c HI = 16.2 mm
3 **a** 4 units **b** 6 units
   c 6 units **d** 12 units

Practice 11.3A

1 **a** The vertical distance from 2 to 9 is 7 units
   b The horizontal distance from 2 to 7 is 5 units
   c $\sqrt{7^2 + 5^2} = 8.6$ units
2 **a** 3 units **b** 6 units **c** 6.7 units
3 **a** **i** 5.0 units **ii** 4.2 units
      iii 10.3 units **iv** 8.1 units
   b 14.9 units **c** 11.7 units
4 **a** 3.6 units **b** 10.8 units
   c 5.1 units **d** 5.7 units
5 **a** **i** 10.2 units **ii** 17.0 units
      iii 10.0 units **iv** 6.4 units
   b 10 units **c** 23.3 units
6 **a** 7.8 units **b** 10.3 units
   c 3.2 units **d** 9.4 units
7 7.8 units

What do you think?

1 **a** 5.4 units
   b Faith is correct as the vector describes the lengths of the two shorter sides of the right-angled triangle for which XX' is the hypotenuse
2 34.5 units

Practice 11.3B

1 **a** **i**

   D — 8 cm — C
   6 cm   6 cm
   H — 8 cm — G

   ii $\sqrt{6^2 + 8^2} = 10$ cm

b i

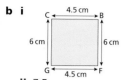

ii 7.5 cm

c i

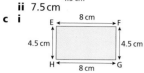

ii 9.2 cm

d i

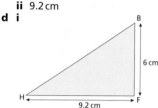

ii 11.0 cm

2 a i 9.5 m **ii** 4.2 m **iii** 9.5 m
b 9.9 m
3 11.4 cm
4 13.9 mm

What do you think?

1 a 8.5 cm
b i 3 cm **ii** 9.5 cm **iii** 9.1 cm
2 Yes, the longest length in the pencil case is 16.6 cm
3 216 mm

Consolidate

1 a 4.5 units **b** 5 units
c 9.9 units
2 a 8.2 units **b** 10.6 units
c 22.8 units
3 a 5 m **b** 16.5 m
c 16.3 m
4 a 4.2 cm **b** 5.2 cm

Stretch

1 $a = 12.94$ or $a = -4.94$
2 11.7 units
3 a $\sqrt{x^2 + y^2}$
b $\sqrt{(x_2 - x_1)^2 + (y_2 - y_1)^2}$
4 a 1539.6 m³ **b** 800 m²
5 $\sqrt{a^2 + b^2 + c^2}$

Check my understanding

1 a 81 **b** 3 **c** 25 **d** 6
2 a BC **b** DE **c** GI
3 a $12^2 + 16^2 = 20^2$
b $20^2 + 30^2 \neq 50^2$
4 a 8.1 cm **b** 10.9 cm **c** 11.4 cm
5 9 units
6 25.98 mm

Block 12 Enlargement and similarity

Chapter 12.1

Are you ready?

1 a 15 **b** 15 **c** 6
d 6 **e** 10 **f** 10
2 a 18 **b** 50

3 a 2:3 **b** 5:3 **c** 2:1
d 3:4 **e** 6:5 **f** 2:3
4 a $\frac{2}{3}$ **b** $\frac{3}{5}$ **c** $\frac{1}{2}$
d $\frac{3}{4}$ **e** $\frac{5}{6}$ **f** $\frac{2}{3}$

Practice 12.1A

1 a Each of the side lengths has been multiplied by 2
b Yes, it is the same dimensions as rectangle B
2 The base has been doubled but the height has not
3 a Any two rectangles such that ratio of length:width is 3:2
b Any two right-angled triangles such that ratio of height:base is 3:2
c Any two correct shapes such that the side lengths are in the correct ratios
4 a 2 **b** 3 **c** $\frac{1}{2}$
5 a 3, because $5 \times 3 = 15$
b $x = 12$
6 a 3 **b** 1:3 **c** 6
7 a 1.5 **b** $z = 6$
c 40°
8 a 5 cm and 3 cm
b 30 cm and 7.5 cm
c Yes, 30:7.5 = 4:1
9 a 9.6 cm **b** 20 cm

What do you think?

1 a 37 cm by 20 cm
b 370 cm by 200 cm
c 3700 cm by 2000 cm

2
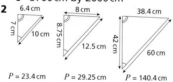

$P = 23.4\,cm$ $P = 29.25\,cm$ $P = 140.4\,cm$

3 a 1.5, $\frac{1}{2}$ or $\frac{2}{3}$
b Three sketches. One with side lengths 6 cm, 13.5 cm and 18 cm
One with side lengths 2.67 cm, 6 cm and 8 cm
One with side lengths 2 cm, 4.5 cm and 6 cm

Consolidate

1 For example

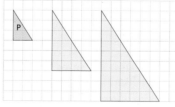

2 All of them
3 a 27 cm **b** 30 mm
c 60 mm **d** 9 cm

Stretch

1 15 mm
2 138.6 cm

3 a Angle BAC = angle CED and angle ABC = angle CDE because alternate angles are equal
Angle ACB = angle DCE because vertically opposite angles are equal
All three angles are equal; therefore the triangles are similar
b i CD = 12.8 cm
ii The wrong pair of sides or the wrong side length have been used
4 a $\frac{4x}{5}$
b $17.6 - \frac{4x}{5}$
c $x = 8.95$ cm so EF = 7.16 cm and FG = 10.44 cm

Chapter 12.2

Are you ready?

1 Correct drawing
2 4 by 6 rectangle drawn
3 A shape in which each side has been tripled has been enlarged by scale factor 3

Practice 12.2A

1 a He has only multiplied the height
b 3 by 6 rectangle drawn

2
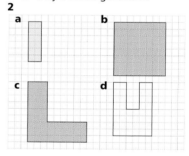

3 a Various answers, for example identify the perpendicular height of the triangle
b
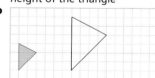

4 a
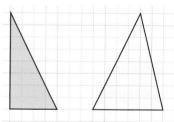

b The first one, because the base and height are perpendicular

5

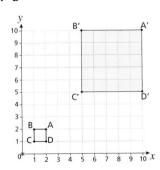

a

b

c

d

6 a 2 **b** 4

What do you think?
1 Yes, it will be the same size
2 Yes

Practice 12.2B
1 a Faith's enlargement is the correct size but it is in the wrong position
b Yes

2

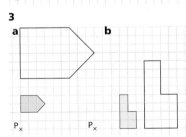

a

b

c

d

3

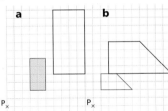

a

b

4 a

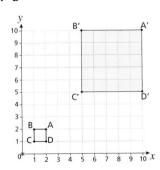

b

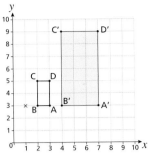

c

d

e

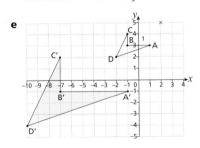

5 a An enlargement by scale factor 3 centre (0, 0)
b An enlargement by scale factor 2 centre (0, 0)
c An enlargement by scale factor 3 centre (2, 3)
d An enlargement by scale factor 3 centre (2, 3)

What do you think?
1

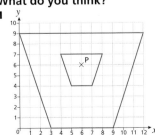

2 (1, 3), (4, 3), (4, 1), (1, 1)

Consolidate
1 a i ii

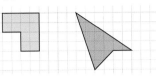

b i ii

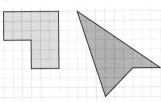

c i ii

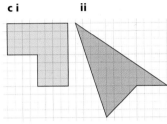

2 a 3 **b** 10
3 a i ii

iii iv

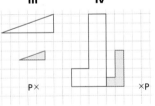

b i ii

iii iv

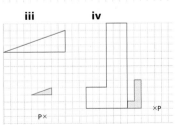

Answers

4 a

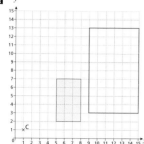

b

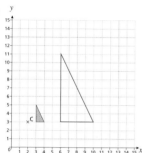

c

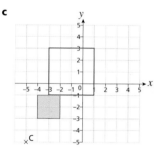

d

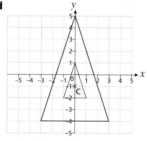

5 a An enlargement by scale factor 4 centre (−5, −4)
b An enlargement by scale factor 2 centre (−5, −2)
c An enlargement by scale factor 2 centre (4, −1)

Stretch

1 DEF is an enlargement of ABC by scale factor 3 centre (−6, 2)
2 (−16, 40)
3 a The original parallelogram has base 3 units and perpendicular height 3 units so the area is 9 square units
The enlarged parallelogram has base 12 units and perpendicular height 12 units so the area is 144 square units
144 is not four times as big as 9
b 6.25%

4 a i $(p+2, q-4)$ $(p+4, q+2)$
$(p+4, q-10)$ $(p+6, q-2)$
ii $(p+5, q-10)$ $(p+10, q+5)$
$(p+10, q-25)$ $(p+15, q-5)$
iii $(p+m, q-2m)$
$(p+2m, q+m)$
$(p+2m, q-5m)$
$(p+3m, q-m)$
b $p=-21, q=49$

Chapter 12.3

Are you ready?
1 a 5 **b** 5 **c** 6 **d** 18 **e** 18
2 a $\frac{1}{2}$ **b** $\frac{1}{3}$ **c** 4
3 a $\frac{1}{2}$ **b** $\frac{1}{5}$ **c** $\frac{1}{10}$

Practice 12.3A
1 a

b

c

2 a

b

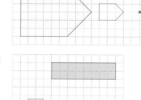

c

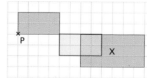

3 a The shape is the correct size
b Junaid has drawn it in the wrong place
c

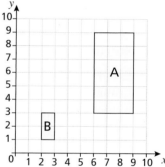

4 a

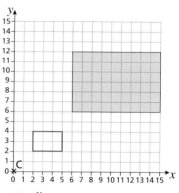

b

5 a B is an enlargement of A by scale factor $\frac{1}{2}$ centre (5, 5)
b B is an enlargement of A by scale factor $\frac{3}{4}$ centre (0, 3)

What do you think?
1 No, $\frac{3}{2}$ is equivalent to 1.5 so the shape will be larger
2 a–b

c i 3:1 **ii** 3:1 **iii** 9:1
d The length ratio is equal to the perimeter ratio because they are both length measurements. The area ratio is the length ratio squared

Practice 12.3B
1 a 1
b The shape has got bigger
c 3
d The shape has got smaller
e 2
f Orientation changes and the direction from the centre changes
2 Jackson; the scale factor is −1

716

3 a

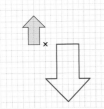

b

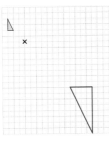

c

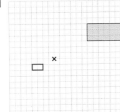

d

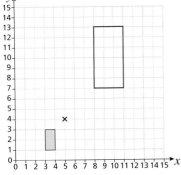

4 a

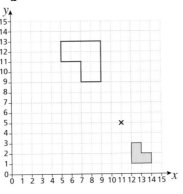

b

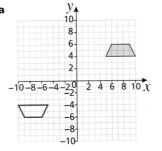

c

d

5 a

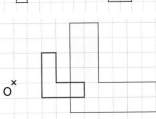

b

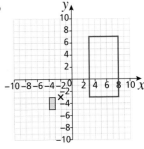

c

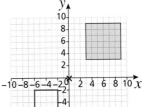

d

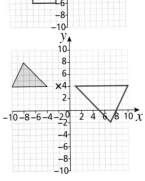

6 B is an enlargement of A by scale factor −2 centre (0, 3)

What do you think?

1 Enlargement by scale factor −1 centre (0, 0) or rotation 180° about (0, 0)

2 a i $\binom{1}{2}$ **ii** $\binom{2}{5}$ **iii** $\binom{5}{2}$

b i $\binom{-2}{-4}$ **ii** $\binom{-4}{-10}$ **iii** $\binom{-10}{-4}$

c The numbers in the vectors for **a** have been multiplied by −2 to give the vectors in **b**

3 $\binom{-20}{-35}$

Consolidate

1 a **b**

c **d**

2

Answers

3 a B is an enlargement of A by scale factor $\frac{1}{4}$

b B is an enlargement of A by scale factor $\frac{3}{4}$

4 a

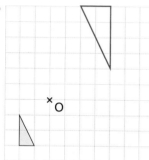

b

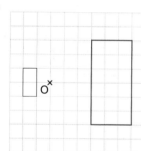

5 a

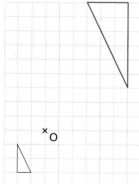

b

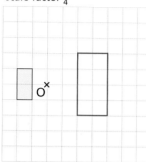

Stretch

1 C is an enlargement of B by scale factor 4 centre (0, 0)

2 a Sometimes; if the scale factor >1, the shape increases in size. If the scale <1, the shape decreases in size. If the scale factor = 1 the shape stays the same size

b Sometimes; a negative scale factor can change the orientation

c Sometimes; if the shape is symmetrical and the scale factor is 1

3 a $3 \times -4 = -12$ and $12^2 = 144$

b $(-12)^2 = 144$ not -144, and an area cannot be negative

4 a i $\binom{14}{-10}$ **ii** $\binom{12}{-16}$ **iii** $\binom{18}{-2}$

b i $\binom{-28}{20}$ **ii** $\binom{-24}{32}$ **iii** $\binom{-36}{4}$

c i $\binom{10.5}{-7.5}$ **ii** $\binom{9}{-12}$ **iii** $\binom{13.5}{-1.5}$

d i $\binom{-17.5}{12.5}$ **ii** $\binom{-15}{20}$ **iii** $\binom{-22.5}{2.5}$

5 a Chloe is correct

b i Any example where $a = b$
ii Any other example
Faith is only correct when the scale factor is 1

6 a i

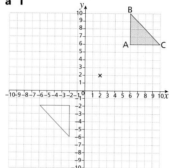

ii

Vertex	Before	After
A	(6, 6)	(−2, −2)
B	(6, 10)	(−2, −6)
C	(10, 6)	(−6, −2)

b i

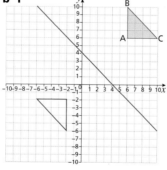

ii

Vertex	Before	After
A	(6, 6)	(−2, −2)
B	(6, 10)	(−6, −2)
C	(10, 6)	(−2, −6)

c The two transformations appear to have the same result but the positions of B and C have swapped

Chapter 12.4

Are you ready?

1 30 cm, 18 cm, 27 cm

2 50°, 85°, 45°

3 $x = 9$, $y = 4$, $z = 34$

Practice 12.4A

1 a C **b** Y **c** B

2 a DEF, JKL
b ABC, ZXY
c PRQ, NML
d ABC, DFE

3 a Angles in a triangle add up to 180° so the third angles must also be equal
b 3 **c** BC **d** 21
e Triangles ABC and MLN are similar

4 XYZ is an enlargement of PQR by scale factor 1.5

5 a $a = 12$ **b** $b = 6$
c $c = 100$, $d = 9$ **d** $e = 40$, $f = 48$

6 a 3.36 cm **b** 3.2 cm

7 a Angle ABC = angle CDE and angle BAC = angle CED because alternate angles are equal.
Angle BCA = angle ECD because vertically opposite angles are equal
All three angles are equal so the triangles are similar
b 7.5 cm **c** 9 cm

8 a 7 cm **b** 14 cm

What do you think?

1 Jackson

2 12

3 a 0.6 **b** 0.75 **c** 0.8
They are equal in each part.

4 Yes, the third angle must be 50° and all three angles are equal

Practice 12.4B

1 a The third angle must be 30° so, yes, all will be similar
b The ratio is 0.5 in each
c Students should notice that the ratio is always 0.5

2 a $x = 6$ **b** $x = 6$ **c** $x = 8.4$

3 B, D, E

4 a i 6.9 cm **ii** 6.1 cm **iii** 8.7 cm
b 0.87 and 1.73

5 a $\sin 30° = 0.5$ **b** $\cos 30° = \frac{\sqrt{3}}{2}$
c $\tan 30° = \frac{1}{\sqrt{3}}$

What do you think?

1 Any correct diagram illustrating this, for example

$\sin 30° = \frac{O}{H}$
$\cos 60° = \frac{A}{H}$
A = O so
$\sin 30° = \cos 60°$

2 Jakub is correct; they all have the same three interior angles of 90°, 45° and 45°

3 a 1 **b** $\sqrt{2}$
c i and ii $\sin 45° = \cos 45° = \frac{1}{\sqrt{2}}$

Consolidate

1 a Angles CAB and QRP; ABC and RPQ; BCA and PQR
Sides BC and PQ; AB and RP; AC and RQ

b Angles DEF and LKJ; DFE and LJK; EDF and KLJ
Sides DE and KL; EF and JK; DF and JL

c Angles MNP and UTV; NMP and TVU; MPN and TUV
Sides MN and TV; NP and TU; MP and UV

2 a 1.5 **b** XZ
c i 30 cm **ii** 24 cm

3 a 14.4 cm **b** 11.1 cm

4 No, the sides are not in the same ratio. The ratio of side lengths in PQR is $1:1.8:2.4$ but in XYZ it is $1:1.7:2.4$

5 a $a = 7.5$ cm **b** $b = 8.4$ cm
c $c = \frac{5\sqrt{3}}{3}$ **d** $d = 3$ cm

Stretch

1

2 88.125 m

3 a Angle PQR = angle RST and angle QPR = angle RTS because alternate angles are equal
Angle QRP = angle TRS because vertically opposite angles are equal
All three angles are equal; therefore the triangles are similar

b If PQ = ST, then the triangles are congruent by ASA

4 Perimeter = $3 + 15z + 4y$

5 a

$\sin 30° = \frac{1}{2}$ $\quad \sin 60° = \frac{\sqrt{3}}{2}$
$\cos 30° = \frac{\sqrt{3}}{2}$ $\quad \cos 60° = \frac{1}{2}$
$\tan 30° = \frac{1}{\sqrt{3}}$ $\quad \tan 60° = \sqrt{3}$

b

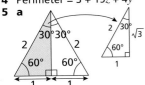

$\sin 45° = \cos 45° = \frac{x}{H}$

No, it is only true because the triangle is isosceles

Check my understanding

1 Length 64 cm and width 32 cm

2 a

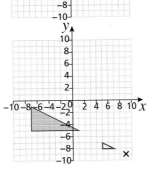

b

3 a All three angles are equal because angles in a triangle add up to 180°

b i 45 mm **ii** 20 mm

4 80 cm

Block 13 Solving ratio and proportion problems

Chapter 13.1

Are you ready?

1 a 12 **b** 6 **c** 35 **d** 2.1
2 a £1.70 **b** £4.25
c £10.20 **d** £14.45
3 A(3, 5), B(0, 4), C(6, 0), D(5, 5)
4 Check answers with a partner

Practice 13.1A

1 a £1.60 **b** £3.84 **c** 32p
2 £2.40
3 a £5.20 **b** 15 pencils
c 25p
4 a i 90 **ii** 37.5
b 43.2
5 a 67.5 g butter, 270 g puffed rice, 450 g marshmallows
b £4.20
6 a Approximately 22–23 cookies
b 35 cookies
7 a £60 **b** 9 hours
c Rob's hourly rate of pay is £12 so his wages are 12 × number of hours worked, $w = 12t$

What do you think?

1 6.4 litres
2 a 4 adults
b None; since there will be 19 children, they still only need 4 adults

c Darius needs to remember that both quantities need to increase or decrease at the same rate and here they do not. The number of adults only increases after five children join the group.

Practice 13.1B

(Note: allow a small tolerance when reading from graphs)

1 a Missing values: £4.50, £9, £18, £22.50, £27
b Compare answers with a partner
c i £6.75 **ii** 4.4 metres
2 a £10.50 **b** 1.4 kg **c** £1.25
3 a €34.50 **b** £17.50
c Convert £35 (to get €40) and then multiply by 10 to get €400
d Amina is right, because the graph is a straight line and goes through (0, 0)
4 Pounds to kilograms is direct proportion but the Celsius to Fahrenheit is not since it does not go through (0, 0) and one quantity is not a multiple of the other
5 Only A since it is a straight line going through the origin

What do you think?

1 £1 is approximately 145 yen
2 150 gallons is about $15 \times 45 = 675$ litres. The volume of the milk in the container is $\pi \times 50^2 \times 150 = 1\,178\,097$ cm³ which is approximately 1178 litres, so it is more than half full

Consolidate

1 a i £15 **ii** £18.75
b 40 litres
2 a £8.40 **b** £9.20
3 45 g butter, 3 garlic cloves, 675 g mushrooms, 3 small onions and 1.5 litres vegetable stock
4 a Missing values are: 0, £2.20, £6.60, £8.80, £11.00
b Compare graphs with a partner
c £7.70
d £17.60 (for example, adding up the price of 3 kg and 5 kg of oranges)
5 a €33 **b** $36 **c** €250

Stretch

1 2.4 litres
2 Yes, a is directly proportional to c
3 a Both would be straight lines. Emily's graph would go through the origin but Seb's graph would go through (0, £8). Emily's graph would be steeper as her cost per unit is higher
b No. Only Emily's example involves direct proportion
c It depends on the number of units used. At first, Emily's tariff is better, but above 200 units Seb's tariff will be cheaper.

Chapter 13.2

Are you ready?

1 a divide by 4   b multiply by 3
   c divide by 6   d multiply by 7
2 a 1 × 16, 2 × 8, 4 × 4
   b 1 × 24, 2 × 12, 3 × 8, 4 × 6
   c 1 × 36, 2 × 18, 3 × 12, 4 × 9,
      6 × 6
   d 1 × 50, 2 × 25, 5 × 10
3 A (4, 4), B (0, 3), C (4, 0), D (6, 1)
4 Compare answers with a partner

Practice 13.2A

1 a 150 days   b 25 days
2 a 1.5 hours   b 18 hours
   c 3 hose pipes
   d Any pair of values that have a
      product of 18
3 24 days
4 a 4 hours and 48 minutes
   b 80 miles per hour
5 a 4 hours   b 8 workers
6 1 hour and 4 minutes
7 Chloe is wrong because the rates
   of increase and decrease are not
   the same. For example, when you
   double the number of people, the
   time taken does not halve.
8 a Missing values (left to right):
      6, 9.6, 40
   b They multiply to give 96
   c $xy = 96$
9 a Inversely proportional
   b Neither
   c Directly proportional

What do you think?

1 a i 6 hours   ii 1 hour
   b 5 workers
2 21.5 days

Practice 13.2B

(Note: allow a small tolerance when
reading from graphs)
1 a As one variable increases,
      the other decreases at the
      same rate
   b Compare answers as a class
   c 9.6 cm
2 a Missing values: 8, 6, 4, 2, 1
   b As one variable increases, the
      other decreases at the same
      rate
   c Compare answers as a class
   d Approx. 2.2 hours (approx.
      2 hours 11 minutes)
3 a i 2 cm   ii 1.6 cm iii 2.7 cm
   b x is the base and y is the
      height and the formula for
      area of a parallelogram is
      base × height, so $x × y = 16$
4 a i 1.7 m³   ii 6.25 Pa
   b $xy = 25$
5 Benji is wrong. Inverse proportion
   graphs are curved and do not
   touch the x- or y-axis

What do you think?

1 a Missing values: 3, 1
   b Because the values do not
      increase/decrease at the same
      rate
   c It is a straight line rather than
      a curve

Consolidate

1 a i 24 days   ii 96 days
   b 6 people
2 6 hours and 18 minutes
3 Missing values (left to right):
   4, 8, 20
4 18 minutes
5 a Missing values (left to right):
      8, 4
   b Compare answers with a
      partner
   c i 5.3 cm   ii 3.2 cm

Stretch

1 240 boxes
2 8 workers take two days,
   so £1440
3 2 hours and 24 minutes

Chapter 13.3

Are you ready?

1 a $\frac{2}{3}$
   b i 1:2   ii 2:1
2 a $\frac{3}{8}$ b $\frac{5}{8}$
3 a 1:3 b 3:2 c 5:7 d 2:5
4 a 6   b 18   c 33

Practice 13.3A

1 20 litres
2 1:5
3 £10 and £35
4 a Marta 18 and Faith 12
   b 50   c 45   d 150
5 a 12   b 60   c £48
6 A and D are true
7 72
8 a i 1:$\frac{3}{7}$ ii 1:$\frac{2}{5}$
   b $\frac{3}{7} > \frac{2}{5}$ so each person in group
      A gets more chocolate
9 100 cm²
10 £32

What do you think?

1 If they are both prime numbers,
   they will only be divisible by 1
   and themselves so they will not
   have any factors in common
2 If the ratio is 3:5, then the total
   number of counters should be
   divisible by 3 + 5 = 8, and 25 is
   not divisible by 8
3 9:12:20

Practice 13.3B

1 B and D are correct
2 $x = \frac{1}{4}y$ or $y = 4x$
3 a $x = \frac{2}{3}y$ (or $3x = 2y$ or $y = \frac{3}{2}x$)
   b 2 + 6:12 − 2 = 8:10 = 4:5
      as required
4 a $x = 7.2$ b $x = 3.75$ c $x = 2.4$
5 $x = 6$
6 1:5:2
7 3:2

8 9 sides
9 a 2 cm, 6 cm and 10 cm
   b 184 cm²
10 75 counters
11 16 and 20 years old

What do you think?

1 28 in box A and 12 in box B
2 8:2:2:3

Consolidate

1 18 girls
2 1:8
3 £30, £18 and £12
4 20 and 12
5 a 80   b 64
6 a 36°, 90° and 54°
   b Right-angled triangle

Stretch

1 36 chocolates
2 £420
3 24 oranges
4 Any pair as long as $b = \frac{3}{4}a$
5 a 21 and 28 years old
   b In 14 years

Chapter 13.4

Are you ready?

1 a 57p b £3.12 c 50p d £2.74
2 a 25p b 31p   c 54p d 60p
3 a 3.22 b 0.42 c 1.35 d 0.15

Practice 13.4A

1 a i 42.5p and 43.75p
      ii 2.35 eggs and 2.29 eggs
      iii £10.20 and £10.50
   b 6 eggs for £2.55
2 a i £1.75 and £1.79
      ii 0.57 kg and 0.56 kg
      iii £24.50 and £25.00
   b 2 kg for £3.50
3 500 g for £3.70
4 300 g for £1.25
5 He has forgotten to convert
   £1.55 into pence. 1p buys
   2000 ÷ 155 = 12.9 ml
6 Buy 21 get 9 free (30 bars for
   £14.70) plus buy 3 get 1 free
   (4 bars for £2.10) plus 1 at full
   price (70p); total cost is £17.50
   OR five lots of the buy 5 get 2
   free (5 × 5 × 70p); total cost is
   also £17.50
7 a 500 ml for £5.75
   b 4 × 300 ml cartons
8 2-litre bottles at 2 for £2.50
9 Ed is working out the cost of 60
   cupcakes in each case. The
   cheapest answer is best value, C

What do you think?

1 a Cost (in pence) per teabag
   b Cost (in pounds) per 10 teabags
   c The number of teabags that
      can be bought for £1
2 France
3 a 20% off offer (£48)
   b Buy 2 for £30 (£90)
   c 5 DVDs (£80 in all scenarios)

Consolidate

1 B
2 a 10 doughnuts for £3.75
  b Two packets of 10 and 1 packet of 6, which gives 26 doughnuts for £9.80. (She could also buy three packets of 6 and two packets of 4, which gives 26 doughnuts for £10 exactly.)
3 a 20.4p, 21p and 20p
  b 4.90 pencils, 4.76 pencils, 5 pencils
  c 35 pencils for £7.00
4 400 g for £1.20
5 a 10 plates for £52
  b Yes, 6 plates is now better value
6 a 1 kg for 75 pence
  b One big bag and one small bag
  c So that you don't buy much more than you need

Stretch

1 Offer 1 (50% extra free)
2 150 g for £2.70 and 250 g for £3.50
3 Spain

Check my understanding

1 a i 85p **ii** £2.13
  b 4.2 kg (1.d.p)
2 a 16 km
  b 7.5 miles
  c Approximately 94 miles
3 a i 24 days **ii** 2 days
  b 6 workers
4 £25 and £10
5 £42 and £24

Block 14 Rates

Chapter 14.1

Are you ready?

1 a $x = 50$ **b** $m = 2400$
  c $t = 5$
2 a $d = \frac{a}{6}$ **b** $d = 4b$ **c** $d = \frac{10}{c}$
3 a 60 **b** 240 **c** 20
4 a 12.9 **b** 1.28 **c** 2.36 **d** 35.4

Practice 14.1A

1 a 60 m **b** 300 m
  c 1800 m **d** 15 m
2 a 360 miles **b** 240 miles
  c 60 miles **d** 30 miles
  e 180 miles
3 a A = 80 B = 200 C = 6
    D = 400 E = 12
  b 40 kilometres per hour
4 a 50 mph
  b No; because of traffic, hills etc.
5 a kilometres per hour
  b 5 hours
6 a 250 m
  b 250 metres per minute
7 a 8.47 m/s **b** 9.03 m/s
  c 8.29 m/s **d** 8.50 m/s
8 a 100 s **b** 125 s **c** $133\frac{1}{3}$ s
9 a 1 hour 30 mins = 1.5 hours, not 1.3 hours
  b 43.3 mph
  c i 52 mph **ii** 61.9 mph

What do you think?

1 a Discuss as a class; they are both correct
  b i 120 mph **ii** 150 mph
    iii 100 mph **iv** 96 mph
2 100 mph as 100 miles > 100 km
3 131 km/h
4 Discuss as a class

Practice 14.1B

1 a 50 mph **b** 1 hour
  c 46.7 mph **d** 5 hours
  e 40 mph **f** 48 mph
2 a One straight line only
  b 16.7 km/h **c** 15 mins
  d 12 km/h **e** 10 km/h
3 a

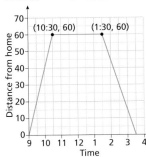

  b i 120 km **ii** 3.5 hours
    iii 34.3 km/h
4 a 78 km **b** 19.5 km/h
5 a 8.62 m/s **b** 12.1 m/s **c** 10.4 m/s
6 5.33 mph
7 60 mph
8 a 4 mph **b** 2.4 mph
  c

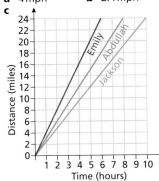

What do you think?

1 a A and B
  b Compare answers as a class
2 a The speed is increasing all the time
  b Compare answers as a class

Consolidate

1 a 960 miles **b** 1200 miles
  c 1560 miles **d** 160 miles
2 5 hours
3 a i 18 km **ii** 9 km
    iii 12 km **iv** 90 km
  b i 3 hours **ii** 45 mins
4 a 42 m **b** 2520 m
  c 4000 s = 1 hour 6 mins 40 s
5 a 30 mins **b** 1 hour **c** 1.5 km/h
  d Faster on her own (3 km/h; the graph line is steeper)
6 a 10 m/s **b** 8 m/s **c** 8.89 m/s

Stretch

1 a 5.87×10^{12} miles
  b about 1600
2 a C
  b A shows constant speed then stopped
    B shows accelerated from rest
    D shows slowing down then stopped
3 a For example, a might be much greater than b
  b $\frac{ax + by}{a + b}$
  c $a = b$
4 Discuss as a class

Chapter 14.2

Are you ready?

1 1000
2 400 cm³
3 a $p = 125$ **b** $h = 420$ **c** $k = 5$
4 A and C

Practice 14.2A

1 11.3 g/cm³
2 1.18 g/ml
3 Quartz sand = 7 g/cm³, silica sand = 2.6 g/cm³, 7 > 2 × 2.6, so yes Faith is correct
4 3510 g
5 a 0.777 cm³ **b** 6.92 g
6 1240 g
7 a 3200 cm³ **b** 2.88 kg
8 A and B

What do you think?

1 Discuss as a class
2 Mass is the amount of substance; weight is a force/the force exerted on an object by gravity. Density calculations involve mass, not weight
3 a Sometimes true
  b Sometimes true
  c Always true
  d Always true

Practice 14.2B

1 a i 600 **ii** 36 000
  b i 0.6 **ii** 36
  c 36 km/h
2 a 0.8 km per min
  b 48 km/h
  c 48 000 m per hour
  d 800 m per minute
  e 13.3 m/s
3 43.2 km/h
4 33.3 m/s
5 20 m/s = 72 km/h > 70 km/h
6 a 6.80 **b** 10.1
7 0.9 g/cm³
8 35.2 feet per second

What do you think?

1 a Beca, as 1 mile = $\frac{8}{5}$ km
  b 30 mph ≈ 48 km/h < 50 km/h, so 50 km/h is faster
  c Australian speed limit is 18 km/h or 11.25 mph faster
2 3.6
3 a 17.8 m/s **b** 22.5 mph

Consolidate

1 a 8.8 g/cm³   **b** 1320 g
2 a 379.5 g   **b** 575 g
3 A = 1.25 g/cm³   B = 1.875 g/ml
   C = 350 g   D = 245 g
   E = 64 cm³   F = 130 ml
4 7.5 g/cm³
5 2000 cm³ or 2 litres
6 a 800 cm³   **b** 0.85 g/cm³
7 54 km/h
8 a 0.0012 N/cm²   **b** 0.08 N/mm²
   c 400 N/m²

Stretch

1 a 1.152 g/cm³   **b** 1.150 g/cm³
2 7.5 g/cm³
3 8.5 g/cm³

Chapter 14.3

Are you ready?

1 a US$98   **b** £50
2 a 480 m   **b** 2 min 5 s
3 5

Practice 14.3A

1 a 3000
  b 3 hours 20 minutes
  c Discuss as a class
2 a 50 000 litres per day
  b On the fourth day after
3 a 40   **b** 57 600
  c No, more people will be
    asleep at some times
4 12.8 miles/litre
5 a 118.9p, 119.6p
  b 42p
6 a 2.94 × 10⁹   **b** 9.9
7 a 12.5 minutes
  b Both are correct
8 a 40 mins per room
  b 8

What do you think?

1 a For example, it must take less
   than 3 hours
  b $1\frac{7}{8}$ hours
2 3.69 and 3.81
3 a i $60x$ **ii** $60xy$
  b i $\frac{3000}{x}$ minutes
    ii $\frac{50}{x}$ hours

Practice 14.3B

1 a 12 litres/min   **b** 720 litres
  c 150 minutes
2 a

Time (s)	0	1	2	3	4	5
Volume of water (litres)	0	8	16	24	32	40

b
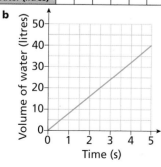

c 8; the rate of flow in litres per
   second
d 2400 litres
e The line will be steeper
3 a i 12.5 litres/min
   ii 375 litres **iii** 80 minutes
  b i 20 litres/min
   ii 600 litres **iii** 50 minutes
  c i 16 litres/min
   ii 480 litres **iii** 62.5 minutes
  d i 6 litres/min
   ii 180 litres
   iii 2 hours 46 minutes
    40 seconds
4 a 4.5 g/cm³
  b The density of titanium
5 a The cost of land per m²
  b The distance travelled per litre
   of petrol
  c The speed in m/s
6 a The steepest part of the graph
  b i 5 litres/min
   ii 10 litres/min
   iii 4 litres/min
  c 6 litres/min
7 The graph is not a straight line

What do you think?

1 a
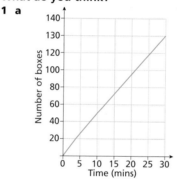

b He spent longer doing 4 boxes
   per minute; correct average is
   $4\frac{1}{3}$ boxes per minute
2 a A and C as both fill at a
   constant rate but B and D
   do not
  b
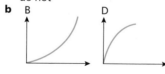

  c Discuss answers as a class

Consolidate

1 a 10   **b** 3 hours
2 a 3200 g **b** 25 m²
3 a 4.5   **b** 32 minutes
  c 6480
4 a 50   **b** 10
5 a

Length of call (mins)	0	1	2	3	4	5
Cost (pence)	0	6	12	18	24	30

b
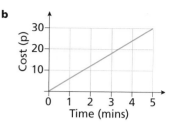

c 6; the cost in pence per
   minute of a call
d 50 minutes

Stretch

1 a €8.21   **b** US$$\frac{28x}{23}$
2 236 minutes
3 360 minutes
4 a 20 hours
  b

Check my understanding

1 a 1.5 hours   **b** 180 mph
2 Brass, 205 g > 195.8 g
3 2.71 g/cm³
4 a 14 boxes per minute
  b 8 hours 20 minutes
5
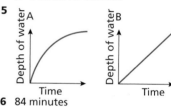

6 84 minutes

Block 15 Probability

Chapter 15.1

Are you ready?

1 1, 2, 3, 4, 5, 6
2 heads, tails
3 a $\frac{3}{7}$ **b** $\frac{4}{7}$ **c** $\frac{0}{7}$ or 0
4 a $\frac{15}{32}$ **b** $\frac{22}{32}$ or $\frac{11}{16}$
  c $\frac{32}{32}$ or 1 whole

Practice 15.1A

1 a The coin is fair; there are two
   possible outcomes and one of
   them is heads
  b $\frac{1}{2}$
  c They sum to 1 because heads
   and tails are the only options
2 a $\frac{1}{6}$ **b** $\frac{1}{6}$ **c** $\frac{3}{6}$ or $\frac{1}{2}$
  d 0   **e** 1
  f 7 is impossible and less than 7
   is certain
3 a $\frac{6}{12}$ or $\frac{1}{2}$ **b** $\frac{2}{12}$ or $\frac{1}{6}$
  c $\frac{4}{12}$ or $\frac{1}{3}$ **d** $\frac{8}{12}$ or $\frac{2}{3}$
4 a $\frac{3}{4}$ **b** 0   **c** $\frac{1}{4}$
5 a $\frac{5}{19}$ **b** $\frac{6}{19}$ **c** $\frac{8}{19}$
  d $\frac{11}{19}$ **e** $\frac{11}{19}$

6 $\frac{4}{30}$ or $\frac{2}{15}$

7 $\frac{14}{29}$

8 $\frac{2}{5}$

9 0.7

10 32%

What do you think?

1 No, the proportion of each spinner that is orange are not both $\frac{1}{2}$

2 Bag A because $\frac{3}{6} > \frac{4}{9}$

3 Various answers, for example sections labelled 1, 4, 16, 25, 49, 2, 6, 12, 21, 30

4 a $\frac{1}{5}$ **b** $\frac{4}{5}$

Practice 15.1B

1 a 250   **b** 250

2 a 100   **b** 100

   c 300

3 a 700   **b** 1000

4

Outcome	Probability	Expected frequency after 800 rolls
orange	$\frac{1}{5}$	160
blue	$\frac{9}{25}$	288
green	0.37	296
yellow	7%	56

5 a $5 \div 20 = 0.25$

   b i 0.35   **ii** 0.4

   c They sum to 1 because they cover all possible outcomes

6 a

Outcome	G	R	A	P	H
Frequency	8	12	7	14	9
Relative frequency	0.16	0.24	0.14	0.28	0.18

   b The relative frequencies aren't too far from the theoretical probabilities (0.2)

   c 36

What do you think?

1 540

2 No, there are not enough trials

3 a

Zach

Outcome	A	B	C	D
Frequency	18	13	19	10
Relative frequency	0.3	0.217	0.317	0.167

Beca

Outcome	A	B	C	D
Frequency	49	53	47	51
Relative frequency	0.245	0.265	0.235	0.255

   b Beca, as she completed more trials

   c No. When more trials are carried out the relative frequency becomes very close to the theoretical probability.

   d More accurate as there are more trials

Consolidate

1 a i $\frac{3}{7}$ **ii** $\frac{4}{7}$

   b There aren't any green counters

   c This covers all outcomes

2 a $\frac{8}{20}$ or $\frac{2}{5}$, 0.4, 40%

   b $\frac{2}{20}$ or $\frac{1}{10}$, 0.1, 10%

   c $\frac{3}{20}$, 0.15, 15%

   d $\frac{7}{20}$, 0.35, 35%

   e $\frac{10}{20}$ or $\frac{1}{2}$, 0.5, 50%

   f $\frac{13}{20}$, 0.65, 65%

   g $\frac{20}{20}$, 1, 100%

3 a i 100   **ii** 200

   b 0.17   **c** 105

Stretch

1 $1 - x$

2 $\frac{a}{p}$

3 a $w + x + y$

   b i $\dfrac{w}{w + x + y}$

     ii $\dfrac{x}{w + x + y}$

     iii $\dfrac{y}{w + x + y}$

     iv $\dfrac{w + x}{w + x + y}$

     v $1 - \dfrac{y}{w + x + y}$

   c $\dfrac{x - 1}{w + x + y - 1}$

4 a 0.2   **b** 0.1

   c 0.15   **d** 0.3

   e 0.125   **f** 0.125

5 $\frac{t}{k}$

6 $500x \div 10(140x + 3) = 7x - 1$
$50x \div (140x + 3) = 7x - 1$
$(7x - 1)(140x + 3) = 50x$
$980x^2 - 140x + 21x - 3 = 50x$
$980x^2 - 169x - 3 = 0$

Chapter 15.2

Are you ready?

1 $\frac{1}{2}$

2 $\frac{1}{6}$

3 a $\frac{1}{2}$ **b** $\frac{1}{8}$ **c** $\frac{3}{4}$

4 a $\frac{1}{12}$ **b** $\frac{1}{8}$ **c** $\frac{1}{4}$ **d** $\frac{5}{24}$

Practice 15.2A

1 a $\frac{1}{2}$ **b** $\frac{1}{6}$

   c

	1	2	3	4	5	6
H	(1, H)	(2, H)	(3, H)	(4, H)	(5, H)	(6, H)
T	(1, T)	(2, T)	(3, T)	(4, T)	(5, T)	(6, T)

   d $\frac{1}{12}$ **e** $\frac{1}{2} \times \frac{1}{6} = \frac{1}{12}$

   f i $\frac{1}{12}$ **ii** $\frac{1}{12}$ **iii** $\frac{3}{12}$ or $\frac{1}{4}$

     iv $\frac{3}{12}$ or $\frac{1}{4}$

2 a

	Plain	Striped	Spotty	Sleeveless	Long-sleeved
Jeans	Je, P	Je, St	Je, Sp	Je, Sl	Je, Ls
Shorts	Sh, P	Sh, St	Sh, Sp	Sh, Sl	Sh, Ls
Leggings	L, P	L, St	L, Sp	L, Sl	L, Ls
Joggers	Jo, P	Jo, St	Jo, Sp	Jo, Sl	Jo, Ls

   b $\frac{1}{20}$ **c** $\frac{1}{5} \times \frac{1}{4} = \frac{1}{20}$

3 $\frac{4}{20}$ or $\frac{1}{5}$

4 a $\frac{1}{10} \times \frac{1}{5} = \frac{1}{50}$ **b** $\frac{7}{10} \times \frac{2}{5} = \frac{14}{50} = \frac{7}{25}$

5 a They are independent events. More than one of the outcomes can happen at the same time.

   b i $\frac{3}{20}$ **ii** $\frac{3}{40}$ **iii** $\frac{1}{8}$

What do you think?

1 No. He has not considered the different orders, heads–tails and tails–heads. The correct answer is $\frac{1}{2}$

2 $\frac{1}{12}$

3 a The spinners aren't both split into two equal-sized sections of each colour

   b P(O) for spinner A = $\frac{1}{2}$. P(O) for spinner B = $\frac{5}{7}$. P(OO) = $\frac{1}{2} \times \frac{5}{7} = \frac{5}{14}$

Practice 15.2B

1 a i $\frac{45}{320}$ or $\frac{9}{64}$ **ii** $\frac{44}{320}$ or $\frac{11}{80}$

     iii $\frac{169}{320}$ **iv** $\frac{48}{320}$ or $\frac{3}{20}$

   b i $\frac{31}{151}$ **ii** $\frac{27}{151}$ **iii** $\frac{48}{151}$ **iv** $\frac{124}{151}$

   c i $\frac{27}{50}$ **ii** $\frac{23}{50}$

2 a i $\frac{57}{100}$ **ii** $\frac{30}{100}$ or $\frac{3}{10}$

     iii $\frac{26}{100}$ or $\frac{13}{50}$

   b $\frac{17}{43}$

3 a i $\frac{30}{120}$ or $\frac{1}{4}$ **ii** $\frac{42}{120}$ or $\frac{7}{20}$

     iii $\frac{71}{120}$

   b $\frac{30}{59}$

4 a

	Under 21	21–40	Over 40	Total
Glasses	105	92	78	275
No glasses	125	178	222	525
Total	230	270	300	800

   b i $\frac{230}{800}$ or $\frac{23}{80}$ **ii** $\frac{275}{800}$ or $\frac{3}{32}$

     iii $\frac{105}{800}$ or $\frac{21}{160}$

   c $\frac{78}{275}$ **d** $\frac{178}{270}$ or $\frac{89}{135}$

5 a

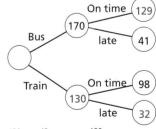

   b $\frac{129}{300}$ or $\frac{43}{100}$ **c** $\frac{129}{170}$

   d $\frac{41}{73}$

6 a There is no number outside the circles

   b i $\frac{16}{100}$ or $\frac{4}{25}$ **ii** $\frac{17}{100}$

     iii $\frac{45}{100}$ or $\frac{9}{20}$

   c $\frac{16}{54}$ or $\frac{8}{27}$

What do you think?

1 a

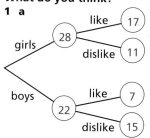

girls 28 — like 17 / dislike 11

boys 22 — like 7 / dislike 15

b $\frac{11}{50}$ **c** $\frac{7}{22}$

2 a

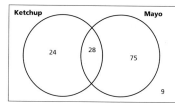

Ketchup 24   28   Mayo 75   9

b $\frac{99}{136}$

3 a $\frac{18}{80}$ or $\frac{9}{40}$

b Compare answers as a class

Consolidate

1 a The outcome of selecting the counter doesn't affect the spinner

b

	Red	Blue	Yellow	Green
1	(1, red)	(1, blue)	(1, yellow)	(1, green)
2	(2, red)	(2, blue)	(2, yellow)	(2, green)
3	(3, red)	(3, blue)	(3, yellow)	(3, green)

c $\frac{1}{12}$ **d** $\frac{1}{12} = \frac{1}{3} \times \frac{1}{4}$

2 $\frac{2}{35}$

3 a i $\frac{35}{100}$ or $\frac{7}{20}$ or 0.35

ii $\frac{57}{100}$ or 0.57

iii $\frac{14}{100}$ or $\frac{7}{50}$ or 0.14

iv $\frac{71}{100}$ or 0.71

v $\frac{28}{100}$ or $\frac{7}{25}$ or 0.28

b i $\frac{14}{57}$ **ii** $\frac{29}{57}$ **iii** $\frac{43}{57}$

iv $\frac{43}{57}$ **v** $\frac{57}{57}$ or 1

Stretch

1 $\frac{141}{1100}$

2 $\frac{34}{240}$ or $\frac{17}{120}$

3 $P(B) = \frac{88}{45}$ which is greater than 1. An individual probability must be between 0 and 1

4 $\frac{3a^3b}{2}$

5 a $\frac{50}{870}$ or $\frac{5}{87}$ **b** $\frac{90}{870}$ or $\frac{9}{87}$

Chapter 15.3

Are you ready?

1 $\frac{8}{13}$

2 12 green and 8 yellow counters

3 a 6

b i 3 cheese and 3 chicken

ii 4 cheese and 2 chicken

Practice 15.3A

1 a

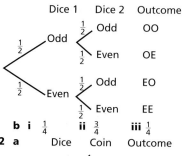

Dice 1	Dice 2	Outcome
$\frac{1}{2}$ Odd	$\frac{1}{2}$ Odd	OO
	$\frac{1}{2}$ Even	OE
$\frac{1}{2}$ Even	$\frac{1}{2}$ Odd	EO
	$\frac{1}{2}$ Even	EE

b i $\frac{1}{4}$ **ii** $\frac{3}{4}$ **iii** $\frac{1}{4}$

2 a

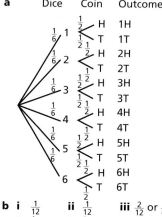

Dice   Coin   Outcome

$\frac{1}{6}$ 1 — $\frac{1}{2}$ H 1H / $\frac{1}{2}$ T 1T

$\frac{1}{6}$ 2 — $\frac{1}{2}$ H 2H / $\frac{1}{2}$ T 2T

$\frac{1}{6}$ 3 — $\frac{1}{2}$ H 3H / $\frac{1}{2}$ T 3T

$\frac{1}{6}$ 4 — $\frac{1}{2}$ H 4H / $\frac{1}{2}$ T 4T

$\frac{1}{6}$ 5 — $\frac{1}{2}$ H 5H / $\frac{1}{2}$ T 5T

$\frac{1}{6}$ 6 — $\frac{1}{2}$ H 6H / $\frac{1}{2}$ T 6T

b i $\frac{1}{12}$ **ii** $\frac{1}{12}$ **iii** $\frac{2}{12}$ or $\frac{1}{6}$

c i $\frac{1}{4}$ **ii** $\frac{3}{4}$

d They add up to 1

3 a

1st pick	2nd pick	Outcome
$\frac{3}{10}$ Red	$\frac{3}{10}$ Red	RR
	$\frac{7}{10}$ Blue	RB
$\frac{7}{10}$ Blue	$\frac{3}{10}$ Red	BR
	$\frac{7}{10}$ Blue	BB

b i $\frac{58}{100}$ or $\frac{29}{50}$ **ii** $\frac{42}{100}$ or $\frac{21}{50}$

4 a

Piano	Spelling	Outcome
$\frac{3}{5}$ Pass	$\frac{6}{7}$ Pass	PP
	$\frac{1}{7}$ Fail	PF
$\frac{2}{5}$ Fail	$\frac{6}{7}$ Pass	FP
	$\frac{1}{7}$ Fail	FF

b $\frac{33}{35}$

5 a

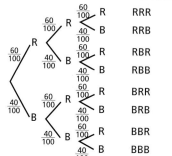

1st	2nd	3rd	Outcome
		$\frac{60}{100}$ R	RRR
	$\frac{60}{100}$ R	$\frac{40}{100}$ B	RRB
$\frac{60}{100}$ R	$\frac{40}{100}$ B	$\frac{60}{100}$ R	RBR
		$\frac{40}{100}$ B	RBB
	$\frac{60}{100}$ R	$\frac{60}{100}$ R	BRR
$\frac{40}{100}$ B		$\frac{40}{100}$ B	BRB
	$\frac{40}{100}$ B	$\frac{60}{100}$ R	BBR
		$\frac{40}{100}$ B	BBB

b i $\frac{280}{1000}$ or $\frac{7}{25}$ **ii** $\frac{288}{1000}$ or $\frac{36}{125}$

6 a

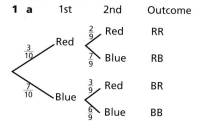

Tennis	Chess	Outcome
0.7 Win	0.9 Win	WW
	0.1 Lose	WL
0.3 Lose	0.9 Win	LW
	0.1 Lose	LL

b 0.97

What do you think?

1 a Yes, all the probabilities are correct

b She needs to find a common denominator before she can add the probabilities. If she had not have simplified the fractions, she wouldn't need this step.

2 No, the probabilities of the outcomes on each pair of branches sum to 1

Practice 15.3B

1 a

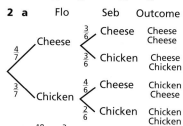

1st	2nd	Outcome
$\frac{3}{10}$ Red	$\frac{2}{9}$ Red	RR
	$\frac{7}{9}$ Blue	RB
$\frac{7}{10}$ Blue	$\frac{3}{9}$ Red	BR
	$\frac{6}{9}$ Blue	BB

b i $\frac{48}{90}$ or $\frac{8}{15}$ **ii** $\frac{42}{90}$ or $\frac{7}{15}$

2 a

Flo	Seb	Outcome
$\frac{4}{7}$ Cheese	$\frac{3}{6}$ Cheese	Cheese Cheese
	$\frac{3}{6}$ Chicken	Cheese Chicken
$\frac{3}{7}$ Chicken	$\frac{4}{6}$ Cheese	Chicken Cheese
	$\frac{2}{6}$ Chicken	Chicken Chicken

b $\frac{18}{42}$ or $\frac{3}{7}$

3 a

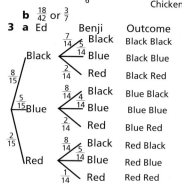

Ed	Benji	Outcome
$\frac{8}{15}$ Black	$\frac{7}{14}$ Black	Black Black
	$\frac{5}{14}$ Blue	Black Blue
	$\frac{2}{14}$ Red	Black Red
$\frac{5}{15}$ Blue	$\frac{8}{14}$ Black	Blue Black
	$\frac{4}{14}$ Blue	Blue Blue
	$\frac{2}{14}$ Red	Blue Red
$\frac{2}{15}$ Red	$\frac{8}{14}$ Black	Red Black
	$\frac{5}{14}$ Blue	Red Blue
	$\frac{1}{14}$ Red	Red Red

b i $\frac{2}{210}$ or $\frac{1}{105}$

ii $\frac{78}{210}$ or $\frac{13}{35}$

iii $\frac{120}{210}$ or $\frac{4}{7}$

4 a

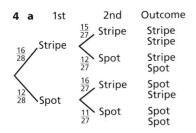

1st	2nd	Outcome

Stripe $\frac{16}{28}$, Spot $\frac{12}{28}$

$\frac{15}{27}$ Stripe → Stripe Stripe
$\frac{12}{27}$ Spot → Stripe Spot
$\frac{16}{27}$ Stripe → Spot Stripe
$\frac{11}{27}$ Spot → Spot Spot

b $\frac{372}{756}$ or $\frac{31}{63}$

5 $\frac{70}{132}$ or $\frac{35}{66}$

6 0.88

What do you think?

1 She has subtracted 2 from the numerator and the denominator between the 1st pick and the 2nd pick rather than subtracting 1. The probability tree diagram is used to model the probability for each pick at a time.

2 0.08

Consolidate

1 a

1st ice lolly	2nd ice lolly	Outcome	Probability
	$\frac{3}{8}$ Lemon	LL	$\frac{3}{8} \times \frac{3}{8} = \frac{9}{64}$
$\frac{3}{8}$ Lemon	$\frac{5}{8}$ Strawberry	LS	$\frac{3}{8} \times \frac{5}{8} = \frac{15}{64}$
$\frac{5}{8}$ Strawberry	$\frac{3}{8}$ Lemon	SL	$\frac{5}{8} \times \frac{3}{8} = \frac{15}{64}$
	$\frac{5}{8}$ Strawberry	SS	$\frac{5}{8} \times \frac{5}{8} = \frac{25}{64}$

b i $\frac{25}{64}$ **ii** $\frac{55}{64}$

2 a

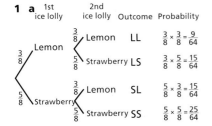

Hockey	Rugby		
	$\frac{2}{5}$ Win	WW	$\frac{1}{3} \times \frac{2}{5} = \frac{2}{15}$
$\frac{1}{3}$ Win	$\frac{3}{5}$ Lose	WL	$\frac{1}{3} \times \frac{3}{5} = \frac{3}{15}$
$\frac{2}{3}$ Lose	$\frac{2}{5}$ Win	LW	$\frac{2}{3} \times \frac{2}{5} = \frac{4}{15}$
	$\frac{3}{5}$ Lose	LL	$\frac{2}{3} \times \frac{3}{5} = \frac{6}{15}$

b i $\frac{7}{15}$
   ii $\frac{6}{15}$ or $\frac{2}{5}$

3 a

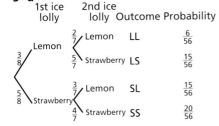

1st ice lolly	2nd ice lolly	Outcome	Probability
	$\frac{2}{7}$ Lemon	LL	$\frac{6}{56}$
$\frac{3}{8}$ Lemon	$\frac{5}{7}$ Strawberry	LS	$\frac{15}{56}$
$\frac{5}{8}$ Strawberry	$\frac{3}{7}$ Lemon	SL	$\frac{15}{56}$
	$\frac{4}{7}$ Strawberry	SS	$\frac{20}{56}$

b i $\frac{20}{56}$ or $\frac{5}{14}$ **ii** $\frac{50}{56}$ or $\frac{25}{28}$

Stretch

1 a $\frac{1}{25}$ **b** 40

   c £100 loss

2 $\frac{66}{196}$ or $\frac{33}{98}$

3 $x = 0.37$

---

4 a

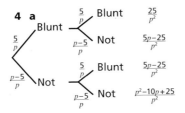

Blunt $\frac{5}{p}$
$\frac{5}{p}$ Blunt → $\frac{25}{p^2}$
$\frac{p-5}{p}$ Not → $\frac{5p-25}{p^2}$
$\frac{p-5}{p}$ Not, $\frac{5}{p}$ Blunt → $\frac{5p-25}{p^2}$
$\frac{p-5}{p}$ Not → $\frac{p^2-10p+25}{p^2}$

$324(p^2 - 10p + 25) = 169p^2$
$324p^2 - 3240p + 8100 = 169p^2$
$155p^2 - 3240p + 8100 = 0$
$31p^2 - 648p + 1620 = 0$

b

Blunt $\frac{4}{p-1}$
$\frac{5}{p}$ Blunt →
$\frac{p-5}{p-1}$ Not → $\frac{5p-25}{p(p-1)}$
$\frac{5}{p-1}$ Blunt → $\frac{5p-25}{p(p-1)}$
$\frac{p-5}{p}$ Not → $\frac{p-6}{p-1}$ Not

$\frac{10p - 50}{p(p-1)} = \frac{65}{153}$
$1530p - 7650 = 65p^2 - 65p$
$0 = 65p^2 - 1595p + 7650$
$0 = 13p^2 - 319p + 1530$

Check my understanding

1 a $\frac{5}{10}$ or $\frac{1}{2}$ **b** $\frac{6}{10}$ or $\frac{3}{5}$

   c $\frac{3}{10}$ **d** 0

2 a $\frac{240}{300}$ or $\frac{4}{5}$ or 0.8

   b 140

3 0.075 or $\frac{3}{40}$

4 $\frac{30}{64}$ or $\frac{15}{32}$

5 $\frac{26}{56}$ or $\frac{13}{28}$

Block 16 Algebraic representations

Chapter 16.1

Are you ready?

1 a

x	-2	-1	0	1	2
$y = 2x + 3$	-1	1	3	5	7

b

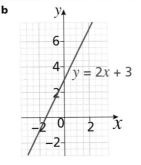

$y = 2x + 3$

2

x	-2	-1	0	1	2
$y = 3 - 2x$	7	5	3	1	-1

3 a i 9   **ii** 9
   b i 12   **ii** 6
   c i 6   **ii** 12
   d i 14   **ii** 2

Practice 16.1A

1 a -4 **b** -1, 4

   c -3.25 **d** -1.2, 4.2

2 b

x	-3	-2	-1	0	1	2	3
$y = x^2$	9	4	1	0	1	4	9

3 a

x	-3	-2	-1	0	1	2	3
$y = x^2 + 1$	10	5	2	1	2	5	10

Graphs for **2c** and **3b**

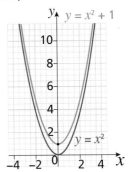

$y = x^2 + 1$
$y = x^2$

3 c, d and **e** Discuss as a class

4 a

x	-3	-2	-1	0	1	2	3
$y = x^2 + 2x + 1$	4	1	0	1	4	9	16

x	-3	-2	-1	0	1	2	3
$y = x^2 - 2x - 1$	14	7	2	-1	-2	-1	2

b

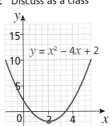

$y = x^2 + 2x + 1$
$y = x^2 - 2x - 1$

c Discuss as a class

5

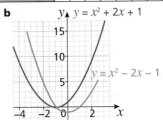

$y = x^2 - 4x + 2$

6 a $x = 2.5$, $y = -3.25$

   b

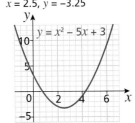

$y = x^2 - 5x + 3$

What do you think?

1 The graph of $y = -x^2$ is a reflection of the graph of $y = x^2$ in the x-axis

2 A, C and F are straight lines; B, D, E, G and H are parabolas

3 Benji is wrong – see Example 2

Practice 16.1B

1 a Compare answers as a class
  b i 0.5   **ii** –0.5   **iii** –1
  c Discuss as a class
2 Discuss as a class
3 b Discuss as a class
  c i 1.4   **ii** 5.7
  d
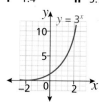

4 a i £1.50   **ii** £2.50   **iii** £3
  b 3 hours
5 a i £1.25   **ii** £1.25   **iii** £1.75
  b Compare answers as a class
6 Discuss as a class

What do you think?

1
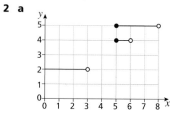

2 a
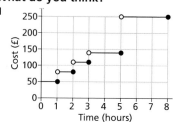

  b
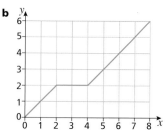

3 Compare answers as a class

Consolidate

1 b

x	–2	–1	0	1	2
$y = 2x^2 + 1$	9	3	1	3	9

c
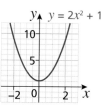

2 b

x	–5	–4	–3	–2	–1	0	1
$x^2 + 4x - 2$	3	–2	–5	–6	–5	–2	3

c
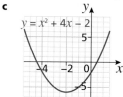

3 a

x	–4	–2	–1	–0.5	0.5	1	2	4
$y = \frac{4}{x}$	–1	–2	–4	–8	8	4	2	1

b
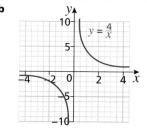

4 a i £9   **ii** £9   **iii** £11
  b 8 mins = £21 is cheaper than 2 × 4 mins = £26 by £5

Stretch

1 a $y = x^2 + ax + b$ has line of symmetry $x = -\frac{1}{2}a$
  b i $x = -3$ **ii** $x = 3$
    iii $x = -1.5$ **iv** $x = -5$
  c $y = x$ and $y = -x$
2 and 3 Compare answers as a class

Chapter 16.2

Are you ready?

1
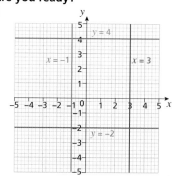

2 a

x	–2	–1	0	1	2
$y = 3x - 1$	–7	–4	–1	2	5

b
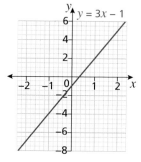

3 a $x = a$ for any a
  b $y = b$ for any b
  c $y = 2x + c$ for any c

Practice 16.2A

1 a A $x = 2$, B $y = -1$
  b $(2, -1)$
2 a
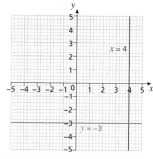

  b $(4, -3)$
3 a $(1, 4)$ **b** $(-2, 3)$
  c $(-1, 4)$ **d** (a, b)
4 a $(3, 7)$ **b** $(1, 3)$
  c The values satisfy the equation
5 $x = -2, y = -5$
6 a $x = 1, y = 5$ **b** $x = 4, y = -2$

What do you think?

1 The lines are parallel and so do not intersect
2 a The values satisfy the equation $y = 5 - x$ but substituting $x = 2.7$ into $y = 2x - 3$ gives a y-value of 2.4 not 2.3
  c Not reliable for non-integer solutions

Practice 16.2B

1 a $(0, 3)$ and $(6, 0)$
  b $(0, 8)$ and $(4, 0)$
  c $(0, 2)$ and $(-3, 0)$
  d $(0, -4)$ and $(2, 0)$
2 a $x = 1, y = 4$ **b** $x = 3, y = 2$
  c $x = -1, y = -3$
3 a The numbers are x and y
  b 4 and 2
  c They have the same solution, 2 and 4
4 5.5 and 2.5
5 a Rearrange to $y = 2x$ and use, for example, $(1, 2)$ or $(2, 4)$
  b $x = -2, y = -4$
6 a $x = 2, y = 1$
  b $x = 3, y = 2$
  c $x = -2, y = 2$

What do you think?

1 and **2** Discuss as a class

Consolidate

1 $x = -3$, $y = 1$

2 a

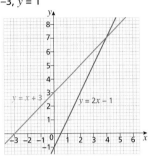

b $x = 4$, $y = 7$

3 a (0, 4) and (8, 0)
b (0, 2) and (4, 0)
c (0, 3) and (−2, 0)
d (0, −9) and (3, 0)

4 a $x = 4$, $y = -2$ **b** $x = 3$, $y = 2$
c $x = -2$, $y = 3$

Stretch

1 Discuss as a class (solution is
$x = -1$, $y = -2$)

2 $x = -2$, $y = -3$

3 a $b = 7$, $a = 5$ **b** $p = 3$, $q = 2.5$
c $c = 5$, $d = 4$

Chapter 16.3

Are you ready?

1 a $t > 7$ **b** $p \leqslant 5$ **c** $x > 48$

2 $x + 3 > 6$, $x + 3 \geqslant 7$, $2x < 9$

3 a

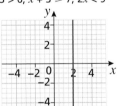

b

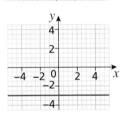

c

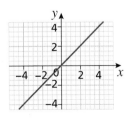

Practice 16.3A

1 a $x \geqslant -4$ **b** $y > -2$
c $y < 1$ **d** $x < 3$

2 a

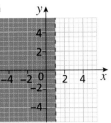

b

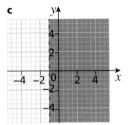

c

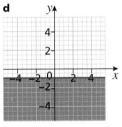

d

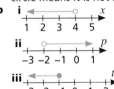

3 a A filled circle means that the
number is included; an open
circle means it is not included

b i

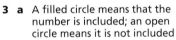

ii

iii

iv

4 a A and D
b

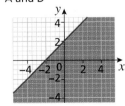

5 a

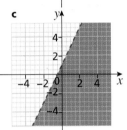

b

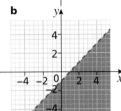

c

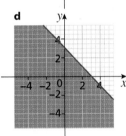

d

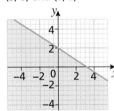

6 a (3, 0) and (0, 2)
b

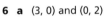

7 a

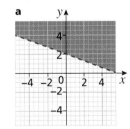

b

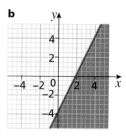

What do you think?

1 **a**

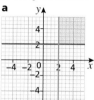

b

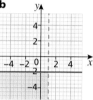

c

2 **a** x can take any value between −1 and 2 (but it cannot be −1 or 2)

b

i

ii

iii

3 $y < x$, $-2 \leqslant x \leqslant 4$, $-2 \leqslant y < 4$

Consolidate

1 **a** $y > 3$

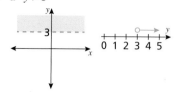

b x is less than or equal to 1

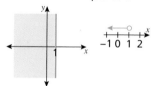

c $y > -1$, y is greater than −1

d $y < -2$, y is less than −2

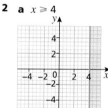

2 **a** $x \geqslant 4$

b $y < -1$

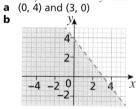

3 **a** $x < 2$ **b** $2 < y \leqslant 4$
    c $x + y \geqslant 4$ **d** $y < 3x + 1$
4 **a** $(0, 4)$ and $(3, 0)$
    b

Stretch

1 **a**

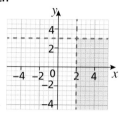

b Compare answers as a class
2 **a** x = number of revision guides and y = number of workbooks. The region satisfies $6x + 8y \leqslant 72$, $x > 2$ and $y > 3$
    b 3 revision guides and 4 or 5 or 6 workbooks, 4 revision guides and 4, 5 or 6 workbooks, 5 revision guides and 4 or 5 workbooks, 6 revision guides and 4 workbooks

Check my understanding

1 **a** −1.25   **b** −0.2, 3.2
2 **a** £4.00   **b** 4 miles
3

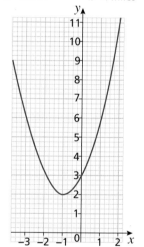

4 **a** **i**

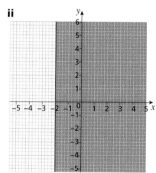

    ii

b i

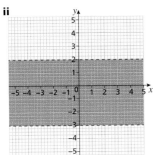

ii

5 $x = -1$, $y = -2$

Block 17 Getting ready for Key Stage 4

Chapter 17.1

Practice 17.1A

1 a Mean = 9.2, median = 10, mode = 12, range = 7
  b Mean = 25.25, median = 22, mode = 12, range = 42

2 a 120
  b 120 is a factor of 360
  c 0
  d The 60th item is 1 and the 61st is 2. Median = 60.5th value = $\frac{1+2}{2}$ = 1.5
  e 1.625

3 a i 1 **ii** 1 **iii** 1.39 **iv** 4
  b i 8 **ii** 8 **iii** 8.21 **iv** 4

4 a $(200 < w \leqslant 250)$ g
  b $(200 < w \leqslant 250)$ g

5 a

Time (hours)	Frequency	Midpoint	Subtotals
$0 < t \leqslant 1$	63	0.5	31.5
$1 < t \leqslant 3$	15	2	30
$3 < t \leqslant 7$	27	5	135
$7 < t \leqslant 10$	24	8.5	204
$10 < t \leqslant 15$	21	12.5	262.5
Total	150		663

  Mean ≈ $\frac{663}{150}$ = 4.42 hours
  b $(1 < t \leqslant 3)$ hours

6 a i $(10 < l \leqslant 30)$ cm
  ii $(10 < l \leqslant 30)$ cm
  iii 32.4 cm
  b i $(3 < t \leqslant 5)$ hours
  ii $(3 < t \leqslant 5)$ hours
  iii 3.60 hours

7 161.4 cm; discuss as a class
8 22

Challenges

1 Work out the fraction of the way through the class to the position of the median
2 Compare answers as a class
3 This is bivariate data and you will want to include a scatter graph to look for correlation

Chapter 17.2

Practice 17.2A

1 8, 13, 18, 23; 4, 9, 14, 19; 16, 21, 26, 31
2 a A and C
  b A is $5n + 2$, B is $5n - 4$
3 a $3n + 2$ **b** $4n - 1$ **c** $6n$
  d $7n - 2$ **e** $9n + 1$ **f** $10n - 3$
  g $n + 11$
4 $3n + 7$, $3n + 10$
5 a A is ascending and B is descending; the step is 6 for both sequences
  c i $60 - 10n$
  ii $77 - 2n$
  iii $61 - 3n$
6 218th
7 a 1001 **b** 29

Challenges

1 a $4n + 1$
  b 4 sticks are added each time ($4n$), then 1 more is needed to make the first pentagon (+ 1)
  c $8n + 1$
  d $(p - 1)n + 1$
  e Compare answers as a class
2 a The first sequence increases indefinitely, the second converges to 2
  b Compare answers as a class
3 Flo is correct
4 Compare answers as a class
5 a The first differences are 6, 8, 10, 12 for all three sequences; the second differences are 2 for all the sequences
  b i $n^2 + 6n - 3$
  ii $n^2 + 2n - 7$
  iii $n^2 - 10n$
6 a 1, 7, 21, 35, 35, 21, 7, 1
  b Compare answers as a class

Chapter 17.3

Practice 17.3A

1 b i 0.643 **ii** 0.669
  iii 1.88 **iv** 44.4°
  v 45.6° **vi** 35.0°
  c 0.940, 0.342, 2.75
  d i 23.6° **ii** 66.4° **iii** 21.8°

2 a

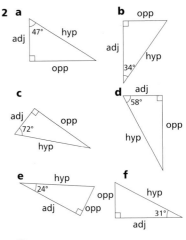

  c

  d

  e

  f

  g

3 a $t = 10.8$ **b** $g = 4.8$
  c $j = 30$
4 a 8.03 cm **b** 6.81 cm
  c 6.16 cm **d** 12.0 cm
  e 7.22 cm **f** 12.6 cm
5 a 35.0° **b** 44.4°
  c 73.0° **d** 25.4°
  e 51.8° **f** 21.8°

6 a $\tan 54 = \dfrac{12}{x}$
    $x = \dfrac{12}{\tan 54}$
    $= 8.72$ cm

  b $\tan(90 - 54) = \dfrac{x}{12}$
    $x = \tan 36 \times 12$
    $= 8.71$ cm

7 25.9°
8 32.0°
9 36.7 cm²
10 WY = 12.9 cm, XZ = 15.3 cm

Challenges

1 a Discuss as a class
  b 45°
2 a i 0.342 **ii** 0.342
  iii 0.766 **iv** 0.766
  v 0.174 **vi** 0.174
  b For example,
    $\sin x° = \cos(90 - x)°$
  c $\dfrac{\sin x}{\cos x} = \dfrac{\text{opp}}{\text{hyp}} \div \dfrac{\text{adj}}{\text{hyp}} = \dfrac{\text{opp}}{\text{hyp}} \times \dfrac{\text{hyp}}{\text{adj}}$
    $= \dfrac{\text{opp}}{\text{adj}} = \tan x$
3 Use the fact that the height of the triangle is $a \sin 60°$
4 Compare answers as a class
5 Compare answers as a class